Two-Point Boundary Value Problems:
Lower and Upper Solutions

This is volume 205 in
MATHEMATICS IN SCIENCE AND ENGINEERING
Edited by C.K. Chui, *Stanford University*

A list of recent titles in this series appears at the end of this volume.

Two-Point Boundary Value Problems: Lower and Upper Solutions

Colette De Coster
UNIVERSITÉ DU LITTORAL
CÔTE D'OPALE
FRANCE

Patrick Habets
UNIVERSITÉ CATHOLIQUE DE LOUVAIN
BELGIUM

ELSEVIER
Amsterdam – Boston – Heidelberg – London – New York – Oxford
Paris – San Diego – San Francisco – Singapore – Sydney – Tokyo

Elsevier
Radarweg 29, PO Box 211, 1000 AE Amsterdam, The Netherlands
The Boulevard, Langford Lane, Kidlington, Oxford OX5 1GB, UK

First edition 2006

Library of Congress Cataloging-in-Publication Data
A catalog record for this book is available from the Library of Congress

British Library Cataloguing in Publication Data
A catalogue record for this book is available from the British Library

ISBN-13: 978-0-444-52200-9
ISBN-10: 0-444-52200-X
ISSN: 0076-5392

For information on all Elsevier publications
visit our website at books.elsevier.com

Printed and bound in The Netherlands
06 07 08 09 10 10 9 8 7 6 5 4 3 2 1

Contents

Preface

The theory of lower and upper solutions is known to be an easy, elementary method to deal with second order boundary value problems. This book aims both to introduce the method for ordinary differential equations and to describe some recent and more involved results on this subject. Most of the theory we present was obtained for PDE or can be extended to deal with such problems. We propose a hint to such extensions at the end of chapter two. However, in this work we restricted our concern to the ODE setting to simplify the underlying techniques and because the difficulties inherent to the method already appear in this framework. As usual, the ODE approach allows a more intuitive interpretation of the theory.

In working applications, the use of lower and upper solutions method is hampered by the difficulty to exhibit such functions. This method replaces a difficult problem, "How to find a solution of a boundary value problem?", by a no easier problem, "How to find lower and upper solutions?". This replacement reminds us of the Liapounov's second method. In both instances, working cases is the key to developing the intuition related to these auxiliary functions. This observation was our primary motivation in starting this work.

In many theorems the assumptions at hand provide lower and upper solutions and their use simplifies the argument. The questions are then, "Is it easy to recognize that a set of assumptions provides such lower and upper solutions?", "Is it easy to find them?"

Keeping these questions in mind we divided the book into two parts. The first part, grouping the first five chapters, is theoretical. It develops the theory in various directions. The second part, consisting of the last five chapters, describes applications and shows how to build lower and upper solutions.

We first provide the basic ideas of the method. The two first chapters deal with the periodic problem and the separated boundary value problem for second order differential equations. Next we describe relations with other methods of nonlinear analysis. Chapter three studies connections with degree theory and is a first key to multiplicity results. The fourth chapter discusses the variational methods and provides a means to distinguish solutions and hence leads to further multiplicity results. Throughout the four first chapters, we focus on problems that satisfy Carathéodory conditions which makes our analysis fairly general. At last, chapter five describes connections with monotonicity which leads to iterative methods to approximate solutions. At this point we assume the nonlinearity to be continuous so as to simplify the analysis.

The second part of the book successively works parametric multiplicity problems, resonance and non-resonance problems, positive solutions, system with singularities, and singular perturbations. This examination provides a somewhat arbitrary selection of mathematical problems which can be investigated using the lower and upper solutions method and reflects the authors' interests. We hope that the study of these applications will give the reader a better understanding of how to use the method and how to find the auxiliary functions. We did not work these problems with maximum generality but rather tried to describe the scope of the method, i.e. both its power and its limitations.

We have tried to present the material of this book in such a way that a graduate student can understand both the formal developments and the basic ideas underlying the method. Only a limited background is needed to read the book and to work the material we present.

To write a book is a long road to travel from the first drafts to the final text. This effort could not have been possible without the numerous collaborations and encouragements of friends and co-workers. We enriched our knowledge of the subject from the remarks, results, and references they discussed with us. We wish to thank them all for the considerable help they gave us. Among them, we are especially grateful to Professor Jean Mawhin who started the work with us but, due to other activities, was forced to let us follow our way alone.

Louvain-la-Neuve, February 2005

Colette De Coster and Patrick Habets

Notations

u^+	$\max\{u, 0\}$
u^-	$\max\{-u, 0\}$
$\mathbb{R}^+$	non-negative real numbers
$\mathbb{R}_0^+$	positive real numbers
$\mathbb{R}^-$	non-positive real numbers
$\mathbb{R}_0^-$	negative real numbers
$\mathcal{C}(A, B)$	continuous maps of A to B
$\mathcal{C}([a, b])$	continuous maps of $[a, b]$ to $\mathbb{R}$
$\mathcal{C}_0([a, b])$	continuous maps u of $[a, b]$ to $\mathbb{R}$ such that $u(a) = u(b) = 0$
$\mathcal{C}^k([a, b])$	maps of $[a, b]$ to $\mathbb{R}$ with continuous k^{th} derivative
$\mathcal{C}_0^1([a, b])$	continuously differentiable maps u of $[a, b]$ to $\mathbb{R}$ such that $u(a) = u(b) = 0$
$\mathcal{B}C(\mathbb{R})$	bounded continuous maps of $\mathbb{R}$ to $\mathbb{R}$
$\mathcal{B}C^2(\mathbb{R})$	bounded continuous maps of $\mathbb{R}$ to $\mathbb{R}$ with bounded continuous first and second derivative
$L^p(a, b)$	with $p \in [1, +\infty[$, maps u of $[a, b]$ to $\mathbb{R}$ so that $\Vert u\Vert_{L^p}^p = \int_a^b \vert u(t)\vert^p \, dt < \infty$
$L^\infty(a, b)$	measurable maps u of $[a, b]$ to $\mathbb{R}$ so that $\Vert u\Vert_{L^\infty} = \sup \vert u(t)\vert < \infty$
$L^1_{loc}(a, b)$	maps u of $[a, b]$ to $\mathbb{R}$ integrable on any compact subset of $]a, b[$
$\mathcal{A}(a, b)$	maps u of $[a, b]$ to $\mathbb{R}$ so that $(\cdot - a)(b - \cdot)u(\cdot) \in L^1(a, b)$
$H^1(a, b)$	maps of $[a, b]$ to $\mathbb{R}$ with square integrable weak derivative
$H^1_{per}(a, b)$	maps in $H^1(a, b)$ such that $u(a) = u(b)$
$W^{1,1}(a, b)$	maps of $[a, b]$ to $\mathbb{R}$ with integrable weak derivative
$W^{2,p}(a, b)$	maps of $[a, b]$ to $\mathbb{R}$ with second weak derivative in $L^p(a, b)$
$W^{2,1}_{loc}(a, b)$	maps of $[a, b]$ to $\mathbb{R}$ with second weak derivative integrable on any compact subset of $]a, b[$
$W^{2,\mathcal{A}}(a, b)$	maps of $[a, b]$ to $\mathbb{R}$ with second weak derivative in $\mathcal{A}$
$W_0^{2,p}(\Omega)$	maps $u : \Omega \subset \mathbb{R}^N \to \mathbb{R}$ such that $u \in W^{2,p}(\Omega)$ whose trace is zero on $\partial\Omega$

$u \succ v$	the functions u and $v : [a, b] \to \mathbb{R}$ are such that for some $\epsilon > 0$, $u(t) - v(t) \geq \epsilon \sin \frac{\pi(t-a)}{b-a}$
$u \prec v$	equivalent to $v \succ u$
$u(t) \gneqq 0$	the function u is nonnegative and positive on a set with positive measure
$D_l u$, $D_r u$	left and right derivative of a function
$D_- u$, $D^- u$	left Dini derivatives of a continuous function
$D_+ u$, $D^+ u$	right Dini derivatives of a continuous function

Introduction – The History

Dans beaucoup d'autres cas, l'emploi des approximations successives pour obtenir certaines intégrales d'une équation différentielle peut conduire à des résultats curieux; il y aurait là, ce semble, un sujet de recherches qui présenterait quelque intérêt.

E. Picard

1 The First Steps

Lower and upper solutions can be traced back into the use of successive approximation. As early as 1893, E. Picard [250] had looked for solutions of the ODE boundary value problem

$$u'' + f(t,u) = 0, \quad u(a) = 0,\ u(b) = 0, \tag{1.1}$$

assuming $u = 0$ is a solution and $f(t,.)$ is increasing. The idea was to build a sequence of successive approximations $(\alpha_n)_n$ which is monotone, i.e.

$$\alpha_0 \leq \alpha_1 \leq \alpha_2 \leq \dots ,$$

that satisfies $\alpha_0(t) > 0$ on $]a,b[$ so as to avoid the trivial solution and that converges to a solution u of (1.1). To this end, using assumptions related to the now called *sublinear case*, he had proved the existence of a first approximation α_0, a positive function such that

$$\alpha_0'' + f(t,\alpha_0) > 0, \text{ on }]a,b[, \quad \alpha_0(a) = 0,\ \alpha_0(b) = 0.$$

Today, such a function is called a *lower solution* and the method used by Picard the *monotone iterative method.*

Independently, some of the basic ideas of the method, i.e. comparison between solutions of differential inequalities, have appeared in the study of first order Cauchy problems

$$u' + f(t,u) = 0, \quad u(0) = u_0,$$

made in 1915 by O. Perron [246] and in its extension to systems worked out by M. Müller [223] in 1926. These authors have deduced existence of solutions together with their localization between functions α and β assuming these functions are ordered $\alpha \leq \beta$, verify $\alpha(0) = \beta(0) = u_0$, have left and right derivatives ($D_l\alpha$, $D_r\alpha$, $D_l\beta$ and $D_r\beta$), and satisfy differential inequalities

$$D_{l,r}\alpha(t) + f(t,\alpha(t)) \leq 0, \quad D_{l,r}\beta(t) + f(t,\beta(t)) \geq 0.$$

A good account of this theory can be found in J. Szarski [295] or W. Walter [304].

The major breakthrough was due to G. Scorza Dragoni in 1931. In a first paper [280], extended and improved in [281], this author considers the Dirichlet boundary value problem

$$u'' = f(t,u,u'), \quad u(a) = A,\ u(b) = B. \tag{1.2}$$

He assumes the existence of functions α and $\beta \in \mathcal{C}^2([a,b])$ such that

$$\alpha(t) \leq \beta(t) \quad \text{on} \quad [a,b]$$

and

$$\begin{gathered}\alpha''(t) + f(t,\alpha(t),y) \geq 0 \ \text{ if } t \in [a,b], y \leq \alpha'(t) \quad (\text{resp. } y \geq \alpha'(t)),\\ \alpha(a) \leq A,\ \alpha(b) \leq B,\\ \beta''(t) + f(t,\beta(t),y) \leq 0 \ \text{ if } t \in [a,b], y \geq \beta'(t) \quad (\text{resp. } y \leq \beta'(t)),\\ \beta(a) \geq A,\ \beta(b) \geq B.\end{gathered}$$

He then obtains existence of a solution u of (1.2) together with its localization $\alpha \leq u \leq \beta$. The regularity assumptions were that f is continuous and bounded on

$$E := \{(t,u,v) \in [a,b] \times \mathbb{R}^2 \mid \alpha(t) \leq u \leq \beta(t)\}. \tag{1.3}$$

These papers are probably the first ones that recognize explicitly the central role of the functions α and β, which we today call *lower and upper solutions.* In that sense they can be considered as founding the lower and upper solutions method.

A drawback of Scorza Dragoni's approach is that, assuming the existence of ordered lower and upper solutions α and β, hides the difficulty. In practical problems there is no clue to finding these functions. This drawback motivated further work to indicate how they can be found. Constructions of lower and upper solutions appear in various proofs (see for example H. Epheser [108]). In a series of papers in 1967 and 1968, K. Akô [7, 8, 9], and later R.E. Gaines [122] and others (see also [123]), make such constructions explicit. They state conditions that ensure that a function is a lower or an upper solution. A large part of the present book is intended to answer this problem. Chapters VI to X deal with applications and show how to build in specific cases appropriate lower and upper solutions. This search for lower and upper solutions also follows from the whole string of examples in Chapters I to V.

Also the approach used by Scorza Dragoni to prove his results turns out to be basic. First, he introduces a modified problem

$$u'' = \overline{f}(t, u, u'), \quad u(a) = A, \ u(b) = B,$$

where $\overline{f}$ is bounded, equal to f on E and increasing for $u < \alpha(t)$ and $u > \beta(t)$. For this modified problem, he obtains then existence of a solution u_0. Second, he proves, using a maximum principle argument, that the differential inequalities imply that the solution u of the modified problem is such that neither $u - \alpha$ can have a negative minimum nor $u - \beta$ a positive maximum on $]a, b[$. In particular, $\alpha \le u_0 \le \beta$ and u_0 solves (1.2). This method of proof has been widely used in relation to lower and upper solutions. The difficulty is to define a modified problem so that the associated flow has the appropriate geometry. Going through this book, mainly in Chapters I and II, the reader will meet a large variety of such modified problems after which guiding lines hopefully will become apparent.

A next step is due in 1937 to M. Nagumo [224]. In this paper, the author proves the existence of at least one solution for (1.2) assuming first that there exist two functions α and $\beta \in \mathcal{C}^1([a, b])$ such that $\alpha < \beta$ on $[a, b]$, α' and β' have left and right derivatives ($D_l\alpha'$, $D_r\alpha'$, $D_l\beta'$ and $D_r\beta'$) and

$$\begin{gathered} D_{l,r}\alpha'(t) + f(t, \alpha(t), \alpha'(t)) > 0 \ \text{ on } [a, b], \\ \alpha(a) \le A, \ \alpha(b) \le B, \end{gathered}$$

$$\begin{gathered} D_{l,r}\beta'(t) + f(t, \beta(t), \beta'(t)) < 0 \ \text{ on } [a, b], \\ \beta(a) \ge A, \ \beta(b) \ge B. \end{gathered}$$

Secondly, Nagumo assumes $f : E \to \mathbb{R}$ is continuous together with the partial derivatives $\frac{\partial f}{\partial u}$, $\frac{\partial f}{\partial v}$, and satisfies what we call today a *Nagumo condition* on E

$$\forall(t,u,v) \in E, \quad |f(t,u,v)| \leq \varphi(|v|)$$

for some positive continuous function $\varphi : \mathbb{R}^+ \to \mathbb{R}$ that satisfies

$$\int_0^\infty \frac{s\,ds}{\varphi(s)} = \infty.$$

One important feature is that the differential inequalities must be satisfied along the functions α and β and not, as in Scorza Dragoni's papers, for sets of values of u'. Also, using functions which are not $\mathcal{C}^2$ but rather $\mathcal{C}^1$ with left and right second derivatives, Nagumo enlarges the class of functions to consider as lower and upper solutions. This process has been extended since then and contributes to diminish the difficulty of finding such functions. At last, introducing the Nagumo condition, this last author generalizes earlier ideas of S. Bernstein [35] to control the derivative of the solution. Nagumo's condition, as well as Bernstein's, control the growth of f as a function of the derivative. Although Nagumo's condition is not optimal, it was quite successful due to the fact that it is both general and simple.

In 1954 Nagumo [228], considering an elliptic problem, went back on his idea to enlarge the class of functions α and β and defined *quasi-subsolution* and *quasi-supersolution*. In a neighbourhood of each point, these are the maximum of a finite number of lower solutions and the minimum of a finite number of upper solutions which allows them to have angles. This idea was applied in 1964 by K. Akô and T. Kusano [12]. Much later, P. Habets and M. Laloy [144] have used such generalizations to deal with singular perturbation problems (see also [156, 169]). As it can be seen in Chapter X, these problems lead naturally to lower and upper solutions with angles. It is interesting to notice that the possibility of using angular functions was already mentioned by E. Picard [250] in 1893 in relation to the monotone iteration method. This idea was taken over in 1963 by H. Knobloch [190] who considers angles explicitly as he imposes at a finite number of points t_i conditions similar to $D_l\alpha(t_i) < D_r\alpha(t_i)$ (resp. $D_l\beta(t_i) > D_r\beta(t_i)$). At such points t_i, the modified problem has the appropriate geometry : $u - \alpha$ cannot have a negative minimum (resp. $u - \beta$ a positive maximum). The definitions of lower and upper solutions that we use take such angles into account.

In 1938, Nagumo's paper had suggested to G. Scorza Dragoni [282] a further generalization. With respect to this paper, he had weakened the continuity assumptions on f assuming it is locally lipschitzian in (u,v), continuous over E, and satisfies a Nagumo condition. He also had assumed that the differential inequalities are not strict, i.e. the functions α and $\beta \in C^2([a,b])$ are such that

$$\alpha''(t) + f(t, \alpha(t), \alpha'(t)) \geq 0 \quad \text{on } [a,b]$$

and

$$\beta''(t) + f(t, \beta(t), \beta'(t)) \leq 0 \quad \text{on } [a,b].$$

The same year, G. Scorza Dragoni [283, 284], looking for minimal continuity assumptions on the nonlinearity f, had considered the boundary value problem (1.2) assuming only f to satisfy Carathéodory conditions. Solutions are in $W^{2,1}(a,b)$ and the notion of lower and upper solutions has to be generalized. He assumed that α, $\beta \in C^1([a,b])$ and the functions

$$-\alpha'(t) + \int_a^t f(s, \alpha(s), \alpha'(s))\, ds \quad \text{and} \quad \beta'(t) - \int_a^t f(s, \beta(s), \beta'(s))\, ds$$

are non-increasing on $[a,b]$. In such a case, the Nagumo condition also has to be generalized. He required that for some positive continuous functions $\varphi : \mathbb{R}^+ \to \mathbb{R}$ and $\chi \in L^1(a,b;\mathbb{R}^+)$ we have

$$\forall (t,u,v) \in E, \quad |f(t,u,v)| \leq \varphi(|v|) + \chi(t),$$

where

$$\int_0^\infty \frac{s\, ds}{\varphi(s)} = \infty$$

and

$$\left| \frac{s}{\varphi(s)} \right| < K \quad \text{except if} \quad \chi \equiv 0.$$

Such an assumption mimics the common situation of a system driven by an autonomous force and an L^1-forcing. Further work on lower and upper solutions to investigate $W^{2,1}$-solutions had been done by H. Epheser [108] in 1955. Since then a variety of definitions of lower and upper solutions has been proposed to deal with such solutions [173, 185, 187, 139, 144, 159, 114, 2, 149, 89]. All of these are strongly related and express basically the same geometry of the vector field. In practice, there is no obvious reason to choose one rather than the other. Our choice (see [85, 87, 88, 89, 91, 92]) tends to be general enough for applications and simple enough to model the geometric intuition built into the concept.

2 Other Boundary Value Problems

Working a contraction method, it had been noticed by A. Rosenblatt in 1933 [262] that Dirichlet problems can be studied for more singular nonlinearities than Carathéodory functions. Roughly, it is sufficient for the nonlinearity to be bounded on bounded sets by a positive function h so that

$$(s-a)(b-s)h(s) \in L^1(a,b).$$

This condition allows for some non-L^1 singularities near the boundary points a and b. In 1953, G. Prodi [256] had used lower and upper solutions to work a parabolic problem with such singularities. Further works were done in 1968 by I.T. Kiguradze [185] (see also [153, 154, 87]). Extensions to derivative dependent problems were done in I.T. Kiguradze [185], I.T. Kiguradze and B.L. Shekhter [187] and C. De Coster [86]. These problems are described in Section II-4.

Boundary conditions such as $a_1u(a) + a_2u'(a) = 0$ were already considered in 1954 for singular perturbation problems by N.I. Briš [46]. This author had used a shooting method together with lower and upper solutions for an associated Cauchy problem. In 1955, H. Epheser [108] had developed a direct lower and upper solutions method for problems with mixed boundary conditions $u(a) = 0$, $u'(b) = 0$.

Due to the difficulty of finding an appropriate equivalent fixed point problem, the periodic problem and the Neumann problem were studied quite late. In 1963, H. Knobloch [190] had considered the periodic problem and used constant lower and upper solutions. The very same year, in [191], he had extended his ideas and worked out the lower and upper solution method (see also [27], [182], [184] and [269]). Neumann type of boundary conditions were investigated by Akô [10] in 1969 (see also [9]).

In 1970, K. Schmitt [271] had considered the general separated boundary value problem

$$\begin{aligned} u'' &= f(t,u,u'), \\ a_1u(a) - a_2u'(a) &= A_0, \\ b_1u(b) + b_2u'(b) &= B_0, \end{aligned}$$

assuming $a_1 + a_2 > 0$, $a_i \geq 0$, $b_1 + b_2 > 0$, $b_i \geq 0$. This problem contains both the Dirichlet and the Neumann problem. General nonlinear boundary conditions containing the case $a_2 \geq 0$, $a_1^2 + a_2^2 > 0$, $b_2 \geq 0$, $b_1^2 + b_2^2 > 0$ were considered by L. Erbe [109]. J.W. Heidel [157] had given a more direct proof together with examples proving the above sign conditions cannot be omitted.

The study of the elliptic problem had started much later. In 1954, M. Nagumo [228] had considered the problem

$$\begin{aligned} \Delta u + f(x,u,\nabla u) &= 0, && \text{in } \Omega, \\ u &= 0, && \text{on } \partial\Omega, \end{aligned}$$

assuming α and $\beta \in \mathcal{C}^2(\overline{\Omega})$, f Hölder continuous such that

$$|f(x,u,v)| \le B\|v\|^2 + C,$$

but with the restriction

$$16MB < 1, \quad \text{where } M = \max\{\|\alpha\|_\infty, \|\beta\|_\infty\}.$$

Several authors such as K. Akô [5] have extended this result but it seems that we had to wait until F. Tomi [299] in 1969 obtained the result with the restriction $16MB < 1$ removed completely. In [228], M. Nagumo had observed that the result is false if f has a superquadratic growth in v.

An alternative setting consists of considering weak solutions of the problem. This approach was first developed by P. Hess [159] and J. Deuel and P. Hess [105] in 1974-1976 (see also G. Stampacchia [292]). The extension of Tomi's result in the weak solution framework is probably due to L. Boccardo, F. Murat and J.P. Puel [38, 39] in 1983. Curiously enough, $W^{2,p}$-solutions were less studied. We refer the reader to [145] and [93] for such results.

Chapter I considers the periodic problem and Chapter II handles the separated boundary conditions.

3 Maximal and Minimal Solution

Non-uniqueness of solutions produces the problem of finding one of them larger than any other between the lower and upper solutions, α and β. Similarly, it is reasonable to look for a solution smaller than any other one. Such solutions are called *maximal and minimal solutions.* For first order ODE, existence of such extremal solutions were already studied in 1885 by G. Peano [244] and in 1915 by O. Perron [246]. They have considered the Cauchy problem

$$u' = f(t,u), \quad u(t_0) = u_0,$$

with f continuous and have assumed the existence of lower and upper solutions, i.e. continuous functions α and β with left and right derivatives so that $\alpha(0) = u_0$, $\beta(0) = u_0$, $\alpha \le \beta$ and

$$D_{l,r}\alpha(t) \le f(t,\alpha(t)), \quad D_{l,r}\beta(t) \ge f(t,\beta(t)).$$

They have showed then the existence of a solution between α and β and have defined

$$u_{max} = \sup\{u_i \mid u_i \text{ is a lower solution with } \alpha \le u_i \le \beta\},$$

$$u_{min} = \inf\{u_i \mid u_i \text{ is an upper solution with } \alpha \le u_i \le \beta\}.$$

At last, they have proved according to their definitions that the maximum of two lower solutions is a lower solution, u_{max} and u_{min} are solutions, and every solution between α and β is between u_{min} and u_{max}. The formulation of G. Peano is slightly different but the idea is the same. In a second work [247] in 1923, O. Perron developed similar results for the Laplace problem.

In 1954, T. Satô [266] had studied this problem for nonlinear boundary value problems. He had considered a nonlinear elliptic problem but had assumed some monotonicity assumption on f. An abstract formulation of these results can be found in A. Tarski [297] and A. Pelczar [245]. T. Satô [267] had completed his result proving in case the maximal and minimal solutions are not equal that a continuum of solutions takes place (see also I. Hirai and K. Akô [163]).

The idea of Perron that the maximal solution can be obtained without extra monotonicity assumption as the supremum of the lower solutions was used by W. Mlak [221] in 1960 for the parabolic problem and independently by K. Akô [5] in 1961 for an elliptic problem. In 1963, W. Mlak [222] had provided an example for the parabolic case where the minimal solution and the maximal solutions are distinct.

4 The Nagumo Condition

To examine derivative dependent nonlinearities, we need $\mathcal{C}^1$-estimates on the solutions of the modified problem. In the method of lower and upper solutions, we know that such a solution lies between α and β, i.e. it already satisfies a $\mathcal{C}^0$-bound. The idea to extend it into a $\mathcal{C}^1$-bound begins with S. Bernstein [35]. He has considered the case where f is continuous and satisfies the growth condition

$$|f(t,u,v)| \leq A + Bv^2.$$

He has proved then that given $r > 0$ there exists $R > 0$ such that any solution u of

$$u'' = f(t,u,u')$$

on $[a,b]$ with $\|u\|_\infty \leq r$ satisfies the *a priori* bound

$$\|u'\|_\infty \leq R.$$

As we already noticed it, in 1937, M. Nagumo [224] generalized these ideas introducing the so-called *Nagumo condition*

$$\forall (t,u,v) \in E, \quad |f(t,u,v)| \leq \varphi(|v|),$$

where $\varphi : \mathbb{R}^+ \to \mathbb{R}$ is some positive continuous function such that

$$\int_0^\infty \frac{s\,ds}{\varphi(s)} = \infty.$$

He then applied his result to the Dirichlet problem. K. Akô [7] has observed that the condition

$$\int_r^\infty \frac{s\,ds}{\varphi(s)} > \max_t \beta(t) - \min_t \alpha(t),$$

where $r = \max\{\frac{\beta(b)-\alpha(a)}{b-a}, \frac{\beta(a)-\alpha(b)}{b-a}\}$, was enough to obtain a C^1-bound. Some authors have also used a simplification of Nagumo condition

$$\lim_{s\to\infty} \frac{s^2}{\varphi(s)} = +\infty$$

which turns out to be useful in looking at vector problems (see J. Mawhin [209] or P. Habets and K. Schmitt [151]). For the elliptic equation, we see that F. Tomi [299] in 1969 proves the existence of a solution in presence of well ordered lower and upper solutions together with the Bernstein condition.

K.W. Schrader [276] in 1968 (see also H.W. Knobloch [191]) had observed that in some cases the Nagumo condition can be restricted to be one-sided. For example, for the Dirichlet problem, he has proved that we can assume only

$$\forall (t,u,v) \in E, \quad \operatorname{sgn}(v) f(t,u,v) \le \bar{\varphi}(|v|),$$

provided $\alpha(a) = \beta(a)$. Related results can already be found in H. Epheser [108] in 1955 and in I.T. Kiguradze [183] in 1967 (see also [185]).

In his 1967 paper [183], I.T. Kiguradze has extended the Nagumo condition so as to deal with $W^{2,1}$-solutions. In this paper the nonlinearity is controlled by a condition

$$\forall (t,u,v) \in E, \quad |f(t,u,v)| \le \psi(t)\varphi(|v|),$$

where $\varphi \in C(\mathbb{R}^+, \mathbb{R}_0^+)$ and $\psi \in L^p(a,b)$ satisfy

$$\int_0^\infty \frac{s^{1/q}}{\varphi(s)}\,ds > \|\psi\|_{L^p} (\max_t \beta(t) - \min_t \alpha(t))^{1/q}.$$

Such a condition gives the desired C^1-bound and is general enough to allow cases where the solutions are in $W^{2,1}(a,b)$ but not in $C^2([a,b])$.

If the nonlinearity f is independant of u', then the existence of a solution follows from the existence of ordered lower and upper solutions alone. Also, if the nonlinearity f depends of u', it is easy to see that the Nagumo condition is not necessary. However, it was not clear whether the existence of ordered lower and upper solutions alone would not be sufficient to ensure the existence of solutions. In 1954, M. Nagumo [228] has given an example of a Dirichlet problem which has no solution although ordered lower and upper solutions can be found. Examples for other boundary value problems such as the periodic problem are more difficult to find since we know that the result is valid with only one of the one-sided Nagumo condition. Such examples were given much later in [148].

In 1941, M. Nagumo [226] had replaced the Nagumo condition by the more geometric condition of *bounding functions* (a related notion already appears in G. Scorza-Dragoni paper [279] concerning first order systems). These are functions $\Omega_1(t,u) \leq \Omega_2(t,u)$ such that the vector field points one way along each of the curves

$$u' = \Omega_1(t,u) \quad \text{and} \quad u' = \Omega_2(t,u).$$

It follows then under additional assumptions that these functions bound the derivative u' of the solutions, i.e.

$$\Omega_1(t,u(t)) \leq u'(t) \leq \Omega_2(t,u(t)).$$

In the same paper, Nagumo has made clear how bounding functions can be deduced from one-sided Nagumo conditions which relates these conditions to the geometry of the flow.

This idea can be further developed assuming the vector field points one way along the curve $u' = \Omega_i(t,u)$ if $u < \mu_i(t)$ and the other way if $u > \mu_i(t)$. The curves $u = \mu_i(t)$, which are called *diagonals*, were introduced by H. Okamura in 1941 [233] (see also L. Tonelli [300]). These were used by H. Epheser [108] in a particular case and were developed in full generality by F.Z. Sadyrbaev [264]. A combined use of bounding functions and diagonals can be found in C. Fabry and P. Habets [114].

An alternative to the Nagumo condition is to assume some global existence of the solutions, i.e. every solution u of

$$u'' = f(t,u,u'),$$

such that $\alpha(t) \leq u(t) \leq \beta(t)$ on its maximal interval of existence exists on the whole interval $[a,b]$. Such an approach has been used in 1969 by K.W. Schrader [277] and later by J.W. Heidel [157].

5 Degree Theory

The idea to associate a degree to a pair of lower and upper solutions goes back to H. Amann [14, 15] in 1972 (see also [67]). He has used additional assumptions such as a one-sided Lipschitz condition on the nonlinearity. An alternative is to define strict lower and upper solutions. The easiest approach assumes that the inequalities that are satisfied by the lower and upper solutions are strict. In this book we use a more refined definition (see Sections III-1.1 and III-2.1). Basically, a lower solution α is strict if any solution u of the boundary value problem with $u \geq \alpha$ is such that $u > \alpha$. Such concepts appear in [93].

The relation with degree theory produces new proofs and further improvements of known results such as multiplicity results. For example in 1972, H. Amann [15] has given a simple degree theoretic proof of the Three Solutions Theorem. This theorem proves the existence of three solutions in case there are two pairs of lower and upper solutions with appropriate ordering. Such a result goes back to Y.S. Kolesov [194] in 1970 and was extended by H. Amann [13] in 1971. For derivative dependent equations, this Three Solutions Theorem was studied by H. Amann [16] without degree theory and later by H. Amann and M. Crandall [20] using degree arguments. Extensions of the order relations between the two pairs of lower and upper solutions were worked out by Bongsoo Ko [192].

A second type of multiplicity result can be obtained in case we have a well-ordered pair of lower and upper solutions, i.e. $\alpha \leq \beta$, and an *a priori* bound on the upper solutions. This idea goes back to K.J. Brown and H. Budin [47]. A more precise study of it can be found for the Dirichlet p-Laplacian and f independent of u' in C. De Coster [84].

Until recently, these types of results were obtained only for continuous nonlinearities f. Since 1994 several authors [84, 87, 145, 88, 31] studied various Dirichlet problems with f independent of u' and satisfying Carathéodory conditions. A study of the derivative dependent case for a Rayleigh equation can be found in [152], for the Liénard equation in [41] and for the general case in [91, 92]. The elliptic case can be found in [145] and [93].

The two first paragraphs of Chapter III concern the relation between lower and upper solutions and degree theory for both the periodic and the Dirichlet problems. In each of these sections, we present as applications some multiplicity results in the line of the Three Solutions Theorem and of the Brown-Budin Theorem.

6 Non Well-ordered Lower and Upper Solutions

In 1972, D.H. Sattinger [268] has presented as an open problem the question of existence of a solution for the problem

$$-\Delta u = f(x,u) \text{ in } \Omega, \qquad u = 0 \text{ on } \partial\Omega, \tag{6.1}$$

in presence of lower and upper solutions which do not satisfy the ordering relation $\alpha \leq \beta$.

In 1976, H. Amann [17] gave a counter-example of the type

$$-\Delta u = \lambda_k u + \varphi_k(x) \text{ in } \Omega, \qquad u = 0 \text{ on } \partial\Omega,$$

where λ_k denotes the k-th eigenvalue of the problem and φ_k the corresponding eigenfunction. For $k \geq 2$ this problem has no solution although $\alpha(x) = K\varphi_1(x)$ and $\beta(x) = -K\varphi_1(x)$ are lower and upper solutions for large enough positive values of K.

A first important contribution to this question was given by H. Amann, A. Ambrosetti and G. Mancini [19] in 1978 when they considered the problem

$$-\Delta u = f(x,u,\nabla u) \text{ in } \Omega, \quad u = 0 \text{ on } \partial\Omega,$$

under the assumption

$$\sup_{\Omega\times\mathbb{R}\times\mathbb{R}^n} |f(x,u,\nabla u) - \lambda_1 u| < \infty.$$

This assumption implies that for large values of u the slope $\frac{f(x,u,\nabla u)}{u}$ remains near the first eigenvalue which means that the only possible resonance is at the first eigenvalue.

Various assumptions to rule out resonance with any eigenvalue but the first one have been worked out by several authors for different boundary value problems. In 1991, J.P. Gossez and P. Omari [130] (see also [117, 118, 147, 131, 133, 238]) have considered a nonlinearity $f(u)$ for a periodic or a Neumann problem. They have assumed there exist constant lower and upper solutions and have given a condition on $F(u) = \int_0^u f(s)\,ds$ so as to keep resonance away from all eigenvalues but the first one. Related results were already obtained by P. Omari [235] in 1988. Notice that because of the inequalities that need to be satisfied at the boundary points, the use of non well-ordered constant lower and upper solutions rules out the possibility of dealing with the Dirichlet problem. In 1994, J.P. Gossez and P. Omari [132] have considered the elliptic Dirichlet problem with a nonlinearity $f(x,u)$.

They have assumed existence of non-ordered lower and upper solutions α and β and that, as $|u|$ goes to infinity, the slope $\frac{f(x,u)}{u}$ becomes larger or equal than the first eigenvalue. To avoid resonance, they also have assumed that the slope $\frac{f(x,u)}{u}$ keeps away from the second eigenvalue, i.e. that it is bounded above by a function $\gamma(x) \lneqq \lambda_2$. This result has been improved by P. Habets and P. Omari [145] who supposed that $\alpha > \beta$ but only needed the slope $\frac{f(x,u)}{u}$ to be bounded below. Further they have given localization information which is that the solution has to be at some point x_1 over β and at some point x_2 below α, i.e. as $\alpha > \beta$ at some point x between these two functions. Further work has replaced the non-resonance with the second eigenvalue by a non-resonance with the second Fučík curve (see [130, 133, 145]).

In 1998, C. De Coster and M. Henrard [93] have stated a general theorem which provides a solution in case there exists lower and upper solutions without order. It also provides the same localization as found in [145]. These authors gave an elegant proof using the Three Solution Theorem. Other results in this direction for the parabolic case and the periodic ODE can be found in [94], [260] and [95].

Another type of result is far older and concerns lower and upper solutions in the reverse order, $\alpha \geq \beta$. For the periodic and the Neumann problem, existence of a solution can be obtained between the lower and upper solutions. This feature is essentially due to the fact that for these problems a uniform anti-maximum principle holds. We find already a contribution in this direction in 1965 with the paper of A.Ya. Khokhryakov and B.M. Arkhipov [182]. We refer also to Y.S. Kolesov [193] and K. Schmitt [272]. This idea has also been revisited by P. Omari and M. Trombetta [236] for the periodic problem and A. Cabada and L. Sanchez [53] for the Neumann problem (see Section III-3.5).

Non-ordered lower and upper solutions are considered in Section III-3.

7 Variational Methods

It is well-known that solutions of conservative boundary value problems can be obtained as stationary points of an associated functional. Existence of lower and upper solutions imposes geometrical properties on this functional. This feature relates variational methods with lower and upper solutions and gives rise to new results.

In 1983 K.C. Chang [58, 59] and in 1984 D.G. de Figueiredo and S. Solimini [102] have noticed independently that, for an elliptic problem between well-ordered lower and upper solutions, the related functional has a minimum. In the particular case of the ODE Dirichlet problem, the functional

$$\phi : H_0^1(a,b) \to \mathbb{R}$$

is such that for some $u_0 \in H_0^1(a,b)$ with $\alpha \le u_0 \le \beta$,

$$\phi(u_0) = \min_{\substack{v \in H_0^1(a,b) \\ \alpha \le v \le \beta}} \phi(v).$$

Furthermore, u_0 is a solution of the Dirichlet problem. Such a result can be used to obtain a second solution u_1 of a problem which has a known solution u_0 between α and β provided

$$\phi(u_0) > \min_{\substack{v \in H_0^1(a,b) \\ \alpha \le v \le \beta}} \phi(v).$$

K.C. Chang [58, 59] as well as D.G. de Figueiredo and S. Solimini [102] have noticed that under appropriate conditions, the minimum u_0 obtained between strict lower and upper solutions is valid in the H^1-topology. For example, considering the Dirichlet problem, there exists some $r > 0$ such that

$$\phi(u_0) = \min_{\substack{v \in H_0^1(a,b) \\ \|v-u_0\|_{H^1} \le r}} \phi(v).$$

This remark was pointed out and extended by H. Brezis and L. Nirenberg [44]. It is essential to use variational methods such as a mountain pass theorem whose natural framework uses a local minimum of the functional in the H^1-topology.

A fundamental step occurred with K.C. Chang [58, 59] and was further developed by Z. Liu and J. Sun [203, 204]. They have worked a minimax method and to this end have considered the minus gradient flow as defined by the Cauchy problem

$$\tfrac{d}{dt}u = -\nabla\phi(u), \quad u(0) = u_0.$$

In the one-dimensional case, their approach amounts to considering this flow in $C_0^1([a,b])$ so that the cones

$$C_\alpha = \{u \in C_0^1([a,b]) \mid \exists\epsilon > 0,\ u \ge \alpha + \epsilon \sin \tfrac{t-a}{b-a}\pi\}$$

and

$$C^\beta = \{u \in C_0^1([a,b]) \mid \exists\epsilon > 0,\ u \le \beta - \epsilon \sin \tfrac{t-a}{b-a}\pi\},$$

are positively invariant and with non-empty interior. This remark turned out to be essential to realize that non-ordered lower and upper solutions impose a mountain pass geometry. Various situations involving ordered and non-ordered lower and upper solutions were worked out in [28, 60, 82, 164, 200, 92]. Some of them are presented in Section IV-3 for an ODE problem. A slightly different point of view was used by M. Conti, L. Merizzi, and S. Terracini [74] who worked with flows in H_0^1. This framework is more natural for variational methods but with this topology the interior of the cones C_α and C^β are empty, which complicates the analysis.

A quite different use of variational methods was introduced by E. Serra and M. Tarallo [287, 288] in 1998. Given a functional ϕ, these authors have considered the real function

$$\varphi(\xi) = \min_{\bar{u}=\xi} \phi(u).$$

They have noticed then that in case φ is not monotone, we have a pair of weak lower and upper solutions but without order relation. Applications to boundary value problems follow. This idea was developed in C. De Coster and M. Tarallo [95] and can be found in Section IV-4. At last, let us mention that some traces of this approach can be found in A. Castro [56].

The use of the variational method to build lower and upper solutions appears in various problems. Chapter VIII presents such examples.

8 Monotone Methods

As we have mentioned, monotone iterative methods based on lower and upper solutions go back to E. Picard's work on boundary value problems such as

$$u'' + f(t,u) = 0, \quad u(a) = 0, \ u(b) = 0. \tag{8.1}$$

In case $f(t,\cdot)$ is increasing, he has proved in 1893 [250] the existence of a solution using an increasing sequence of lower solutions defined from the iteration scheme

$$-\alpha_n'' = f(t, \alpha_{n-1}), \quad \alpha_n(a) = 0, \ \alpha_n(b) = 0,$$

and converging to the solution. In 1890, he worked the case where $f(t,\cdot)$ is decreasing [249] (see also [253, 254]). To this end he used sequences $(\alpha_n)_n$ and $(\beta_n)_n$ defined from the problems

$$\begin{aligned} \beta_n'' + f(t, \alpha_{n-1}) &= 0, \quad && \beta_n(a) = 0, \ \beta_n(b) = 0, \\ \alpha_n'' + f(t, \beta_n) &= 0, && \alpha_n(a) = 0, \ \alpha_n(b) = 0. \end{aligned}$$

Here $\alpha_n \leq \beta_n$; $(\alpha_n)_n$ is increasing; $(\beta_n)_n$ is decreasing; and the sequences $(\alpha_n)_n$ and $(\beta_n)_n$ converge to functions u and $v \geq u$ which bound a solution. Picard has given an example where these bounds are different [251, 252] and also has provided conditions that imply that these bounds are equal and hence are solutions to (8.1).

Following S.A. Chaplygin's work in 1935 [62], the Russian school has further developed the method. In 1954, B.N. Babkin [25] has considered the problem (8.1) assuming that the function $f(t,u) + Ku$ is increasing in u for some $K > 0$. Taking as first approximations some lower and upper solutions α_0 and $\beta_0 \geq \alpha_0$, he has defined monotone sequences of approximations from

$$\begin{aligned} -\alpha_n'' + K\alpha_n &= f(t,\alpha_{n-1}) + K\alpha_{n-1}, \quad & \alpha_n(a) = 0,\ \alpha_n(b) = 0, \\ -\beta_n'' + K\beta_n &= f(t,\beta_{n-1}) + K\beta_{n-1}, \quad & \beta_n(a) = 0,\ \beta_n(b) = 0. \end{aligned}$$

Each of these sequences converges to a solution to (8.1) which, under his assumptions on $f(t,u)$, is unique.

In 1958 S. Slugin [291] and in 1960 W. Mlak [221] have noticed that under weaker assumptions the solutions are not unique but that the sequences $(\alpha_n)_n$ and $(\beta_n)_n$ converge to extremal solutions.

An important step is due to L. Kantorovich [175] in 1939. He had observed that the first approximation scheme, used for the Cauchy problem as well as for other boundary value problems, has a common structure related to positive operators. He then developed an abstract formulation of the method. This idea was further developed by M.A. Krasnosel'skii [195], M.A. Krasnosel'skii, G.M. Vainniko, P.P. Zabreiko, Ya.B. Rutitskii and V.Ya. Stetsenko [196], H. Amann [15, 17], and E. Zeidler [309].

Independently of the Russian school, R. Courant and D. Hilbert [77] proposed in 1962 a monotone iterative scheme similar to Babkin's. The main problem was to find appropriate conditions on the function f to apply the method. In [72], Cohen had obtained a maximizing sequence assuming the nonlinearity f is increasing and strictly concave. In 1968, L.F. Shampine [289] (see also L.F. Shampine and G.M. Wing [290]) had introduced a one-sided Lipschitz condition.

$$f(x,v) - f(x,u) \leq k(x)(u - v) \quad \text{if } u \geq v. \tag{8.2}$$

This condition unifies R. Courant and D. Hilbert [77], H.B. Keller and D.S. Cohen [178], D.S. Cohen [72] and S.V. Parter [243] approaches.

In 1974, H. Amann [13] has generalized the one-sided Lipschitz condition by imposing a Hölder condition on f. This generalization is a particular case of the condition used in 1960 by W. Mlak [221] for a parabolic problem. On the other side, in 1978, C.A. Stuart [293] has assumed that f is of bounded variation on compact intervals which implies $f = g - h$, where g and h are increasing functions. This assumption was also used in 1992 by S. Carl [55]. In such a case, the approximation sequences are obtained from the nonlinear problems

$$\begin{aligned} -\alpha_n'' + h(t, \alpha_n) &= g(t, \alpha_{n-1}), \quad && \alpha_n(a) = 0, \ \alpha_n(b) = 0, \\ -\beta_n'' + h(t, \beta_n) &= g(t, \beta_{n-1}), \quad && \beta_n(a) = 0, \ \beta_n(b) = 0. \end{aligned}$$

In general, solutions cannot be made explicit which reduces considerably the interest of the approach.

The study of the monotone iterative methods for nonlinearities depending on the derivative was initiated in 1964 by G.V. Gendzhoyan [127] who had considered the problem

$$u'' + f(t, u, u') = 0, \quad u(a) = 0, \ u(b) = 0.$$

Using lower and upper solutions α_0 and $\beta_0 \geq \alpha_0$ as first approximations, he has defined sequences of approximations $(\alpha_n)_n$ and $(\beta_n)_n$ as solutions of

$$\begin{gathered} -\alpha_n'' + l(t)\alpha_n' + k(t)\alpha_n = f(t, \alpha_{n-1}, \alpha_{n-1}') + l(t)\alpha_{n-1}' + k(t)\alpha_{n-1}, \\ \alpha_n(a) = 0, \ \alpha_n(b) = 0, \\ -\beta_n'' + l(t)\beta_n' + k(t)\beta_n = f(t, \beta_{n-1}, \beta_{n-1}') + l(t)\beta_{n-1}' + k(t)\beta_{n-1}, \\ \beta_n(a) = 0, \ \beta_n(b) = 0, \end{gathered}$$

where $k(t)$ and $l(t)$ are functions related to the assumptions on f. Here also, the convergence is monotone and gives approximations of the solution (which is unique under his assumptions) together with some error bounds.

Another approach to deal with derivative dependent problems was used in S.R. Bernfeld and J. Chandra [34]. For the Dirichlet problem, these authors had defined monotone sequences of approximations from

$$\begin{aligned} -\alpha_n'' + K\alpha_n &= f(t, \alpha_{n-1}, \alpha_n') + K\alpha_{n-1}, \quad && \alpha_n(a) = 0, \ \alpha_n(b) = 0, \\ -\beta_n'' + K\beta_n &= f(t, \beta_{n-1}, \beta_n') + K\beta_{n-1}, \quad && \beta_n(a) = 0, \ \beta_n(b) = 0. \end{aligned}$$

Such a scheme which can be used for other problems imposes a solution to nonlinear boundary value problems which makes the method implicit.

An alternative was proposed by P. Omari [234] in 1986 who has assumed $f(t,u,u')$ to be one-sided Lipschitz in u and Lipschitz in u'. For the Dirichlet problem, the author has considered the iterative process defined from the piecewise linear problems

$$\alpha_n'' - L|\alpha_n' - \alpha_{n-1}'| - K\alpha_n = f(t, \alpha_{n-1}, \alpha_{n-1}') - K\alpha_{n-1}',$$
$$\alpha_n(a) = 0, \ \alpha_n(b) = 0,$$
$$\beta_n'' + L|\beta_n' - \beta_{n-1}'| - K\beta_n = f(t, \beta_{n-1}, \beta_{n-1}') - K\beta_{n-1},$$
$$\beta_n(a) = 0, \ \beta_n(b) = 0.$$

He has considered in the same way periodic and Neumann problems.

The monotone iterative method was also developed in case the lower and upper solutions appear in the reverse order i.e. $\alpha \geq \beta$. A first contribution was made by P. Omari and M. Trombetta [236] in 1992. They have considered in particular the periodic problem

$$-u'' + cu' + f(t,u) = 0, \qquad u(a) = u(b), \ u'(a) = u'(b),$$

and have proved the convergence of approximations $(\alpha_n)_n$ and $(\beta_n)_n$ as defined by

$$-\alpha_n'' + c\alpha_n' + K\alpha_n = -f(t, \alpha_{n-1}) + K\alpha_{n-1},$$
$$\alpha_n(a) = \alpha_n(b), \ \alpha_n'(a) = \alpha_n'(b),$$
$$-\beta_n'' + c\beta_n' + K\beta_n = -f(t, \beta_{n-1}) + K\beta_{n-1},$$
$$\beta_n(a) = \beta_n(b), \ \beta_n'(a) = \beta_n'(b).$$

The key assumptions are that the function $f(t,u) - Ku$ is non-decreasing in u for some $K < 0$ and that this K is such that the operator $-u'' + cu' + Ku$ is inverse negative on the space of periodic functions, i.e. that an anti-maximum principle holds. Analogous results for the Neumann problem were obtained by A. Cabada and L. Sanchez [53]. We also refer to [52] for other results in this direction. In case f depends nonlinearly on u', we can quote A. Bellen [30] for the periodic problem, A. Cabada, P. Habets and S. Lois [51] for the Neumann BVP and M. Cherpion, C. De Coster and P. Habets [65] for both problems.

Chapter V gives an account of the principal results using monotone methods together with lower and upper solutions.

The second iterative scheme introduced by E. Picard

$$\beta_n'' + f(t, \alpha_{n-1}) = 0, \quad \beta_n(a) = 0, \ \beta_n(b) = 0,$$
$$\alpha_n'' + f(t, \beta_n) = 0, \qquad \alpha_n(a) = 0, \ \alpha_n(b) = 0,$$

was further developed. In 1959, L. Collatz and J. Schröder [73] have given an abstract formulation of this process. In 1960, J. Schröder [278] has shown that it can be reduced to

$$u_n'' + F(t, u_{n-1}) = 0, \quad u_n(a) = 0, \ u_n(b) = 0,$$

where $u_n(t) \in \mathbb{R}^2$. Further work can be found in [306, 179, 285, 140, 141, 181, 303, 134]. Section V-4 describes results obtained on this second scheme by M. Cherpion, C. De Coster and P. Habets [64].

9 Applications

The lower and upper solutions method has been used in relation with several problems of the theory of boundary value problems. Chapters VI to X of this book present a selection of them.

Chapter VI mainly concerns Ambrosetti-Prodi problems. In 1972, these authors [24] have considered the problem

$$\Delta u + f(u) = h(x) \text{ in } \Omega, \quad u = 0 \text{ on } \partial\Omega,$$

and have proved that under appropriate conditions on f there is a manifold M which separates the space $\mathcal{C}^{0,\alpha}(\overline{\Omega})$ in two regions O_0 and O_2 such that the above problem has zero, exactly one, or exactly two solutions according to $h \in O_0$, $h \in M$ or $h \in O_2$. In 1975, M.S. Berger and E. Podolak [33] have used the decomposition of h in terms of the first eigenfunction φ, i.e. $h(x) = s\varphi(x) + \widetilde{h}(x)$, where $\int_\Omega \widetilde{h}(x)\varphi(x)\,dx = 0$. They characterize the manifold M to be of the type $M = \{s\varphi + \widetilde{h} \mid s = s_0\}$ for some $s_0 = s_0(\widetilde{h})$. The same year, J.L. Kazdan and F.W. Warner [177] have used the lower and upper solutions method to study this problem but they were only able to prove the existence of one solution if $h \in O_2$. The multiplicity result combining lower and upper solutions technique with degree theory has been obtained independently in 1978 by E.N. Dancer [81] and in 1979 by H. Amann and P. Hess [21]. The corresponding ODE problem

$$u'' + u + f(t, u) = s\varphi(t), \quad u(0) = 0, \ u(\pi) = 0,$$

where $\varphi(t) = \sin t$ is the first eigenfunction, was studied by R. Chiappinelli, J. Mawhin and R. Nugari [66]. They have weakened the assumptions but were merely able to localize the set M within $\{s\varphi + \widetilde{h} \mid s \in [s_0, s_1]\}$. On the other hand, J.L. Kazdan and F.W. Warner have noticed that their result still holds if the function h is decomposed using a positive function φ rather than an eigenfunction. In order to better understand the relation between these two decompositions of the forcing h, C. De Coster and P. Habets [88] have studied an ODE problem with two parameters, i.e. where $h(t) = r + s\sin t + \tilde{h}(t)$. This is presented in Section VI-3. Similar results were obtained by C. Fabry, J. Mawhin and M.N. Nkashama [115] for periodic solutions of the one parameter equation

$$u'' + f(t, u, u') = s,$$

with application to the Liénard equation. Periodic solutions of the Rayleigh equation can be found in P. Habets and P. Torres [152]. Both these papers combine degree theory and lower and upper solutions. Sections VI-1 and VI-2 concern these problems.

Some first important results concerning non-resonance problems are due to C.L. Dolph [106]. Resonance problems were analyzed by E.M. Landesman and A.C. Lazer [198] who introduced the so-called Landesman-Lazer conditions. Here, a boundary value problem associated with the equation

$$u'' + f(t, u) = h(t),$$

is said to be non-resonant for cases where it has a solution for every h and resonant if the problem is solvable only for some h. The use of lower and upper solutions can be useful to treat such situations when the nonlinearity interacts with the first eigenvalue. This matter is the subject of Chapter VII. The idea to use lower and upper solutions to treat such problems seems to go back to J.L. Kazdan and F.W. Warner [177] in 1975 for the results on the left of the first eigenvalue. Extensions to problems with nonlinearities between the two first eigenvalues were considered mainly by J.P. Gossez and P. Omari [132] and P. Habets and P. Omari [145] with the help of non-ordered lower and upper solutions.

In Sections VII-1 and VII-3 we present different kinds of non-resonance conditions at the left of the first eigenvalue and between the two first ones. Sections VII-2 and VII-4 are devoted to resonance conditions closely related to the so-called Landesman-Lazer conditions at the left of the first eigenvalue and between the first two eigenvalues for the periodic and the Dirichlet problems. Extensions to derivative dependent Dirichlet problems are presented in Section VII-7.

In Section VII-5 we present further resonance results where the Landesman-Lazer conditions fail. These are sign conditions as introduced by D.G. de Figueiredo and W.N. Ni [101] and what is now called a strong resonance condition, i.e. situations where $f(t, u)$ tends to 0 at $\pm\infty$ but not too quickly. A first important contribution on these kind of problems was obtained by

A. Ambrosetti and G. Mancini [23] although some previous steps are due to J.L. Kazdan and F.W. Warner [177]. Other results, including the so-called Ahmad, Lazer and Paul condition as introduced by these authors in [4] have already been considered in Section IV-4.

The situation is quite different if we consider nonlinearities depending only on the derivative such as

$$\begin{gathered} u'' + u + g(u') = p(t), \\ u(0) = 0, \ \ u(\pi) = 0. \end{gathered}$$

In that case the Landesman-Lazer condition does not apply. A first systematic study of these problems appears in A. Cañada and P. Drábek [54] for Dirichlet, Neumann, and periodic problems. This paper was followed by P. Habets and L. Sanchez [150] for the Dirichlet problem and by J. Mawhin [216] for the Neumann and the periodic problems. Some of these results are presented in Section VI-4.

Section VII-6 presents the use of lower and upper solutions to deal with some Fredholm alternative problems associated with the Fučík spectrum. These were introduced by L. Aguinaldo and K. Schmitt [3] using other methods. In this section we work also problems with monotonicity assumptions.

The problem of finding positive solutions of the boundary value problem

$$u'' + f(t,u) = 0, \quad u(a) = 0, \ u(b) = 0, \tag{9.1}$$

is considered in Chapter VIII. This problem has been widely studied, see for example P.L. Lions [201], D.G. de Figueiredo [96, 97], L.H. Erbe, S. Hu and H. Wang [110], L.H. Erbe and H. Wang [111] and the references therein. The most studied case assumes that $\alpha(t) \equiv 0$ is a lower solution. A necessary condition to obtain positive solutions of (9.1) is that $\frac{f(t,u)}{u}$ crosses the first eigenvalue for u going from 0 to $+\infty$. This approach gives us four typical situations:

(a) The sublinear problem where $\frac{f(t,u)}{u}$ is above the first eigenvalue in 0 and below it at $+\infty$. This case is classically dealt with via lower and upper solutions as in D.G. de Figueiredo [96], D.G. Costa and J.V.A. Goncalves [76]. Preliminary ideas can already be found in the work of E. Picard [250].

(b) The superlinear case, where $\frac{f(t,u)}{u}$ is below the first eigenvalue in 0 and above it at $+\infty$, is generally dealt with by using other methods (see [45, 201, 96, 97, 231, 111]). The treatment by lower and upper solutions can be found in K. Ben-Naoum and C. De Coster [31]. The main difficulty

is that in this case the natural lower and upper solutions are in the reverse order.
(c) The sub-superlinear case, where $\frac{f(t,u)}{u}$ is above the first eigenvalue in 0 and at $+\infty$, and
(d) the super-sublinear case, where $\frac{f(t,u)}{u}$ is below the first eigenvalue in 0 and at $+\infty$, give raise to multiplicity results. In these cases we need an additional condition which forces a "double crossing" of the first eigenvalue. In the sub-superlinear case, the additional condition, introduced for example by K.J. Brown and H. Budin [47] (see also [100, 75, 110]), is related to the existence of a strict upper solution. P. Rabinowitz [258] (see also [110]) has considered also super-sub nonlinearities.
A complete study of these four problems using lower and upper solutions was done by K. Ben-Naoum and C. De Coster [31] (see also C. De Coster [84] for the sub-superlinear case) and is presented in Sections VIII-1.1, VIII-1.2, VIII-2.1 and VIII-2.2. In case the nonlinearity crosses the two first eigenvalues, we can prove the existence of more solutions, positive, negative, or sign-changing as observed in Section IV-3.

Section VIII-1.3 is concerned with the less studied semipositone problem where $\alpha(t) \equiv 0$ is not necessarily a lower solution. The results presented in that section are related to A. Castro, J.B. Garner and R. Shivaji [57].

Until this point, our assumptions were such that for $f(t,u) = q(t)g(u)$ the function q has to be positive. In the last ten years, a great deal of interest has been devoted to the investigation of equations with an indefinite weight, i.e. the case that q changes sign. This question is the subject of Section VIII-1.4 whose results are due to M. Gaudenzi, P. Habets and F. Zanolin [126].

Perhaps one of the oldest ways to obtain multiplicity results relies on the study of the linearization of the equation along a known solution. Such results, obtained by H. Amann [14, 15] and K. Schmitt [274], are described in Section VIII-2.3.

In Section VIII-3, we study positive solutions of parametric problems

$$u'' + sf(t,u) = 0, \quad u(0) = 0, \; u(1) = 0,$$

where $s > 0$ and $f(t,u) \geq 0$ can be singular in $t = 0$ and 1. The results of this section can be found in M. Gaudenzi and P. Habets [125].

The problem

$$u'' + g(u)u' + f(t,u) = h(t), \quad u(a) = u(b), \; u'(a) = u'(b),$$

where f is singular for $u = 0$, is considered in Section IX-1. Although M. Nagumo [227] has proven in 1944 some very interesting results for this problem, the paper of A.C. Lazer and S. Solimini [199] was the starting point for a large literature on the subject. The attractive case was considered by P. Habets and L. Sanchez [149] using well-ordered lower and upper solutions. The repulsive case gives rise to non-ordered lower and upper solution and has been investigated by D. Bonheure and C. De Coster [41].

Problem

$$u'' + f(t, u) = 0, \quad u(0) = 0, \ u(\pi) = 0,$$

where f is singular at both the end points $t = 0$, $t = \pi$ and $u = 0$, appears in several applied mathematical problems (see [29]). A typical example is the Dirichlet problem for the generalized Emden-Fowler equation

$$u'' + \frac{g(t)}{u^\sigma} = 0, \quad u(0) = 0, \ u(\pi) = 0,$$

which was investigated by S.D. Taliaferro [296]. The general case was considered by several authors [29, 37, 124, 142, 298]. Most of these results rely on the fact that, $f(t, u)$ being positive, the solution u is concave. This assumption was given up in J. Janus and J. Myjak [174] for nonlinearities $f(t, u) = \frac{g(t)}{u^\sigma} + h(t)$. The general case was considered, using lower and upper solutions, by A.G. Lomtatidze [206], I.T. Kiguradze and B.L. Shekhter [187] and P. Habets and F. Zanolin [153, 154]. We present in Section IX-2 the results of P. Habets and F. Zanolin [154]. In 2001, C. De Coster [86] generalized these results to the derivative dependent case. This work is described in Section IX-3. Existence of two positive solutions for singular problems in the sub-superlinear case, which extends Section VIII-2.1, was considered by C. De Coster, M.R. Grossinho and P. Habets [87]. Section IX-4 describes these results.

The use of lower and upper solutions in the singular perturbed Cauchy problem was introduced in 1939 by M. Nagumo [225]. In 1955, using Nagumo's work on lower and upper solutions, N.I. Briš [46] considered various boundary value problems. This method has not received much attention until the works of P. Habets and M. Laloy [144] and F.A. Howes [166, 167]. Since then a family of classical problems in singular perturbation theory was revisited using lower and upper solutions (see [170, 61] and the references therein). We present some of them in Chapter X. We had already considered in Section II-1.3 the model example

$$\epsilon u'' + uu' - u = 0, \quad u(0) = A, \ u(1) = B,$$

due to W.A. Harris [156].

The Dirichlet problem with a boundary layer at one end point had already been considered in 1952 by E.A. Coddington and N. Levinson [71] and had further been developed in A. Erdélyi [112]. General separated boundary conditions were first considered by N.I. Briš [46] using lower and upper solutions. Dirichlet problems with boundary layers at both end points were introduced by N.I. Briš [46] and further extended using lower and upper solutions by P. Habets and M. Laloy [144] and F. Howes [166]. S. Haber and N. Levinson [143] have noticed in 1955 the possibility for the reduced problem to have angular solutions. Lower and upper solutions with angles have turned out to be well-adapted to this problem and were used by F. Howes [169]. Several recent results on angular solutions can be found in V.F. Butuzov, N.N. Nefedov, and K.R. Schneider [49]. Problems with an algebraic boundary layer appear together with moving singularities and are related to turning point problems. Some results of this type can be found in F.A. Howes [168] or in H. Gingold and S. Rosenblat [128].

Chapter I

The Periodic Problem

1 A First Approach of Lower and Upper Solutions

Consider the periodic boundary value problem

$$\begin{aligned} u'' &= f(t,u), \\ u(a) = u(b), &\ u'(a) = u'(b), \end{aligned} \tag{1.1}$$

where $a < b$ and f is a continuous function. The simplest approach to the method of lower and upper solutions for this problem uses the following definitions.

Definitions 1.1. *A function $\alpha \in C^2(]a,b[) \cap C^1([a,b])$ is a* lower solution *of the periodic problem* (1.1) *if*
(a) for all $t \in]a,b[$, $\alpha''(t) \geq f(t,\alpha(t))$;
(b) $\alpha(a) = \alpha(b)$, $\alpha'(a) \geq \alpha'(b)$.

A function $\beta \in C^2(]a,b[) \cap C^1([a,b])$ is an upper solution *of* (1.1) *if*
(a) for all $t \in]a,b[$, $\beta''(t) \leq f(t,\beta(t))$;
(b) $\beta(a) = \beta(b)$, $\beta'(a) \leq \beta'(b)$.

The main result of the method is an intermediate value theorem. It proves that if we can find a lower solution which is smaller than an upper one, there is a solution wedged between these two.

Theorem 1.1. *Let α and β be lower and upper solutions of* (1.1) *such that $\alpha \leq \beta$, define $E = \{(t,u) \in [a,b] \times \mathbb{R} \mid \alpha(t) \leq u \leq \beta(t)\}$ and assume $f : E \to \mathbb{R}$ is continuous.*

Then the problem (1.1) *has at least one solution $u \in C^2([a,b])$ such that for all $t \in [a,b]$*

$$\alpha(t) \leq u(t) \leq \beta(t).$$

Proof. Consider the modified problem

$$\begin{aligned} u'' - u &= f(t, \gamma(t,u)) - \gamma(t,u), \\ u(a) &= u(b), \ u'(a) = u'(b), \end{aligned} \tag{1.2}$$

where $\gamma : [a,b] \times \mathbb{R} \to \mathbb{R}$ is defined by

$$\begin{aligned} \gamma(t,u) &= \alpha(t), && \text{if } u < \alpha(t), \\ &= u, && \text{if } \alpha(t) \le u \le \beta(t), \\ &= \beta(t), && \text{if } u > \beta(t). \end{aligned} \tag{1.3}$$

Claim 1 – The problem (1.2) *has at least one solution.* Let us write (1.2) as an integral equation

$$u(t) = \int_a^b G(t,s)[f(s,\gamma(s,u(s))) - \gamma(s,u(s))]\, ds,$$

where $G(t,s)$ is the Green's function corresponding to the problem

$$\begin{aligned} u'' - u &= f(t), \\ u(a) &= u(b), \ u'(a) = u'(b). \end{aligned} \tag{1.4}$$

The operator

$$T : C([a,b]) \to C([a,b]),$$

defined by

$$(Tu)(t) = \int_a^b G(t,s)[f(s,\gamma(s,u(s))) - \gamma(s,u(s))]\, ds, \tag{1.5}$$

is completely continuous and bounded. By Schauder's Theorem, T has a fixed point which is a solution of (1.2).

Claim 2 – All solutions u of (1.2) *satisfy on* $[a,b]$

$$\alpha(t) \le u(t) \le \beta(t). \tag{1.6}$$

Let us assume on the contrary that, for some $t_0 \in [a,b]$,

$$\min_t (u(t) - \alpha(t)) = u(t_0) - \alpha(t_0) < 0.$$

If $t_0 \in \,]a,b[$, we obtain the contradiction

$$0 \le u''(t_0) - \alpha''(t_0) = f(t_0, \alpha(t_0)) + u(t_0) - \alpha(t_0) - \alpha''(t_0) < 0.$$

In case

$$\min_t (u(t) - \alpha(t)) = u(a) - \alpha(a) = u(b) - \alpha(b) < 0,$$

we obtain

$$u'(a) - \alpha'(a) \geq 0 \geq u'(b) - \alpha'(b)$$

and from the definition of a lower solution

$$u'(a) - \alpha'(a) \leq u'(b) - \alpha'(b).$$

Hence, $u'(a) - \alpha'(a) = 0$ and for $t > a$ small enough

$$u'(t) - \alpha'(t) = \int_a^t [f(s, \alpha(s)) + u(s) - \alpha(s) - \alpha''(s)]\, ds < 0$$

which contradicts $u - \alpha$ to be minimum in a. This proves $\alpha \leq u$.

In a similar way, we prove $u \leq \beta$.

Conclusion – From Claim 1, (1.2) has a solution which, from Claim 2, satisfies (1.6). Therefore, it is also a solution of (1.1). □

Remark The proof of Claim 2 relies on a maximum principle. Here, we have used two different arguments to obtain such a conclusion. For $t_0 \in]a, b[$, we deduce the claim from the pointwise condition

$$u''(t_0) - \alpha''(t_0) < 0$$

and for the end points we used a first order condition

$$u'(t) - \alpha'(t) < 0$$

valid on an interval $[a, a + \varepsilon]$. These are two basic ways to obtain the conclusion. Notice however that the interval condition is more general and can be used in both cases. More generally the lower and upper solution method works for a large class of problems where a maximum principle holds. This remark will be exemplified in section II-5.

Theorem 1.1 furnishes two kinds of information. It is an existence result but it also gives a localization of the solution. In the following example, such a localization provides an asymptotic estimate on the solution.

Example 1.1. Consider the problem

$$\begin{gathered}\epsilon^2 u'' = u^3 - \sin^3 t,\\ u(0) = u(2\pi),\ u'(0) = u'(2\pi),\end{gathered}$$

where $\epsilon > 0$ is a parameter. It is easy to see that

$$\alpha(t) = \sin t - 2\epsilon \quad \text{and} \quad \beta(t) = \sin t + 2\epsilon$$

are lower and upper solutions. From Theorem 1.1, we deduce the existence of a solution u such that $\alpha(t) \leq u(t) \leq \beta(t)$ on $[0, 2\pi]$ which implies for ϵ small the asymptotic estimate

$$u(t) = \sin t + O(\epsilon).$$

The localization result contained in Theorem 1.1 can also give multiplicity results.

Example 1.2. Consider the problem

$$\begin{gathered} u'' + u^4 - u^2 = h(t), \\ u(a) = u(b), \ u'(a) = u'(b), \end{gathered}$$

where $h \in \mathcal{C}([a,b])$ is such that for all $t \in [a,b]$, $-1/4 \leq h(t) \leq 0$. It is easy to observe that $\alpha_1(t) = -1$ and $\alpha_2(t) = 0$ are lower solutions and $\beta_1(t) = -1/\sqrt{2}$ and $\beta_2(t) = 1/\sqrt{2}$ are upper solutions. Hence, we have at least two solutions u_1 and u_2 such that on $[a,b]$

$$-1 \leq u_1(t) \leq -1/\sqrt{2}, \quad 0 \leq u_2(t) \leq 1/\sqrt{2}.$$

Notice that in this example, it is possible, using further results on lower and upper solutions and reinforcing the assumptions, to prove existence of two more solutions (see Example III-1.4).

Remarks Theorem 1.1 can be interpreted as an intermediate value theorem for the operator T defined by (1.5). It is easy to see that

$$\begin{aligned} T\alpha &= \int_a^b G(\cdot, s)[f(s, \alpha(s)) - \alpha(s)]\, ds \\ &\geq \int_a^b G(\cdot, s)[\alpha''(s) - \alpha(s)]\, ds \geq \alpha \end{aligned}$$

and

$$T\beta \leq \beta.$$

Theorem 1.1 proves then the existence of an intermediate value $u \in [\alpha, \beta]$ such that $Tu = u$.

This intermediate value property depends strongly on the ordering $\alpha \leq \beta$. In case this ordering is not satisfied, the result does not hold as such. Consider for example the problem

$$\begin{gathered} u'' + u = \sin t, \\ u(0) = u(2\pi), \ u'(0) = u'(2\pi). \end{gathered}$$

Multiplying by $\sin t$ and integrating, it follows right away that this problem has no solution although

$$\alpha(t) = 1 \quad \text{and} \quad \beta(t) = -1$$

are lower and upper solutions.

Another remark of the same type is that the result is no longer true if we reverse the inequalities in the boundary conditions. Consider for example the problem

$$u'' = 1,$$
$$u(-1) = u(1),\ u'(-1) = u'(1).$$

It has no solution although $\alpha(t) = t^2 - 1$ is almost a lower solution (i.e. $\alpha''(t) = 2 > 1$, $\alpha(-1) = \alpha(1)$), $\beta(t) = 1$ is an upper solution ($\beta''(t) \leq 1$, $\beta(-1) = \beta(1)$, $\beta'(-1) = \beta'(1)$) and $\alpha(t) < \beta(t)$. Clearly, Theorem 1.1 does not apply here since $\alpha'(-1) < \alpha'(1)$ which means that α is not a lower solution.

Exercise 1.1. (Alternative proof) Assume α and $\beta \in C^2([a,b])$, $\alpha(t) < \beta(t)$ on $[a,b]$ and prove the previous result from the homotopy

$$u'' = \lambda f(t,u) + (1-\lambda)\,[k^2(u - \tfrac{\alpha(t)+\beta(t)}{2}) + \tfrac{\alpha''(t)+\beta''(t)}{2}],$$
$$u(a) = u(b),\ u'(a) = u'(b),$$

where k is large enough. Use degree theory and the set

$$\Omega = \{u \in C([a,b]) \mid \forall t \in [a,b],\ \alpha(t) < u(t) < \beta(t)\}.$$

Hint : See [113].

The main difficulty in applying Theorem 1.1 is to find appropriate lower and upper solutions. The first and simplest idea is to try constant functions. This gives the following corollary.

Corollary 1.2. *Let $f : [a,b] \times \mathbb{R} \to \mathbb{R}$ be a continuous function such that for some $r_1 \leq r_2$ and all $t \in [a,b]$,*

$$f(t,r_1) \leq 0 \leq f(t,r_2).$$

Then the problem (1.1) *has at least one solution $u \in C^2([a,b])$ such that for all $t \in [a,b]$,*

$$r_1 \leq u(t) \leq r_2.$$

Theorem 1.1 and Corollary 1.2 can be used in various situations.

Example 1.3. Consider the problem

$$\begin{gathered} u'' = ku^{2n+1} + h(t), \\ u(0) = u(2\pi), \ u'(0) = u'(2\pi), \end{gathered}$$

where $k > 0$, $n \in \mathbb{N}$ and $h \in \mathcal{C}([0, 2\pi])$. It is easy to see that

$$\alpha(t) = -\left[\frac{\|h\|_\infty}{k}\right]^{\frac{1}{2n+1}} \quad \text{and} \quad \beta(t) = \left[\frac{\|h\|_\infty}{k}\right]^{\frac{1}{2n+1}}$$

are lower and upper solutions. Hence, there is a solution u such that

$$-\left[\frac{\|h\|_\infty}{k}\right]^{\frac{1}{2n+1}} \le u \le \left[\frac{\|h\|_\infty}{k}\right]^{\frac{1}{2n+1}}.$$

Example 1.4. Consider the forced pendulum equation

$$\begin{gathered} u'' + \sin u = h(t), \\ u(0) = u(2\pi), \ u'(0) = u'(2\pi), \end{gathered} \tag{1.7}$$

where $h \in \mathcal{C}([0, 2\pi])$. Then, if $\|h\|_\infty \le 1$, $\alpha(t) = \pi/2$ and $\beta(t) = 3\pi/2$ are respectively lower and upper solutions. Existence of a solution in $[\frac{\pi}{2}, \frac{3\pi}{2}]$ follows.

A more elaborate approach uses a different construction of the lower and upper solutions. Write $h(t) = \bar{h} + \tilde{h}(t)$, where $\bar{h} = \frac{1}{2\pi}\int_0^{2\pi} h(t)\,dt$ is the constant term in the Fourier expansion of h. We then look for a lower solution using the same type of decomposition, i.e.

$$\alpha(t) = A + w(t)$$

with $\int_0^{2\pi} w(t)\,dt = 0$. This function has to satisfy the inequality

$$\alpha''(t) + \sin\alpha(t) = w''(t) + \sin\alpha(t) \ge \bar{h} + \tilde{h}(t).$$

In case $\bar{h}$ lower bounds $\sin\alpha(t)$, it is natural to choose w to be the solution of

$$w'' = \tilde{h}(t), \quad w(0) = w(2\pi), \ w'(0) = w'(2\pi), \ \bar{w} = 0.$$

We know from Proposition A-4.2 that

$$\|w\|_\infty \le \tfrac{\pi}{6}\|\tilde{h}\|_{L^1}.$$

Hence, if we assume

$$\|\tilde{h}\|_{L^1} \le 3 \quad \text{and} \quad \bar{h} \le \cos(\tfrac{\pi}{6}\|\tilde{h}\|_{L^1}),$$

it is easy to observe that

$$\|w\|_\infty \le \tfrac{\pi}{2}$$

and

$$\alpha(t) = \tfrac{\pi}{2} + w(t)$$

is a lower solution of (1.7). If further $\bar{h} \ge -\cos(\frac{\pi}{6}\|\tilde{h}\|_{L^1})$, we prove similarly that

$$\beta(t) = \tfrac{3\pi}{2} + w(t) > \alpha(t)$$

is an upper solution and existence of a solution follows from Theorem 1.1.

The existence of a second solution, different from u mod 2π, can be obtained reinforcing the assumptions and using further results on lower and upper solutions (see Example III-1.1).

2 Existence of C^2-Solutions

2.1 Definitions

Dealing with applications, it is useful to generalize the concept of lower and upper solutions. The following definitions take this observation into account and try to capture what is essential in Definitions 1.1.

Definitions 2.1. *A function $\alpha \in C([a,b])$ such that $\alpha(a) = \alpha(b)$ is a C^2-*lower solution *of* (1.1) *if its periodic extension on $\mathbb{R}$, defined by $\alpha(t) = \alpha(t+b-a)$, is such that for any $t_0 \in \mathbb{R}$*
either $D_-\alpha(t_0) < D^+\alpha(t_0)$;
or there exist an open interval I_0 with $t_0 \in I_0$ and a function $\alpha_0 \in C^1(I_0)$ such that
(a) $\alpha(t_0) = \alpha_0(t_0)$ and $\alpha(t) \ge \alpha_0(t)$ for all $t \in I_0$;
(b) $\alpha_0''(t_0)$ exists and $\alpha_0''(t_0) \ge f(t_0, \alpha_0(t_0))$.

*A function $\beta \in C([a,b])$ such that $\beta(a) = \beta(b)$ is a C^2-*upper solution *of* (1.1) *if its periodic extension on $\mathbb{R}$ is such that for any $t_0 \in \mathbb{R}$*
either $D^-\beta(t_0) > D_+\beta(t_0)$;
or there exist an open interval I_0 with $t_0 \in I_0$ and a function $\beta_0 \in C^1(I_0)$ such that
(a) $\beta(t_0) = \beta_0(t_0)$ and $\beta(t) \le \beta_0(t)$ for all $t \in I_0$;
(b) $\beta_0''(t_0)$ exists and $\beta_0''(t_0) \le f(t_0, \beta_0(t_0))$.

When there is no possible confusion with Definitions 1.1, we will speak of lower and upper solutions of (1.1) particularly when the definitions of lower (resp. upper) solutions are equivalent. For example, Definitions 2.1 reduce to Definitions 1.1 if f is continuous and the lower and upper solutions α and β are in $C^2([a,b])$.

Geometric interpretations The definitions of lower and upper solutions can be understood geometrically. To this end, let us reinforce these notions and assume the inequalities (b) in Definitions 2.1 are strict. The graphs of α and β are then curves with a definite geometry with respect to the vector field. Solutions cannot be tangent to a lower solution from above or tangent to an upper solution from below.

The existence of α_0 and β_0 such that $\alpha_0''(t_0) \geq f(t_0, \alpha_0(t_0))$ and $\beta_0''(t_0) \leq f(t_0, \beta_0(t_0))$ is a "second order" condition. In most applications, the functions α_0 and β_0 will be chosen as restrictions of α and β.

On the other hand, the conditions $D_-\alpha(t_0) < D^+\alpha(t_0)$ and $D^-\beta(t_0) > D_+\beta(t_0)$ are of "first order". Geometrically, the first one means that the curve $u = \alpha(t)$ can have angles but with opening from above. In such a case, solutions cannot be tangent from above at the vertex $(t = t_0)$ of the angle. This angle condition already appears in Definitions 1.1 if we consider the boundary points a and b after extending α by periodicity. Similarly, the condition $D^-\beta(t_0) > D_+\beta(t_0)$ means that the curve $u = \beta(t)$ can have angles with an opening from below. The same geometry appears if we consider as the lower solution the maximum of a finite number of $\mathcal{C}^2$-lower solutions or as the upper solution the minimum of a finite number of $\mathcal{C}^2$-upper solutions. The following results concretize this idea.

Proposition 2.1. *Let $\alpha_i \in \mathcal{C}([a,b])$ $(i = 1, ..., n)$ be $\mathcal{C}^2$-lower solutions of* (1.1)*. Then the function*

$$\alpha(t) = \max_{1 \leq i \leq n} \alpha_i(t), \quad t \in [a,b]$$

is a $\mathcal{C}^2$-lower solution of (1.1).

Proof. The proof is straightforward since for any $t_0 \in [a,b]$, there exists α_k such that $\alpha(t_0) = \alpha_k(t_0)$, $\alpha(t) \geq \alpha_k(t)$. Either $D_-\alpha_k(t_0) < D^+\alpha_k(t_0)$, which implies $D_-\alpha(t_0) < D^+\alpha(t_0)$, or there exist an open interval I_0 with $t_0 \in I_0$ and a function $\alpha_{k0} \in \mathcal{C}^1(I_0)$ which satisfies (a) and (b) in Definition 2.1. These facts imply that α is a $\mathcal{C}^2$-lower solution. □

In a similar way, we can prove the corresponding result for $\mathcal{C}^2$-upper solutions.

Proposition 2.2. *Let $\beta_j \in \mathcal{C}([a,b])$ $(j = 1, ..., m)$ be $\mathcal{C}^2$-upper solutions of* (1.1)*. Then the function*

$$\beta(t) = \min_{1 \leq j \leq m} \beta_j(t), \quad t \in [a,b]$$

is a $\mathcal{C}^2$-upper solution of (1.1).

2.2 Existence of Solutions

Our first approach to periodic solutions based on lower and upper solutions can be extended to $\mathcal{C}^2$-lower and upper solutions.

Theorem 2.3. *Let α and β be $\mathcal{C}^2$-lower and upper solutions of* (1.1) *such that $\alpha \le \beta$, define $E = \{(t,u) \in [a,b] \times \mathbb{R} \mid \alpha(t) \le u \le \beta(t)\}$ and assume $f : E \to \mathbb{R}$ is continuous.*

Then the problem (1.1) *has at least one solution $u \in \mathcal{C}^2([a,b])$ such that for all $t \in [a,b]$*

$$\alpha(t) \le u(t) \le \beta(t).$$

Proof. As in the proof of Theorem 1.1 we consider the modified problem (1.2), where γ is defined by (1.3). From the proof of this theorem, we know that problem (1.2) has a solution and we only have to prove that this solution satisfies

$$\alpha(t) \le u(t) \le \beta(t) \quad \text{on } [a,b].$$

Extend α and u by periodicity. If

$$\min_t (u(t) - \alpha(t)) < 0,$$

we can find $t_0 \in [a,b[$ such that

$$u(t_0) - \alpha(t_0) = \min_t (u(t) - \alpha(t))$$

and we have

$$u'(t_0) - D_-\alpha(t_0) \le u'(t_0) - D^+\alpha(t_0).$$

Hence, by definition of a $\mathcal{C}^2$-lower solution, there exist an open interval I_0 with $t_0 \in I_0$ and a function $\alpha_0 \in \mathcal{C}^1(I_0)$ such that
(i) $\alpha(t_0) = \alpha_0(t_0)$ and $\alpha(t) \ge \alpha_0(t)$ for all $t \in I_0$;
(ii) $\alpha_0''(t_0)$ exists and $\alpha_0''(t_0) \ge f(t_0, \alpha_0(t_0))$.
It follows that t_0 is a minimum of $u - \alpha_0$, $(u - \alpha_0)'(t_0) = 0$ and $(u - \alpha_0)''(t_0) \ge 0$, thus we obtain the contradiction

$$0 \le u''(t_0) - \alpha_0''(t_0) \le f(t_0, \alpha_0(t_0)) - \alpha_0(t_0) + u(t_0) - f(t_0, \alpha_0(t_0)) < 0.$$

In a similar way we prove that $u \le \beta$. Hence, the solution u of (1.2) is a solution of (1.1). □

The following elementary example illustrates the interest of Theorem 2.3. It exemplifies a class of problems that use "angular" lower and upper solutions.

Example 2.1. Consider the problem

$$\epsilon^2 u'' = \varphi(u) - \varphi(|t-\pi|),$$
$$u(0) = u(2\pi), \; u'(0) = u'(2\pi),$$

where $\epsilon > 0$ is a parameter and $\varphi \in C^1(\mathbb{R})$ is such that $\varphi'(t) \geq a^2 > 0$. Notice that the function $\alpha \in C([0, 2\pi])$, defined by $\alpha(t) = \max\{\alpha_1(t), \alpha_2(t)\}$ where

$$\alpha_1(t) = -(t-\pi) - \tfrac{\epsilon}{a} e^{-\frac{at}{\epsilon}} \quad \text{and} \quad \alpha_2(t) = (t-\pi) - \tfrac{\epsilon}{a} e^{\frac{a(t-2\pi)}{\epsilon}},$$

is a C^2-lower solution. Next we can check that the function

$$\beta(t) = |t-\pi| + \tfrac{\epsilon}{a} e^{-\frac{a|t-\pi|}{\epsilon}}$$

is a C^2-upper solution. From Theorem 2.3, we deduce the existence of a solution u together with the asymptotic estimate

$$u(t) = |t-\pi| + O(\epsilon).$$

2.3 Structure of the Set of Solutions

The set of solutions between α and β admits extremal elements in the following sense.

Theorem 2.4. *Let α and β be C^2-lower and upper solutions of* (1.1) *such that $\alpha \leq \beta$, define $E = \{(t,u) \in [a,b] \times \mathbb{R} \mid \alpha(t) \leq u \leq \beta(t)\}$ and assume $f : E \to \mathbb{R}$ is continuous.*

Then the problem (1.1) *has solutions u_{min}, $u_{max} \in C^2([a,b])$ such that*

$$\alpha \leq u_{min} \leq u_{max} \leq \beta,$$

and any other solution u of (1.1) *such that $\alpha \leq u \leq \beta$ satisfies*

$$u_{min} \leq u \leq u_{max}.$$

Solutions u_{min} and u_{max} of (1.1), which lie between α and β and are such that any other solution u of (1.1) with $\alpha \leq u \leq \beta$ satisfies $u_{min} \leq u \leq u_{max}$ are called *minimal* and *maximal solutions in* $[\alpha, \beta]$ or more simply *minimal* and *maximal solutions.*

Proof. Notice first that solutions of (1.1) are fixed points of the operator

$$T : C([a,b]) \to C([a,b])$$

defined by

$$(Tu)(t) = \int_a^b G(t,s)[f(s,u(s)) - u(s)]\,ds,$$

where $G(t,s)$ is the Green's function of (1.4). Define then the set

$$S = \{u \in \mathcal{C}([a,b]) \mid u = Tu,\ \alpha \le u \le \beta\}.$$

From Theorem 2.3, $S \neq \emptyset$. Further, S is compact as T is completely continuous. Consider next the family of sets

$$F_x = \{u \in S \mid u \ge x\},$$

where $x \in S$. This family has the finite intersection property as follows from Proposition 2.1 and Theorem 2.3. Hence, it is known (see [180, Theorem 5.1]) that there exists

$$u_{max} \in \bigcap_{x \in S} F_x,$$

which is a maximal solution.

Similarly, we prove existence of a minimal solution. □

The structure of the set of solutions is richer if f is non-decreasing with respect to u. In such a case, we have a continuum of solutions as follows from the following theorem.

Theorem 2.5. *Let α and β be $\mathcal{C}^2$-lower and upper solutions of (1.1) with $\alpha \le \beta$, define $E = \{(t,u) \in [a,b] \times \mathbb{R} \mid \alpha(t) \le u \le \beta(t)\}$ and assume $f : E \to \mathbb{R}$ is continuous, non-decreasing with respect to u.*

Let $t_0 \in [a,b]$ and u^* be a real number with $u_{min}(t_0) \le u^* \le u_{max}(t_0)$, where u_{min} and u_{max} are the minimal and maximal solutions defined in Theorem 2.4.

Then there exists a solution $u \in \mathcal{C}^2([a,b])$ of (1.1) such that $u_{min} \le u \le u_{max}$ and $u(t_0) = u^*$.

Proof. Let $t_0 \in [a,b]$ and u^* be such that $u_{min}(t_0) \le u^* \le u_{max}(t_0)$. Choose $\epsilon > 0$ such that $u_{max} - \epsilon \le u_{min} + \epsilon$ and define

$$\alpha_1(t) = \max\{u_{min}(t), u_{max}(t) - \epsilon\}, \quad \beta_1(t) = \min\{u_{max}(t), u_{min}(t) + \epsilon\}.$$

Observe that α_1 and β_1 are respectively $\mathcal{C}^2$-lower and upper solutions of (1.1). By Theorem 2.3, the problem (1.1) has a solution u_1 such that for all $t \in [a,b]$,

$$u_{min}(t) \le u_1(t) \le u_{min}(t) + \epsilon, \quad u_{max}(t) - \epsilon \le u_1(t) \le u_{max}(t).$$

In case $u^* \in [u_{min}(t_0), u_1(t_0)]$ we define

$$\alpha_2(t) = \max\{u_{min}(t), u_1(t) - \epsilon/2\}, \quad \beta_2(t) = \min\{u_1(t), u_{min}(t) + \epsilon/2\}$$

and obtain from Theorem 2.3 a solution u_2 such that on $[a, b]$

$$u_{min}(t) \le u_2(t) \le u_{min}(t) + \epsilon/2, \quad u_1(t) - \epsilon/2 \le u_2(t) \le u_1(t).$$

If $u^* \in]u_1(t_0), u_{max}(t_0)]$, we proceed in a similar way. This procedure defines a sequence of solutions $(u_k)_k$ that satisfies $|u_k(t_0) - u^*| \le \epsilon/2^{k-1}$. Next, from the Arzelà-Ascoli Theorem, there is a subsequence $(u_{k_n})_n$ such that for some $u \in \mathcal{C}([a, b])$, u_{k_n} converges to u in $\mathcal{C}([a, b])$. It follows then that u is a fixed point of T defined by (1.5) and therefore a solution of (1.1). Further, we have $u(t_0) = \lim_{n\to\infty} u_{k_n}(t_0) = u^*$. □

Existence of a continuum of solutions depends strongly on f being non-decreasing. Such a continuum does not exist in the following example.

Example 2.2. Consider the problem

$$\begin{gathered} u'' = u^3 - u^2, \\ u(0) = u(1), \; u'(0) = u'(1). \end{gathered}$$

Notice first that $\alpha(t) = -2$ and $\beta(t) = 2$ are lower and upper solutions. Also, it is straightforward from a phase plane analysis that this problem has only two solutions $u_1 = 0$, $u_2 = 1$, which are between α and β.

Notice also that solutions between lower and upper solutions are not necessarily ordered as shown in the example that follows.

Example 2.3. The piecewise linear problem

$$\begin{gathered} u'' = \min\{u + 2, \max\{-u, u - 2\}\}, \\ u(0) = u(2\pi), \; u'(0) = u'(2\pi) \end{gathered}$$

is such that $u_1(t) = \sin t$ and $u_2(t) = -\sin t$ are non-ordered solutions which lie between the lower solution $\alpha(t) = -3$ and the upper one $\beta(t) = 3$. In this problem, it follows from the phase plane analysis that the minimal and maximal solutions are respectively $u_{min}(t) = -2$ and $u_{max}(t) = 2$.

Exercise 2.1. Prove that under the assumptions of Theorem 2.4, if for some $M > 0$ and all $(t, u_1), (t, u_2) \in E$

$$u_1 \le u_2 \Rightarrow f(t, u_1) - f(t, u_2) \le M(u_1 - u_2),$$

then there exists one and only one solution u of (1.1) such that $\alpha \le u \le \beta$.

3 Existence of $W^{2,1}$-Solutions

3.1 Definitions

In this section, we consider the periodic boundary value problem (1.1), where f is L^1-Carathéodory. A function $f : D \subset [a,b] \times \mathbb{R}^n \to \mathbb{R}$ is said to satisfy *Carathéodory conditions* or to be a *Carathéodory function* if

(a) for a.e. $t \in [a,b]$, the function $f(t,\cdot)$ with domain $\{z \in \mathbb{R}^n \mid (t,z) \in D\}$ is continuous;

(b) for all $z \in \mathbb{R}^n$, the function $f(\cdot, z)$ with domain $\{t \in [a,b] \mid (t,z) \in D\}$ is measurable.

If furthermore, for some $p \in [1,\infty]$, the Carathéodory function f satisfies

(c) for all $r > 0$, there exists $h \in L^p(a,b)$ such that for all $(t,z) \in D$ with $|z| \leq r$, $|f(t,z)| \leq h(t)$,

we say that f is an *L^p-Carathéodory function* or that it satisfies *L^p-Carathéodory conditions.*

If f is an L^1-Carathéodory function, solutions of (1.1) are in $W^{2,1}(a,b)$. It is then natural to look for lower and upper solutions which are in this space or at least which are piecewise $W^{2,1}$. This observation motivates the definitions we present here. To simplify the notations, we extend $f(t,u)$ by periodicity $f(t,u) = f(t+T,u)$, where $T = b - a$.

Definitions 3.1. *A function $\alpha \in C([a,b])$ such that $\alpha(a) = \alpha(b)$ is a* $W^{2,1}$-lower solution *of* (1.1) *if its periodic extension on $\mathbb{R}$, defined by $\alpha(t) = \alpha(t+b-a)$ is such that for any $t_0 \in \mathbb{R}$*
either $D_-\alpha(t_0) < D^+\alpha(t_0)$
or there exists an open interval I_0 such that $t_0 \in I_0$, $\alpha \in W^{2,1}(I_0)$ and for a.e. $t \in I_0$,

$$\alpha''(t) \geq f(t,\alpha(t)).$$

A function $\beta \in C([a,b])$ such that $\beta(a) = \beta(b)$ is a $W^{2,1}$-upper solution *of* (1.1) *if its periodic extension on $\mathbb{R}$, defined by $\beta(t) = \beta(t+b-a)$ is such that for any $t_0 \in \mathbb{R}$*
either $D^-\beta(t_0) > D_+\beta(t_0)$
or there exists an open interval I_0 such that $t_0 \in I_0$, $\beta \in W^{2,1}(I_0)$ and for a.e. $t \in I_0$,

$$\beta''(t) \leq f(t,\beta(t)).$$

In the definition of $W^{2,1}$-lower solution we did not use, as for C^2-lower solutions, auxiliary functions α_0. This feature is due to the fact that for $W^{2,1}$-lower solutions we cannot work pointwise and the natural modification to be used in the proof does not allow to use information along the curve $u = \alpha_0(t)$.

In practical problems the "natural" lower and upper solutions are often more regular than needed in Definitions 3.1. For example, α will often be defined as the solution of some auxiliary problem so that it will be in $W^{2,1}(a,b)$ and satisfy for a.e. $t \in [a,b]$

$$\alpha''(t) \geq f(t, \alpha(t)),$$
$$\alpha(a) = \alpha(b),\ \alpha'(a) \geq \alpha'(b).$$

In such a case, α is a $W^{2,1}$-lower solution of (1.1).

In applications the various notions of lower and upper solution will frequently coincide. This observation holds, for instance, if f is continuous and $\alpha \in \mathcal{C}^1([a,b]) \cap \mathcal{C}^2(]a,b[)$. With these assumptions, α is a $W^{2,1}$-lower solution if and only if it satisfies Definition 1.1. Therefore and if there is no possible confusion, we will often call $W^{2,1}$-lower and upper solutions of (1.1), lower and upper solutions.

3.2 Existence of Solutions

In this paragraph, we present the basic existence results for solutions $u \in W^{2,1}(a,b)$ of (1.1).

Theorem 3.1. *Let α and β be $W^{2,1}$-lower and upper solutions of* (1.1) *such that $\alpha \leq \beta$, $E = \{(t,u) \in [a,b] \times \mathbb{R} \mid \alpha(t) \leq u \leq \beta(t)\}$ and $f : E \to \mathbb{R}$ be an L^1-Carathéodory function.*

Then the problem (1.1) *has at least one solution $u \in W^{2,1}(a,b)$ such that for all $t \in [a,b]$*

$$\alpha(t) \leq u(t) \leq \beta(t).$$

Proof. As in the proof of Theorem 1.1, we consider the modified problem (1.2) where γ is defined by (1.3). Repeating Claim 1 of this proof shows that problem (1.2) has a solution and we only have to verify that this solution satisfies on $[a,b]$

$$\alpha(t) \leq u(t) \leq \beta(t).$$

Let us assume on the contrary that, for some $t_0 \in \mathbb{R}$

$$\min_t (u(t) - \alpha(t)) = u(t_0) - \alpha(t_0) < 0.$$

Then we have

$$u'(t_0) - D_-\alpha(t_0) \leq u'(t_0) - D^+\alpha(t_0)$$

and, by definition of a $W^{2,1}$-lower solution, there exists an open interval I_0, with $t_0 \in I_0, \alpha \in W^{2,1}(I_0)$ and for almost every $t \in I_0$

$$\alpha''(t) \geq f(t, \alpha(t)).$$

Further $u'(t_0) - \alpha'(t_0) = 0$ and for $t \geq t_0$, near enough t_0,

$$\begin{aligned} u'(t) - \alpha'(t) &= \int_{t_0}^{t} (u''(s) - \alpha''(s))\, ds \\ &\leq \int_{t_0}^{t} [f(s, \alpha(s)) + u(s) - \alpha(s) - f(s, \alpha(s))]\, ds < 0. \end{aligned}$$

This inequality proves that $u(t_0) - \alpha(t_0)$ is not a minimum of $u - \alpha$ which is a contradiction. A similar argument holds to prove $u(t) \leq \beta(t)$. □

A simple illustration of Theorem 3.1 is the following.

Example 3.1. Consider the problem

$$\begin{gathered} u'' - \frac{1}{\sqrt{t}} u^2 = q(t), \\ u(0) = u(2\pi),\ u'(0) = u'(2\pi), \end{gathered}$$

where $q \in L^1(0, 2\pi)$. To build a lower solution, we write $q(t) = \bar{q} + \tilde{q}(t)$, with $\bar{q} = \frac{1}{2\pi} \int_0^{2\pi} q(s)\, ds$ and look for a lower solution $\alpha(t) = A + w(t)$ such that $\bar{w} = \frac{1}{2\pi} \int_0^{2\pi} w(s)\, ds = 0$. We wish

$$\alpha'' - \frac{1}{\sqrt{t}} \alpha^2 = w'' - \frac{1}{\sqrt{t}} \alpha^2 \geq \bar{q} + \tilde{q}(t).$$

To this end we choose w to be the solution of

$$\begin{gathered} w'' = \tilde{q}(t), \\ w(0) = w(2\pi),\ w'(0) = w'(2\pi),\ \bar{w} = 0. \end{gathered}$$

We have $\|\alpha'\|_\infty = \|w'\|_\infty \leq \|\tilde{q}\|_{L^1}$ and if we choose $A = -w(0)$, we can write for all $t \in [0, 2\pi]$, $|\alpha(t)| \leq \|\tilde{q}\|_{L^1} t$. We compute then

$$\alpha'' - \frac{1}{\sqrt{t}} \alpha^2 - \bar{q} - \tilde{q}(t) \geq -\|\tilde{q}\|_{L^1}^2 (2\pi)^{3/2} - \bar{q}$$

and $\alpha(t) = w(t) - w(0)$ is a lower solution if

$$\bar{q} + (2\pi)^{3/2} \|\tilde{q}\|_{L^1}^2 \leq 0.$$

If B is a large enough positive constant then $\beta(t) = \alpha(t) + B \geq \alpha(t)$ is an upper solution and we can conclude from Theorem 3.1.

With our definition, it is not obvious that maxima of lower solutions and minima of upper solutions are still lower and upper solutions. However, it is easy to prove existence of solutions between such maxima and minima. Such a result is a substitute to Propositions 2.1 and 2.2.

Theorem 3.2. *Let α_i $(i = 1, \ldots, n)$ be $W^{2,1}$-lower solutions and β_j $(j = 1, \ldots, m)$ be $W^{2,1}$-upper solutions of* (1.1), $\alpha := \max\limits_{1\le i\le n} \alpha_i$ *and* $\beta := \min\limits_{1\le j\le m} \beta_j$ *be such that* $\alpha \le \beta$. *Define* $E = \{(t,u) \in [a,b] \times \mathbb{R} \mid \alpha(t) \le u \le \beta(t)\}$ *and let* $f : E \to \mathbb{R}$ *be an* L^1*-Carathéodory function.*

Then the problem (1.1) *has at least one solution* $u \in W^{2,1}(a,b)$ *such that for all* $t \in [a,b]$

$$\alpha(t) \le u(t) \le \beta(t).$$

Proof. Consider the modified problem

$$\begin{aligned} u'' - u &= \bar{f}(t,u) - \gamma(t,u), \\ u(a) &= u(b), \quad u'(a) = u'(b), \end{aligned} \tag{3.1}$$

where $\gamma(t,u)$ is defined in (1.3) and

$$\begin{aligned} \bar{f}(t,u) &= \min_{1\le i\le n} f(t, \max\{\alpha_i(t), u\}), && \text{if } u \le \alpha(t), \\ &= f(t,u), && \text{if } \alpha(t) < u < \beta(t), \\ &= \max_{1\le j\le m} f(t, \min\{\beta_j(t), u\}), && \text{if } \beta(t) \le u. \end{aligned}$$

First, we prove as in Theorem 1.1 that problem (3.1) has a solution. Next, we show that on $[a,b]$

$$\alpha(t) \le u(t) \le \beta(t).$$

Extend α and u by periodicity and assume by contradiction that

$$\min_t (u(t) - \alpha(t)) < 0.$$

It follows that for some t_0 and $i \in \{1, \ldots, n\}$ we have

$$\min_t (u(t) - \alpha(t)) = u(t_0) - \alpha_i(t_0) = \min_t (u(t) - \alpha_i(t)) < 0.$$

A contradiction follows now as in the proof of Theorem 3.1. At last, $u \le \beta$ follows from the same argument. □

Using Theorem 3.1, we can also obtain infinitely many solutions as in the following result.

Proposition 3.3. *Let $f : [a,b] \times \mathbb{R} \to \mathbb{R}$ satisfy L^1-Carathéodory conditions, let $(p_k)_k$, $(q_k)_k$, $(r_k)_k$, $(s_k)_k \subset \mathbb{R}$ be sequences such that $p_k < q_k \leq r_k < s_k \leq p_{k+1}$ and let $(g_{0,k})_k$, $(g_{1,k})_k \subset L^1(a,b)$. Define $g_{i,k} = \bar{g}_{i,k} + \tilde{g}_{i,k}$, where $\bar{g}_{i,k} = \frac{1}{b-a}\int_a^b g_{i,k}(t)\,dt$. Assume*

(a) for a.e. $t \in \mathbb{R}$ and all $u \in]p_k, q_k[$,

$$f(t,u) \leq g_{0,k}(t) \quad \text{and} \quad \bar{g}_{0,k} \leq 0;$$

(b) for a.e. $t \in [a,b]$ and all $u \in]r_k, s_k[$,

$$f(t,u) \geq g_{1,k}(t) \quad \text{and} \quad \bar{g}_{1,k} \geq 0;$$

(c) $q_k - p_k \geq \frac{b-a}{6}\|\tilde{g}_{0,k}\|_{L^1}$ and $s_k - r_k \geq \frac{b-a}{6}\|\tilde{g}_{1,k}\|_{L^1}$.

Then the problem (1.1) has infinitely many solutions $u_k \in [p_k, s_k]$.

Proof. Let us prove the existence of a solution in each interval $[p_k, s_k]$. To this end we construct first an upper solution $\beta_k \in [r_k, s_k]$. Let w be the solution of the problem

$$u'' + \tilde{g}_{1,k}(t) = 0,$$
$$u(a) = u(b), \;\; u'(a) = u'(b),$$

such that $\bar{w} = \frac{1}{b-a}\int_a^b w(t)\,dt = 0$. The following estimate

$$\|w\|_\infty \leq \frac{b-a}{12}\,\|\tilde{g}_{1,k}\|_{L^1} \leq \frac{s_k - r_k}{2}$$

follows from Proposition A-4.2. Hence, $\beta_k(t) = \frac{s_k + r_k}{2} + w(t)$ is a $W^{2,1}$-upper solution such that $\beta_k \in [r_k, s_k]$.

In the same way, we can construct a $W^{2,1}$-lower solution $\alpha_k \in [p_k, q_k]$ and the result follows from Theorem 3.1. □

Example 3.2. A model example is

$$u'' + \sin\sqrt{|u|} = h(t),$$
$$u(a) = u(b), \;\; u'(a) = u'(b),$$

where $h \in L^1(a,b)$ is such that $\bar{h} \in]-1,1[$. Existence of infinitely many solutions follows from Proposition 3.3 if we choose, for k large enough, $p_k = (2k\pi + \arcsin|\bar{h}|)^2$, $q_k = ((2k+1)\pi - \arcsin|\bar{h}|)^2$, $r_k = ((2k+1)\pi + \arcsin|\bar{h}|)^2$, $s_k = (2(k+1)\pi - \arcsin|\bar{h}|)^2$, $g_{0,k} = h - |\bar{h}|$ and $g_{1,k} = h + |\bar{h}|$.

3.3 Structure of the Set of Solutions

As in the continuous case, we can prove existence of minimal and maximal solutions. Also, there is a continuum of solutions if f is non-decreasing with respect to u.

Theorem 3.4. *Let α and β be $W^{2,1}$-lower and upper solutions of* (1.1) *such that $\alpha \leq \beta$, $E = \{(t,u) \in [a,b] \times \mathbb{R} \mid \alpha(t) \leq u \leq \beta(t)\}$ and $f : E \to \mathbb{R}$ satisfy L^1-Carathéodory conditions.*

Then the problem (1.1) *has a minimal solution $u_{min} \in W^{2,1}(a,b)$ and a maximal solution $u_{max} \in W^{2,1}(a,b)$ in $[\alpha, \beta]$, i.e.*

$$\alpha \leq u_{min} \leq u_{max} \leq \beta$$

and any other solution u of (1.1) *with $\alpha \leq u \leq \beta$ satisfies*

$$u_{min} \leq u \leq u_{max}.$$

Theorem 3.5. *Assume the hypotheses of Theorem 3.4 hold and f is non-decreasing with respect to u.*

Then for any $t_0 \in [a,b]$ and $u^* \in [u_{min}(t_0), u_{max}(t_0)]$, there exists a solution $u \in W^{2,1}(a,b)$ of (1.1) *such that $u_{min} \leq u \leq u_{max}$ and $u(t_0) = u^*$.*

Exercise 3.1. Prove the above results adapting the proofs of Theorems 2.4 and 2.5 and using Theorem 3.2.

4 *A Priori* Bound on the Derivative

Consider the problem

$$\begin{gathered} u'' = f(t,u,u'), \\ u(a) = u(b), \;\; u'(a) = u'(b). \end{gathered} \tag{4.1}$$

The Nemytskii operator $Nu := f(\cdot, u, u')$ and therefore the fixed point problem to be considered are defined now on $C^1([a,b])$. Upper and lower solutions will give *a priori* bounds on u. In order to apply the Schauder Fixed Point Theorem or degree theory we shall also need *a priori* bounds on the derivative u'. In some cases, the special structure of the nonlinearity f gives this information. In others, this information follows from a Nagumo condition. The following example shows that in any case some condition is necessary since the existence of solutions does not follow from the existence of ordered lower and upper solutions.

Example 4.1. Consider the problem

$$\begin{gathered} u'' = (1+u'^2)^2(u-p(t)), \\ u(0)=u(T), \;\; u'(0)=u'(T), \end{gathered} \tag{4.2}$$

where p is a continuous function such that $p(t)=2$ on $[0,T/3[$, $p(t)\in[-2,2]$ on $[T/3,2T/3[$ and $p(t)=-2$ on $[2T/3,T]$. For such a problem, we define lower and upper solutions from Definitions 5.1 or 6.1 below. It follows that $\alpha=-3$ is a lower solution and $\beta=3$ an upper one. We are going to prove nevertheless that if $T>0$ is large enough, problem (4.2) has no solution.

Proof. Assume there is a solution u with $u(0)=u(T)\le 0$. On the interval $[0,T/3]$, the differential equation (4.2) is autonomous

$$u'' = (1+u'^2)^2(u-2) \tag{4.3}$$

and admits an energy integral

$$\mathcal{E}(u,u') = \frac{1}{1+u'^2} + (u-2)^2.$$

If $u(0)\le 0$, we have $E := \mathcal{E}(u(0),u'(0)) \ge 4$.

Claim : The solutions $\tilde u$ of (4.3) *with $\tilde u(0)\le 0$ are defined on a maximum interval of length*

$$\tau \le 2\cosh^{-1}\frac{2}{\sqrt{3}}.$$

From a phase plane analysis, we can see that the orbit of $\tilde u$ goes through some point $(u_0,0)$. Consider then the solution v of (4.3) such that

$$v(0) = 2-\sqrt{E-1}, \;\; v'(0)=0,$$

where $E=\mathcal{E}(u(0),u'(0))$. Notice that $(v(t)-2)^2 \le E \le 1+(v(t)-2)^2$ and $v'(t)\le 0$ for $t\ge 0$. Hence, we compute for such t

$$-v' = \sqrt{\tfrac{(v-2)^2-(E-1)}{E-(v-2)^2}} \ge \sqrt{(v-2)^2-(E-1)}$$

and

$$t \le -\int_{v(0)}^{v(t)} \frac{dv}{\sqrt{(v-2)^2-(E-1)}} = \int_{w_0}^{w(t)} \frac{dr}{\sqrt{r^2-1}} = \cosh^{-1} r \Big|_{w_0}^{w(t)},$$

where $w_0 = \frac{2-v(0)}{\sqrt{E-1}} = 1$ and $w(t) = \frac{2-v(t)}{\sqrt{E-1}} \le \frac{\sqrt{E}}{\sqrt{E-1}} \le \frac{2}{\sqrt{3}}$. Hence, we have

$$t \le \cosh^{-1}\tfrac{2}{\sqrt{3}}.$$

Repeating the computations in reverse time, the claim follows.

Conclusion – If $\frac{T}{3} > 2\cosh^{-1}\frac{2}{\sqrt{3}}$, it follows from the claim that there is no solution of (4.2) such that $u(0) = u(T) \leq 0$.

A similar result holds if $u(0) = u(T) \geq 0$. Here, we have to compute the travelling time of solutions along orbits of the autonomous equation defined by (4.2) for $t \in [2T/3, T]$ and notice that again the travelling time along orbits is too small. □

4.1 The Rayleigh Equation

Consider the Rayleigh equation

$$\begin{aligned} &u'' + g(u') + h(t, u, u') = 0, \\ &u(a) = u(b), \ u'(a) = u'(b), \end{aligned} \tag{4.4}$$

where $g \in \mathcal{C}(\mathbb{R})$ and h is a Carathéodory function.

Proposition 4.1. *Let $g \in \mathcal{C}(\mathbb{R})$ and let h be a Carathéodory function. Assume that for all $s > 0$ and some $h_s \in L^2(a,b)$, h satisfies*

$$\text{for a.e. } t \in [a,b], \forall (u,v) \in \mathbb{R}^2, \quad |u| \leq s \Rightarrow |h(t,u,v)| \leq h_s(t).$$

Then for all $r > 0$ there exists $R > 0$ such that every solution u of (4.4) with $\|u\|_\infty \leq r$ satisfies $\|u'\|_\infty < R$.

Proof. Let $r > 0$ be given and u be a solution of (4.4) with $\|u\|_\infty \leq r$. Multiplying (4.4) by u'' and integrating we obtain

$$\begin{aligned} \|u''\|_{L^2}^2 &= -\int_a^b g(u'(t))u''(t)\,dt - \int_a^b h(t,u(t),u'(t))u''(t)\,dt \\ &\leq \|h_r\|_{L^2}\|u''\|_{L^2}. \end{aligned}$$

Now it is easy to see that for some $t_0 \in [a,b]$ and all $t \in [a,b]$

$$|u'(t)| = |\int_{t_0}^t u''(s)\,ds| \leq \sqrt{b-a}\,\|h_r\|_{L^2}.$$

Hence, $R := \sqrt{b-a}\,\|h_r\|_{L^2} + 1 > 0$ satisfies the claim. □

We can obtain a bound on u' independent of r if we reinforce the bound on h.

Proposition 4.2. *Let $g \in \mathcal{C}(\mathbb{R})$ and h be a Carathéodory function. Assume there exists $h_0 \in L^2(a,b)$ such that*

$$\text{for a.e. } t \in [a,b] \text{ and all } (u,v) \in \mathbb{R}^2, \quad |h(t,u,v)| \leq h_0(t).$$

Then there exists $R > 0$ such that every solution u of (4.4) satisfies $\|u'\|_\infty < R$.

Proof. If we repeat the argument of Proposition 4.1, the proof follows with $R := \sqrt{b-a}\,\|h_0\|_{L^2} + 1 > 0$. □

4.2 The Liénard Equation

As for Rayleigh equation, we can prove *a priori* bounds on the derivative for the Liénard equation

$$\begin{aligned} &u'' + g(u)u' + h(t,u) = 0, \\ &u(a) = u(b), \;\; u'(a) = u'(b), \end{aligned} \tag{4.5}$$

where $g \in \mathcal{C}(\mathbb{R})$ and h is L^1-Carathéodory.

Proposition 4.3. *Let $g \in \mathcal{C}(\mathbb{R})$ and h be an L^1-Carathéodory function.*

Then, for all $r > 0$, there exists $R > 0$ such that for every solution u of (4.5) with $\|u\|_\infty \le r$ we have $\|u'\|_\infty < R$.

Proof. Let $r > 0$. As h is L^1-Carathéodory, there exists $h_r \in L^1(a,b)$ such that for a.e. $t \in [a,b]$ and all $u \in [-r,r]$, $|h(t,u)| \le h_r(t)$. Next we choose a solution u of (4.5) with $\|u\|_\infty \le r$. Then multiplying equation (4.5) by u and integrating we obtain

$$\|u'\|_{L^2}^2 = \int_a^b g(u(t))u(t)u'(t)\,dt + \int_a^b h(t,u(t))u(t)\,dt \le r\|h_r\|_{L^1}.$$

It is now easy to bound $\|u''\|_{L^1}$ and next $\|u'\|_\infty$. □

4.3 The Nagumo Condition: The Continuous Case

In case the equation does not have any special structure, *a priori* bounds on the derivative can still be obtained for nonlinearities which do not grow too quickly with respect to the derivative.

A first result of this type is due to S. Bernstein [35], [36] who considered nonlinearities that satisfy the growth condition

$$|f(t,u,v)| \le A + Bv^2,$$

known as the *Bernstein condition.* He proved that for any $r > 0$ there exists $R > 0$ such that any solution u of

$$u'' = f(t,u,u') \tag{4.6}$$

on $[a,b]$ with $\|u\|_\infty \le r$ satisfies the *a priori* bound

$$\|u'\|_\infty \le R.$$

Bernstein's result was generalized by M. Nagumo [224] who introduced the so-called Nagumo condition which became a classical way to obtain *a priori* bounds on the derivative. To describe this condition, we consider α and $\beta \in \mathcal{C}([a,b])$ such that $\alpha \le \beta$, define

$$E := \{(t,u,v) \in [a,b] \times \mathbb{R}^2 \mid \alpha(t) \le u \le \beta(t)\} \tag{4.7}$$

and suppose $f : E \to \mathbb{R}$ satisfies

$$\forall (t,u,v) \in E, \qquad |f(t,u,v)| \le \varphi(|v|), \tag{4.8}$$

where $\varphi : \mathbb{R}^+ \to \mathbb{R}$ is some positive continuous function such that

$$\int_0^\infty \frac{s\,ds}{\varphi(s)} = \infty. \tag{4.9}$$

Condition (4.8) is then said to be the *Nagumo condition.*

Notice that if a function f satisfies a Bernstein condition,

$$\forall (t,u,v) \in E, \quad |f(t,u,v)| \le A + Bv^2,$$

it satisfies a Nagumo condition with $\varphi(v) = A + Bv^2$. Also, the Bernstein condition is equivalent to assuming the existence of a positive continuous function $\varphi : \mathbb{R}^+ \to \mathbb{R}$ such that

$$\liminf_{s \to +\infty} \frac{s^2}{\varphi(s)} > 0$$

and

$$\forall (t,u,v) \in E, \quad |f(t,u,v)| \le \varphi(|v|).$$

However, the Bernstein condition is not equivalent to the Nagumo condition where φ satisfies (4.9). Indeed, the function $f(t,u,v) = (v^2+1)\ln(v^2+1)$ satisfies (4.8) and (4.9), with $\varphi(v) = f(t,u,v)$, but it does not satisfy the Bernstein condition.

It is easy to see that in case h is an L^∞-Carathéodory function, the function $f(t,u,u') = g(u)u' + h(t,u)$ in the Liénard equation (4.5) satisfies a Nagumo condition together with (4.9). However, this fact does not hold true for the Rayleigh equation if, for example, $g(u') = u'^3$.

In applications, we might consider nonlinearities $f+g$ such that f and g satisfy the Nagumo condition together with (4.9). This assumption does not imply that $f+g$ satisfies such conditions. A counter-example can be build as follows. Let $A = \cup_1^\infty [n, n + \frac{1}{n^3}]$, $B = [1,\infty[\setminus A$ and define the function $f(v) = \varphi(v)$ by

$$\begin{aligned} \varphi(s) &= 1/n, \quad \text{if } s \in [n, n+1/n^3], \\ &= s^3, \quad \text{if } s \in B. \end{aligned}$$

We have

$$\int_1^\infty \frac{s}{\varphi(s)}\,ds \geq \int_A \frac{s}{\varphi(s)}\,ds \geq \sum_{n=1}^{\infty} \frac{n}{1/n}\frac{1}{n^3} = \sum_{n=1}^{\infty}\frac{1}{n} = \infty.$$

On the other hand, if $g = 1$, $f+g$ does not satisfy a Nagumo condition. One computes

$$\int_A \frac{s}{\varphi(s)+1}\,ds \leq \sum_{n=1}^{\infty} \frac{n+1}{\frac{1}{n}+1}\frac{1}{n^3} = \sum_{n=1}^{\infty}\frac{1}{n^2} < \infty,$$

and

$$\int_B \frac{s}{s^3+1}\,ds < \infty,$$

which shows that

$$\int_1^\infty \frac{s}{\varphi(s)+1}\,ds < \infty.$$

It is now easy to see that we can regularize $f = \varphi$ to obtain a continuous function with the same property.

In many cases however, $f+g$ still satisfies a Nagumo condition with (4.9). For example, if f satisfies such conditions with $\varphi \geq \epsilon > 0$ and g is bounded then the same holds true for $f+g$, which can be deduced from the fact that for some $k \geq 0$

$$|f+g| \leq \varphi + k \leq \frac{k+\epsilon}{\epsilon}\varphi.$$

A similar result holds if f and g satisfy a Nagumo condition with the same function $\varphi(v)$.

The following proposition deduces *a priori* bounds on the derivative from a Nagumo condition.

Proposition 4.4. *Let $\bar\alpha$ and $\bar\beta \in C([a,b])$ be such that $\bar\alpha \leq \bar\beta$. Define $E \subset [a,b]\times\mathbb{R}^2$ from (4.7) (with $\alpha = \bar\alpha$ and $\beta = \bar\beta$) and let $\bar\varphi : \mathbb{R}^+ \to \mathbb{R}$ be a positive continuous function satisfying*

$$\int_r^\infty \frac{s\,ds}{\bar\varphi(s)} > \max_t \bar\beta(t) - \min_t \bar\alpha(t), \tag{4.10}$$

where $r = \max\{\frac{\bar\beta(b)-\bar\alpha(a)}{b-a}, \frac{\bar\beta(a)-\bar\alpha(b)}{b-a}\} \geq 0.$

Then there exists $R > 0$ such that for every continuous function $f : E \to \mathbb{R}$ which satisfies

$$\forall (t,u,v) \in E, \qquad |f(t,u,v)| \leq \bar\varphi(|v|) \tag{4.11}$$

and every solution u *of* (4.6) *on* $[a,b]$ *such that* $\bar\alpha \le u \le \bar\beta$, *we have*

$$\|u'\|_\infty < R.$$

Proof. Define R to be such that

$$\int_r^R \frac{s\,ds}{\bar\varphi(s)} > \max_t \bar\beta(t) - \min_t \bar\alpha(t). \tag{4.12}$$

Let u be a solution of (4.6) such that $\bar\alpha \le u \le \bar\beta$. Observe that

$$-r \le \frac{\bar\alpha(b) - \bar\beta(a)}{b-a} \le \frac{u(b)-u(a)}{b-a} \le \frac{\bar\beta(b) - \bar\alpha(a)}{b-a} \le r$$

and using the Lagrange Theorem, there exists $\tau \in [a,b]$ with

$$|u'(\tau)| = |\frac{u(b)-u(a)}{b-a}| \le r.$$

Now consider an interval $I = [t_0, t_1]$ or $[t_1, t_0]$ such that $u'(t) \ge 0$ on I, $u'(t_0) = r$, $u'(t_1) > r$. Then we have

$$\begin{aligned}\int_{u'(t_0)}^{u'(t_1)} \frac{s\,ds}{\bar\varphi(s)} &= \int_{t_0}^{t_1} \frac{u'(t)u''(t)}{\bar\varphi(u'(t))}\,dt = \int_{t_0}^{t_1} \frac{u'(t)f(t,u(t),u'(t))}{\bar\varphi(u'(t))}\,dt \\ &\le |\int_{t_0}^{t_1} u'(t)\,dt| = |u(t_1) - u(t_0)| \le \max_t \bar\beta(t) - \min_t \bar\alpha(t).\end{aligned}$$

From (4.12) we deduce that $u'(t_1) < R$.

In the same way, we prove that for any $t \in [a,b]$, $u'(t) > -R$ and the result follows. □

A fundamental generalization of this proposition concerns one-sided Nagumo conditions. This applies to solutions u such that some *a priori* bound on the derivative is known at the end points of the interval of definition, which is clearly the case for the Neumann problem

$$\begin{aligned} u'' &= f(t,u,u'), \\ u'(\bar a) &= A, \ u'(\bar b) = B. \end{aligned} \tag{4.13}$$

Notice also that if u solves the periodic problem (4.1), there exists $\bar a \in [a,b]$ such that $u'(\bar a) = 0$ and the periodic extension of $u(t)$ solves (4.13) with $\bar a$, $\bar b = \bar a + b - a$ and $A = B = 0$.

Our first result concerns solutions such that a one-sided *a priori* bound on the derivative is known at both end points.

Proposition 4.5. *Let $\bar{\alpha}$ and $\bar{\beta} \in C([\bar{a},\bar{b}])$ be such that $\bar{\alpha} \le \bar{\beta}$. Define $E \subset [\bar{a},\bar{b}] \times \mathbb{R}^2$ from (4.7) (with $a = \bar{a}$, $b = \bar{b}$, $\alpha = \bar{\alpha}$ and $\beta = \bar{\beta}$). Let $r \ge 0$, $R > r$ and $\bar{\varphi} : \mathbb{R}^+ \to \mathbb{R}$ be a positive continuous function satisfying (4.12).*

Then for every continuous function $f : E \to \mathbb{R}$ which satisfies

$$\forall (t,u,v) \in E, \qquad f(t,u,v) \le \bar{\varphi}(|v|) \tag{4.14}$$

and for every solution u of (4.6) on $[\bar{a},\bar{b}]$ which is such that $\bar{\alpha} \le u \le \bar{\beta}$, $u'(\bar{a}) \le r$ and $u'(\bar{b}) \ge -r$ we have

$$\|u'\|_\infty < R.$$

Similarly, for every continuous function $f : E \to \mathbb{R}$ which satisfies

$$\forall (t,u,v) \in E, \qquad f(t,u,v) \ge -\bar{\varphi}(|v|) \tag{4.15}$$

and for every solution u of (4.6) on $[\bar{a},\bar{b}]$ which is such that $\bar{\alpha} \le u \le \bar{\beta}$, $u'(\bar{a}) \ge -r$ and $u'(\bar{b}) \le r$ we have

$$\|u'\|_\infty < R.$$

Proof. Let u be a solution as above and assume by contradiction that $\min_t u'(t) \le -R$. Hence, there exists an interval $I = [t_0, t_1]$ such that $u'(t) \le 0$ on I, $u'(t_0) \le -R$ and $u'(t_1) = -r$. We deduce then the contradiction

$$\begin{aligned}
\int_r^R \frac{s\,ds}{\bar{\varphi}(s)} &\le \int_{-u'(t_1)}^{-u'(t_0)} \frac{s\,ds}{\bar{\varphi}(s)} \\
&= -\int_{t_0}^{t_1} \frac{(-u'(t))(-u''(t))}{\bar{\varphi}(-u'(t))}\,dt = \int_{t_0}^{t_1} \frac{(-u'(t))f(t,u(t),u'(t))}{\bar{\varphi}(|u'(t)|)}\,dt \\
&\le -u(t_1) + u(t_0) \le \max_t \bar{\beta}(t) - \min_t \bar{\alpha}(t) < \int_r^R \frac{s\,ds}{\bar{\varphi}(s)}.
\end{aligned}$$

A similar argument holds if $\max_t u'(t) \ge R$.

The second part of the proposition is proved in the same way. □

Example 4.2. As an example, notice that condition (4.14) applies for a problem such as

$$\begin{gathered}
u'' + u'^{32} + g(t,u) = 0, \\
u(a) = u(b), \ u'(a) = u'(b).
\end{gathered}$$

Remark In the previous proposition, the condition

$$f(t,u,v) \leq \bar{\varphi}(|v|)$$

uses the same function φ for $v \geq 0$ and $v \leq 0$. A trivial generalization would be

$$\begin{aligned} f(t,u,v) &\leq \bar{\varphi}_1(v), && \text{if } v \geq 0, \\ f(t,u,v) &\leq \bar{\varphi}_2(-v), && \text{if } v \leq 0. \end{aligned}$$

Another variant concerns solutions such that a two-sided *a priori* bound is known but only in one of the boundary points.

Proposition 4.6. *Let $\bar{\alpha}$ and $\bar{\beta} \in \mathcal{C}([\bar{a},\bar{b}])$ be such that $\bar{\alpha} \leq \bar{\beta}$. Define $E \subset [\bar{a},\bar{b}] \times \mathbb{R}^2$ from* (4.7) *(with $a = \bar{a}$, $b = \bar{b}$, $\alpha = \bar{\alpha}$ and $\beta = \bar{\beta}$). Let $r \geq 0$, $R > r$ and $\bar{\varphi} : \mathbb{R}^+ \to \mathbb{R}$ be a positive continuous function satisfying* (4.12).

Then for every continuous function $f : E \to \mathbb{R}$ which satisfies

$$\forall (t,u,v) \in E, \qquad \operatorname{sgn}(v) f(t,u,v) \leq \bar{\varphi}(|v|) \tag{4.16}$$

and for every solution u of (4.6) *on $[\bar{a},\bar{b}]$, which is such that $\bar{\alpha} \leq u \leq \bar{\beta}$ and $|u'(\bar{a})| \leq r$, we have*

$$\|u'\|_\infty < R.$$

Similarly, for every continuous function $f : E \to \mathbb{R}$ which satisfies

$$\forall (t,u,v) \in E, \qquad \operatorname{sgn}(v) f(t,u,v) \geq -\bar{\varphi}(|v|) \tag{4.17}$$

and for every solution u of (4.6) *on $[\bar{a},\bar{b}]$, which is such that $\bar{\alpha} \leq u \leq \bar{\beta}$ and $|u'(\bar{b})| \leq r$, we have*

$$\|u'\|_\infty < R.$$

The reader will easily find examples of functions $f(t,u,v)$ which satisfy one and only one of these one-sided conditions.

Exercise 4.1. Prove Proposition 4.6.

Each of the conditions (4.14), (4.15), (4.16) and (4.17) is called a *one-sided Nagumo condition.*

Remark Nagumo conditions and more specifically one-sided Nagumo conditions have a nice geometric interpretation. Assume for simplicity $\bar{\alpha}$ and $\bar{\beta}$ are constant and $f(t,\bar{\alpha},0) < 0 < f(t,\bar{\beta},0)$. Consider then the assumption

$$f(t,u,v) < \bar{\varphi}(v) \quad \text{for } v > 0.$$

In the phase space (u, u') we define the curve $u' = \nu(u)$, where ν is the solution of the problem

$$v' = \frac{\bar{\varphi}(v)}{v}, \quad v(\bar{\alpha}) = v_0.$$

If we assume (4.10) and choose $v_0 \in [0, r]$ it is easy to see that ν is defined on $[\bar{\alpha}, \bar{\beta}]$. Also, if $u \in [\bar{\alpha}, \bar{\beta}]$, the vector field corresponding to the differential equation (4.6) points downward along that curve. An assumption such as

$$f(t, u, v) > \bar{\varphi}(v) \qquad \text{for } v < 0$$

furnishes a curve $u' = \mu(u) < 0$ so that the vector field points upward. More generally, the vector field cannot be tangent from the inside along the boundary of the set

$$\mathcal{E} := \{(u, u') \mid \bar{\alpha} \leq u \leq \bar{\beta},\ \mu(u) \leq u' \leq \nu(u)\}.$$

This geometry is essential to obtain existence of solutions from a degree argument or a modification method as we did in this chapter.

Notice that it is not essential for the vector field to point upward along the curve $u' = \mu(u)$ and downward along the curve $u' = \nu(u)$. Other geometries are possible. These underlie the various one-sided Nagumo conditions we mentioned above.

4.4 A Generalization: the Carathéodory Case

Observe that if φ is continuous, any L^p-Carathéodory function f that satisfies the Nagumo condition (4.8) is L^∞-Carathéodory. In the next result, we extend the Nagumo condition so as to deal with L^p-Carathéodory functions which are not L^∞-Carathéodory.

Proposition 4.7. *Let $\bar{\alpha}$, $\bar{\beta} \in C([a,b])$ be such that $\bar{\alpha} \leq \bar{\beta}$ and define $r = \max\{\bar{\beta}(b) - \bar{\alpha}(a), \bar{\beta}(a) - \bar{\alpha}(b)\}/(b-a)$. Let p, $q \in [1, \infty]$ be such that $\frac{1}{p} + \frac{1}{q} = 1$. Assume there exist $\bar{\varphi} \in C(\mathbb{R}^+, \mathbb{R}_0^+)$, $\bar{\psi} \in L^p(a,b)$ and $R > r$ such that*

$$\int_r^R \frac{s^{1/q}}{\bar{\varphi}(s)}\, ds > \|\bar{\psi}\|_{L^p} (\max_t \bar{\beta}(t) - \min_t \bar{\alpha}(t))^{1/q}. \tag{4.18}$$

Define $E \subset [a,b] \times \mathbb{R}^2$ from (4.7) (with $\alpha = \bar{\alpha}$ and $\beta = \bar{\beta}$).

Then, for every L^p-Carathéodory function $f : E \to \mathbb{R}$ such that

for a.e. $t \in [a,b]$ and all $(u,v) \in \mathbb{R}^2$, with $(t,u,v) \in E$,

$$|f(t,u,v)| \leq \bar{\psi}(t)\bar{\varphi}(|v|),$$

and for every solution u of (4.6) such that $\bar{\alpha} \leq u \leq \bar{\beta}$, we have

$$\|u'\|_\infty < R.$$

Proof. Let u be a solution of (4.6) and $t \in [a,b]$ be such that $u'(t) \geq R$. We can choose, as in the proof of Proposition 4.4, $t_0 < t_1$ (or $t_0 > t_1$) such that $u'(t_0) = r$, $u'(t_1) = R$ and $r \leq u'(s) \leq R$ on $[t_0, t_1]$ (or $[t_1, t_0]$). Next, we write

$$\begin{aligned}\int_r^R \frac{r^{1/q}}{\bar{\varphi}(r)}\,dr &= \int_{t_0}^{t_1} \frac{u'^{1/q}(s)u''(s)}{\bar{\varphi}(u'(s))}\,ds = \int_{t_0}^{t_1} \frac{u'^{1/q}(s)f(s,u(s),u'(s))}{\bar{\varphi}(u'(s))}\,ds \\ &\leq \left|\int_{t_0}^{t_1} \bar{\psi}(s)u'^{1/q}(s)\,ds\right| \leq \|\bar{\psi}\|_{L^p} \left|\int_{t_0}^{t_1} u'(s)\,ds\right|^{1/q} \\ &\leq \|\bar{\psi}\|_{L^p}(\max_t \bar{\beta}(t) - \min_t \bar{\alpha}(t))^{1/q}.\end{aligned}$$

From (4.18) we obtain a contradiction and deduce that $u'(t) < R$. In the same way we prove that $u'(t) > -R$. □

As a first remark, notice that (4.18) is obviously implied by

$$\int_0^\infty \frac{s^{1/q}}{\varphi(s)}\,ds = \infty.$$

Also one-sided Nagumo conditions can be worked out for the Carathéodory case just as in Proposition 4.5.

Proposition 4.8. *Let $\bar{\alpha}$ and $\bar{\beta} \in C([\bar{a},\bar{b}])$ be such that $\bar{\alpha} \leq \bar{\beta}$ and p, $q \in [1,\infty]$ such that $\frac{1}{p} + \frac{1}{q} = 1$. Assume there exist $\bar{\varphi} \in C(\mathbb{R}^+, \mathbb{R}_0^+)$, $\bar{\psi} \in L^p(\bar{a},\bar{b})$, $r \geq 0$ and $R > r$ that satisfy (4.18). Define $E \subset [\bar{a},\bar{b}] \times \mathbb{R}^2$ from (4.7) (with $a = \bar{a}$, $b = \bar{b}$, $\alpha = \bar{\alpha}$ and $\beta = \bar{\beta}$).*

Then, for every L^p-Carathéodory function $f : E \to \mathbb{R}$ such that

for a.e. $t \in [\bar{a},\bar{b}]$ and all $(u,v) \in \mathbb{R}^2$, with $(t,u,v) \in E$,
$$f(t,u,v) \leq \bar{\psi}(t)\bar{\varphi}(|v|),$$

and for every solution u of (4.6) on $[\bar{a},\bar{b}]$ such that $\bar{\alpha} \leq u \leq \bar{\beta}$, $u'(\bar{a}) \leq r$ and $u'(\bar{b}) \geq -r$, we have
$$\|u'\|_\infty < R.$$

Similarly, for every L^p-Carathéodory function $f : E \to \mathbb{R}$ such that

for a.e. $t \in [\bar{a},\bar{b}]$ and all $(u,v) \in \mathbb{R}^2$, with $(t,u,v) \in E$,
$$f(t,u,v) \geq -\bar{\psi}(t)\bar{\varphi}(|v|),$$

and for every solution u of (4.6) on $[\bar{a},\bar{b}]$ such that $\bar{\alpha} \leq u \leq \bar{\beta}$, $u'(\bar{a}) \geq -r$ and $u'(\bar{b}) \leq r$, we have
$$\|u'\|_\infty < R.$$

As in Proposition 4.6, we can deal with solutions u such that $u'(\bar{a})$ only (or $u'(\bar{b})$ only) is *a priori* bounded by some constant r.

Proposition 4.9. *Let $\bar{\alpha}$ and $\bar{\beta} \in C([\bar{a}, \bar{b}])$ be such that $\bar{\alpha} \leq \bar{\beta}$ and p, $q \in [1, \infty]$ such that $\frac{1}{p} + \frac{1}{q} = 1$. Assume there exist $\bar{\varphi} \in C(\mathbb{R}^+, \mathbb{R}_0^+)$, $\bar{\psi} \in L^p(\bar{a}, \bar{b})$, $r \geq 0$ and $R > r$ that satisfy (4.18). Define $E \subset [\bar{a}, \bar{b}] \times \mathbb{R}^2$ from (4.7) (with $a = \bar{a}$, $b = \bar{b}$, $\alpha = \bar{\alpha}$ and $\beta = \bar{\beta}$).*

Then, for every L^p-Carathéodory function $f : E \to \mathbb{R}$ such that

$$\text{for a.e. } t \in [\bar{a}, \bar{b}] \text{ and all } (u, v) \in \mathbb{R}^2, \text{ with } (t, u, v) \in E,$$
$$\operatorname{sgn}(v) f(t, u, v) \leq \bar{\psi}(t)\bar{\varphi}(|v|), \quad (\text{resp. } \operatorname{sgn}(v) f(t, u, v) \geq -\bar{\psi}(t)\bar{\varphi}(|v|)),$$

and for every solution u of (4.6) on $[\bar{a}, \bar{b}]$ such that $\bar{\alpha} \leq u \leq \bar{\beta}$ and $|u'(\bar{a})| \leq r$ (resp. $|u'(\bar{b})| \leq r$), we have

$$\|u'\|_\infty < R.$$

Exercise 4.2. Prove Propositions 4.8 and 4.9.

5 BVP with Derivative Dependence : C^2-Solutions

5.1 Definitions

Consider the periodic boundary value problem

$$\begin{aligned} u'' &= f(t, u, u'), \\ u(a) &= u(b), \ u'(a) = u'(b). \end{aligned} \tag{5.1}$$

The dependence of f in the derivative u' does not really change the definitions of lower and upper solutions. These are straightforward extensions of Definitions 2.1.

Definitions 5.1. *A function $\alpha \in C([a, b])$ such that $\alpha(a) = \alpha(b)$ is a C^2-*lower solution *of* (5.1) *if its periodic extension on $\mathbb{R}$, defined by $\alpha(t) = \alpha(t + b - a)$ is such that for any $t_0 \in \mathbb{R}$*
either $D_-\alpha(t_0) < D^+\alpha(t_0)$
or there exist an open interval I_0 with $t_0 \in I_0$ and a function $\alpha_0 \in C^1(I_0)$ such that
(a) $\alpha(t_0) = \alpha_0(t_0)$ and $\alpha(t) \geq \alpha_0(t)$ for all $t \in I_0$;
(b) $\alpha_0''(t_0)$ exists and $\alpha_0''(t_0) \geq f(t_0, \alpha_0(t_0), \alpha_0'(t_0))$.

*A function $\beta \in C([a, b])$ such that $\beta(a) = \beta(b)$ is a C^2-*upper solution *of* (5.1) *if its periodic extension on $\mathbb{R}$ is such that for any $t_0 \in \mathbb{R}$*
either $D^-\beta(t_0) > D_+\beta(t_0)$
or there exist an open interval I_0 with $t_0 \in I_0$ and a function $\beta_0 \in C^1(I_0)$ such that
(a) $\beta(t_0) = \beta_0(t_0)$ and $\beta(t) \leq \beta_0(t)$ for all $t \in I_0$;
(b) $\beta_0''(t_0)$ exists and $\beta_0''(t_0) \leq f(t_0, \beta_0(t_0), \beta_0'(t_0))$.

Remark Let α be a lower solution and assume t_0 is such that

$$D_-\alpha(t_0) \geq D^+\alpha(t_0).$$

From the definition, there exists then $\alpha_0 \in \mathcal{C}^1(I_0)$ such that $\alpha(t_0) = \alpha_0(t_0)$ and $\alpha(t) \geq \alpha_0(t)$ on I_0. It follows that

$$D_-\alpha(t_0) \leq D^-\alpha(t_0) \leq \alpha_0'(t_0) \leq D_+\alpha(t_0) \leq D^+\alpha(t_0) \leq D_-\alpha(t_0),$$

which implies α has a derivative at t_0 and $\alpha'(t_0) = \alpha_0'(t_0)$.

As in Section 2, we can build lower and upper solutions from maxima of lower solutions and minima of upper solutions.

Proposition 5.1. *Let $\alpha_i \in \mathcal{C}([a,b])$ $(i = 1, ..., n)$ be $\mathcal{C}^2$-lower solutions of* (5.1). *Then the function*

$$\alpha(t) = \max_{1\leq i\leq n} \alpha_i(t), \quad t \in [a,b]$$

is a $\mathcal{C}^2$-lower solution of (5.1).

Proposition 5.2. *Let $\beta_j \in \mathcal{C}([a,b])$ $(j = 1, ..., m)$ be $\mathcal{C}^2$-upper solutions of* (5.1). *Then the function*

$$\beta(t) = \min_{1\leq j\leq m} \beta_j(t), \quad t \in [a,b]$$

is a $\mathcal{C}^2$-upper solution of (5.1).

Exercise 5.1. Prove the above propositions.

5.2 Existence of Solutions

For a first and simple approach, we state a theorem which uses a two-sided Nagumo condition, which as we will see later, can be generalized in various directions.

Theorem 5.3. *Let α and β be $\mathcal{C}^2$-lower and upper solutions of the problem* (5.1) *such that $\alpha \leq \beta$, E be defined in* (4.7), *$\varphi : \mathbb{R}^+ \to \mathbb{R}$ be a positive continuous function satisfying* (4.9) *and $f : E \to \mathbb{R}$ be a continuous function which satisfies* (4.8).

Then the problem (5.1) *has at least one solution $u \in \mathcal{C}^2([a,b])$ such that for all $t \in [a,b]$*

$$\alpha(t) \leq u(t) \leq \beta(t).$$

Proof. Consider the modified problem

$$\begin{gathered} u'' = \lambda f(t, \gamma(t,u), u') + \varphi(|u'|)(u - \lambda\gamma(t,u)), \\ u(a) = u(b), \ u'(a) = u'(b), \end{gathered} \tag{5.2}$$

where $\gamma : [a,b] \times \mathbb{R} \to \mathbb{R}$ is defined by (1.3). Choose $\rho > 0$ such that

$$\begin{gathered} -\rho < \alpha(t) \leq \beta(t) < \rho, \\ f(t, \alpha(t), 0) + \varphi(0)(-\rho - \alpha(t)) < 0, \\ f(t, \beta(t), 0) + \varphi(0)(\rho - \beta(t)) > 0. \end{gathered}$$

Claim 1 – Every solution u of (5.2) with $\lambda \in [0,1]$ is such that $-\rho < u(t) < \rho$ on $[a,b]$. Assume there exists t_0 such that $u(t_0) = \min_t u(t) \leq -\rho$. This assumption leads to a contradiction since in such a case

$$0 \leq u''(t_0) = \lambda[f(t_0, \alpha(t_0), 0) + \varphi(0)(u(t_0) - \alpha(t_0))] + (1-\lambda)\varphi(0)u(t_0) < 0.$$

Similarly, we prove that $u(t) < \rho$ on $[a,b]$.

Claim 2 – There exists $R > 0$ such that every solution u of (5.2) with $\lambda \in [0,1]$ satisfies $\|u'\|_\infty < R$. The claim follows choosing $R > 0$ from Proposition 4.4, where $\bar{\alpha} = -\rho$, $\bar{\beta} = \rho$ and $\bar{\varphi}(v) = (1+2\rho)\varphi(v)$.

Claim 3 – Existence of solutions of (5.2) for $\lambda = 1$. Let us define the operators

$$\begin{aligned} L &: \operatorname{Dom} L \subset \mathcal{C}^1([a,b]) \to \mathcal{C}([a,b]), u \mapsto u'' - u, \\ N_\lambda &: \mathcal{C}^1([a,b]) \to \mathcal{C}([a,b]), \\ &\qquad u \mapsto \lambda f(t, \gamma(t,u), u') + \varphi(|u'|)(u - \lambda\gamma(t,u)) - u, \end{aligned}$$

where $\operatorname{Dom} L = \{u \in \mathcal{C}^2([a,b]) \mid u(a) = u(b),\ u'(a) = u'(b)\}$. Observe that L has a compact inverse. Hence, we can define the completely continuous operator

$$T_\lambda(u) = L^{-1}N_\lambda(u).$$

From degree theory, we have that

$$\deg(T_0, \Omega) = \deg(T_1, \Omega),$$

where

$$\Omega = \{u \in \mathcal{C}^1([a,b]) \mid \|u\|_\infty < \rho,\ \|u'\|_\infty < R\}.$$

Using the Odd Mapping Theorem A-1.5, we compute that

$$\deg(T_0, \Omega) \neq 0$$

and the problem (5.2) with $\lambda = 1$ has a solution u.

Conclusion – Using the argument in Theorem 2.3, we prove that this solution u is such that for all $t \in [a,b]$

$$\alpha(t) \le u(t) \le \beta(t).$$

Hence u is also a solution of (5.1). □

Notice that we can try to prove this result using the modified problem

$$\begin{gathered} u'' - u = f(t, \gamma(t,u), \delta(u')) - \gamma(t,u), \\ u(a) = u(b), \; u'(a) = u'(b), \end{gathered} \tag{5.3}$$

where $\gamma(t,u)$ is defined from (1.3) and $\delta(v) = \min\{R, \max\{-R, v\}\}$. The proof follows as in Theorem 2.3 provided α and β are in $W^{1,\infty}(a,b)$. On the other hand, with this last assumption we can weaken the condition (4.9) on φ and assume

$$\int_0^\infty \frac{s\,ds}{\varphi(s)} > \max_t \beta(t) - \min_t \alpha(t). \tag{5.4}$$

As always for periodic problems, we can also replace the two-sided Nagumo condition by a one-sided one.

Theorem 5.4. *Let α, $\beta \in W^{1,\infty}(a,b)$ be C^2-lower and upper solutions of the problem* (5.1) *such that $\alpha \le \beta$, E be defined in* (4.7), *$\varphi : \mathbb{R}^+ \to \mathbb{R}$ be a positive continuous function satisfying* (5.4) *and $f : E \to \mathbb{R}$ a continuous function which satisfies* (4.14) *(with $\bar{\varphi} = \varphi$).*

Then the problem (5.1) *has at least one solution $u \in C^2([a,b])$ such that for all $t \in [a,b]$*

$$\alpha(t) \le u(t) \le \beta(t).$$

Proof. Let R be large enough so that

$$\int_0^R \frac{s\,ds}{\varphi(s)} > \max_t \beta(t) - \min_t \alpha(t).$$

and

$$\max\{\|D^+\alpha\|_\infty, \|D_-\alpha\|_\infty, \|D_-\beta\|_\infty, \|D^+\beta\|_\infty\} \le R.$$

Consider then the modified problem (5.3).

Step 1 – Using Schauder's Theorem, we prove that problem (5.3) has a solution u.

Step 2 – *The solution u is such that $\alpha(t) \le u(t) \le \beta(t)$ on $[a,b]$.* This result follows from the argument in the proof of Theorem 2.3. As a consequence, u satisfies

$$\begin{gathered} u'' = f(t, u, \delta(u')), \\ u(a) = u(b), \; u'(a) = u'(b). \end{gathered} \tag{5.5}$$

Step 3 – The solution u is such that $\|u'\|_\infty \le R$. As u solves a periodic problem, there exists $\bar a \in [a,b]$ such that $u'(\bar a) = 0$. Let $\bar b = \bar a + b - a$ and consider the periodic extension of u on $[\bar a, \bar b]$. Observe that for all $(t,u,v) \in E$ (with $a = \bar a$ and $b = \bar b$),

$$f(t,u,\delta(v)) \le \bar\varphi(|v|),$$

where $\bar\varphi(v) := \varphi(\delta(v))$ satisfies (4.12) with $r = 0$. From Proposition 4.5, every solution $u \in [\alpha,\beta]$ of (5.5) satisfies

$$\|u'\|_\infty < R.$$

It follows that the function u is a solution of (5.1). □

The bound on the derivative of the lower and upper solutions can be replaced by one-sided bound on $f(t,\alpha(t),\alpha'(t))$ and $f(t,\beta(t),\beta'(t))$. The key of the proof of this generalization is to use an appropriate modified problem.

Theorem 5.5. *Let α, $\beta \in C([a,b])$ be C^2-lower and upper solutions of the problem* (5.1) *such that $\alpha \le \beta$. Define $A \subset [a,b]$ (resp. $B \subset [a,b]$) to be the set of points where α (resp. β) is derivable.*

Let E be defined in (4.7), *$\varphi : \mathbb{R}^+ \to \mathbb{R}$ be a positive continuous function satisfying* (5.4) *and $f : E \to \mathbb{R}$ a continuous function which satisfies* (4.14) *(with $\bar\varphi = \varphi$). Assume there exists $N > 0$ such that for all $t \in A$ (resp. for all $t \in B$),*

$$f(t,\alpha(t),\alpha'(t)) \ge -N \quad (\textit{resp. } f(t,\beta(t),\beta'(t)) \le N). \tag{5.6}$$

Then the problem (5.1) *has at least one solution $u \in C^2([a,b])$ such that for all $t \in [a,b]$*

$$\alpha(t) \le u(t) \le \beta(t).$$

Proof. Let R be large enough so that

$$\int_0^R \frac{s\,ds}{\varphi(s)} > \max_t \beta(t) - \min_t \alpha(t).$$

Increasing the value of N if necessary, we can assume

$$N \ge \max\{|f(t,u,v)| \mid t \in [a,b],\ \alpha(t) \le u \le \beta(t),\ |v| \le R\}.$$

Consider then the modified problem

$$\begin{aligned} u'' - u &= \bar f(t,u,u') - \gamma(t,u), \\ u(a) &= u(b),\ u'(a) = u'(b), \end{aligned} \tag{5.7}$$

where $\bar f(t,u,v) := \max\{\min\{f(t,\gamma(t,u),v), N\}, -N\}$ and $\gamma(t,u)$ is defined in (1.3).

Step 1 – Using Schauder's Theorem we prove that problem (5.7) has a solution u.

Step 2 – *The solution u is such that $\alpha \le u \le \beta$.* Extend α and u by periodicity. Let $t_0 \in [a, b[$ be such that

$$u(t_0) - \alpha(t_0) = \min_t (u(t) - \alpha(t)) < 0,$$

which implies

$$D_-\alpha(t_0) \ge D^+\alpha(t_0)$$

and there exists $\alpha_0 \in \mathcal{C}^1(I_0)$ as in Definition 5.1. It follows that t_0 is a minimum of $u - \alpha_0$, $(u - \alpha_0)'(t_0) = 0$ and $(u - \alpha_0)''(t_0) \ge 0$. From the remark that follows Definition 5.1, we have $\alpha'(t_0) = \alpha_0'(t_0)$. Hence $t_0 \in A$ and

$$\bar{f}(t_0, u(t_0), u'(t_0)) \le f(t_0, \alpha(t_0), \alpha'(t_0)) = f(t_0, \alpha_0(t_0), \alpha_0'(t_0)).$$

We compute now

$$\begin{aligned} 0 \le u''(t_0) - \alpha_0''(t_0) &= \bar{f}(t_0, u(t_0), u'(t_0)) + u(t_0) - \alpha_0(t_0) - \alpha_0''(t_0) \\ &\le f(t_0, \alpha_0(t_0), \alpha_0'(t_0)) - \alpha_0''(t_0) + u(t_0) - \alpha_0(t_0) < 0 \end{aligned}$$

which is a contradiction.

In a similar way we prove that $u \le \beta$.

As a consequence, u satisfies

$$\begin{aligned} u'' &= \max\{\min\{f(t, u, u'), N\}, -N\}, \\ u(a) &= u(b), \ u'(a) = u'(b). \end{aligned} \tag{5.8}$$

Step 3 – *The solution u is such that $|u'(t)| \le R$ on $[a, b]$.* Observe that, for all $(t, u, v) \in E$,

$$\max\{\min\{f(t, u, v), N\}, -N\} \le \varphi(|v|).$$

From Proposition 4.5 (with $r = 0$ and $\bar{\varphi} = \varphi$), every solution $u \in [\alpha, \beta]$ of (5.8) satisfies

$$\|u'\|_\infty < R.$$

Hence $|f(t, u(t), u'(t))| \le N$ and the function u is a solution of (5.1). □

Notice the idea of the modification (5.7). Here f is truncated rather than u and u'.

Remark also that condition (5.6) is satisfied if α, $\beta \in W^{1,\infty}(a, b)$ or if f does not depend on u'.

Remark The Theorems 5.3, 5.4, and 5.5 are still valid if we replace the Nagumo condition (4.8) or the one-sided Nagumo condition (4.14) by this last one-sided condition or any of the following one:

(a) for all $(t, u, v) \in E$, $f(t, u, v) \geq -\varphi(|v|)$;

(b) for all $(t, u, v) \in E$, $\operatorname{sgn}(v) f(t, u, v) \leq \varphi(|v|)$;

(c) for all $(t, u, v) \in E$, $\operatorname{sgn}(v) f(t, u, v) \geq -\varphi(|v|)$.

5.3 Structure of the Set of Solutions

The following results concern existence of maximal and minimal solutions between α and β as well as existence of a continuum of solutions.

Theorem 5.6. *Let α and β be C^2-lower and upper solutions of the problem* (5.1) *such that $\alpha \leq \beta$, E be defined in* (4.7), *$\varphi : \mathbb{R}^+ \to \mathbb{R}$ be a positive continuous function satisfying* (4.9) *and $f : E \to \mathbb{R}$ be a continuous function which satisfies* (4.8).

Then the problem (5.1) *has a minimal solution $u_{min} \in C^2([a, b])$ and a maximal solution $u_{max} \in C^2([a, b])$ in $[\alpha, \beta]$, i.e.*

$$\alpha \leq u_{min} \leq u_{max} \leq \beta,$$

and any other solution u of (5.1) *such that $\alpha \leq u \leq \beta$ satisfies*

$$u_{min} \leq u \leq u_{max}.$$

Theorem 5.7. *Assume the hypothesis of Theorem 5.6 hold and f is nondecreasing with respect to u.*

Then for any $t_0 \in [a, b]$ and $u^* \in \mathbb{R}$ with $u_{min}(t_0) \leq u^* \leq u_{max}(t_0)$, there exists a solution $u \in C^2([a, b])$ of (5.1) *such that $u_{min} \leq u \leq u_{max}$ and $u(t_0) = u^*$.*

Exercise 5.2. Prove the above results adapting the proofs of Theorems 2.4 and 2.5.

Exercise 5.3. State and prove the counterpart of Theorem 5.6 under the assumptions of Theorem 5.5.

6 BVP with Derivative Dependence : $W^{2,1}$-Solutions

For problems with a nonlinearity that depends on the derivative and satisfies L^1-Carathéodory conditions, the definitions of lower and upper solutions are the obvious extensions of Definitions 3.1.

Definitions 6.1. *A function* $\alpha \in C([a,b])$ *such that* $\alpha(a) = \alpha(b)$ *is a* $W^{2,1}$-lower solution *of* (5.1) *if its periodic extension on* $\mathbb{R}$, *defined by* $\alpha(t) = \alpha(t+b-a)$, *is such that for any* $t_0 \in \mathbb{R}$
either $D_-\alpha(t_0) < D^+\alpha(t_0)$
or there exists an open interval I_0 *such that* $t_0 \in I_0$, $\alpha \in W^{2,1}(I_0)$ *and for a.e.* $t \in I_0$,

$$\alpha''(t) \geq f(t, \alpha(t), \alpha'(t)).$$

A function $\beta \in C([a,b])$ *such that* $\beta(a) = \beta(b)$ *is a* $W^{2,1}$-upper solution *of* (5.1) *if its periodic extension on* $\mathbb{R}$, *defined by* $\beta(t) = \beta(t+b-a)$, *is such that for any* $t_0 \in \mathbb{R}$
either $D^-\beta(t_0) > D_+\beta(t_0)$
or there exists an open interval I_0 *such that* $t_0 \in I_0$, $\beta \in W^{2,1}(I_0)$ *and for a.e.* $t \in I_0$,

$$\beta''(t) \leq f(t, \beta(t), \beta'(t)).$$

Our first result uses lower and upper solutions which are in $W^{1,\infty}$. This assumption is not essential but it allows a method of proof which is interesting by itself.

Theorem 6.1. *Let* α *and* $\beta \in W^{1,\infty}(a,b)$ *be* $W^{2,1}$*-lower and upper solutions of* (5.1) *such that* $\alpha \leq \beta$.

Let E *be defined from* (4.7) *and* p, $q \in [1,\infty]$ *be such that* $\frac{1}{p} + \frac{1}{q} = 1$. *Assume* $f : E \to \mathbb{R}$ *satisfies* L^p*-Carathéodory conditions.*

Assume moreover there exist $\varphi \in C(\mathbb{R}^+, \mathbb{R}_0^+)$ *and* $\psi \in L^p(a,b)$ *with*

$$\int_0^\infty \frac{s^{1/q}}{\varphi(s)}\, ds > \|\psi\|_{L^p} (\max_t \beta(t) - \min_t \alpha(t))^{1/q} \tag{6.1}$$

and that the function f *satisfies one of the following one-sided Nagumo conditions*

(a) for a.e. $t \in [a,b]$ *and all* $(u,v) \in \mathbb{R}^2$ *such that* $\alpha(t) \leq u \leq \beta(t)$,
$$f(t,u,v) \leq \psi(t)\varphi(|v|);$$
(b) for a.e. $t \in [a,b]$ *and all* $(u,v) \in \mathbb{R}^2$ *such that* $\alpha(t) \leq u \leq \beta(t)$,
$$f(t,u,v) \geq -\psi(t)\varphi(|v|);$$
(c) for a.e. $t \in [a,b]$ *and all* $(u,v) \in \mathbb{R}^2$ *such that* $\alpha(t) \leq u \leq \beta(t)$,
$$\operatorname{sgn}(v) f(t,u,v) \leq \psi(t)\varphi(|v|);$$
(d) for a.e. $t \in [a,b]$ *and all* $(u,v) \in \mathbb{R}^2$ *such that* $\alpha(t) \leq u \leq \beta(t)$,
$$\operatorname{sgn}(v) f(t,u,v) \geq -\psi(t)\varphi(|v|).$$

Then the problem (5.1) *has at least one solution* $u \in W^{2,p}(a,b)$ *such that for all* $t \in [a,b]$

$$\alpha(t) \leq u(t) \leq \beta(t).$$

A natural idea to prove this theorem is to use a modified problem. For example, we could consider (5.3) as in Theorem 5.4. To prove then that solutions of the modified problem lie between α and β requires however that f satisfies an extra assumption such as Condition (A) in Proposition III-1.5. To avoid such an unnecessary condition, we will use another modified problem, which will need several auxiliary results.

Lemma 6.2. *Let $u \in W^{1,\infty}(a,b)$. Then $u^+ = \max\{u,0\} \in W^{1,\infty}(a,b)$ and*

$$\frac{d}{dt}u^+(t) = \frac{d}{dt}u(t), \quad \text{if } u(t) > 0,$$
$$\frac{d}{dt}u^+(t) = 0, \quad \text{if } u(t) \leq 0.$$

Proof. See [188, Theorem A1 (p.50)]. □

Lemma 6.3. *Let $u \in W^{1,\infty}(a,b)$ and $(u_n)_n \subset W^{1,\infty}(a,b)$ be such that $u_n \to u$ in $W^{1,\infty}(a,b)$. Then $u_n^+ \to u^+$ in $\mathcal{C}([a,b])$ and for almost every $t \in [a,b]$*

$$\lim_{n\to\infty} \frac{d}{dt}u_n^+(t) = \frac{d}{dt}u^+(t).$$

Proof. The first part of the proof comes from the continuity of the application $P : u \mapsto u^+$ as follows from

$$||u_n^+ - u^+||_\infty \leq ||u_n - u||_\infty.$$

To prove the second part, notice that for almost every t, the derivatives $\frac{d}{dt}u^+(t)$, $\frac{d}{dt}u(t)$, $\frac{d}{dt}u_n^+(t)$ and $\frac{d}{dt}u_n(t)$ exist and $\lim_{n\to\infty}\frac{d}{dt}u_n(t) = \frac{d}{dt}u(t)$. Let t be such a point. If $u(t) > 0$, we have $\frac{d}{dt}u^+(t) = \frac{d}{dt}u(t)$ and, for n large enough, $u_n(t) > 0$, which implies that for such an n,

$$\frac{d}{dt}u_n^+(t) = \frac{d}{dt}u_n(t) \quad \text{and} \quad \lim_{n\to\infty} \frac{d}{dt}u_n^+(t) = \frac{d}{dt}u^+(t).$$

Similarly, if $u(t) < 0$, we have $\frac{d}{dt}u^+(t) = 0$ and for n large enough,

$$\frac{d}{dt}u_n^+(t) = 0.$$

At last if $u(t) = 0$ and as at point t both derivatives $\frac{d}{dt}u^+(t)$ and $\frac{d}{dt}u(t)$ exist, we have $\frac{d}{dt}u^+(t) = 0$ and also $\frac{d}{dt}u(t) = 0$. As $\frac{d}{dt}u_n^+(t)$ is $\frac{d}{dt}u_n(t)$ or 0, this implies

$$\lim_{n\to\infty} \frac{d}{dt}u_n^+(t) = 0 = \frac{d}{dt}u(t).$$

□

Corollary 6.4. *Let α and $\beta \in W^{1,\infty}(a,b)$ be such that $\alpha \leq \beta$, and define $\gamma(t,u)$ from (1.3).*

Then, for any $u \in W^{1,\infty}(a,b)$, we have $\gamma(.,u) \in W^{1,\infty}(a,b)$ and

$$\begin{aligned} \tfrac{d}{dt}\gamma(t,u(t)) &= \alpha'(t), \quad \text{if } u(t) < \alpha(t), \\ &= u'(t), \quad \text{if } \alpha(t) \leq u(t) \leq \beta(t), \\ &= \beta'(t), \quad \text{if } \beta(t) < u(t). \end{aligned}$$

Furthermore, if $(u_n)_n \subset W^{1,\infty}(a,b)$ is such that $u_n \to u$ in $W^{1,\infty}(a,b)$, then

$$\gamma(.,u_n) \to \gamma(.,u)$$

in $C([a,b])$ and for almost every $t \in [a,b]$

$$\lim_{n\to\infty} \frac{d}{dt}\gamma(t,u_n(t)) = \frac{d}{dt}\gamma(t,u(t)).$$

Proof. One can notice that

$$\gamma(t,u(t)) = u(t) + (\alpha - u)^+(t) - (u - \beta)^+(t)$$

and use Lemmas 6.2 and 6.3. □

Lemma 6.5. *Let α and $\beta \in W^{1,\infty}(a,b)$ be such that $\alpha \leq \beta$, define $\gamma(t,u)$ from (1.3) and E from (4.7), and assume $f : E \to \mathbb{R}$ satisfies L^1-Carathéodory conditions.*

Then, for any $u \in C^1([a,b])$, the function

$$(Tu)(t) = \int_a^b G(t,s) f\Big(s, \gamma(s,u(s)), \frac{d}{ds}\gamma(s,u(s))\Big)\, ds, \tag{6.2}$$

where $G(t,s)$ is the Green's function corresponding to the problem (1.4), is in $C^1([a,b])$. Furthermore, the operator

$$T : C^1([a,b]) \to C^1([a,b])$$

defined by (6.2) is completely continuous.

Proof. Notice that given $u \in C^1([a,b])$, it follows from Corollary 6.4 that $f(s,\gamma(s,u(s)), \frac{d}{ds}\gamma(s,u(s)))$ is in $L^1(a,b)$. It is then straightforward to see that Tu is in $C^1([a,b])$.

Next, let $u_n \to u$ in $C^1([a,b])$. From

$$\begin{aligned}|T(u_n)(t) - &T(u)(t)| \\ &= |\int_a^b G(t,s) f(s,\gamma(s,u_n(s)), \tfrac{d}{ds}\gamma(s,u_n(s)))\,ds \\ &\qquad - \int_a^b G(t,s) f(s,\gamma(s,u(s)), \tfrac{d}{ds}\gamma(s,u(s)))\,ds| \\ &\le \max_{t,s}|G(t,s)| \int_a^b |f(s,\gamma(s,u_n(s)), \tfrac{d}{ds}\gamma(s,u_n(s))) \\ &\qquad\qquad - f(s,\gamma(s,u(s)), \tfrac{d}{ds}\gamma(s,u(s)))\,|ds,\end{aligned}$$

Lebesgue's Dominated Convergence Theorem, and Corollary 6.4, we have

$$T(u_n) \to T(u)$$

in $C([a,b])$. Similarly, one proves $\frac{d}{dt}T(u_n) \to \frac{d}{dt}T(u)$ in $C([a,b])$ as $u_n \to u$ in $C^1([a,b])$. The continuity of T follows.

At last, the complete continuity of T follows as usually from Arzelà-Ascoli Theorem. □

Proof of Theorem 6.1 : Let $R > \max\{\|\alpha'\|_\infty, \|\beta'\|_\infty\}$ be large enough so that

$$\int_0^R \frac{s^{1/q}}{\varphi(s)}\,ds > \|\psi\|_{L^p}(\max_t \beta(t) - \min_t \alpha(t))^{1/q}.$$

Consider the modified problem

$$\begin{aligned} u'' - u = \bar{f}(t,\gamma(t,u), \tfrac{d}{dt}\gamma(\cdot,u)) - \gamma(t,u), \\ u(a) = u(b),\ u'(a) = u'(b), \end{aligned} \tag{6.3}$$

where $\bar{f}(t,u,v) := f(t,u,\max\{\min\{v,R\},-R\})$ and $\gamma(t,u)$ is defined by (1.3). Notice that (6.3) is a functional equation.

Claim 1 – The problem (6.3) *has at least one solution.* We can write (6.3) as a fixed point problem

$$u = Tu,$$

where

$$(Tu)(t) = \int_a^b G(t,s)(\bar{f}(t,\gamma(t,u), \tfrac{d}{dt}\gamma(\cdot,u)) - \gamma(t,u))\,ds.$$

This operator is clearly bounded and from Lemma 6.5 we know it is completely continuous. Hence by Schauder's Theorem, T has a fixed point which is a solution of (6.3).

Claim 2 – The solution u of (6.3) *satisfies on* $[a,b]$

$$\alpha(t) \le u(t) \le \beta(t).$$

This claim follows from the argument used to prove Theorem 3.1.

Claim 3 – The solution u of (6.3) *is such that* $\|u'\|_\infty \le R$. This result is proved from the argument used in Theorem 5.4 with Proposition 4.5 replaced by Proposition 4.8 or 4.9.

Conclusion – It follows now from Claims 2 and 3 that the solution u of (6.3) solves (5.1). □

The following theorem extends Theorem 6.1 to lower and upper solutions whose derivative is not in $L^\infty(a,b)$. As in Theorem 5.5, we have to assume a one-sided bound on $f(t,\alpha(t),\alpha'(t))$ and $f(t,\beta(t),\beta'(t))$. Notice that such a condition holds if α and $\beta \in W^{1,\infty}(a,b)$.

Theorem 6.6. *Let α and $\beta \in C([a,b])$ be $W^{2,1}$-lower and upper solutions of* (5.1) *such that $\alpha \le \beta$. Define $A \subset [a,b]$ (resp. $B \subset [a,b]$) to be the set of points where α (resp. β) is derivable.*

Let E be defined from (4.7) *and p, $q \in [1,\infty]$ be such that $\frac{1}{p}+\frac{1}{q}=1$. Assume $f : E \to \mathbb{R}$ satisfies L^p-Carathéodory conditions and there exists $N \in L^1(a,b)$, $N > 0$ such that for a.e. $t \in A$ (resp. for a.e. $t \in B$)*

$$f(t,\alpha(t),\alpha'(t)) \ge -N(t) \quad (\text{resp. } f(t,\beta(t),\beta'(t)) \le N(t)).$$

Assume moreover there exist $\varphi \in C(\mathbb{R}^+,\mathbb{R}_0^+)$ and $\psi \in L^p(a,b)$ satisfying (6.1) *and that the function f satisfies one of the one-sided Nagumo conditions (a), (b), (c), or (d) in Theorem 6.1.*

Then the problem (5.1) *has at least one solution $u \in W^{2,p}(a,b)$ such that for all $t \in [a,b]$*

$$\alpha(t) \le u(t) \le \beta(t).$$

Proof. The proof proceeds in several steps.

Step 1: The modified problem. Let $R > 0$ be large enough so that

$$\int_0^R \frac{s^{1/q}}{\varphi(s)} ds > \|\psi\|_{L^p} (\max_t \beta(t) - \min_t \alpha(t))^{1/q}.$$

Increasing N if necessary, we can assume $N(t) \ge |f(t,u,v)|$ if $t \in [a,b]$, $\alpha(t) \le u \le \beta(t)$ and $|v| \le R$. Define then

$$\begin{aligned}
\bar f(t,u,v) &= \max\{\min\{f(t,\gamma(t,u),v),N(t)\},-N(t)\},\\
\omega_1(t,\delta) &= \chi_A(t) \max_{|v|\le\delta} |\bar f(t,\alpha(t),\alpha'(t)+v) - \bar f(t,\alpha(t),\alpha'(t))|,\\
\omega_2(t,\delta) &= \chi_B(t) \max_{|v|\le\delta} |\bar f(t,\beta(t),\beta'(t)+v) - \bar f(t,\beta(t),\beta'(t))|,
\end{aligned}$$

where γ is defined from (1.3), χ_A and χ_B are the characteristic functions of the sets A and B. It is clear that ω_i are L^1-Carathéodory functions, non-decreasing in δ, such that $\omega_i(t,0) = 0$ and $|\omega_i(t,\delta)| \le 2N(t)$.

We consider now the modified problem

$$\begin{aligned} u'' - u &= \bar{f}(t,u,u') - \omega(t,u), \\ u(a) &= u(b),\ u'(a) = u'(b), \end{aligned} \tag{6.4}$$

where

$$\begin{aligned} \omega(t,u) &= \beta(t) - \omega_2(t, u - \beta(t)), && \text{if } u > \beta(t), \\ &= u, && \text{if } \alpha(t) \le u \le \beta(t), \\ &= \alpha(t) + \omega_1(t, \alpha(t) - u), && \text{if } u < \alpha(t). \end{aligned}$$

Step 2: Existence of a solution of (6.4). Notice that the operator

$$T : C^1([a,b]) \to C^1([a,b])$$

defined by

$$(Tu)(t) = \int_a^b G(t,s)[\bar{f}(s,u(s),u'(s)) - \omega(s,u(s))]\,ds,$$

where $G(t,s)$ is the Green's function corresponding to (1.4), is completely continuous and bounded. Hence, by Schauder's Theorem, T has a fixed point u which is a solution of (6.4).

Step 3: The solution u of (6.4) *is such that $\alpha \le u \le \beta$.* Let us assume on the contrary that for some $t_0 \in \mathbb{R}$

$$\min_t (u(t) - \alpha(t)) = u(t_0) - \alpha(t_0) < 0.$$

Then, as in Theorem 3.1, there exists an open interval I_0 with $t_0 \in I_0$, $\alpha \in W^{2,1}(I_0)$ and, for a.e. $t \in I_0$,

$$\alpha''(t) \ge f(t,\alpha(t),\alpha'(t)).$$

Furthermore $u'(t_0) - \alpha'(t_0) = 0$ and for $t \ge t_0$ near enough t_0

$$|u'(t) - \alpha'(t)| \le \alpha(t) - u(t).$$

As ω_1 is non-decreasing and $\bar{f}(t,\alpha(t),\alpha'(t)) \le f(t,\alpha(t),\alpha'(t))$, we have for $t \ge t_0$ near enough t_0

$$\begin{aligned} u'(t) - \alpha'(t) &= \int_{t_0}^t (u''(s) - \alpha''(s))\,ds \\ &\le \int_{t_0}^t [\bar{f}(s,\alpha(s),u'(s)) - \bar{f}(s,\alpha(s),\alpha'(s)) \\ &\qquad + u(s) - \alpha(s) - \omega_1(s,\alpha(s) - u(s))]\,ds < 0. \end{aligned}$$

This inequality proves $u(t_0) - \alpha(t_0)$ is not a minimum of $u - \alpha$ which is a contradiction.

A similar argument holds to prove $u \leq \beta$.

Step 4: The solution u of (6.4) *is such that* $\|u'\|_\infty < R$. Observe that for all $(t, u, v) \in E$, one of the conditions (a), (b), (c), or (d) in Theorem 6.1 is satisfied with $f(t, u, v)$ replaced by $\bar{f}(t, u, v)$. Extending u by periodicity and using Propositions 4.8 or 4.9, we prove then that every solution $u \in [\alpha, \beta]$ of (6.4) satisfies

$$\|u'\|_\infty < R.$$

Conclusion. It follows from Step 3 and 4 that the solution u of (6.4) solves (5.1). □

We can improve Theorems 6.1 and 6.6 by assuming that f is an L^r-Carathéodory function and that conditions (a) to (d) are satisfied with $\psi \in L^p(a, b)$, where r is not necessarily equal to p. Observe that a Nagumo condition $|f(t, u, v)| \leq \psi(t)\varphi(|v|)$ would imply $r \geq p$ and the solution would be in $W^{2,r}(a, b)$.

As a first illustration consider the following example where Theorem 5.5 does not apply.

Example 6.1. Consider the boundary value problem

$$u'' = \frac{1}{\sqrt{t}}|u'|^a + u + t,$$
$$u(0) = u(1), \; u'(0) = u'(1),$$

where $1 \leq a < 3/2$. Existence of a solution in $W^{2,2-\epsilon}(0, 1)$, for $\epsilon > 0$ small enough, follows from Theorem 6.1 with $\alpha(t) = -1$, $\beta(t) = 0$, $p = \frac{1}{2-a} < 2$, $\psi(t) = \frac{1}{\sqrt{t}}$ and $\varphi(y) = y^a + 1$. Notice that

$$\int_0^\infty \frac{s^{1/q}}{\varphi(s)}\, ds = \int_0^\infty \frac{s^{a-1}}{s^a + 1}\, ds = \infty.$$

In Theorem 6.6, the limit cases $p = \infty$, $q = 1$, and $p = 1$, $q = \infty$ are of interest. If $p = \infty$, $q = 1$, it is natural to choose $\psi = 1$ and (6.1) reads

$$\int_0^\infty \frac{s\, ds}{\varphi(s)} > \max_t \beta(t) - \min_t \alpha(t).$$

In this case, Theorems 6.6 and 5.5 differ from the regularity assumptions. If $p = 1$, $q = \infty$, condition (6.1) becomes

$$\int_0^\infty \frac{ds}{\varphi(s)} > \|\psi\|_{L^1}$$

and we obtain the following corollary.

Corollary 6.7. *Let α and $\beta \in C([a,b])$ be $W^{2,1}$-lower and upper solutions of* (5.1) *such that $\alpha \le \beta$. Define $A \subset [a,b]$ (resp. $B \subset [a,b]$) to be the set of points where α (resp. β) is derivable.*

Let E be defined by (4.7). *Assume $f : E \to \mathbb{R}$ satisfies L^1-Carathéodory conditions and there exists $N \in L^1(a,b)$, $N > 0$ such that for a.e. $t \in A$ (resp. for a.e. $t \in B$)*

$$f(t,\alpha(t),\alpha'(t)) \ge -N(t) \quad (\textit{resp. } f(t,\beta(t),\beta'(t)) \le N(t)).$$

Assume moreover there exist $\varphi \in C(\mathbb{R}^+,\mathbb{R}_0^+)$ and $\psi \in L^1(a,b)$ with

$$\int_0^\infty \frac{ds}{\varphi(s)} > \|\psi\|_{L^1}$$

and that the function f satisfies one of the one-sided Nagumo conditions (a), (b), (c), or (d) in Theorem 6.1.

Then the problem (5.1) *has at least one solution $u \in W^{2,1}(a,b)$ such that for all $t \in [a,b]$*

$$\alpha(t) \le u(t) \le \beta(t).$$

The conditions of the corollary are satisfied if $f : E \to \mathbb{R}$ satisfies a L^1-Lipschitz condition in v, i.e. there exists $L \in L^1(a,b)$ such that for a.e. $t \in [a,b]$ and all (u,v_1), $(u,v_2) \in \mathbb{R}^2$

$$(t,u,v_1),(t,u,v_2) \in E \;\Rightarrow\; |f(t,u,v_1) - f(t,u,v_2)| \le L(t)|v_1 - v_2|.$$

Exercise 6.1. Prove a result similar to Corollary 6.7 assuming there exist $\varphi \in C(\mathbb{R}^+,\mathbb{R}_0^+)$ and $\psi \in L^1(a,b)$ with

$$\int_0^\infty \frac{ds}{\varphi(s)} = \infty$$

and that the function f satisfies one of the one-sided Nagumo conditions

(a) for a.e. $t \in [a,b]$ and all $(u,v) \in \mathbb{R}^2$ such that $\alpha(t) \le u \le \beta(t)$,
$$f(t,u,v) \le (\psi(t) + |v|)\varphi(|v|);$$

(b) for a.e. $t \in [a,b]$ and all $(u,v) \in \mathbb{R}^2$ such that $\alpha(t) \le u \le \beta(t)$,
$$f(t,u,v) \ge -(\psi(t) + |v|)\varphi(|v|);$$

(c) for a.e. $t \in [a,b]$ and all $(u,v) \in \mathbb{R}^2$ such that $\alpha(t) \le u \le \beta(t)$,
$$\operatorname{sgn}(v)f(t,u,v) \le (\psi(t) + |v|)\varphi(|v|);$$

(d) for a.e. $t \in [a,b]$ and all $(u,v) \in \mathbb{R}^2$ such that $\alpha(t) \le u \le \beta(t)$,
$$\operatorname{sgn}(v)f(t,u,v) \ge -(\psi(t) + |v|)\varphi(|v|).$$

Hint : See [187].

Notice that the Example 4.1 shows we cannot replace the Nagumo condition by the assumption that solutions of the given problem (5.1) between α and β are *a priori* bounded in $\mathcal{C}^1([a,b])$.

We can however replace the Nagumo condition by any condition so that an *a priori* bound in $\mathcal{C}([a,b])$ on solutions of the corresponding modified problem implies an *a priori* bound in $\mathcal{C}^1([a,b])$. In particular, we have the following result for the Rayleigh equation.

Theorem 6.8. *Let α and $\beta \in \mathcal{C}([a,b])$ be $W^{2,1}$-lower and upper solutions of (4.4) such that $\alpha \le \beta$. Let E be defined by (4.7), $g \in \mathcal{C}(\mathbb{R})$ and $h : E \to \mathbb{R}$ be a Carathéodory function such that for some $H \in L^2(a,b)$, for a.e. $t \in [a,b]$ and all $(u,v) \in \mathbb{R}^2$ with $(t,u,v) \in E$,*

$$|h(t,u,v)| \le H(t).$$

Then the problem (4.4) has at least one solution $u \in W^{2,2}(a,b)$ such that for all $t \in [a,b]$

$$\alpha(t) \le u(t) \le \beta(t).$$

Proof. Define $A \subset [a,b]$ (resp. $B \subset [a,b]$) to be the set of points where α (resp. β) is derivable and

$$\begin{aligned}\omega_1(t,\delta) &= \chi_A(t)\max_{|v|\le\delta}|h(t,\alpha(t),\alpha'(t)+v) - h(t,\alpha(t),\alpha'(t))|,\\ \omega_2(t,\delta) &= \chi_B(t)\max_{|v|\le\delta}|h(t,\beta(t),\beta'(t)+v) - h(t,\beta(t),\beta'(t))|,\end{aligned}$$

where χ_A and χ_B are the characteristic functions of the sets A and B. It is clear that ω_i are Carathéodory functions, non-decreasing in δ, such that $\omega_i(t,0) = 0$ and $|\omega_i(t,\delta)| \le 2H(t)$.

Consider the family of modified problems

$$\begin{gathered}u'' - C(t)u = -[\lambda g(u') + h(t,\gamma(t,u),u') + \omega(t,u)],\\ u(a) = u(b),\ \ u'(a) = u'(b),\end{gathered} \tag{6.5}$$

where $C \in L^1(a,b)$ is chosen such that $C(t) > |g(0)| + 1 + 3H(t)$ on $[a,b]$, $\lambda \in [0,1]$, $\gamma(t,u)$ is defined from (1.3), and

$$\begin{aligned}\omega(t,u) &= C(t)\beta(t) - \omega_2(t,u-\beta(t)), && \text{if } u > \beta(t),\\ &= C(t)u, && \text{if } \alpha(t) \le u \le \beta(t),\\ &= C(t)\alpha(t) + \omega_1(t,\alpha(t)-u), && \text{if } u < \alpha(t).\end{aligned}$$

Claim 1 – Define $\rho = \max\{\|\alpha\|_\infty, \|\beta\|_\infty\} + 2$. *Then every solution* u *of* (6.5) *satisfies* $\|u\|_\infty < \rho$. Let us assume on the contrary that for some $t_0 \in \mathbb{R}$

$$\min_t u(t) = u(t_0) \leq -\rho.$$

Hence, for $t \geq t_0$ close enough to t_0, we have

$$|u'(t)| \leq \alpha(t) - u(t),$$

and

$$\begin{aligned} u'(t) &= \int_{t_0}^t u''(s)\,ds \\ &\leq \int_{t_0}^t [C(s)(u(s) - \alpha(s)) - \lambda g(u'(s)) \\ &\qquad\qquad - h(s, \alpha(s), u'(s)) - \omega_1(s, \alpha(s) - u(s))]\,ds \\ &\leq -\int_{t_0}^t [C(s) - |g(u'(s))| - 3H(s)]\,ds < 0. \end{aligned}$$

This inequality proves that $u(t_0)$ is not a minimum of u which is a contradiction. A similar argument holds to prove that $u \leq \rho$.

Claim 2 – There exists $R > 0$ *such that every solution* u *of* (6.5) *satisfies* $\|u'\|_\infty < R$. The proof follows the argument in Proposition 4.1.

Claim 3 – There exists a solution u *of* (6.5) *with* $\lambda = 1$. Define the operator $T_\lambda : C^1([a,b]) \to C^1([a,b])$ by

$$T_\lambda(u) = -\int_a^b G(t,s)[\lambda g(u'(s)) + h(s, \gamma(s, u(s)), u'(s)) + \omega(s, u(s))]\,ds,$$

where $G(t,s)$ is the Green's function of

$$\begin{aligned} u'' - C(t)u &= f(t), \\ u(a) = u(b),\ u'(a) &= u'(b). \end{aligned}$$

Observe that there exists $R_0 > 0$ such that $T_0(C^1([a,b])) \subset B(0, R_0)$. Hence,

$$\deg(I - T_0, B(0, R_0)) = 1$$

and by the properties of the degree, we prove easily that (6.5) with $\lambda = 1$ has a solution.

Claim 4 – The solution u *of* (6.5) *with* $\lambda = 1$ *is such that* $\alpha(t) \leq u(t) \leq \beta(t)$ *on* $[a,b]$. Extend α and u by periodicity. Let $t_0 \in [a, b[$ be such that

$$u(t_0) - \alpha(t_0) = \min_t (u(t) - \alpha(t)) < 0.$$

From the definition of lower solution there exists an open interval I_0 such that $t_0 \in I_0$, $\alpha \in W^{2,1}(I_0)$ and for a.e. $t \in I_0$,

$$\alpha''(t) + g(\alpha'(t)) + h(t, \alpha(t), \alpha'(t)) \geq 0.$$

Further, for $t \geq t_0$ near enough t_0,

$$|u'(t) - \alpha'(t)| \leq \alpha(t) - u(t)$$

and we compute

$$\begin{aligned} u'(t) - \alpha'(t) &= \int_{t_0}^{t} (u''(s) - \alpha''(s))\, ds \\ &\leq \int_{t_0}^{t} [h(s, \alpha(s), \alpha'(s)) - h(s, \alpha(s), u'(s)) - \omega_1(s, \alpha(s) - u(s)) \\ &\qquad\qquad + g(\alpha'(s)) - g(u'(s)) + C(s)(u(s) - \alpha(s))]\, ds \\ &< 0. \end{aligned}$$

This result follows as $g(\alpha'(s)) - g(u'(s))$ is small and $C(s)(\alpha(s) - u(s)) > \alpha(s) - u(s) \geq k$ for some $k > 0$. Hence, we contradict the minimality of $u - \alpha$ at t_0.

In a similar way we prove that $u \leq \beta$. Hence u is also a solution of (4.4). □

Using the same ideas, we can work the Liénard equation.

Theorem 6.9. *Let α and $\beta \in C([a, b])$ be $W^{2,1}$-lower and upper solutions of* (4.5) *such that $\alpha \leq \beta$. Let $E = \{(t, u) \in [a, b] \times \mathbb{R} \mid \alpha(t) \leq u \leq \beta(t)\}$, $g \in C(\mathbb{R})$ and $h : E \to \mathbb{R}$ be an L^1-Carathéodory function.*

Then the problem (4.5) *has at least one solution $u \in W^{2,1}(a, b)$ such that for all $t \in [a, b]$*

$$\alpha(t) \leq u(t) \leq \beta(t).$$

Proof. Consider the family of modified problems

$$\begin{aligned} u'' - C(t)u = -[\lambda g(\gamma(t, u))u' + h(t, \gamma(t, u)) + C(t)\gamma(t, u)], \\ u(a) = u(b),\ u'(a) = u'(b), \end{aligned} \tag{6.6}$$

where $C \in L^1(a, b)$ is chosen such that for a.e. $t \in [a, b]$, $|g(\alpha(t))| < C(t)$, $|g(\beta(t))| < C(t)$ and for every $(t, u) \in E$, $|h(t, u)| < C(t)$.

Claim 1 – Let $\rho = \max\{\|\alpha\|_\infty, \|\beta\|_\infty\} + 3$. Then every solution u of (6.6) *satisfies $\|u\|_\infty < \rho$.* This result follows as Claim 1 in Theorem 6.8.

Claim 2 – There exists $R > 0$ such that any solution u of (6.6) *is such that $\|u'\|_\infty < R$.* Use the argument of Proposition 4.3.

Claim 3 – There exists a solution u of (6.6) *with $\lambda = 1$.* The problem (6.6) is equivalent to the fixed point problem

$$u = T_\lambda u,$$

where $T_\lambda : \mathcal{C}^1([a,b]) \to \mathcal{C}^1([a,b])$ is defined by

$$(T_\lambda u)(t) = -\int_a^b G(t,s)[\lambda g(\gamma(t,u))u' + h(t,\gamma(t,u)) + C(t)\gamma(t,u)]\, ds,$$

with $G(t,s)$ the Green's function corresponding to

$$\begin{aligned} u'' - C(t)u &= f(t), \\ u(a) = u(b),\ u'(a) &= u'(b). \end{aligned}$$

Observe that for some $R_0 > 0$ we have $T_0(\mathcal{C}^1([a,b])) \subset B(0,R_0)$. Hence, by the properties of the degree, we prove easily that (6.6) with $\lambda = 1$ has a solution u.

Claim 4 – The solution u of (6.6) *with $\lambda = 1$ is such that $\alpha(t) \le u(t) \le \beta(t)$ on* $[a,b]$. The proof repeats the argument of Claim 4 in Theorem 6.8. □

Existence of solutions between the maximum of lower solutions and the minimum of upper solutions can be proved along the lines of Theorem 3.2. We state such a result for (5.1) with a Nagumo condition. A similar statement is valid for (4.4) or (4.5) under the assumptions of Theorems 6.8 or 6.9.

Theorem 6.10. *Let $\alpha_i \in \mathcal{C}([a,b])$ $(i = 1,\dots,n)$ and $\beta_j \in \mathcal{C}([a,b])$ $(j = 1,\dots,m)$ be respectively $W^{2,1}$-lower solutions and $W^{2,1}$-upper solutions of* (5.1). *Define $A_i \subset [a,b]$ (resp. $B_j \subset [a,b]$) to be the set of points where α_i (resp. β_j) is derivable. Assume $\alpha := \max_{1\le i\le n} \alpha_i \le \beta := \min_{1\le j\le m} \beta_j$.*

Define $E = \{(t,u,v) \in [a,b] \times \mathbb{R}^2 \mid \min_i \alpha_i \le u \le \max_j \beta_j\}$ and let p, $q \in [1,\infty]$ be such that $\frac{1}{p} + \frac{1}{q} = 1$. Assume $f : E \to \mathbb{R}$ satisfies L^p-Carathéodory conditions and there exists $N \in L^1(a,b)$, $N > 0$ such that for all i and a.e. $t \in A_i$ (resp. for all j and a.e. $t \in B_j$)

$$f(t,\alpha_i(t),\alpha_i'(t)) \ge -N(t) \quad (\textit{resp. } f(t,\beta_j(t),\beta_j'(t)) \le N(t)).$$

Assume moreover there exist $\varphi \in \mathcal{C}(\mathbb{R}^+,\mathbb{R}_0^+)$ and $\psi \in L^p(a,b)$ that satisfy (6.1) *and assume that the function f satisfies one of the one-sided Nagumo conditions (a), (b), (c), or (d) in Theorem 6.1.*

Then the problem (5.1) *has at least one solution $u \in W^{2,p}(a,b)$ such that for all $t \in [a,b]$*

$$\alpha(t) \le u(t) \le \beta(t).$$

Proof. Let R and N be chosen as in the proof of Theorem 6.6. Define then

$$\begin{aligned}
\bar{f}(t,u,v) &= \max\{\min\{f(t,u,v),N(t)\},-N(t)\},\\
g_i(t,u,v) &= \bar{f}(t,\alpha_i(t),v)-\alpha_i(t)-\omega_{1i}(t,\alpha_i(t)-u), \text{ if } u<\alpha_i(t),\\
&= \bar{f}(t,u,v)-u, \qquad \text{if } u\geq\alpha_i(t),\\
h_j(t,u,v) &= \bar{f}(t,\beta_j(t),v)-\beta_j(t)+\omega_{2j}(t,u-\beta_j(t)), \text{ if } u>\beta_j(t),\\
&= \bar{f}(t,u,v)-u, \qquad \text{if } u\leq\beta_j(t),
\end{aligned}$$

where

$$\begin{aligned}
\omega_{1i}(t,\delta) &= \chi_{A_i}(t)\max_{|v|\leq\delta}|\bar{f}(t,\alpha_i(t),\alpha_i'(t)+v)-\bar{f}(t,\alpha_i(t),\alpha_i'(t))|,\\
\omega_{2j}(t,\delta) &= \chi_{B_j}(t)\max_{|v|\leq\delta}|\bar{f}(t,\beta_j(t),\beta_j'(t)+v)-\bar{f}(t,\beta_j(t),\beta_j'(t))|,
\end{aligned}$$

χ_{A_i} and χ_{B_j} are the characteristic functions of the sets A_i and B_j.

We consider now the modified problem

$$\begin{aligned}
u''-u &= F(t,u,u'),\\
u(a) &= u(b),\ u'(a)=u'(b),
\end{aligned} \tag{6.7}$$

where

$$\begin{aligned}
F(t,u,v) &= \min_{1\leq i\leq n} g_i(t,u,v), \quad \text{if } u\leq\alpha(t),\\
&= \bar{f}(t,u,v)-u, \quad \text{if } \alpha(t)<u<\beta(t),\\
&= \max_{1\leq j\leq m} h_j(t,u,v), \text{ if } \beta(t)\leq u.
\end{aligned}$$

Adapting the argument in Theorem 3.2 and 6.6, we prove (6.7) has a solution such that $\alpha(t)\leq u(t)\leq\beta(t)$ and $\|u'\|_\infty<R$. The proof follows. □

The structure of the set of $W^{2,1}$-solutions is similar to the structure of C^2-solutions. In particular, the following result concerns existence of extremal solutions.

Theorem 6.11. *Let α and $\beta\in C([a,b])$ be respectively $W^{2,1}$-lower and upper solutions of* (5.1) *such that $\alpha\leq\beta$. Define $A\subset[a,b]$ (resp. $B\subset[a,b]$) to be the set of points where α (resp. β) is derivable.*

Let E be defined from (4.7) *and p, $q\in[1,\infty]$ be such that $\frac{1}{p}+\frac{1}{q}=1$. Assume $f:E\to\mathbb{R}$ satisfies L^p-Carathéodory conditions and there exists $N\in L^1(a,b)$, $N>0$ such that for a.e. $t\in A$ (resp. for a.e. $t\in B$)*

$$f(t,\alpha(t),\alpha'(t))\geq -N(t) \quad (\text{resp. } f(t,\beta(t),\beta'(t))\leq N(t)).$$

Assume moreover there exist $\varphi\in C(\mathbb{R}^+,\mathbb{R}_0^+)$ and $\psi\in L^p(a,b)$ that satisfy (6.1) *and assume that the function f satisfies one of the one-sided Nagumo conditions (a), (b), (c), or (d) in Theorem 6.1.*

Then the problem (5.1) *has a minimal solution* $u_{min} \in W^{2,1}(a,b)$ *and a maximal solution* $u_{max} \in W^{2,1}(a,b)$ *in* $[\alpha, \beta]$, *i.e.*

$$\alpha \leq u_{min} \leq u_{max} \leq \beta$$

and any other solution u *of* (5.1) *with* $\alpha \leq u \leq \beta$ *satisfies*

$$u_{min} \leq u \leq u_{max}.$$

Exercise 6.2. Prove the above theorem.

Exercise 6.3. As in Theorem 3.5, prove that if f is non-decreasing with respect to u we have a continuum of solutions.

Exercise 6.4. As an application of Theorem 6.6, we can study the influence of a parameter on the number of solutions. Consider the problem

$$\begin{array}{c} u'' + g(u)\,\mathrm{sgn}(u')\,|u'|^r + f(t,u) = s(t), \\ u(0) = u(2\pi), \;\; u'(0) = u'(2\pi) \end{array} \tag{6.8}$$

where f is an L^p-Carathéodory function, g is continuous, and $s \in L^p(0, 2\pi)$. It is easy to observe that the Nagumo condition is satisfied if $0 < r \leq 2 - 1/p$. Assume further

$$\lim_{u \to +\infty} f(t,u) = -\infty, \qquad \lim_{u \to -\infty} f(t,u) = +\infty,$$

uniformly in t.

(a) Prove that for all $s \in L^\infty(0, 2\pi)$, the problem (6.8) has at least one solution.

(b) Suppose moreover, there exist real numbers $v < u$ such that $f(t,v) < f(t,u)$. Prove then that for any $s \in L^\infty(0, 2\pi)$ with $f(t,v) \leq s(t) \leq f(t,u)$ problem (6.8) has at least two solutions.

Hint : Use constant lower and upper solutions : $\alpha_1 < \beta_1 = v < \alpha_2 = u < \beta_2$.

Combining the lower and upper solution method with other methods, we can prove the existence of a third solution of (6.8) for every s with $f(t,v) \leq s(t) \leq f(t,u)$. We refer the interested reader to Exercise III-1.4.

Observe that using one-sided Nagumo conditions, we can replace the restriction on r by a sign condition on g.

Exercise 6.5. Consider the singular problem

$$u'' + g(u)u' + \frac{1}{u} = h(t),$$
$$u(a) = u(b),\ u'(a) = u'(b),$$

where g is continuous and $h \in L^1(a,b)$ is such that $\int_a^b h(t)\,dt > 0$ and has an essential upper bound. Prove existence of a positive solution.

Hint : We can choose $\alpha > 0$ constant small enough and $\beta(t) = c + w(t)$, where c is a large enough constant and w is solution of

$$u'' + g(u+c)u' = h(t) - \frac{1}{b-a}\int_a^b h(t)\,dt,$$
$$u(a) = u(b),\ u'(a) = u'(b),\ \int_a^b u(t)\,dt = 0.$$

The idea is to prove an *a priori* bound on w which is independent of c and then to pick c so that β is an upper solution and $\beta \geq \alpha$. For more details see [149].

Chapter II

The Separated BVP

1 $\mathcal{C}^2$-Solutions

1.1 The method of lower and upper solutions

Consider the problem

$$\begin{aligned} u'' &= f(t, u, u'), \\ a_1 u(a) - a_2 u'(a) &= A_0, \\ b_1 u(b) + b_2 u'(b) &= B_0, \end{aligned} \tag{1.1}$$

where A_0, $B_0 \in \mathbb{R}$, a_1, $b_1 \in \mathbb{R}$, a_2, $b_2 \in \mathbb{R}^+$, $a_1^2 + a_2^2 > 0$ and $b_1^2 + b_2^2 > 0$. This problem is called the separated boundary value problem and contains as special cases the Dirichlet problem

$$\begin{aligned} u'' &= f(t, u, u'), \\ u(a) &= A_0, \;\; u(b) = B_0, \end{aligned}$$

the Neumann problem

$$\begin{aligned} u'' &= f(t, u, u'), \\ u'(a) &= A_0, \;\; u'(b) = B_0, \end{aligned}$$

and the Robin problem

$$\begin{aligned} u'' &= f(t, u, u'), \\ u'(a) &= A_0, \;\; u(b) = B_0. \end{aligned}$$

If we are looking for $\mathcal{C}^2$-solutions the Definitions I-5.1 can be adapted to problem (1.1) as follows.

Definitions 1.1. *A function* $\alpha \in \mathcal{C}([a,b])$ *is a* $\mathcal{C}^2$-lower solution *of* (1.1) *if*
(a) for any $t_0 \in]a,b[$, *either* $D_-\alpha(t_0) < D^+\alpha(t_0)$,
or there exist an open interval $I_0 \subset]a,b[$ *with* $t_0 \in I_0$ *and a function* $\alpha_0 \in \mathcal{C}^1(I_0)$ *such that*
(i) $\alpha(t_0) = \alpha_0(t_0)$ *and* $\alpha(t) \geq \alpha_0(t)$ *for all* $t \in I_0$;
(ii) $\alpha_0''(t_0)$ *exists and* $\alpha_0''(t_0) \geq f(t_0, \alpha_0(t_0), \alpha_0'(t_0))$;
(b) $a_1\alpha(a) - a_2 D^+\alpha(a) \leq A_0$, $b_1\alpha(b) + b_2 D_-\alpha(b) \leq B_0$.

A function $\beta \in \mathcal{C}([a,b])$ *is a* $\mathcal{C}^2$-upper solution *of* (1.1) *if*
(a) for any $t_0 \in]a,b[$, *either* $D^-\beta(t_0) > D_+\beta(t_0)$,
or there exist an open interval $I_0 \subset]a,b[$ *with* $t_0 \in I_0$ *and a function* $\beta_0 \in \mathcal{C}^1(I_0)$ *such that*
(i) $\beta(t_0) = \beta_0(t_0)$ *and* $\beta(t) \leq \beta_0(t)$ *for all* $t \in I_0$;
(ii) $\beta_0''(t_0)$ *exists and* $\beta_0''(t_0) \leq f(t_0, \beta_0(t_0), \beta_0'(t_0))$;
(b) $a_1\beta(a) - a_2 D_+\beta(a) \geq A_0$, $b_1\beta(b) + b_2 D^-\beta(b) \geq B_0$.

Remark Let α be a lower solution and assume t_0 is such that

$$D_-\alpha(t_0) \geq D^+\alpha(t_0).$$

As in Chapter I, we can observe that in that case α has a derivative at t_0 and $\alpha'(t_0) = \alpha_0'(t_0)$.

As in Chapter I, we can build lower and upper solutions from maxima of lower solutions and minima of upper solutions. The proofs of these propositions are left as exercises.

Proposition 1.1. *Let* $\alpha_i \in \mathcal{C}([a,b])$ $(i = 1, ..., n)$ *be* $\mathcal{C}^2$*-lower solutions of* (1.1). *Then the function*

$$\alpha(t) = \max_{1 \leq i \leq n} \alpha_i(t), \quad t \in [a,b]$$

is a $\mathcal{C}^2$*-lower solution of* (1.1).

Proposition 1.2. *Let* $\beta_j \in \mathcal{C}([a,b])$ $(j = 1, ..., m)$ *be* $\mathcal{C}^2$*-upper solutions of* (1.1). *Then the function*

$$\beta(t) = \min_{1 \leq j \leq m} \beta_j(t), \quad t \in [a,b]$$

is a $\mathcal{C}^2$*-upper solution of* (1.1).

The main existence results for solutions of (1.1) parallel the corresponding ones for the periodic problem.

Theorem 1.3. *Let A_0, $B_0 \in \mathbb{R}$, a_1, $b_1 \in \mathbb{R}$, a_2, $b_2 \in \mathbb{R}^+$ with $a_1^2 + a_2^2 > 0$, $b_1^2 + b_2^2 > 0$.*

Assume α and $\beta \in C([a,b])$ are C^2-lower and upper solutions of the problem (1.1) such that $\alpha \le \beta$.

Let

$$E := \{(t,u,v) \in [a,b] \times \mathbb{R}^2 \mid \alpha(t) \le u \le \beta(t)\}, \tag{1.2}$$

$\varphi : \mathbb{R}^+ \to \mathbb{R}$ be a positive continuous function satisfying

$$\int_0^\infty \frac{s\,ds}{\varphi(s)} = \infty, \tag{1.3}$$

and assume $f : E \to \mathbb{R}$ is a continuous function which satisfies

$$\forall (t,u,v) \in E, \qquad |f(t,u,v)| \le \varphi(|v|). \tag{1.4}$$

Then the problem (1.1) has at least one solution $u \in C^2([a,b])$ such that for all $t \in [a,b]$

$$\alpha(t) \le u(t) \le \beta(t).$$

Proof. The proof can be adapted from the proof of Theorem I-5.3. Consider the modified problem

$$\begin{aligned} u'' &= \lambda f(t, \gamma(t,u), u') + \varphi(|u'|)\,(u - \lambda\gamma(t,u)),\\ u(a) &= \lambda[A_0 - a_1\gamma(a,u(a)) + a_2 u'(a) + \gamma(a,u(a))],\\ u(b) &= \lambda[B_0 - b_1\gamma(b,u(b)) - b_2 u'(b) + \gamma(b,u(b))], \end{aligned} \tag{1.5}$$

where $\gamma : [a,b] \times \mathbb{R} \to \mathbb{R}$ is defined by

$$\begin{aligned} \gamma(t,u) &= \alpha(t), && \text{if } u < \alpha(t),\\ &= u, && \text{if } \alpha(t) \le u \le \beta(t),\\ &= \beta(t), && \text{if } u > \beta(t). \end{aligned} \tag{1.6}$$

Let $\rho > 0$ be such that, for all $t \in [a,b]$,

$$\begin{gathered} -\rho < \alpha(t) \le \beta(t) < \rho,\\ f(t,\alpha(t),0) + \varphi(0)\,(-\rho - \alpha(t)) < 0,\\ f(t,\beta(t),0) + \varphi(0)\,(\rho - \beta(t)) > 0,\\ A_0 - a_1\alpha(a) + \alpha(a) > -\rho, \quad B_0 - b_1\alpha(b) + \alpha(b) > -\rho,\\ A_0 - a_1\beta(a) + \beta(a) < \rho, \quad B_0 - b_1\beta(b) + \beta(b) < \rho. \end{gathered}$$

Claim 1 – Every solution u of (1.5) with $\lambda \in [0,1]$ is such that $\|u\|_\infty < \rho$. Assume there exists $t_0 \in]a,b[$ such that $\min_t u(t) = u(t_0) \le -\rho$. We have $u'(t_0) = 0$ and obtain the contradiction

$$0 \le u''(t_0) = \lambda[f(t_0,\alpha(t_0),0) + \varphi(0)\,(u(t_0) - \alpha(t_0))] + (1-\lambda)\varphi(0)u(t_0) < 0.$$

If the minimum is achieved for $t_0 = a$, we have the contradiction

$$-\rho \geq u(a) = \lambda[A_0 - a_1\alpha(a) + a_2 u'(a) + \alpha(a)] \geq \lambda[A_0 - a_1\alpha(a) + \alpha(a)] > -\rho.$$

In case the minimum is for $t_0 = b$, we come to the same contradiction. At last, we prove in a similar way that $u(t) < \rho$.

Claim 2 – *There exists* $R > 0$ *such that every solution* u *of* (1.5) *with* $\lambda \in [0,1]$ *satisfies* $\|u'\|_\infty < R$. The claim follows choosing $R > 0$ from Proposition I-4.4, where $\bar{\alpha} = -\rho$, $\bar{\beta} = \rho$ and $\bar{\varphi}(v) = (1+2\rho)\varphi(v)$.

Claim 3 – *Existence of solutions of* (1.5) *for* $\lambda = 1$. Let us define the operators

$$\begin{aligned} L &: \mathcal{C}^2([a,b]) \subset \mathcal{C}^1([a,b]) \to \mathcal{C}([a,b]) \times \mathbb{R}^2 \\ N_i &: \mathcal{C}^1([a,b]) \to \mathcal{C}([a,b]) \times \mathbb{R}^2, \quad i = 1,2, \end{aligned}$$

where

$$\begin{aligned} Lu &= (u'', u(a), u(b)), \\ N_0(u) &= (\varphi(|u'|)u, 0, 0), \\ N_1(u) &= (f(t, \gamma(t,u), u') + \varphi(|u'|)(u - \gamma(t,u)), \\ &\qquad A_0 - a_1\gamma(a, u(a)) + a_2 u'(a) + \gamma(a, u(a)), \\ &\qquad B_0 - b_1\gamma(b, u(b)) - b_2 u'(b) + \gamma(b, u(b))). \end{aligned}$$

Observe that L has a compact inverse. Hence, we can define the completely continuous operator

$$T_\lambda : \ \mathcal{C}^1([a,b]) \to \mathcal{C}^1([a,b])$$

by $T_\lambda(u) = L^{-1}(\lambda N_1 + (1-\lambda)N_0)(u)$. From degree theory, we have that

$$\deg(T_0, \Omega) = \deg(T_1, \Omega),$$

where $\Omega = \{u \in \mathcal{C}^1([a,b]) \mid \|u\|_\infty < \rho, \ \|u'\|_\infty < R\}$. Using the Odd Mapping Theorem A-1.5, we compute that

$$\deg(T_0, \Omega) \neq 0$$

and the problem (1.5) with $\lambda = 1$ has a solution u.

Claim 4 – *The solution* u *of* (1.5) *with* $\lambda = 1$ *is such that for all* $t \in [a,b]$,

$$\alpha(t) \leq u(t) \leq \beta(t).$$

If $u(t_0) - \alpha(t_0) = \min_t(u(t) - \alpha(t)) < 0$ for $t_0 \in]a,b[$, then this result follows from the argument in the proof of Theorem I-2.3. If $t_0 = a$, we have $u'(a) - D^+\alpha(a) \geq 0$. This fact together with (1.5) and the definition of a $\mathcal{C}^2$-lower solution provides the contradiction

$$A_0 \geq a_1\alpha(a) - a_2 D^+\alpha(a) \geq a_1\alpha(a) - a_2 u'(a) = A_0 - u(a) + \alpha(a) > A_0.$$

A similar argument holds if $t_0 = b$ and the rest of the claim follows as well.

Conclusion – It follows from the preceding claims that u is also a solution of (1.1). □

As in Theorem I-5.4, we can improve condition (1.3) in case α and β are in $W^{1,\infty}(a,b)$. More generally, we can prove a result similar to Theorem I-5.5 which uses one-sided bounds on $f(t,\alpha(t),\alpha'(t))$ and $f(t,\beta(t),\beta'(t))$.

Theorem 1.4. *Let A_0, $B_0 \in \mathbb{R}$, a_1, $b_1 \in \mathbb{R}$, a_2, $b_2 \in \mathbb{R}^+$ with $a_1^2 + a_2^2 > 0$, $b_1^2 + b_2^2 > 0$.*

Let α and $\beta \in C([a,b])$ be C^2-lower and upper solutions of the problem (1.1) such that $\alpha \le \beta$. Define $A \subset [a,b]$ (resp. $B \subset [a,b]$) to be the set of points where α (resp. β) is derivable.

Let E be defined by (1.2), $\varphi : \mathbb{R}^+ \to \mathbb{R}$ be a positive continuous function satisfying

$$\int_r^\infty \frac{s\,ds}{\varphi(s)} > \max_t \beta(t) - \min_t \alpha(t), \tag{1.7}$$

where $r = \max\{\frac{\beta(b)-\alpha(a)}{b-a}, \frac{\beta(a)-\alpha(b)}{b-a}\}$, and let $f : E \to \mathbb{R}$ be a continuous function which satisfies (1.4). Assume there exists $N > 0$ such that for all $t \in A$ (resp. for all $t \in B$)

$$f(t,\alpha(t),\alpha'(t)) \ge -N \quad (\textit{resp. } f(t,\beta(t),\beta'(t)) \le N).$$

Then the problem (1.1) has at least one solution $u \in C^2([a,b])$ such that for all $t \in [a,b]$

$$\alpha(t) \le u(t) \le \beta(t).$$

Proof. The proof of this theorem follows the proof of Theorem I-5.5. Let R be large enough so that

$$\int_r^R \frac{s\,ds}{\varphi(s)} > \max_t \beta(t) - \min_t \alpha(t).$$

Increasing the value of N if necessary, we can assume $N \ge \max_{[0,R]} \varphi(v)$. Consider then the modified problem

$$\begin{aligned} u'' &= \bar{f}(t,u,u') + \arctan(u - \gamma(t,u)),\\ u(a) - a_2 u'(a) &= A_0 - a_1\gamma(a,u(a)) + \gamma(a,u(a)),\\ u(b) + b_2 u'(b) &= B_0 - b_1\gamma(b,u(b)) + \gamma(b,u(b)), \end{aligned} \tag{1.8}$$

where $\bar{f}(t,u,v) := \max\{\min\{f(t,\gamma(t,u),v),N\},-N\}$ and $\gamma(t,u)$ is defined in (1.6).

Step 1 – Taking the inverse of the operator $Lu = (u'', u(a)-a_2u'(a), u(b)+b_2u'(b))$, we transform (1.8) into a fixed point problem in C^1. Next, using Schauder's Theorem we prove existence of a fixed point u.

Step 2 – The solution u is such that on $[a,b]$, $\alpha(t) \le u(t) \le \beta(t)$. The claim follows from the argument in the proof of Theorem 1.3 and I-5.5. As a consequence, u satisfies

$$\begin{aligned} u'' &= \max\{\min\{f(t,u,u'),N\},-N\},\\ a_1u(a) - a_2u'(a) &= A_0,\\ b_1u(b) + b_2u'(b) &= B_0. \end{aligned} \tag{1.9}$$

Step 3 – The solution u is such that $\|u'\|_\infty < R$. Observe that for all $(t,u,v) \in E$,

$$|\max\{\min\{f(t,u,v),N\},-N\}| \le \varphi(|v|).$$

From Proposition I-4.4, every solution $u \in [\alpha,\beta]$ of (1.9) satisfies

$$\|u'\|_\infty < R.$$

Hence $|f(t,u(t),u'(t))| \le N$ and the function u is a solution of (1.1). □

In the proof of this theorem, we could have used the modified problem

$$\begin{aligned} u'' - u &= \bar{f}(t,u,u') - \gamma(t,u),\\ u(a) - a_2u'(a) &= A_0 - a_1\gamma(a,u(a)) + \gamma(a,u(a)),\\ u(b) + b_2u'(b) &= B_0 - b_1\gamma(b,u(b)) + \gamma(b,u(b)), \end{aligned}$$

equally for the periodic problem. In this last problem, the aim of the modification was two-fold. First, we needed the linear problem to be invertible which is why we replaced u'' by $u'' - u$. Second, we needed to confine solutions between α and β which was obtained by the modification of f outside α and β adding a term as $u - \gamma(t,u)$. For the separated boundary value problem, we do not need to modify the linear problem. However, we still need to modify f outside α and β which has been done using the term $\arctan(u - \gamma(t,u))$; however, the alternative $u - \gamma(t,u)$ works as well.

Example 1.1. Consider the Dirichlet boundary value problem

$$\begin{gathered} u'' = u'^2 \ln(u'^2+1) - t,\\ u(0) = 0, \; u(1) = 1. \end{gathered}$$

It is easy to see that $\alpha(t) = 0$ is a lower solution and that $\beta(t) = 2t$ is an upper solution. Notice that

$$|f(t,u,v)| \le (v^2+1)\ln(v^2+1) \quad \text{if } v^2 \ge e-1$$

and

$$\int^\infty \frac{s\,ds}{(s^2+1)\ln(s^2+1)} = \tfrac{1}{2}\ln(\ln(y^2+1))\Big|^\infty = \infty.$$

Hence $f(t, u, v)$ satisfies a Nagumo condition. From Theorem 1.4, this problem has a solution u such that for all $t \in [0, 1]$,

$$0 \leq u(t) \leq 2t.$$

As in the periodic case, if the nonlinearity depends on u' the existence of a well ordered pair of lower and upper solutions alone, i.e. without a Nagumo condition does not guarantee the existence of a solution. Such a situation holds true modifying Example I-4.1. A simpler time independent example is the following.

Example 1.2. Consider the problem

$$\begin{aligned} u'' &= g(u') + u + 1, \\ u(0) &= 0, \ u(T) = 0, \end{aligned}$$

where g is continuous, $g(0) = 0$, $g(v) > 0$ if $v \neq 0$ and $1/g \in L^1(\mathbb{R})$. Observe that -1 and 0 are lower and upper solutions of this problem. Further every solution is a convex function such that $u \in [-1, 0]$. Hence, $u'' \geq g(u')$ and $\int_0^T \frac{u''}{g(u')}\, dt \geq T$. As $1/g \in L^1(\mathbb{R})$ we have a contradiction for T large enough.

As for the periodic case, such examples prove we cannot replace the Nagumo condition assuming that solutions between α and β are *a priori* bounded in $C^1([a, b])$.

In case the nonlinearity is independent of u', as lower and upper solutions are continuous and f is continuous, we automatically have the existence of N such that

$$f(t, \alpha(t)) > -N, \qquad f(t, \beta(t)) < N.$$

Hence, for the problem

$$\begin{aligned} u'' &= f(t, u), \\ a_1 u(a) - a_2 u'(a) &= A_0, \\ b_1 u(b) + b_2 u'(b) &= B_0, \end{aligned} \tag{1.10}$$

the following result is a consequence of Theorem 1.4.

Theorem 1.5. *Let A_0, $B_0 \in \mathbb{R}$, a_1, $b_1 \in \mathbb{R}$, a_2, $b_2 \in \mathbb{R}^+$ with $a_1^2 + a_2^2 > 0$, $b_1^2 + b_2^2 > 0$. Assume α and $\beta \in C([a, b])$ are C^2-lower and upper solutions of the problem* (1.10) *such that $\alpha \leq \beta$. Define $E = \{(t, u) \in [a, b] \times \mathbb{R} \mid \alpha(t) \leq u \leq \beta(t)\}$ and suppose $f : E \to \mathbb{R}$ is a continuous function.*

Then the problem (1.10) *has at least one solution $u \in C^2([a, b])$ such that for all $t \in [a, b]$*

$$\alpha(t) \leq u(t) \leq \beta(t).$$

As for the periodic problem, we have the existence of maximal and minimal solutions between lower and upper solutions.

Theorem 1.6. *Let A_0, $B_0 \in \mathbb{R}$, a_1, $b_1 \in \mathbb{R}$, a_2, $b_2 \in \mathbb{R}^+$ with $a_1^2 + a_2^2 > 0$, $b_1^2 + b_2^2 > 0$.*

Assume α and $\beta \in \mathcal{C}([a,b])$ are $\mathcal{C}^2$-lower and upper solutions of the problem (1.1) *such that $\alpha \le \beta$. Define $A \subset [a,b]$ (resp. $B \subset [a,b]$) to be the set of points where α (resp. β) is derivable.*

Define E from (1.2), *let $\varphi : \mathbb{R}^+ \to \mathbb{R}$ be a positive continuous function satisfying* (1.7) *(with r defined as in Theorem 1.4) and suppose $f : E \to \mathbb{R}$ is a continuous function which satisfies* (1.4). *Assume further there exists $N > 0$ such that for all $t \in A$ (resp. for all $t \in B$),*

$$f(t, \alpha(t), \alpha'(t)) \ge -N, \quad (resp.\ f(t, \beta(t), \beta'(t)) \le N).$$

Then the problem (1.1) *has a minimal solution u_{min} and a maximal solution $u_{max} \in \mathcal{C}^2([a,b])$ in $[\alpha, \beta]$, i.e.*

$$\alpha \le u_{min} \le u_{max} \le \beta,$$

and any other solution u of (1.1) *with $\alpha \le u \le \beta$ satisfies*

$$u_{min} \le u \le u_{max}.$$

Exercise 1.1. Prove this theorem adapting the proof of Theorem I-2.4.

1.2 A First Application

The following proposition is an example of a possible application.

Proposition 1.7. *Let $f \in \mathcal{C}([0,1] \times \mathbb{R}^2)$ be such that $\frac{\partial f}{\partial u} \in \mathcal{C}([0,1] \times \mathbb{R}^2)$ and assume there exists $\varphi : \mathbb{R}^+ \to \mathbb{R}$ a positive continuous function satisfying*

$$\int_0^\infty \frac{s\,ds}{\varphi(s)} = \infty$$

such that for any $(t,u,v) \in [0,1] \times \mathbb{R}^2$, $|f(t,u,v)| \le \varphi(|v|)$. Let $r, s \in \mathbb{R}$ and assume there exists $k > 0$ such that for all $(t,u,v) \in [0,1] \times \mathbb{R} \times [\min(r,s), \max(r,s)]$,

$$\frac{\partial f}{\partial u}(t,u,v) \ge k.$$

Then the problem

$$u'' = f(t,u,u'),$$
$$u'(0) = r, \ \ u'(1) = s$$

has at least one solution.

Proof. Let $M = \max(r,s)$ and $m = \min(r,s)$, define $v(t) = m\frac{t^2}{2} - M\frac{(1-t)^2}{2}$ and choose $A > 0$ large enough so that $v(t) - A < 0$ and $v'' - f(t,0,v') - k(v(t) - A) > 0$. It is easy to see that $\alpha(t) = v(t) - A$ is a lower solution. Similarly, taking B large enough, we can build an upper solution $\beta(t) = B + w(t) \geq \alpha(t)$, where $w(t) = M\frac{t^2}{2} - m\frac{(1-t)^2}{2}$. The proposition follows then from Theorem 1.3. □

1.3 A Model Example in Singular Perturbations

Another field of applications concerns singular perturbation problems. The problem

$$\begin{aligned} \epsilon u'' + uu' - u &= 0, \\ u(0) = A, \ u(1) &= B, \end{aligned} \tag{1.11}$$

where $\epsilon > 0$, A, $B \in \mathbb{R}$ is a model example whose solutions exhibit a large variety of interesting behaviours as ϵ goes to zero. These can be shown off through lower and upper solutions. To this end we shall assume here that $\epsilon > 0$ is as small as needed. Observe also that $f(t,u,v) = (u - uv)/\epsilon$ satisfies a Nagumo condition.

Case 1 : $0 < B - 1 \leq A$ – *There exists a solution of* (1.11) *with a boundary layer at* $t = 0$. The functions

$$\alpha(t) = t + B - 1 \quad \text{and} \quad \beta(t) = t + B - 1 + (A - B + 1)\exp((1 - B)\frac{t}{\epsilon})$$

are lower and upper solutions of (1.11). Hence, we have a solution u_ϵ of (1.11) between α and β.

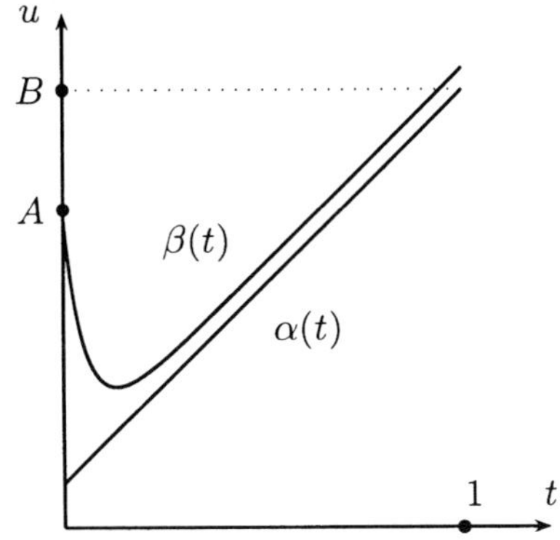

Fig. 1 : $0 < B - 1 \leq A$

Further, as the solution u_ϵ lies between the lower and upper solutions, it is easy to see that the solution $u_0(t) = t + B - 1$ of the reduced problem

$$uu' - u = 0, \;\; u(1) = B$$

approximates u_ϵ as $\epsilon \to 0$ uniformly on compact subsets of $]0,1]$ but not on $[0,1]$. In such a case we say that there is a *boundary layer at* $t = 0$. The case $B - 1 > A > 0$ is similar.

Case 2 : $0 < A$, $0 < B < 1$ *– There exists a solution of* (1.11) *with a boundary layer at* $t = 0$ *and a transition layer in the derivative at* $t = \lambda = 1 - B$. The function

$$\begin{aligned} \alpha(t) &= 0, &\quad &\text{if } t \in [0, \lambda[, \\ &= t - \lambda, &\quad &\text{if } t \in [\lambda, 1], \end{aligned}$$

is a lower solution and

$$\begin{aligned} \beta(t) &= A\exp(-t/\sqrt{\epsilon}) + \sqrt{\epsilon}, &\quad &\text{if } t \in [0, \lambda[, \\ &= t - \lambda + K\exp(-(t-\lambda)/\sqrt{\epsilon}), &\quad &\text{if } t \in [\lambda, 1], \end{aligned}$$

where $K = A\exp(-\lambda/\sqrt{\epsilon}) + \sqrt{\epsilon}$, is an upper solution.

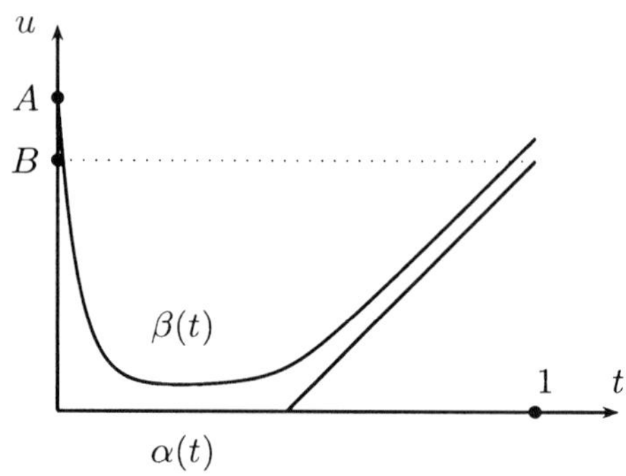

Fig. 2 : $0 < A$, $0 < B < 1$

To verify that β is an upper solution we check that

$$\beta(t) \le \beta_0(t) := t - \lambda + K\exp(-(t-\lambda)/\sqrt{\epsilon})$$

for all $t \in I_0 = [0,1]$ and

$$\epsilon\,\beta_0''(\lambda) + \beta_0(\lambda)\beta_0'(\lambda) - \beta_0(\lambda) \le 0.$$

Notice that we use the full generality of Definitions 1.1 for $t = \lambda$, since $\beta \neq \beta_0$ in a neighbourhood of this point. It follows that we have a solution u_ϵ of (1.11) between α and β and we see that

$$\begin{aligned} \lim_{\epsilon \to 0^+} u_\epsilon(t) &= 0, && \text{if } t \in]0, \lambda[, \\ &= t - \lambda, && \text{if } t \in [\lambda, 1], \end{aligned}$$

uniformly on compact subsets of $]0, 1]$.

In this problem, two solutions of the reduced problem

$$uu' - u = 0,$$

are needed to approximate u_ϵ, i.e. $u_0 = 0$ if $t \in]0, \lambda[$ and $u_0 = t - \lambda$ if $t \in [\lambda, 1]$. There is a boundary layer at $t = 0$ and it is clear that a non-uniformity in the limit of the derivative u'_ϵ takes place at the interior point $t = \lambda$. In such a case, we speak of a *transition* or *interior layer* in the derivative at $t = \lambda$.

Case 3 : $0 < B < A + 1 < 1$ – *There exists a solution of* (1.11) *with a transition layer in the derivatives.* Here, three solutions of the reduced equation are needed to approximate the solution, i.e.

$$\begin{aligned} \lim_{\epsilon \to 0^+} u_\epsilon(t) &= t + A, && \text{on } [0, -A[, \\ &= 0, && \text{on } [-A, 1 - B[, \\ &= t + B - 1, && \text{on } [1 - B, 1]. \end{aligned}$$

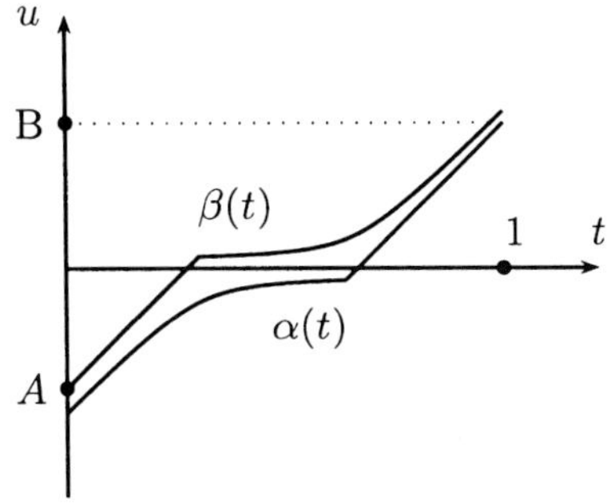

Fig. 3 : $0 < B < A + 1 < 1$

For ϵ small, a lower solution reads

$$\begin{aligned} \alpha(t) &= \tfrac{1}{2}\left(t + A - \sqrt{(t + A)^2 + 4\epsilon^{1/2}}\right), && \text{if } t < t_0 = 1 - B - \tfrac{\epsilon^{1/2}}{1 - B + A}, \\ &= t - 1 + B, && \text{if } t \geq t_0, \end{aligned}$$

and an upper solution

$$\begin{aligned}\beta(t) &= t + A, & &\text{if } t \le t_1 = -A + \tfrac{\epsilon^{1/2}}{1-B+A},\\ &= \tfrac{1}{2}\left(t - 1 + B + \sqrt{(t-1+B)^2 + 4\epsilon^{1/2}}\right), & &\text{if } t > t_1.\end{aligned}$$

Case 4 : $1 - B < A \le 0$ *– there exists a solution of* (1.11) *with a boundary layer at* $t = 0$. Define $\lambda = \tanh^{-1}(-A/(B-1))$. The functions

$$\alpha(t) = t + (B-1)\tanh\left(\tfrac{B-1}{2\epsilon}t - \lambda\right) \quad \text{and} \quad \beta(t) = t + B - 1,$$

are lower and upper solutions. It follows there exists a solution u_ϵ of (1.11) between α and β and

$$\lim_{\epsilon \to 0^+} u_\epsilon(t) = t + B - 1, \quad \text{if } t \in]0,1].$$

The convergence is uniform on compact subsets of $]0,1]$ and we have a boundary layer at $t = 0$.

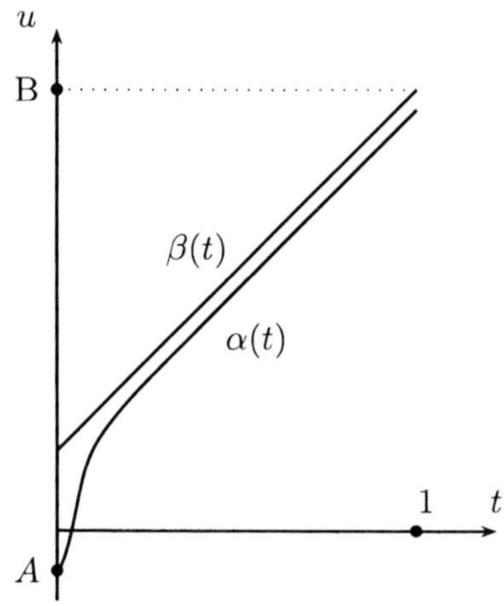

Fig. 4 : $1 - B < A \le 0$

Case 5 : $A < 0$, $0 < B$, $A + 1 < B$, $|A + B| \le 1$ *– There exists a solution of* (1.11) *with a transition or interior layer at* $t = \frac{1-A-B}{2}$. Let $\epsilon > 0$,

$$t_0 = \tfrac{1-A-B}{2} + \epsilon$$

and $t_1 > t_0$ be such that

$$\tanh \tfrac{B-A-1}{4\epsilon}(t_0 - t_1) = 2\,\tfrac{t_0 + A}{B-A-1}.$$

It can be seen that $t_1 - t_0 \to 0$ as $\epsilon \to 0$ and that for ϵ small enough,

$$\begin{aligned} \alpha(t) &= t + A, && \text{if } t \in [0, t_0[, \\ &= t - t_0 + \tfrac{B-A-1}{2} \tanh \tfrac{B-A-1}{4\epsilon}(t - t_1), && \text{if } t \in [t_0, 1], \end{aligned}$$

is a lower solution.

In the same way, we consider $\epsilon > 0$, $t_2 = \frac{1-A-B}{2} - \epsilon$ and $t_3 < t_2$ that satisfy

$$\tanh \tfrac{B-A-1}{4\epsilon}(t_2 - t_3) = 2\,\tfrac{t_2+B-1}{B-A-1}.$$

We prove then that $t_3 - t_2 \to 0$ as $\epsilon \to 0$ and that

$$\begin{aligned} \beta(t) &= t - t_2 + \tfrac{B-A-1}{2} \tanh \tfrac{B-A-1}{4\epsilon}(t - t_3), && \text{if } t \in [0, t_2[, \\ &= t + B - 1, && \text{if } t \in [t_2, 1], \end{aligned}$$

is an upper solution.

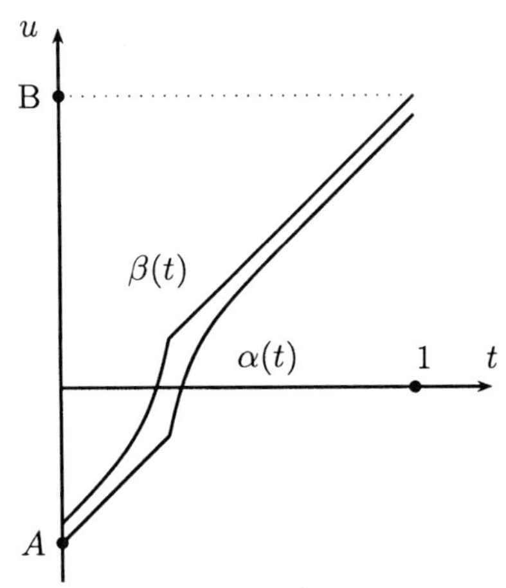

Fig. 5 : $A < 0$, $0 < B$, $A + 1 < B$, $|A + B| \leq 1$

Hence, we have a solution u_ϵ of (1.11) between α and β and we see that

$$\begin{aligned} \lim_{\epsilon \to 0^+} u_\epsilon(t) &= t + A, && \text{on } [0, \tfrac{1-A-B}{2}[, \\ &= t + B - 1, && \text{on }]\tfrac{1-A-B}{2}, 1]. \end{aligned}$$

Here, we have a transition layer at $t = \frac{1-A-B}{2}$.

2 $W^{2,1}$-Solutions

2.1 The Method of Lower and Upper Solutions

In this section we consider nonlinearities f which are L^1-Carathéodory. As in Sections I-3 and I-6, we adapt accordingly the definitions of lower and upper solutions.

Definitions 2.1. *A function $\alpha \in C([a,b])$ is a $W^{2,1}$-*lower solution *of* (1.1) *if*
(a) for any $t_0 \in]a,b[$, either $D_-\alpha(t_0) < D^+\alpha(t_0)$,
or there exists an open interval $I_0 \subset]a,b[$ such that $t_0 \in I_0$, $\alpha \in W^{2,1}(I_0)$ and, for a.e. $t \in I_0$,

$$\alpha''(t) \geq f(t, \alpha(t), \alpha'(t));$$

(b) $a_1\alpha(a) - a_2 D^+\alpha(a) \leq A_0$, $b_1\alpha(b) + b_2 D_-\alpha(b) \leq B_0$.

*A function $\beta \in C([a,b])$ is a $W^{2,1}$-*upper solution *of* (1.1) *if*
(a) for any $t_0 \in]a,b[$, either $D^-\beta(t_0) > D_+\beta(t_0)$,
or there exists an open interval $I_0 \subset]a,b[$ such that $t_0 \in I_0$, $\beta \in W^{2,1}(I_0)$ and, for a.e. $t \in I_0$,

$$\beta''(t) \leq f(t, \beta(t), \beta'(t));$$

(b) $a_1\beta(a) - a_2 D_+\beta(a) \geq A_0$, $b_1\beta(b) + b_2 D^-\beta(b) \geq B_0$.

Theorem 2.1. *Let A_0, $B_0 \in \mathbb{R}$, a_1, $b_1 \in \mathbb{R}$, a_2, $b_2 \in \mathbb{R}^+$ with $a_1^2 + a_2^2 > 0$ and $b_1^2 + b_2^2 > 0$.*

Assume α and $\beta \in C([a,b])$ are $W^{2,1}$-lower and upper solutions of problem (1.1) *such that $\alpha \leq \beta$. Define $A \subset [a,b]$ (resp. $B \subset [a,b]$) to be the set of points where α (resp. β) is derivable.*

Let E be defined in (1.2)*, $p \in [1,\infty]$ and $f : E \to \mathbb{R}$ be an L^p-Carathéodory function. Suppose there exists $N \in L^1(a,b)$, $N > 0$ such that for a.e. $t \in A$ (resp. for a.e. $t \in B$),*

$$f(t, \alpha(t), \alpha'(t)) \geq -N(t), \quad (\textit{resp. } f(t, \beta(t), \beta'(t)) \leq N(t)).$$

Assume moreover there exist $\varphi \in C(\mathbb{R}^+, \mathbb{R}_0^+)$ and $\psi \in L^p(a,b)$ satisfying
(a) $\displaystyle\int_r^\infty \frac{s^{1/q}}{\varphi(s)}\, ds > \|\psi\|_{L^p}(\max_t \beta(t) - \min_t \alpha(t))^{1/q}$,
where $r = \max\{\frac{\beta(a)-\alpha(b)}{b-a}, \frac{\beta(b)-\alpha(a)}{b-a}\}$ and $q = \frac{p}{p-1}$;
(b) for a.e. $t \in [a,b]$ and all (u,v) with $(t,u,v) \in E$,

$$|f(t,u,v)| \leq \psi(t)\varphi(|v|).$$

Then the problem (1.1) *has at least one solution* $u \in W^{2,p}(a,b)$ *such that for all* $t \in [a,b]$

$$\alpha(t) \le u(t) \le \beta(t).$$

Proof. The proof proceeds in several steps.

Step 1 – The modified problem. Let R be large enough so that

$$\int_r^R \frac{s^{1/q}}{\varphi(s)}\, ds > \|\psi\|_{L^p} (\max_t \beta(t) - \min_t \alpha(t))^{1/q}.$$

Increasing N if necessary, we can assume $N(t) \ge |f(t,u,v)|$ if $t \in [a,b]$, $\alpha(t) \le u \le \beta(t)$ and $|v| \le R$. Define then

$$\begin{aligned}
\bar f(t,u,v) &= \max\{\min\{f(t,\gamma(t,u),v), N(t)\}, -N(t)\},\\
\omega_1(t,\delta) &= \chi_A(t) \max_{|v|\le\delta} |\bar f(t,\alpha(t),\alpha'(t)+v) - \bar f(t,\alpha(t),\alpha'(t))|,\\
\omega_2(t,\delta) &= \chi_B(t) \max_{|v|\le\delta} |\bar f(t,\beta(t),\beta'(t)+v) - \bar f(t,\beta(t),\beta'(t))|,
\end{aligned}$$

where γ is defined from (1.6), χ_A and χ_B are the characteristic functions of the sets A and B. It is clear that ω_i are L^1-Carathéodory functions, non-decreasing in δ, such that $\omega_i(t,0) = 0$ and $|\omega_i(t,\delta)| \le 2N(t)$.

We consider now the modified problem

$$\begin{aligned}
u'' &= \bar f(t,u,u') - \omega(t,u),\\
u(a) - a_2 u'(a) &= A_0 - a_1\gamma(a,u(a)) + \gamma(a,u(a)),\\
u(b) + b_2 u'(b) &= B_0 - b_1\gamma(b,u(b)) + \gamma(b,u(b)),
\end{aligned} \tag{2.1}$$

where

$$\begin{aligned}
\omega(t,u) &= -\omega_2(t, u-\beta(t)), && \text{if } u > \beta(t),\\
&= 0, && \text{if } \alpha(t) \le u \le \beta(t),\\
&= \omega_1(t,\alpha(t)-u), && \text{if } u < \alpha(t).
\end{aligned}$$

Step 2 – Existence of a solution u of (2.1). We argue as in Theorem 1.4.

Step 3 – The solution u of (2.1) *satisfies on* $[a,b]$

$$\alpha(t) \le u(t) \le \beta(t).$$

Assume $u - \alpha$ has a negative minimum and let $t_0 = \sup\{t \in [a,b] \mid u(t) - \alpha(t) = \min_{s\in[a,b]}(u(s) - \alpha(s))\}$.

If $t_0 = a$, we have $u'(a) - D^+\alpha(a) \ge 0$. This observation together with (2.1) and the definition of $W^{2,1}$-lower solution provides the contradiction

$$A_0 \ge a_1\alpha(a) - a_2 D^+\alpha(a) \ge a_1\alpha(a) - a_2 u'(a) = A_0 - u(a) + \alpha(a) > A_0.$$

A similar argument holds if $t_0 = b$.

If $t_0 \in]a, b[$, there exist an open interval I_0 and a $t_1 \in I_0$, with $t_1 > t_0$, such that $\alpha \in W^{2,1}(I_0)$, $t_0 \in I_0$, $u'(t_1) - \alpha'(t_1) > 0$ and for a.e. $t \in I_0$

$$\alpha''(t) \geq f(t, \alpha(t), \alpha'(t)).$$

Further $u'(t_0) - \alpha'(t_0) = 0$ and, for t near enough t_0,

$$|u'(t) - \alpha'(t)| \leq \alpha(t) - u(t).$$

As ω_1 is non-decreasing and $\bar{f}(t, \alpha(t), \alpha'(t)) \leq f(t, \alpha(t), \alpha'(t))$, we come to the contradiction

$$\begin{aligned} 0 < u'(t_1) - \alpha'(t_1) &= \int_{t_0}^{t_1} (u''(s) - \alpha''(s))ds \\ &\leq \int_{t_0}^{t_1} [\bar{f}(s, \alpha(s), u'(s)) - \bar{f}(s, \alpha(s), \alpha'(s)) - \omega_1(s, \alpha(s) - u(s))]ds \leq 0. \end{aligned}$$

Step 4 – The solution u of (2.1) *is such that* $\|u'\|_\infty < R$. Observe that for any $(t, u, v) \in E$, $|\bar{f}(t, u, v) - \omega(t, u)| = |\bar{f}(t, u, v)| \leq \psi(t)\varphi(|v|)$ so that the claim follows from Proposition I-4.7.

Conclusion. A consequence of Steps 3 and 4 is that along the solution u of (2.1), $|f(t, u(t), u'(t))| \leq N(t)$. Hence, $\bar{f}(t, u(t), u'(t)) = f(t, u(t), u'(t))$ and u solves (1.1). □

Remark 2.1. Consider the Dirichlet problem

$$\begin{aligned} u'' &= f(t, u, u'), \\ u(a) &= 0, \quad u(b) = 0. \end{aligned}$$

As $u \in W^{2,1}(a, b) \subset C^1([a, b])$, the assumptions of Theorem 2.1 imply some one-sided bounds on α and β,

$$\begin{aligned} \alpha(t) \leq u(t) &\leq C \sin(\pi \tfrac{t-a}{b-a}), \\ \beta(t) \geq u(t) &\geq -C \sin(\pi \tfrac{t-a}{b-a}), \end{aligned}$$

for some $C > 0$.

In the applications, we can use the following simplification of Theorem 2.1.

Corollary 2.2. *Let A_0, $B_0 \in \mathbb{R}$, a_1, $b_1 \in \mathbb{R}$, a_2, $b_2 \in \mathbb{R}^+$ with $a_1^2 + a_2^2 > 0$ and $b_1^2 + b_2^2 > 0$.*

Assume α and $\beta \in W^{1,\infty}(a, b)$ are $W^{2,1}$-lower and upper solutions of problem (1.1) *such that $\alpha \leq \beta$.*

Let E be defined in (1.2), $p \in [1, \infty]$ and $f : E \to \mathbb{R}$ be an L^p-Carathéodory function. Assume moreover there exist $\varphi \in C(\mathbb{R}^+, \mathbb{R}_0^+)$ and $\psi \in L^p(a, b)$ such that

(a) $\displaystyle\int_0^{\infty} \frac{s^{1/q}}{\varphi(s)}\, ds = +\infty$, *where* $q = \frac{p}{p-1}$ *and*

(b) for a.e. $t \in [a, b]$ *and all* (u, v) *with* $(t, u, v) \in E$, $|f(t, u, v)| \le \psi(t)\varphi(|v|)$.

Then the problem (1.1) has at least one solution $u \in W^{2,p}(a, b)$ *such that for all* $t \in [a, b]$

$$\alpha(t) \le u(t) \le \beta(t).$$

An even more specific result uses constant lower and upper solutions.

Corollary 2.3. *Let A_0, $B_0 \in \mathbb{R}$, a_1, $b_1 \in \mathbb{R}$, a_2, $b_2 \in \mathbb{R}^+$ with $a_1^2 + a_2^2 > 0$ and $b_1^2 + b_2^2 > 0$.*

Assume $\alpha \le \beta$ are constants, let E be defined in (1.2), $p \in [1, \infty]$ and $f : E \to \mathbb{R}$ be an L^p-Carathéodory function such that

(a) for a.e. $t \in [a, b]$, $f(t, \alpha, 0) \le 0 \le f(t, \beta, 0)$,

(b) $a_1\alpha \le A_0 \le a_1\beta$, $b_1\alpha \le B_0 \le b_1\beta$.

Assume moreover there exist $\varphi \in C(\mathbb{R}^+, \mathbb{R}_0^+)$ and $\psi \in L^p(a, b)$ such that

$$\int_r^{\infty} \frac{s^{1/q}}{\varphi(s)}\, ds > \|\psi\|_{L^p} (\beta - \alpha)^{1/q},$$

where $r = \frac{\beta - \alpha}{b - a}$, $q = \frac{p}{p-1}$ *and, for a.e.* $t \in [a, b]$ *and all* (u, v) *such that* $(t, u, v) \in E$,

$$|f(t, u, v)| \le \psi(t)\varphi(|v|).$$

Then the problem (1.1) has at least one solution $u \in W^{2,p}(a, b)$ *such that for all* $t \in [a, b]$

$$\alpha \le u(t) \le \beta.$$

If f does not depend on u' as in problem (1.10), Theorem 2.1 reduces to the following one.

Theorem 2.4. *Let A_0, $B_0 \in \mathbb{R}$, a_1, $b_1 \in \mathbb{R}$, a_2, $b_2 \in \mathbb{R}^+$ with $a_1^2 + a_2^2 > 0$, $b_1^2 + b_2^2 > 0$.*

Assume α and $\beta \in C([a, b])$ are $W^{2,1}$-lower and upper solutions of the problem (1.10) such that $\alpha \le \beta$.

Let $E = \{(t, u) \in [a, b] \times \mathbb{R} \mid \alpha(t) \le u \le \beta(t)\}$ and $f : E \to \mathbb{R}$ be an L^1-Carathéodory function.

Then the problem (1.10) has at least one solution $u \in W^{2,1}(a, b)$ such that for all $t \in [a, b]$

$$\alpha(t) \le u(t) \le \beta(t).$$

As for the periodic problem, we have solutions of (1.1) between maxima of lower solutions and minima of upper solutions.

Theorem 2.5. *Let A_0, $B_0 \in \mathbb{R}$, a_1, $b_1 \in \mathbb{R}$, a_2, $b_2 \in \mathbb{R}^+$ with $a_1^2 + a_2^2 > 0$, $b_1^2 + b_2^2 > 0$.*

Let $\alpha_i \in \mathcal{C}([a,b])$ $(i = 1, \ldots, n)$ and $\beta_j \in \mathcal{C}([a,b])$ $(j = 1, \ldots, m)$ be respectively $W^{2,1}$-lower solutions and $W^{2,1}$-upper solutions of (1.1). Define $A_i \subset [a,b]$ (resp. $B_j \subset [a,b]$) to be the set of points where α_i (resp. β_j) is derivable. Assume $\alpha := \max\limits_{1\le i\le n} \alpha_i \le \beta := \min\limits_{1\le j\le m} \beta_j$.

Define $E = \{(t,u,v) \in [a,b] \times \mathbb{R}^2 \mid \min_i \alpha_i \le u \le \max_j \beta_j\}$ and let p, $q \in [1,\infty]$ be such that $\frac{1}{p} + \frac{1}{q} = 1$. Assume $f : E \to \mathbb{R}$ satisfies L^p-Carathéodory conditions and there exists $N \in L^1(a,b)$, $N > 0$ such that for all i and a.e. $t \in A_i$ (resp. for all j and a.e. $t \in B_j$)

$$f(t, \alpha_i(t), \alpha_i'(t)) \ge -N(t) \quad (\textit{resp. } f(t, \beta_j(t), \beta_j'(t)) \le N(t)).$$

Assume moreover there exist $\varphi \in \mathcal{C}(\mathbb{R}^+, \mathbb{R}_0^+)$ and $\psi \in L^p(a,b)$ that satisfy conditions (a) and (b) in Theorem 2.1.

Then the problem (1.1) has at least one solution $u \in W^{2,p}(a,b)$ such that for all $t \in [a,b]$

$$\alpha(t) \le u(t) \le \beta(t).$$

Exercise 2.1. Prove the previous theorem using as a guideline the proof of Theorem I-6.10.

As previously we have the existence of maximal and minimal solutions between α and β in the following sense.

Theorem 2.6. *Let A_0, $B_0 \in \mathbb{R}$, a_1, $b_1 \in \mathbb{R}$, a_2, $b_2 \in \mathbb{R}^+$ with $a_1^2 + a_2^2 > 0$ and $b_1^2 + b_2^2 > 0$.*

Assume α and $\beta \in \mathcal{C}([a,b])$ are $W^{2,1}$-lower and upper solutions of problem (1.1) such that $\alpha \le \beta$. Define $A \subset [a,b]$ (resp. $B \subset [a,b]$) to be the set of points where α (resp. β) is derivable.

Let E be defined in (1.2), $p \in [1,\infty]$ and $f : E \to \mathbb{R}$ be an L^p-Carathéodory function. Suppose there exists $N \in L^1(a,b)$, $N > 0$, such that for a.e. $t \in A$ (resp. for a.e. $t \in B$)

$$f(t, \alpha(t), \alpha'(t)) \ge -N(t), \quad (\textit{resp. } f(t, \beta(t), \beta'(t)) \le N(t)).$$

Assume moreover there exist $\varphi \in \mathcal{C}(\mathbb{R}^+, \mathbb{R}_0^+)$ and $\psi \in L^p(a,b)$ satisfying conditions (a) and (b) in Theorem 2.1.

Then the problem (1.1) has a minimal solution $u_{min} \in W^{2,p}(a,b)$ and a maximal solution $u_{max} \in W^{2,p}(a,b)$ in $[\alpha,\beta]$, i.e.

$$\alpha \le u_{min} \le u_{max} \le \beta,$$

and any other solution u such that $\alpha \leq u \leq \beta$ satisfies

$$u_{min} \leq u \leq u_{max}.$$

Exercise 2.2. Prove Theorem 2.6 adapting the argument used to prove Theorem I-2.4.

Exercise 2.3. Suppose the assumptions of Theorem 2.6 and that there exist
(a) an upper solution $\hat{\beta}$ such that for all $t \in [a, b]$, $\alpha(t) \leq \hat{\beta}(t) < \beta(t)$ and
(b) a continuum of lower solutions $\mathcal{S}$ in $\mathcal{C}([a, b])$ such that $\alpha \in \mathcal{S}$ and for some $\hat{\alpha} \in \mathcal{S}$, $\hat{\alpha} \not\leq \beta$.

Prove there exist then two solutions of (1.1).

Hint : See [22].

2.2 A Multiplicity Result for the Neumann Problem

Consider the problem

$$\begin{aligned} u'' + f(t, u) = 0, \\ u'(a) = 0, \ u'(b) = 0, \end{aligned} \tag{2.2}$$

where $f(t, u)$ is L^1-Carathéodory. Assume that for some sequences $(a_n)_n$ and $(b_n)_n$ of positive numbers such that $a_1 \leq b_1 \leq \ldots \leq a_n \leq b_n \leq \ldots$ we have

$$f(t, a_n) \geq 0 \quad \text{and} \quad f(t, b_n) \leq 0.$$

It is clear then that a_n and b_n are $W^{2,1}$-lower and upper solutions of (2.2). Existence of infinitely many solutions follows.

The problem is more difficult if we add some forcing $h(t)$ which might be unbounded. Consider for example the problem

$$\begin{aligned} u'' + f(t, u) = h(t), \\ u'(0) = 0, \ u'(1) = 0. \end{aligned} \tag{2.3}$$

To deal with this problem, let us recall that the operator

$$L : \text{Dom}\, L \subset \mathcal{C}([0, 1]) \to X \subset L^1(0, 1), u \mapsto u'', \tag{2.4}$$

where $X = \{u \in L^1(0, 1) \mid \int_0^1 u(t)\, dt = 0\}$ and $\text{Dom}\, L = \{u \in W^{2,1}(0, 1) \mid \int_0^1 u(t)\, dt = 0 \text{ and } u'(0) = u'(1) = 0\}$, has a continuous inverse.

We can obtain lower solutions from the following result.

Proposition 2.7. *Let c be the norm of the inverse of L defined in* (2.4), *f be an L^1-Carathéodory function and let $h \in L^1(0,1)$ satisfy $\int_0^1 h(t)\,dt = 0$. Assume that there exists $a > 0$ such that*

$$\text{for a.e. } t \in [0,1] \text{ and all } u \in [a - c\|h\|_{L^1}, a + c\|h\|_{L^1}], \quad f(t,u) \geq 0.$$

Then there exists a $W^{2,1}$-lower solution α of (2.3) *with*

$$\alpha(t) \in [a - c\|h\|_{L^1}, a + c\|h\|_{L^1}] \text{ on } [0,1].$$

Proof. The function $\alpha = a + L^{-1}(h)$ is a lower solution which is in $[a - c\|h\|_{L^1}, a + c\|h\|_{L^1}]$. □

Upper solutions are obtained along the same lines.

Proposition 2.8. *Let c be the norm of the inverse of L defined in* (2.4), *f be an L^1-Carathéodory function and let $h \in L^1(0,1)$ satisfy $\int_0^1 h(t)\,dt = 0$. Assume that there exists $b > 0$ such that*

$$\text{for a.e. } t \in [0,1] \text{ and all } u \in [b - c\|h\|_{L^1}, b + c\|h\|_{L^1}], \quad f(t,u) \leq 0.$$

Then there exists a $W^{2,1}$-upper solution β of (2.3) *with*

$$\beta(t) \in [b - c\|h\|_{L^1}, b + c\|h\|_{L^1}] \text{ on } [0,1].$$

Exercise 2.4. Prove the existence of an infinite number of solutions for the Neumann problem

$$u'' + \sin\sqrt{u} = h(t),$$
$$u'(0) = 0, \; u'(1) = 0,$$

in case $h \in L^1(0,1)$ and $|\int_0^1 h(t)\,dt| < 1$.

In Propositions 2.7 and 2.8, lower and upper solutions are obtained from a sign condition on $f(t,u)$, which holds on intervals long enough. A similar result can be obtained if the intervals are small but $|f(t,u)|$ is large enough.

Proposition 2.9. *Let $f : [0,1] \times \mathbb{R} \to \mathbb{R}$ be an L^2-Carathéodory function and $h \in L^2(0,1)$. Assume there exist sequences $(a_n)_n$, $(b_n)_n$ and $(c_n)_n$ of positive numbers such that*
(a) $\lim_{n\to\infty} a_n = \lim_{n\to\infty} b_n - c_n = +\infty$;
(b) for some $\epsilon > 0$ and any n, $a_n c_n \geq \epsilon$;
(c) for any $u \in [b_n - c_n, b_n + c_n]$, $f(t,u) \geq a_n$.

Then, for any $R > 0$, there exists a function $\alpha(t) \geq R$, which is a $W^{2,1}$-lower solution for (2.3).

Proof. Let $R > 0$ be fixed. Consider the problem

$$\begin{gathered} w'' - dw = h(t), \\ w'(0) = 0,\ w'(1) = 0, \end{gathered}$$

where $d > 0$. Any solution w of this problem is such that

$$\int_0^1 w'^2\,dt + d\int_0^1 w^2\,dt = -\int_0^1 hw\,dt.$$

Hence, we have the *a priori* estimates

$$\|w\|_{L^2} \le \frac{1}{d}\|h\|_{L^2} \quad \text{and} \quad \|w'\|_{L^2} \le \frac{1}{\sqrt{d}}\|h\|_{L^2}.$$

Also, for some t_0,

$$|w(t_0)| \le \frac{1}{d}\,\|h\|_{L^2}.$$

Hence, we obtain

$$w^2(t) = w^2(t_0) + 2\int_{t_0}^t w(s)w'(s)\,ds \le \frac{\|h\|_{L^2}^2}{d^2}(1 + 2d^{\frac{1}{2}}).$$

Let us choose d_0 such that

$$\frac{\|h\|_{L^2}^2}{d_0}(1 + 2d_0^{\frac{1}{2}}) = \epsilon,$$

n so that

$$a_n \ge \|h\|_{L^2}(1 + 2d_0^{\frac{1}{2}})^{\frac{1}{2}}, \quad b_n - c_n \ge R$$

and next $d \ge d_0$ so that $a_n = \|h\|_{L^2}(1+2d^{\frac{1}{2}})^{\frac{1}{2}}$. For this choice of d and n, and all $t \in [0,1]$, we have

$$a_n \ge d|w(t)|$$

and

$$c_n \ge \frac{\epsilon}{a_n} \ge \|h\|_{L^2}\frac{(1 + 2d^{\frac{1}{2}})^{\frac{1}{2}}}{d} \ge |w(t)|.$$

Now, it is easy to see that $\alpha(t) = b_n + w(t) \ge R$ is a lower solution since

$$\alpha'' + f(t,\alpha) - h(t) \ge dw + h(t) + a_n - h(t) \ge 0. \qquad \square$$

A similar result holds for upper solutions.

Proposition 2.10. *Let $f : [0,1] \times \mathbb{R} \to \mathbb{R}$ be an L^2-Carathéodory function and $h \in L^2(0,1)$. Assume there exist sequences $(a_n)_n$, $(b_n)_n$ and $(c_n)_n$ of positive numbers such that*
(a) $\lim_{n\to\infty} a_n = \lim_{n\to\infty} b_n - c_n = +\infty$;
(b) for some $\epsilon > 0$ *and any* n, $a_n c_n \geq \epsilon$;
(c) for any $u \in [b_n - c_n, b_n + c_n]$, $f(t,u) \leq -a_n$.

Then, for any $R > 0$, *there exists a function* $\beta(t) \geq R$, *which is a* $W^{2,1}$*-upper solution for* (2.3).

Exercise 2.5. Prove that the problem

$$\begin{aligned} u'' + u \sin u^2 &= h(t), \\ u'(0) = 0,\ u'(1) &= 0, \end{aligned}$$

where $h \in L^2(0,1)$, has an infinite number of solutions.

2.3 A Multiplicity Result for the Dirichlet Problem

For the Neumann problem, existence of an infinite number of solutions was deduced from sequences of lower and upper solutions $(\alpha_n)_n$ and $(\beta_n)_n$ which are well-ordered

$$\alpha_1 \leq \beta_1 \leq \ldots \leq \alpha_n \leq \beta_n \leq \ldots$$

Such a geometry might seem difficult to find for a Dirichlet problem since we cannot enclose the lower and upper solutions in disjoint intervals. It occurs however if the nonlinearity f oscillates enough around the first eigenvalue. To build such an example, consider the Dirichlet problem

$$\begin{aligned} u'' + u + f(t,u) &= 0, \\ u(0) = 0,\ u(\pi) &= 0, \end{aligned} \tag{2.5}$$

where $f(t,u)$ is an L^1-Carathéodory function defined as follows. First, let $t_1 \in]0, \frac{\pi}{2}[$ be such that

$$\tfrac{1}{2} - \tfrac{t_1(\pi - t_1)}{4} \geq \tfrac{\pi^2}{16} - \tfrac{1}{2},$$

define

$$k = \sin t_1 - \tfrac{t_1(\pi - t_1)}{4} < 1 - \tfrac{\pi^2}{16}$$

and let μ, ν satisfy

$$\tfrac{1}{k}\left(\tfrac{\pi^2}{16} - \tfrac{1}{2}\right) \leq \mu \leq \nu \leq \tfrac{1}{k}\left(\tfrac{1}{2} - \tfrac{t_1(\pi - t_1)}{4}\right).$$

Next, we choose positive sequences $(A_n)_n$ and $(B_n)_n$ such that

$$kA_n < (1 - \tfrac{\pi^2}{16})A_n < kB_n < (1 + \tfrac{\pi^2}{16})B_n < kA_{n+1} \quad \text{and} \quad A_{n+1} \geq \tfrac{4+\pi}{4-\pi} B_n.$$

At last, we choose f so that

$$\begin{aligned}
f(t,u) &\geq \mu u, && \text{on } [0,\pi] \times \bigcup_{n=1}^{\infty} [kA_n, (1 - \tfrac{\pi^2}{16})A_n], \\
f(t,u) &\leq -\mu u, && \text{on } [0,\pi] \times \bigcup_{n=1}^{\infty} [kB_n, (1 + \tfrac{\pi^2}{16})B_n], \\
|f(t,u)| &\leq \nu u, && \text{on } [0,\pi] \times \mathbb{R}^+.
\end{aligned}$$

We shall then consider nested lower and upper solutions

$$\alpha_n(t) = A_n(\sin t - \tfrac{t(\pi - t)}{4}) \geq 0,$$

and

$$\beta_n(t) = B_n(\sin t + \tfrac{t(\pi - t)}{4}) \in [\alpha_n, \alpha_{n+1}].$$

To verify that α_n is a lower solution, notice that in case $\alpha_n(t) \leq kA_n$, i.e. $t \leq t_1$ or $t \geq \pi - t_1$, we have

$$\alpha_n'' + \alpha_n + f(t, \alpha_n) \geq A_n[\tfrac{1}{2} - \tfrac{t_1(\pi - t_1)}{4} - k\nu] \geq 0$$

and if $\alpha_n(t) > kA_n$, we compute

$$\alpha_n'' + \alpha_n + f(t, \alpha_n) \geq A_n[\tfrac{1}{2} - \tfrac{\pi^2}{16} + k\mu] \geq 0.$$

Similarly, we can check that β_n is an upper solution. As a consequence the problem (2.5) has an infinite number of ordered positive solutions.

3 One-Sided Nagumo Condition

One-sided Nagumo conditions were introduced in Sections I-4.3 and I-4.4. These ideas can be applied for the separated boundary value problem (1.1).

Theorem 3.1. *Let A_0, $B_0 \in \mathbb{R}$, a_1, $b_1 \in \mathbb{R}$, a_2, $b_2 \in \mathbb{R}^+$ with $a_1^2 + a_2^2 > 0$, $b_1^2 + b_2^2 > 0$.*

Suppose α and $\beta \in C([a,b])$ are $W^{2,1}$-lower and upper solutions of the problem (1.1) such that $\alpha \leq \beta$. Define $A \subset [a,b]$ (resp. $B \subset [a,b]$) to be the set of points where α (resp. β) is derivable.

Let E be defined in (1.2) *and, for some $p \in [1, \infty]$, $f : E \to \mathbb{R}$ be an L^p-Carathéodory function. Assume there exists $N \in L^1(a,b)$, $N > 0$ such that for a.e. $t \in A$ (resp. for a.e. $t \in B$),*

$$f(t, \alpha(t), \alpha'(t)) \geq -N(t), \quad (\textit{resp. } f(t, \beta(t), \beta'(t)) \leq N(t)).$$

If moreover there exist $r \geq 0$, $\varphi \in C(\mathbb{R}^+, \mathbb{R}_0^+)$ and $\psi \in L^p(a,b)$ such that
(a) for all u solution of (1.1) *with $\alpha \leq u \leq \beta$, we have*

$$u'(a) \leq r \textit{ and } u'(b) \geq -r \quad (\textit{resp. } u'(a) \geq -r \textit{ and } u'(b) \leq r);$$

(b) $\displaystyle\int_r^\infty \frac{s^{1/q}}{\varphi(s)}\, ds > \|\psi\|_{L^p} (\max_t \beta(t) - \min_t \alpha(t))^{1/q}$, *where* $q = \frac{p}{p-1}$;

(c) for a.e. $t \in [a,b]$ and all $(u,v) \in \mathbb{R}^2$ such that $(t,u,v) \in E$, we have

$$f(t,u,v) \leq \psi(t)\varphi(|v|) \quad (\textit{resp. } f(t,u,v) \geq -\psi(t)\varphi(|v|)).$$

Then the problem (1.1) *has at least one solution $u \in W^{2,p}(a,b)$ such that for all $t \in [a,b]$*

$$\alpha(t) \leq u(t) \leq \beta(t).$$

Proof. We proceed as in the proof of Theorem 2.1 except for Step 4 where we use Proposition I-4.8 rather than Proposition I-4.7. □

Theorem 3.2. *Let A_0, $B_0 \in \mathbb{R}$, a_1, $b_1 \in \mathbb{R}$, a_2, $b_2 \in \mathbb{R}^+$ with $a_1^2 + a_2^2 > 0$, $b_1^2 + b_2^2 > 0$.*

Suppose α and $\beta \in C([a,b])$ are $W^{2,1}$-lower and upper solutions of the problem (1.1) *such that $\alpha \leq \beta$. Define $A \subset [a,b]$ (resp. $B \subset [a,b]$) to be the set of points where α (resp. β) is derivable.*

Let E be defined in (1.2) *and for some $p \in [1, \infty]$, $f : E \to \mathbb{R}$ be an L^p-Carathéodory function. Assume there exists $N \in L^1(a,b)$, $N > 0$ such that for a.e. $t \in A$ (resp. for a.e. $t \in B$),*

$$f(t, \alpha(t), \alpha'(t)) \geq -N(t), \quad (\textit{resp. } f(t, \beta(t), \beta'(t)) \leq N(t)).$$

If moreover there exist $r \geq 0$, $\varphi \in C(\mathbb{R}^+, \mathbb{R}_0^+)$ and $\psi \in L^p(a,b)$ such that
(a) for all u solution of (1.1) *with $\alpha \leq u \leq \beta$, we have*

$$|u'(a)| \leq r \quad (\textit{resp. } |u'(b)| \leq r);$$

(b) $\displaystyle\int_r^\infty \frac{s^{1/q}}{\varphi(s)}\, ds > \|\psi\|_{L^p} (\max_t \beta(t) - \min_t \alpha(t))^{1/q}$, *where* $q = \frac{p}{p-1}$;

(c) for a.e. $t \in [a,b]$ and all $(u,v) \in \mathbb{R}^2$ such that $(t,u,v) \in E$, we have

$$\operatorname{sgn}(v) f(t,u,v) \le \psi(t)\varphi(|v|) \quad (\textit{resp. } \operatorname{sgn}(v) f(t,u,v) \ge -\psi(t)\varphi(|v|)).$$

Then the problem (1.1) *has at least one solution $u \in W^{2,p}(a,b)$ such that for all $t \in [a,b]$*

$$\alpha(t) \le u(t) \le \beta(t).$$

Proof. We proceed as in the proof of Theorem 2.1 except for Step 4 where we use Proposition I-4.9 rather than Proposition I-4.7. □

The conditions $|u'(a)| \le r$ or $|u'(b)| \le r$ are automatically satisfied in case of the Neumann problem or of the Robin problem.

Example 3.1. Consider the boundary value problem

$$\begin{gathered} u'' = -u'^{2n+1} + k \sin 2\pi t, \\ u'(0) = 0, \; u(1) = 0, \end{gathered}$$

with $n \ge 0$ and $k \in \mathbb{R}$. It is easy to see that Theorem 3.2 applies with lower and upper solutions

$$\alpha(t) = -\beta(t) = -|k|^{1/2n+1}(1-t).$$

Assumption (c) is satisfied since $f(t,u,v) = -v^{2n+1} + k \sin 2\pi t \le k \sin 2\pi t$, for $v > 0$, and $f(t,u,v) \ge k \sin 2\pi t$ for $v < 0$. Notice that Theorem 2.1 does not apply if n is large enough.

For other problems, the estimate r can sometimes be deduced from the upper and lower solutions if they coincide at $t = a$ or $t = b$.

Example 3.2. Consider the boundary value problem

$$\begin{gathered} u'' = -u'^{2n} \sin \pi t + k \sin 2\pi t, \\ u(0) = 0, \; u(1) = 0, \end{gathered}$$

with $n \ge 0$ and $k \in [-1,1]$. Theorem 3.1 applies now with

$$\beta(t) = t(1-t) = -\alpha(t).$$

Assumption (a) is satisfied since the conditions $\alpha \le u \le \beta$ and $\alpha(0) = \beta(0) = 0$ imply

$$\alpha'(0) = -1 \le u'(0) \le \beta'(0) = 1.$$

We mentioned in Section I-4.3 that the one-sided Nagumo conditions impose that along curves $u' = \mu(u)$ and $u' = \nu(u) > \mu(u)$ the vector field is not tangent from above along $u' = \mu(u)$ and not from below along $u' = \nu(u)$. The following proposition describes such a situation for the Dirichlet problem.

Proposition 3.3. *Let $R > 0$, let α, $\beta \in \mathcal{C}([a,b])$ be such that $\alpha \leq \beta$ and define E from* (1.2).

Then for every L^1-Carathéodory function $f : E \to \mathbb{R}$ which satisfies

$$uf(t,u,v) \geq 0 \quad \text{if } (t,u,v) \in E \text{ and } |v| \geq R, \tag{3.1}$$

and every solution u of

$$\begin{aligned} u'' &= f(t,u,u'), \\ u(a) &= 0, \; u(b) = 0, \end{aligned} \tag{3.2}$$

such that $\alpha \leq u \leq \beta$, we have

$$\|u'\|_\infty \leq R.$$

Proof. Let u be a solution of (3.2) such that $\alpha \leq u \leq \beta$.

Claim 1 – The function u is such that

$$(u(t),u'(t)) \notin F_1 := \{(u,v) \mid u \geq 0, v > R\} \quad \text{if } t \in [a,b[.$$

Assume on the contrary that for some $t_1 \in [a,b[$, $(u(t_1),u'(t_1)) \in F_1$. Notice first that we can find a $t_2 \geq t_1$ such that $u(t_2) > 0$ and $u'(t_2) > R$. As $u(b) = 0$, we can find a $t_3 \in]t_2,b]$ such that $(u(t),u'(t)) \in F_1$ for any $t \in [t_2,t_3[$ and $u'(t_3) = R$. This result gives the contradiction

$$R = u'(t_3) = u'(t_2) + \int_{t_2}^{t_3} f(s,u(s),u'(s))\,ds > R.$$

Claim 2 – The function u is such that

$$(u(t),u'(t)) \notin F_2 := \{(u,v) \mid u < 0, v > R\} \quad \text{if } t \in [a,b].$$

This claim is proved reversing the time and using the arguments of Claim 1.

Conclusion – For $t \in [a,b]$, we deduce from the above claims and the continuity of u' that $(u(t),u'(t)) \notin F_1 \cup F_2$, i.e. $u'(t) \leq R$.

We prove in a similar way that $u'(t) \geq -R$. □

The assumptions of this proposition impose some conditions on the vector field $(u(t),u'(t))$ associated with (3.2); see figure 6.

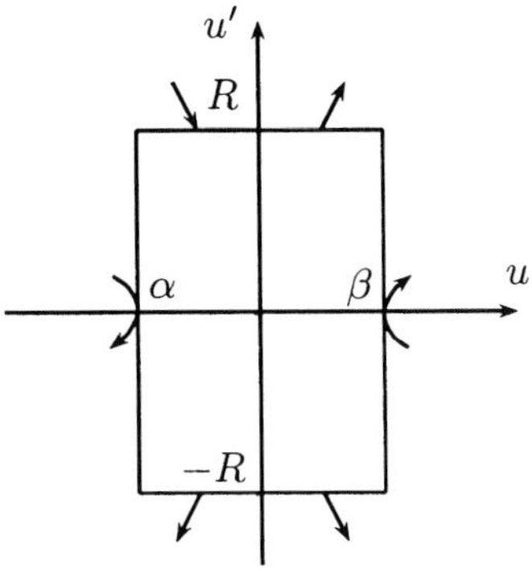

Fig. 6 : Vector field in Proposition 3.3

The above proposition can be improved in two ways. We can replace the sign conditions (3.1) by Nagumo conditions (see the remark after Proposition I-4.6). This procedure amounts to replacing the segments $u' = R$ and $u' = -R$ by curves deduced from these Nagumo conditions. Another improvement concerns the inversion of the vector field on the segments $u' = R$ and $u' = -R$ which can take place at some point $u = \mu_1(t)$ and $u = \mu_2(t)$. Such a generalization is worked out in the next proposition.

Proposition 3.4. *Let $R > 0$, let α, $\beta \in C([a,b])$ be such that $\alpha \leq \beta$ and define E from (1.2). Assume there exist μ_1 and $\mu_2 \in W^{1,1}(a,b)$ such that $-R \leq \mu_1'$, $R \geq \mu_2'$, $\mu_1(a) \geq 0 \geq \mu_2(a)$ and $\mu_1(b) \leq 0 \leq \mu_2(b)$.*

Then for every L^1-Carathéodory function $f : E \to \mathbb{R}$ which satisfies

$$\begin{aligned}(u - \mu_1(t))f(t,u,v) &\geq 0 \quad \text{if } v \leq -R,\\ (u - \mu_2(t))f(t,u,v) &\geq 0 \quad \text{if } v \geq R,\end{aligned}$$

on E and every solution u of (3.2) such that $\alpha \leq u \leq \beta$, we have

$$\|u'\|_\infty \leq R.$$

Functions μ_1 and μ_2 that satisfies such assumptions are often called *diagonals.*

Exercise 3.1. Prove the above result.

Hint : Let u be a solution of (3.2) such that $\alpha \leq u \leq \beta$ and show that the function u is such that
(i) if $(u(t), u'(t)) \in F_1(t) := \{(u,v) \mid u \geq \mu_2(t),\ v > R\}$ for some $t \in [a,b[$, the condition $u(b) = 0$ cannot be satisfied;
(ii) if $(u(t), u'(t)) \in F_2(t) := \{(u,v) \mid u < \mu_2(t),\ v > R\}$ for some $t \in [a,b]$, the condition $u(a) = 0$ cannot be satisfied.

Exercise 3.2. State and prove the counterpart of Theorems 3.1 and 3.2 using Propositions 3.3 or 3.4.

4 Dirichlet Problem

4.1 Derivative Independent Problems

Dirichlet boundary value problems

$$\begin{aligned} u'' &= f(t,u), \\ u(a) = 0,& \;\; u(b) = 0 \end{aligned} \tag{4.1}$$

can be studied for more general nonlinearities than L^1-Carathéodory functions. This remark follows from the analysis of the linear problem

$$\begin{aligned} u'' &= h(t), \\ u(a) = 0,& \;\; u(b) = 0. \end{aligned} \tag{4.2}$$

Its solution makes sense as a function in $\mathcal{C}_0([a,b]) \cap W^{2,1}_{loc}(a,b)$ provided

$$(s-a)(b-s)|h(s)| \in L^1(a,b)$$

(cfr. Proposition 4.1). For example, the problem

$$\begin{aligned} t(1-t)u'' &= 1, \\ u(0) = 0,& \;\; u(1) = 0, \end{aligned}$$

has the solution $u(t) = t\ln t + (1-t)\ln(1-t)$ which is in $\mathcal{C}_0([0,1]) \cap W^{2,1}_{loc}(0,1)$ but not in $\mathcal{C}^1([0,1])$. Notice that in some cases, problems with such singularities turn out to have solutions in $\mathcal{C}^\infty([a,b])$. This is the case in the example

$$\begin{aligned} t(1-t)u'' + u &= -t(1-t), \\ u(0) = 0,& \;\; u(1) = 0, \end{aligned}$$

whose solution is $u(t) = t(1-t)$.

The following proposition makes precise the main result we need concerning the linear problem (4.2).

Proposition 4.1. *If*

$$h \in \mathcal{A} := \{h \mid (s-a)(b-s)h(s) \in L^1(a,b)\},$$

the problem (4.2) *has one and only one solution*

$$u \in W^{2,\mathcal{A}}(a,b) := \{u \in W^{1,1}(a,b) \mid u'' \in \mathcal{A}\} \subset \mathcal{C}([a,b]) \cap \mathcal{C}^1(]a,b[)$$

such that

$$u(t) = \int_a^b G(t,s)\,h(s)\,ds,$$

where $G(t,s)$ is the Green's function corresponding to (4.2). Further, we have

$$\|u\|_\infty \leq \tfrac{1}{b-a}\|h\|_{\mathcal{A}},$$

where $\|h\|_{\mathcal{A}} = \displaystyle\int_a^b (s-a)(b-s)|h(s)|\,ds.$

Proof. Define

$$\begin{aligned} u(t) &= -(b-t)\int_a^t \tfrac{s-a}{b-a}\,h(s)\,ds \\ &\qquad\qquad -(t-a)\int_t^b \tfrac{b-s}{b-a}\,h(s)\,ds, \quad \text{for } t \in]a,b[, \\ &= 0, \qquad\qquad\qquad\qquad\qquad\qquad \text{for } t \in \{a,b\}. \end{aligned} \tag{4.3}$$

Claim 1 – $u \in W^{2,1}_{loc}(a,b) \cap W^{1,1}(a,b)$ *and* $u'' = h$. Let $c \in]a,b[$ be fixed. As $h \in \mathcal{A}$, we have $(\cdot - a)h \in L^1(a,c)$ and $(b-\cdot)h \in L^1(c,b)$. It follows that

$$\int_a^t \tfrac{s-a}{b-a}\,h(s)\,ds \in W^{1,1}(a,c) \quad \text{and} \quad \int_t^b \tfrac{b-s}{b-a}\,h(s)\,ds \in W^{1,1}(c,b).$$

Hence $u \in W^{1,1}_{loc}(a,b)$ and we compute

$$u'(t) = \int_a^t \tfrac{s-a}{b-a}\,h(s)\,ds - \int_t^b \tfrac{b-s}{b-a}\,h(s)\,ds.$$

Using Hake's Theorem and Lebesgue's Dominated Convergence Theorem we obtain $u' \in L^1(a,b)$. At last we deduce from a similar argument that $u' \in W^{1,1}_{loc}(a,b)$ and $u''(t) = h(t)$.

Claim 2 – *The function u is continuous and $u(a) = u(b) = 0$.* From Claim 1, $u \in C(]a,b[)$. Next, as $\displaystyle\int_a^t \tfrac{s-a}{b-a}\,h(s)\,ds \in W^{1,1}(a,c)$, we have

$$\lim_{t\to a}(b-t)\int_a^t \tfrac{s-a}{b-a}\,h(s)\,ds = 0.$$

On the other hand, we can write

$$|(t-a)\int_t^b \tfrac{b-s}{b-a}\,h(s)\,ds| \leq \int_a^b \chi_{[t,b]}(s)\tfrac{(t-a)(b-s)}{b-a}\,|h(s)|\,ds,$$

where $\chi_{[t,b]}$ is the characteristic function of the interval $[t, b]$. Further, as

$$\chi_{[t,b]}(s)\tfrac{(t-a)(b-s)}{b-a}\,|h(s)| \le \tfrac{(s-a)(b-s)}{b-a}\,|h(s)| \in L^1(a,b),$$

the Lebesgue's Dominated Convergence Theorem implies

$$\lim_{t\to a}(t-a)\int_t^b \tfrac{b-s}{b-a}\,h(s)\,ds = 0.$$

This result proves that $\lim\limits_{t\to a} u(t) = 0$. Similarly we obtain $\lim\limits_{t\to b} u(t) = 0$.
Claim 3 – $\|u\|_\infty \le \frac{1}{b-a}\|h\|_{\mathcal{A}}$. We deduce from (4.3) that

$$\begin{aligned} |u(t)| &\le \int_a^t \tfrac{(b-t)(s-a)}{b-a}\,|h(s)|\,ds + \int_t^b \tfrac{(b-s)(t-a)}{b-a}\,|h(s)|\,ds \\ &\le \int_a^b \tfrac{(b-s)(s-a)}{b-a}\,|h(s)|\,ds. \end{aligned}$$

and the claim follows. □

Remark If $h \in L^1_{loc}(a,b)$, any solution u of (4.2) reads

$$u(t) = u(t_0) + \int_{t_0}^t h(s)(t-s)\,ds,$$

where t_0 is chosen such that $u'(t_0) = 0$. If further h is one-signed we can deduce from Levi's Theorem that the integrals

$$\int_a^{t_0} (s-a)h(s)\,ds \quad \text{and} \quad \int_{t_0}^b (b-s)h(s)\,ds$$

exist so that $(s-a)(b-s)h(s) \in L^1(a,b)$.

Notice that this result does not hold anymore if h changes sign. Consider for example the problem

$$u'' = -\tfrac{1}{t^3}\sin\tfrac{1}{t}, \quad u(0) = 0,\ u(\tfrac{1}{\pi}) = 0,$$

which has a continuous solution $u(t) = t\sin\frac{1}{t}$ although $s(\frac{1}{\pi} - s)h(s) \notin L^1(0, \frac{1}{\pi})$.

The Proposition 4.1 suggests how to generalize the Carathéodory conditions in case of the Dirichlet problem. Let us write the boundary value problem (4.1) as an operator equation

$$Lu = Nu.$$

To inverse the linear operator L, we deduce from Proposition 4.1 that the function Nu should be in $\mathcal{A}$. A natural assumption to ensure this property is captured with the following definition. A function $f : D \subset [a,b] \times \mathbb{R} \to \mathbb{R}$ is said to be an *$\mathcal{A}$-Carathéodory function* or to satisfy *$\mathcal{A}$-Carathéodory conditions* if it is a Carathéodory function such that for any $r > 0$ there exists $h \in \mathcal{A}$ and for all $(t,u) \in D$ with $|u| \leq r$

$$|f(t,u)| \leq h(t). \tag{4.4}$$

Notice that in the introductory examples, the functions

$$f(t,u) = \frac{1}{t(1-t)} \quad \text{and} \quad -1 - \frac{u}{t(1-t)}$$

are $\mathcal{A}$-Carathéodory functions. Observe also that $L^1(a,b) \subset \mathcal{A}$ so that this definition generalizes the classical L^1-Carathéodory conditions on f.

For Dirichlet problems, the notion of $W^{2,1}$-lower and upper solutions particularizes as follows.

Definitions 4.1. *A function $\alpha \in \mathcal{C}([a,b])$ is said to be a* $W^{2,1}$-lower solution *of* (4.1) *if*

(a) for any $t_0 \in]a,b[$, either $D_-\alpha(t_0) < D^+\alpha(t_0)$,
or there exists an open interval $I_0 \subset]a,b[$ such that $t_0 \in I_0$, $\alpha \in W^{2,1}(I_0)$, and for a.e. $t \in I_0$,

$$\alpha''(t) \geq f(t,\alpha(t));$$

(b) $\alpha(a) \leq 0$, $\alpha(b) \leq 0$.

A function $\beta \in \mathcal{C}([a,b])$ is a $W^{2,1}$-upper solution *of* (4.1) *if*

(a) for any $t_0 \in]a,b[$, either $D^-\beta(t_0) > D_+\beta(t_0)$,
or there exists an open interval $I_0 \subset]a,b[$ such that $t_0 \in I_0$, $\beta \in W^{2,1}(I_0)$, and for a.e. $t \in I_0$,

$$\beta''(t) \leq f(t,\beta(t));$$

(b) $\beta(a) \geq 0$, $\beta(b) \geq 0$.

We can now state the main result for the Dirichlet problem (4.1).

Theorem 4.2. *Assume that α and β are $W^{2,1}$-lower and upper solutions of* (4.1) *such that $\alpha \leq \beta$. Let $E = \{(t,u) \in [a,b] \times \mathbb{R} \mid \alpha(t) \leq u \leq \beta(t)\}$ and $f : E \to \mathbb{R}$ be an $\mathcal{A}$-Carathéodory function.*

Then the problem (4.1) *has at least one solution $u \in W^{2,\mathcal{A}}(a,b)$ such that for all $t \in [a,b]$*

$$\alpha(t) \leq u(t) \leq \beta(t).$$

Proof. We consider the modified problem

$$\begin{aligned} u'' &= f(t,\gamma(t,u)), \\ u(a) &= 0, \ \ u(b) = 0, \end{aligned} \tag{4.5}$$

where γ is defined by (1.6).

Claim 1 – The problem (4.5) *has at least one solution* $u \in W^{2,\mathcal{A}}(a,b)$. Define the operator $T : \mathcal{C}([a,b]) \to \mathcal{C}([a,b])$ given by

$$(Tu)(t) = \int_a^b G(t,s) f(s,\gamma(s,u(s)))\, ds,$$

where $G(t,s)$ is the Green's function associated with (4.2). We can prove that T is completely continuous and bounded. Hence by Schauder's Fixed Point Theorem, T has a fixed point u which is a solution of (4.5) in $W^{2,\mathcal{A}}(a,b)$.

Claim 2 – Any solution u *of* (4.5) *satisfies* $\alpha \le u \le \beta$. Observe first that $\alpha(a) \le u(a) \le \beta(a)$ and $\alpha(b) \le u(b) \le \beta(b)$. Next, we argue as in Theorem 2.1 to obtain the result.

Conclusion – From Claim 2, the function u, solution of (4.5), solves (4.1). □

In the above proof, the condition $h \in \mathcal{A}$ is used to insure

$$\| \int_a^b G(t,s)\, h(s)\, ds \|_\infty < +\infty \ \text{ and } \ \int_a^b \left| \tfrac{\partial G}{\partial t}(t,s) \right| h(s)\, ds \in L^1(a,b). \tag{4.6}$$

We can see that in the periodic case, the set of functions $h :]a,b[\to \mathbb{R}^+$ measurable that satisfy (4.6) is in fact $L^1(a,b)$. The same holds true for the separated boundary value problem (1.10) if $a_2 b_2 \neq 0$. In that sense, Theorem 4.2 has the same generality as Theorem I-3.1 and Theorem 2.4.

Example 4.1. Consider the boundary value problem

$$\begin{aligned} u'' + |u|^{1/2} - \tfrac{1}{t} &= 0, \\ u(0) = 0, \ \ u(\pi) &= 0. \end{aligned}$$

It is easy to see that $\beta(t) = 0$ is an upper solution and $\alpha(t) = t \ln \frac{t}{\pi} - t$ is a lower solution. Hence we have a solution u such that for all $t \in [0,\pi]$

$$t \ln \frac{t}{\pi} - t \le u(t) \le 0.$$

Notice that in this example the function $f(t,u)$ is not L^1-Carathéodory.

If we want more regularity, we need more restrictive conditions on f. For example, we have the following theorem.

Theorem 4.3. *Assume that α and β are $W^{2,1}$-lower and upper solutions of (4.1) such that $\alpha \le \beta$. Let $E = \{(t,u) \in [a,b] \times \mathbb{R} \mid \alpha(t) \le u \le \beta(t)\}$ and $f : E \to \mathbb{R}$ satisfy Carathéodory conditions. Assume there exists a measurable function h such that $\int_a^b (s-a)h(s)\,ds < \infty$ and for a.e. $t \in [a,b]$ and all $u \in \mathbb{R}$ with $(t,u) \in E$,*

$$|f(t,u)| \le h(t).$$

Then the problem (4.1) has at least one solution $u \in W^{2,\mathcal{A}}(a,b) \cap C^1(]a,b])$ such that for all $t \in [a,b]$

$$\alpha(t) \le u(t) \le \beta(t).$$

Proof. Existence of a solution $u \in W^{2,\mathcal{A}}(a,b)$ follows from Theorem 4.2. Further

$$u(t) = \int_a^b G(t,s) f(s,u(s))\,ds,$$

where $G(t,s)$ is the Green's function corresponding to (4.2). It is now standard to see that $u \in C^1(]a,b])$. □

In the same way, we can consider the problem

$$\begin{aligned} u'' &= f(t,u), \\ u(a) &= 0, \\ b_1 u(b) + b_2 u'(b) &= 0. \end{aligned} \tag{4.7}$$

Theorem 4.4. *Let $b_1 \in \mathbb{R}$, $b_2 \in \mathbb{R}^+$ with $b_1^2 + b_2^2 > 0$ and assume that α and β are $W^{2,1}$-lower and upper solutions of (4.7) such that $\alpha \le \beta$. Let $E = \{(t,u) \in [a,b] \times \mathbb{R} \mid \alpha(t) \le u \le \beta(t)\}$ and $f : E \to \mathbb{R}$ satisfy Carathéodory conditions. Assume there exists a measurable function $h : [a,b] \to \mathbb{R}^+$ such that $\int_a^b (s-a)h(s)\,ds < \infty$ and for a.e. $t \in [a,b]$ and all $u \in \mathbb{R}$ with $(t,u) \in E$,*

$$|f(t,u)| \le h(t).$$

Then the problem (4.7) has at least one solution $u \in W^{2,\mathcal{A}}(a,b) \cap C^1(]a,b])$ such that for all $t \in [a,b]$

$$\alpha(t) \le u(t) \le \beta(t).$$

Exercise 4.1. Prove the above theorem.

4.2 Derivative Dependent Problems

Consider now the Dirichlet problem with derivative dependence

$$\begin{aligned} u'' &= f(t,u,u'), \\ u(a) &= 0, \ u(b) = 0. \end{aligned} \tag{4.8}$$

A natural idea would be to impose a Nagumo condition

$$|f(t,u,v)| \le \psi(t)\varphi(|v|),$$

with $\psi \in L^p(a,b)$ and $\varphi \in C(\mathbb{R}^+, \mathbb{R}_0^+)$ so as to apply Proposition I-4.7. This condition forces f to be L^p-Carathéodory and excludes the type of singularities we considered in the first part of this section. Moreover with such singularities we expect solutions u to be in $W^{2,\mathcal{A}}(a,b)$ and it is not natural to put conditions on f that imply an *a priori* bound on $\|u'\|_\infty$. A more reasonable idea is to look for an *a priori* bound on u' in $L^1(a,b)$. More precisely, we shall approximate the problem on intervals $[a_n,b_n] \subset]a,b[$ and will need an *a priori* bound $|u'(t)| \le h(t)$ valid on $[a_n,b_n]$ but with h independent of n.

To state such results we need the following concept. A function $f : D \subset [a,b] \times \mathbb{R}^2 \to \mathbb{R}$ is said to be an L^1_{loc}*-Carathéodory function* or to satisfy L^1_{loc}*-Carathéodory conditions* if it is a Carathéodory function and for all $r > 0$, there exists $h \in L^1_{loc}(a,b)$ such that for all $(t,u,v) \in D$ with $|u| + |v| \le r$, we have

$$|f(t,u,v)| \le h(t).$$

Proposition 4.5. *Let α and $\beta \in C([a,b])$ be such that $\alpha \le \beta$ and define E from (1.2). Assume there exist $\psi \in L^1_{loc}(a,b)$ and a non-decreasing function $\varphi \in C(\mathbb{R}^+, \mathbb{R}_0^+)$ such that*

(a) $\displaystyle\int_0^\infty \frac{ds}{\varphi(s)} = \infty$;

(b) for some $\xi \in]a,b[$,

$$\Phi^{-1}(2|\int_\xi^t \psi(s)\,ds|) \in L^1(a,b),$$

where $\displaystyle\Phi(u) = \int_0^u \frac{ds}{\varphi(s)}$.

Then there exists $h \in L^1(a,b) \cap C(]a,b[)$ such that, for every $a \le a_1 \le \frac{a+\xi}{2} < \frac{\xi+b}{2} \le b_1 \le b$, every L^1_{loc}-Carathéodory function $f : E \to \mathbb{R}$ which satisfies

for a.e. $t \in [a_1,b_1]$ and all $(u,v) \in \mathbb{R}^2$ with $(t,u,v) \in E$

$$|f(t,u,v)| \le \psi(t)\varphi(|v|),$$

and every solution u of

$$u'' = f(t, u, u') \tag{4.9}$$

defined on $[a_1, b_1]$ and such that $\alpha \le u \le \beta$, we have

$$\forall t \in [a_1, b_1], \quad |u'(t)| \le h(t).$$

Remark 4.1. Let us comment assumption (b).

(i) If $\psi \in L^1(a, b)$, condition (b) is implied by condition (a);

(ii) In case $\varphi(v) \equiv 1$, condition (b) becomes $\psi \in \mathcal{A}$;

(iii) Condition (b) is implied by $\Phi^{-1}(2(b-a)\psi(\cdot)) \in \mathcal{A}$. This fact can be seen using Jensen inequality (see for example [80, Theorem II-2.2]). For $t \ge \xi$ we obtain

$$\begin{aligned}\Phi^{-1}(2\int_\xi^t \psi(s)\,ds) &= \Phi^{-1}(\tfrac{1}{b-a}\int_a^b 2(b-a)\psi(s)\chi_{[\xi,t]}(s)\,ds) \\ &\le \tfrac{1}{b-a}\int_\xi^t \Phi^{-1}(2(b-a)\psi(s))\,ds.\end{aligned}$$

If $\Phi^{-1}(2(b-a)\psi(\cdot)) \in \mathcal{A}$, we have

$$\int_\xi^t \Phi^{-1}(2(b-a)\psi(s))\,ds \in L^1(\xi, b)$$

and it follows that

$$\Phi^{-1}(2\int_\xi^t \psi(s)\,ds) \in L^1(\xi, b).$$

In the same way, we prove

$$\Phi^{-1}(2\int_t^\xi \psi(s)\,ds) \in L^1(a, \xi).$$

Proof. Step 1 – Existence of a function $h \in C(]a, b[)$ that satisfies the assertions of the proposition. Let c, d be such that $\frac{a+\xi}{2} \le c < \xi < d \le \frac{\xi+b}{2}$ and $M > \frac{1}{d-c}(\max_t \beta - \min_t \alpha)$. Define h_1 to be the solution of

$$h_1' = \psi(t)\varphi(h_1), \ t \in [c, b[, \quad h_1(c) = M,$$

h_2 the solution of

$$h_2' = -\psi(t)\varphi(h_2), \ t \in]a, d], \quad h_2(d) = M,$$

and

$$\begin{aligned} h(t) &= h_2(t), && \text{on }]a,c[, \\ &= \max\{h_1(t), h_2(t)\}, && \text{on } [c,d[, \\ &= h_1(t), && \text{on } [d,b[. \end{aligned}$$

Let a_1, b_1 be such that $a \le a_1 \le \frac{a+\xi}{2} < \frac{\xi+b}{2} \le b_1 \le b$ and u be a solution of (4.9) on $[a_1, b_1]$ with $\alpha \le u \le \beta$.

Observe there exists $\tau \in [c,d]$ with $|u'(\tau)| \le M$. Consider an interval $I = [t_0, t_1]$ such that $u'(t) \ge 0$ on I, $u'(t_0) = M$ and $t_0 \ge c$. We have, for every $t \in [t_0, t_1]$,

$$\begin{aligned} \int_M^{u'(t)} \frac{d\sigma}{\varphi(\sigma)} &= \int_{t_0}^t \frac{u''(s)}{\varphi(u'(s))} ds \le \int_{t_0}^t \psi(s)\, ds \\ &\le \int_c^t \psi(s)\, ds = \int_c^t \frac{h_1'(s)}{\varphi(h_1(s))} ds = \int_M^{h_1(t)} \frac{d\sigma}{\varphi(\sigma)}, \end{aligned}$$

i.e. $u'(t) \le h_1(t)$ on I and hence on $[\tau, b_1]$.

In the same way, we prove $u'(t) \le h_2(t)$ on $[a_1, \tau]$. It follows that $u'(t) \le h(t)$ on $[a_1, b_1]$. At last, we obtain from the same argument that $u'(t) \ge -h(t)$ on $[a_1, b_1]$.

Step 2 – $h \in L^1(a,b)$. We compute

$$\Phi(h_1(t)) - \Phi(M) = \int_M^{h_1(t)} \frac{dr}{\varphi(r)} = \int_c^t \psi(s)\, ds.$$

Hence we write

$$\begin{aligned} h_1(t) &= \Phi^{-1}(\Phi(M) + \int_c^t \psi(s)\, ds) \\ &\le \Phi^{-1}(\Phi(M) + \int_c^d \psi(s)\, ds + \int_\xi^t \psi(s)\, ds) \end{aligned}$$

and as Φ^{-1} is convex

$$h_1(t) \le \frac{1}{2}[\Phi^{-1}(2[\Phi(M) + \int_c^d \psi(s)\, ds]) + \Phi^{-1}(2 \int_\xi^t \psi(s)\, ds)],$$

from which we deduce $h_1 \in L^1(c,b)$. In the same way, we have

$$h_2(t) = \Phi^{-1}(\Phi(M) + \int_t^d \psi(s)\, ds)$$

and deduce $h_2 \in L^1(a,d)$. □

Remark The condition φ non-decreasing is not essential. If it is not satisfied, we replace condition (b) by

$$\Phi^{-1}(\Phi(M) + \int_c^t \psi(s)\,ds) \in L^1(c,b)$$

and

$$\Phi^{-1}(\Phi(M) + \int_t^d \psi(s)\,ds) \in L^1(a,d),$$

where M, c and d are defined in the proof of Proposition 4.5.

Theorem 4.6. *Let α, $\beta \in \mathcal{C}([a,b])$ be $W^{2,1}$-lower and upper solutions of the problem (4.8) such that $\alpha \leq \beta$. Define $A \subset [a,b]$ (resp. $B \subset [a,b]$) to be the set of points where α (resp. β) is derivable.*

Let E be defined by (1.2) and $f : E \to \mathbb{R}$ be an L^1_{loc}-Carathéodory function. Suppose there exists $N \in L^1_{loc}(a,b)$, $N > 0$ such that for a.e. $t \in A$ (resp. for a.e. $t \in B$),

$$f(t,\alpha(t),\alpha'(t)) \geq -N(t), \qquad (\text{resp. } f(t,\beta(t),\beta'(t)) \leq N(t)).$$

Assume moreover there exist $\psi \in L^1_{loc}(a,b)$ and a non-decreasing function $\varphi \in \mathcal{C}(\mathbb{R}^+,\mathbb{R}^+_0)$ such that

(a) $\displaystyle\int_0^\infty \frac{ds}{\varphi(s)} = \infty;$

(b) for some $\xi \in]a,b[$,

$$\Phi^{-1}(2|\int_\xi^t \psi(s)\,ds|) \in L^1(a,b),$$

where $\displaystyle\Phi(u) = \int_0^u \frac{ds}{\varphi(s)};$

(c) for a.e. $t \in [a,b]$ and all (u,v) with $(t,u,v) \in E$,

$$|f(t,u,v)| \leq \psi(t)\varphi(|v|).$$

Then the problem (4.8) has at least one solution

$$u \in \mathcal{C}([a,b]) \cap W^{2,1}_{loc}(a,b)$$

such that for all $t \in [a,b]$,

$$\alpha(t) \leq u(t) \leq \beta(t).$$

Proof. Step 1 – The modified problem. Let $(a_n)_n$, $(b_n)_n \subset]a,b[$, $(A_n)_n$, $(B_n)_n \subset \mathbb{R}$ be such that

$$\lim_{n\to\infty} a_n = a, \ \lim_{n\to\infty} b_n = b, \ \lim_{n\to\infty} A_n = 0, \ \lim_{n\to\infty} B_n = 0,$$
$$\alpha(a_n) \le A_n \le \beta(a_n), \ \alpha(b_n) \le B_n \le \beta(b_n).$$

Consider the modified problem

$$\begin{aligned} u'' &= f(t,u,u'), \\ u(a_n) &= A_n, \ u(b_n) = B_n. \end{aligned} \tag{4.10}$$

We can assume that for any n, $a_n \le \frac{a+\xi}{2} < \frac{\xi+b}{2} \le b_n$. Hence by Theorem 2.1 and Proposition 4.5 problem (4.10) has, for every n, a solution u_n satisfying on $[a_n,b_n]$

$$\alpha(t) \le u_n(t) \le \beta(t), \qquad |u_n'(t)| \le h(t),$$

with h given by Proposition 4.5.

Step 2 – Existence of a solution u of (4.8). Using Arzelà-Ascoli Theorem, we can find a subsequence $(u_n^1)_n \subset (u_n)_n$ that converges in $\mathcal{C}^1([a_1,b_1])$. Proceeding by induction, for any $k \in \mathbb{N}$, we build $(u_n^k)_n$ which is a subsequence of $(u_n^{k-1})_n$ that converges in $\mathcal{C}^1([a_k,b_k])$. It follows that the diagonal sequence $(u_n^n)_n$ converges pointwise to some function u and that for any compact $K \subset]a,b[$, the convergence takes place in $\mathcal{C}^1(K)$. Hence, u satisfies on $]a,b[$

$$u'' = f(t,u,u')$$

and

$$\alpha(t) \le u(t) \le \beta(t), \qquad |u'(t)| \le h(t).$$

Let us prove that $\lim_{t\to a} u(t) = 0$. Fix $\epsilon > 0$ and choose $\delta > 0$ so that $\int_a^{a+\delta} h(s)\,ds \le \epsilon/3$. Let $t \in [a, a+\delta]$ and pick n large enough so that $t \in [a_n,b_n]$, $|u_n(t) - u(t)| \le \epsilon/3$ and $|A_n| \le \epsilon/3$. We compute then

$$\begin{aligned} |u(t)| &\le |u(t) - u_n(t)| + |u_n(t) - u_n(a_n)| + |A_n| \\ &\le |u(t) - u_n(t)| + \int_{a_n}^t h(s)\,ds + |A_n| \le \epsilon. \end{aligned}$$

In the same way, we prove $\lim_{t\to b} u(t) = 0$. □

Remark We can improve condition (a) asking $\displaystyle\int_M^\infty \frac{ds}{\varphi(s)} > \|\psi\|_{L^1}$, where M is defined in the proof of Proposition 4.5. Hence, by the remark following Proposition 4.5, this result generalizes Theorem 2.1 for the Dirichlet case as well as Theorem 4.2.

As a first illustration consider the following example where Theorem 2.1 does not apply.

Example 4.2. Consider the boundary value problem

$$\begin{aligned} u'' &= \tfrac{1}{t^n}|u'|^a + u + t, \\ u(0) &= 0, \; u(1) = 0, \end{aligned}$$

where $0 \le a < 1$ and $1 < n < 2 - a$. Existence of a solution follows from Theorem 4.6 with $\alpha(t) = -1$, $\beta(t) = 0$, $\xi = \frac{1}{2}$, $\psi(t) = \frac{1}{t^n} + 1$, $\varphi(y) = \max\{1, y^a\}$ and explicit computation of $\Phi^{-1}(2|\int_{1/2}^{t} \psi(s)\, ds|)$.

In Theorem 4.6 we do not use the full power of condition (c) so that we can generalize it in the following way.

Theorem 4.7. *Let the assumptions of Theorem 4.6 be satisfied with (c) replaced by*

(c') there exist $a \le c < \xi < d \le b$ such that for a.e. $t \in [a, b]$ and all (u, v) with $(t, u, v) \in E$,

$$\begin{aligned} f(t,u,v)\,sgn(v) &\ge -\psi(t)\varphi(|v|) \quad && \text{if } t \in]a, d[, \\ f(t,u,v)\,sgn(v) &\le \psi(t)\varphi(|v|) \quad && \text{if } t \in]c, b[. \end{aligned}$$

Then the problem (4.8) has at least one solution

$$u \in C([a, b]) \cap W^{2,1}_{loc}(a, b)$$

such that for all $t \in [a, b]$,

$$\alpha(t) \le u(t) \le \beta(t).$$

Exercise 4.2. Prove Theorem 4.7.

Remark In Theorem 4.6 and 4.7, we can assume a control on $f(t, u, v)$ only for $|v| > R$.

Example 4.3. Consider the following modification of Example 4.1

$$\begin{aligned} u'' + |u|^{1/2} - \tfrac{1}{t^n} u'^{1/3} - \tfrac{1}{t} &= 0, \\ u(0) = 0, \; u(\pi) &= 0, \end{aligned}$$

where $n > 0$. We can apply Theorem 4.7 to prove the existence of a solution choosing $\alpha(t) = t \ln \frac{t}{\pi} - t$, $\beta(t) = 0$, $\xi = \pi/2$, $c = \pi/3$, $d = 2\pi/3$, $\varphi(v) = \max\{1, v^{1/3}\}$ and

$$\begin{aligned} \psi(t) &= \tfrac{1}{t} + \pi^{1/2}, && \text{on } [0, \pi/3], \\ &= \tfrac{1}{t} + \pi^{1/2} + \tfrac{1}{t^n}, && \text{on }]\pi/3, \pi]. \end{aligned}$$

Hence, there is a solution u such that for all $t \in [0, \pi]$

$$t \ln \tfrac{t}{\pi} - t \leq u(t) \leq 0.$$

5 Other Boundary Value Problems

In the Chapters VI to X we will present applications of the method of lower and upper solutions to a variety of problems in the theory of differential equations. The ideas of these two first chapters can also be extended to solve other problems. This section presents a selection of such extensions. The first one concerns radial solutions of partial differential equations. Here, we have to rewrite the method so as to include systems with singularities. In the second one, we investigate differential equations with deviating arguments. These two first applications concern second order problems. Therefore, we can use the maximum principle on which the theory was grounded and we just have to adapt the method to the problem at hand. The third application concerns a fourth order problem which imposes a more fundamental generalization of the underlying maximum principle. In the fourth application we consider the existence of bounded solutions. Here the lower and upper solutions method applies as such to solve an auxiliary problem. A more elaborate application is the study of the Ginzburg-Landau model which is a vector problem where the use of lower and upper solutions technique is not straightforward. Reading these applications, the reader will realize how large the scope is of the method.

To end this introduction, let us notice that the existence of lower and upper solutions does not necessarily imply existence of solutions. For example, it is proved in [242] that there exists some $c > 0$ and some function $f(t, u)$, which is continuous and almost periodic so that the equation

$$u'' + cu' = f(t, u)$$

has no almost periodic solution although there exist constant lower and upper solutions, i.e.

$$f(t, \alpha) \leq 0 \leq f(t, \beta)$$

for some $\alpha < \beta$.

In order to concentrate on the methods and to simplify the exposition, we will present in this section problems with a maximum of regularity and consider the simplest possible cases. We do not consider here functions f which satisfy Carathéodory conditions nor upper and lower solutions with corners.

5.1 Radial Solutions

Consider the problem

$$\begin{aligned} -\Delta v &= f(|x|, v), && \text{in } B_R, \\ v &= 0, && \text{on } \partial B_R, \end{aligned} \tag{5.1}$$

where B_R denotes the ball of center 0 and radius R in $\mathbb{R}^n$ (with $n \geq 1$) and $|x|$ the euclidian norm of $x \in \mathbb{R}^n$. It is well known that v is a classical radial solution of (5.1) if and only if $v(x) = u(|x|)$, where u is a solution of

$$\begin{aligned} -(t^{n-1}u')' &= t^{n-1}f(t,u), \\ u'(0) = 0,\ & u(R) = 0. \end{aligned} \tag{5.2}$$

More generally, we consider the problem

$$\begin{aligned} -(p(t)u')' &= q(t)f(t,u), \\ (pu')(0) = 0,\ & u(R) = 0, \end{aligned} \tag{5.3}$$

with f continuous, p, q continuous and p, $q > 0$ on $]0, R]$. Here the condition $(pu')(0) = 0$ means $\lim_{t\to 0}(pu')(t) = 0$. The difficulty is that we must allow the equation to be singular at $t = 0$ since in the model example $p(0) = 0$. However the zero of p is balanced by a zero of q. This is clear from the following lemma which is fundamental in our approach.

Lemma 5.1. *Let p, $q \in \mathcal{C}([0, R])$, p, $q > 0$ on $]0, R]$, define $w(t) = \int_t^R \frac{ds}{p(s)}$ and assume $qw \in L^1(0, R)$.*

Then for any $h \in \mathcal{C}([0, R])$, there exists a unique function $u \in \mathcal{C}([0, R]) \cap \mathcal{C}^1(]0, R])$, with $pu' \in \mathcal{C}^1([0, R])$, solution of

$$\begin{aligned} -(p(t)u')' &= q(t)h(t), \\ (pu')(0) = 0,\ & u(R) = 0. \end{aligned}$$

Further, we have

$$\|u\|_\infty \leq \|qw\|_{L^1}\|h\|_\infty.$$

Proof. The proof follows from the explicit solution

$$u(t) = \int_0^R G(t,s)q(s)h(s)\,ds,$$

where $G(t, s) = w(t)$ if $s \leq t$ and $G(t, s) = w(s)$ if $s > t$. □

Remark 5.1. The condition

$$\lim_{t\to 0} \frac{1}{p(t)} \int_0^t q(s)\,ds = 0 \tag{5.4}$$

implies $u'(0) = 0$ and also $qw \in L^1(0, R)$. Hence, if we reinforce the assumptions of Lemma 5.1 with condition (5.4), we obtain existence of a solution of

$$\begin{aligned} -(p(t)u')' &= q(t)h(t), \\ u'(0) = 0, &\ u(R) = 0. \end{aligned}$$

The main existence result is as follows.

Theorem 5.2. *Let $f \in C([0, R] \times \mathbb{R})$, p, $q \in C([0, R])$, p, $q > 0$ on $]0, R]$, define $w(t) = \int_t^R \frac{ds}{p(s)}$ and assume $qw \in L^1(0, R)$.*

Consider α and $\beta \in C([0, R]) \cap C^1(]0, R])$ with $p\alpha'$ and $p\beta' \in C^1([0, R])$, $\alpha \leq \beta$ and suppose

$$\begin{aligned} -(p\alpha')'(t) &\leq q(t)f(t, \alpha(t)), & -(p\beta')'(t) &\geq q(t)f(t, \beta(t)), \\ (p\alpha')(0) \geq 0, &\ \alpha(R) \leq 0, & (p\beta')(0) \leq 0, &\ \beta(R) \geq 0. \end{aligned}$$

Then the problem (5.3) has at least one solution u such that $\alpha \leq u \leq \beta$ on $[0, R]$.

Proof. Consider the modified problem

$$\begin{aligned} -(p(t)u')' &= q(t)[f(t, \gamma(t, u)) - \arctan(u - \gamma(t, u))], \\ (pu')(0) &= 0, \ u(R) = 0, \end{aligned} \tag{5.5}$$

where $\gamma(t, u)$ is defined from (1.6).

Claim 1 – The problem (5.5) *has at least one solution.* Using Lemma 5.1, we write (5.5) as an integral equation

$$u(t) = \int_0^R G(t, s)q(s)[f(s, \gamma(s, u(s))) - \arctan(u(s) - \gamma(s, u(s)))]\, ds$$

and apply Schauder's Theorem.

Claim 2 – The solution u of (5.5) *is such that $\alpha \leq u \leq \beta$.* Let $t_0 \in [0, R[$ be such that $u(t_0) - \alpha(t_0) = \min_t (u(t) - \alpha(t)) < 0$.

If $t_0 \in]0, R[$, we have $\alpha'(t_0) = u'(t_0)$ and

$$\begin{aligned} -(p\alpha')'(t_0) &\leq q(t_0)f(t_0, \alpha(t_0)) \\ &< q(t_0)[f(t_0, \gamma(t_0, u(t_0))) - \arctan(u(t_0) - \gamma(t_0, u(t_0)))] \\ &= -(pu')'(t_0). \end{aligned}$$

It follows that for $t > t_0$, near enough t_0, $p(t)(u'(t) - \alpha'(t)) < p(t_0)(u'(t_0) - \alpha'(t_0)) = 0$ which contradicts the fact that t_0 is a minimum of $u(t) - \alpha(t)$.

If $t_0 = 0$, we deduce from the boundary conditions $(pu')(0) - (p\alpha')(0) \leq 0$. On the other hand, we can write for all $t \in]0, R]$ small enough

$$\begin{aligned} -(p\alpha')'(t) &\leq q(t)f(t, \alpha(t)) \\ &< q(t)[f(t, \alpha(t)) - \arctan(u(t) - \alpha(t))] = -(pu')'(t). \end{aligned}$$

We deduce then for such t,

$$(pu')(t) - (p\alpha')(t) < (pu')(0) - (p\alpha')(0) \leq 0,$$

which contradicts $u - \alpha$ to be minimum at $t_0 = 0$.

We prove in a similar way that $u \leq \beta$.

Conclusion – As a consequence of Claim 2, the solution u of (5.5) solves (5.3) and is such that $\alpha \leq u \leq \beta$. □

Example 5.1. As an application, consider the problem (5.2). Assume the function f is such that for some $a \geq 0$ and $b \geq 0$,

$$0 \leq f(t, u) \leq au + b,$$

on $[0, R] \times \mathbb{R}^+$. Define w_λ to be the solution of the problem

$$(t^{n-1}w')' + \lambda t^{n-1}w = 0, \quad w(0) = 1,\ w'(0) = 0,$$

and $z_0(\lambda)$ to be the first positive zero of w_λ. Assume there exists $\lambda > a$ such that $z_0(\lambda) > R$. With such assumptions, it is easy to see that if $B > 0$ is large enough, the functions $\alpha(t) = 0$ and $\beta(t) = Bw_\lambda(t)$ are lower and upper solutions of (5.2).

Notice that β is an eigenfunction of an eigenvalue problem

$$(t^{n-1}w')' + \lambda t^{n-1}w = 0, \quad w'(0) = 0,\ w(z_0) = 0,$$

with $z_0 > R$. Such eigenfunctions are often the keys to the construction of lower and upper solutions.

5.2 Differential Equations with Deviating Arguments

Consider the following boundary value problem

$$\begin{aligned} u'' &= f(t, u, u(h(t))), & t &\in [0, 1], \\ a_1 u - a_2 u' &= A_0(t), & t &\in [a, 0], \\ b_1 u + b_2 u' &= B_0(t), & t &\in [1, b], \end{aligned} \tag{5.6}$$

where $a < 0$ and $1 < b$. The functions f, A_0, B_0 and h are continuous but no restriction on the sign of $h(t) - t$ is imposed so that both retarded and advanced shifts are allowed. A solution of (5.6) is supposed to be in $\mathcal{C}([a,b]) \cap \mathcal{C}^1(I) \cap \mathcal{C}^2([0,1])$, where

$$\begin{array}{ll} I = [a,b], & \text{if } a_2 \neq 0 \text{ and } b_2 \neq 0, \\ I = [a,1], & \text{if } a_2 \neq 0 \text{ and } b_2 = 0, \\ I = [0,b], & \text{if } a_2 = 0 \text{ and } b_2 \neq 0, \\ I = [0,1], & \text{if } a_2 = 0 \text{ and } b_2 = 0. \end{array}$$

A first result uses a monotonicity assumption on f.

Theorem 5.3. *Let $a < 0$, $b > 1$, $f : [0,1] \times \mathbb{R}^2 \to \mathbb{R}$, $A_0 : [a,0] \to \mathbb{R}$, $B_0 : [1,b] \to \mathbb{R}$ and $h : [0,1] \to [a,b]$ be continuous functions, a_1, a_2, b_1, b_2 be nonnegative numbers such that $a_1 + a_2 > 0$ and $b_1 + b_2 > 0$.*

Assume $f(t,u,v)$ is non-increasing in v and there exist α and $\beta \in \mathcal{C}([a,b]) \cap \mathcal{C}^1(I) \cap \mathcal{C}^2([0,1])$ such that for any $t \in [a,b]$, $\alpha(t) \leq \beta(t)$ and

$$\begin{array}{lll} \alpha''(t) \geq f(t,\alpha(t),\alpha(h(t))), & \beta''(t) \leq f(t,\beta(t),\beta(h(t))), & \forall t \in [0,1], \\ a_1\alpha(t) - a_2\alpha'(t) \leq A_0(t), & a_1\beta(t) - a_2\beta'(t) \geq A_0(t), & \forall t \in [a,0], \\ b_1\alpha(t) + b_2\alpha'(t) \leq B_0(t), & b_1\beta(t) + b_2\beta'(t) \geq B_0(t), & \forall t \in [1,b]. \end{array}$$

Then the problem (5.6) has at least one solution u such that for all $t \in [a,b]$, $\alpha(t) \leq u(t) \leq \beta(t)$.

Proof. Consider the modified problem

$$\begin{array}{rll} u'' - u = f(t,\gamma(t,u),\gamma(h(t),u(h(t)))) - \gamma(t,u), & t \in [0,1], & \\ a_1 u - a_2 u' = A_0(t), & t \in [a,0], & \\ b_1 u + b_2 u' = B_0(t), & t \in [1,b], & \end{array} \tag{5.7}$$

where $\gamma(t,u)$ is defined in (1.6).

Claim 1 – Problem (5.7) *has at least one solution.* We write (5.7) as an integral equation

$$u = Tu. \tag{5.8}$$

The operator $T : \mathcal{C}([a,b]) \to \mathcal{C}([a,b])$ is defined as follows.

(i) For $t \in [0,1]$, we set

$$(Tu)(t) = \int_0^1 G(t,s)[f(s,\gamma(s,u(s)),\gamma(h(s),u(h(s)))) - \gamma(s,u(s))]\,ds + u_0(t),$$

where $G(t,s)$ is the Green's function associated with

$$u'' - u = f(t), \quad a_1 u(0) - a_2 u'(0) = 0,\ b_1 u(1) + b_2 u'(1) = 0,$$

and $u_0(t)$ is the unique solution of

$$u'' - u = 0, \quad a_1 u(0) - a_2 u'(0) = A_0(0),\ b_1 u(1) + b_2 u'(1) = B_0(1).$$

(ii) For $t \in [a, 0[$ we write

$$\begin{aligned}(Tu)(t) &= (Tu)(0)\exp(\tfrac{a_1}{a_2}t) + \int_t^0 \exp(\tfrac{a_1}{a_2}(t-s))\ \tfrac{A_0(s)}{a_2}\,ds, && \text{if } a_2 \neq 0,\\ &= \frac{A_0(t)}{a_1}, && \text{if } a_2 = 0.\end{aligned}$$

(iii) Similarly, for $t \in]1, b]$ we define

$$\begin{aligned}(Tu)(t) &= (Tu)(1)\exp(-\tfrac{b_1}{b_2}(t-1)) +\\ &\qquad \int_1^t \exp(-\tfrac{b_1}{b_2}(t-s))\ \tfrac{B_0(s)}{b_2}\,ds, && \text{if } b_2 \neq 0,\\ &= \frac{B_0(t)}{b_1}, && \text{if } b_2 = 0.\end{aligned}$$

The operator T is completely continuous and bounded. It follows from Schauder's Theorem that T has a fixed point which is a solution of the modified problem (5.7).

Claim 2 – Solutions u of the modified problem (5.7) *are such that*

$$\alpha \leq u \leq \beta.$$

Let u be a solution of the modified problem (5.7) and assume $\min_t(u(t) - \alpha(t)) = u(t_0) - \alpha(t_0) < 0$.

Assume first $a_2 \neq 0$ and notice that for $t \in [a, 0]$, $u(t) < \alpha(t)$ implies $u'(t) \leq \alpha'(t)$. If it were not the case, we come to the contradiction

$$A_0(t) = a_1 u(t) - a_2 u'(t) < a_1\alpha(t) - a_2\alpha'(t) \leq A_0(t).$$

As a consequence, $t_0 \geq 0$. Similarly, if $b_2 \neq 0$, $t_0 \leq 1$.

In case $a_2 = 0$, we also have for $t \in [a, 0]$,

$$A_0(t) = a_1 u(t) \geq a_1\alpha(t)$$

and $u(t) \geq \alpha(t)$. Hence, $t_0 > 0$. Similarly, if $b_2 = 0$, we deduce $t_0 < 1$.

In all cases, a global negative minimum of $u - \alpha$ takes place at $t_0 \in \operatorname{int} I$. It is such that $u'(t_0) - \alpha'(t_0) = 0$ and we can compute

$$\begin{aligned}0 &\leq u''(t_0) - \alpha''(t_0)\\ &\leq f(t_0, \alpha(t_0), \gamma(h(t_0), u(h(t_0)))) + u(t_0) - \alpha(t_0) - f(t_0, \alpha(t_0), \alpha(h(t_0)))\\ &\leq u(t_0) - \alpha(t_0) < 0,\end{aligned}$$

which is a contradiction.

It follows from these arguments that $u - \alpha$ has no negative minimum, i.e. that $u \geq \alpha$. Similarly, we prove $u \leq \beta$.

Conclusion – It follows from Claim 2 that the solution u of (5.7) solves (5.6). □

Exercise 5.1. Prove that we can improve Theorem 5.3 by replacing the non-increasing assumption on f by the following hypothesis on the lower and the upper solutions :

$$\text{for all } t \in [0,1] \text{ and } y \in [\alpha(h(t)), \beta(h(t))], \quad \alpha''(t) \geq f(t, \alpha(t), y) \text{ and } \beta''(t) \leq f(t, \beta(t), y).$$

If we reinforce the notion of lower and upper solution, we can replace the monotonicity condition by a bound on the shifts $h(t) - t$.

Theorem 5.4. *Let $a < 0$, $b > 1$, $f : [0,1] \times \mathbb{R}^2 \to \mathbb{R}$, $A_0 : [a,0] \to \mathbb{R}$, $B_0 : [1,b] \to \mathbb{R}$ and $h : [0,1] \to [a,b]$ be continuous functions, a_1, a_2, b_1, b_2 be nonnegative numbers such that $a_1 + a_2 > 0$ and $b_1 + b_2 > 0$.*

Assume there exist α and $\beta \geq \alpha$ such that

$$\begin{array}{lll} \alpha''(t) > f(t, \alpha(t), \alpha(h(t))), & \beta''(t) < f(t, \beta(t), \beta(h(t))), & \forall t \in [0,1], \\ a_1\alpha(t) - a_2\alpha'(t) \leq A_0(t), & a_1\beta(t) - a_2\beta'(t) \geq A_0(t), & \forall t \in [a,0], \\ b_1\alpha(t) + b_2\alpha'(t) \leq B_0(t), & b_1\beta(t) + b_2\beta'(t) \geq B_0(t), & \forall t \in [1,b]. \end{array}$$

Then there exists δ such that if $\max_t |h(t) - t| < \delta$, *the problem* (5.6) *has at least one solution u such that for all $t \in [a,b]$, $\alpha(t) \leq u(t) \leq \beta(t)$.*

Exercise 5.2. Prove the above theorem.

Hint : See [171] for the Dirichlet problem and [85] for the general case.

5.3 Fourth Order BVP

Consider the fourth order problem

$$\begin{aligned} u^{(4)} &= f(t,u), \\ u(0) = 0,\ & u(1) = 0, \\ u'(0) = 0,\ & u'(1) = 0. \end{aligned} \tag{5.9}$$

The technique of lower and upper solution can be extended to such boundary value problems provided we can make sure that any solution of the modified problem is between the lower and the upper solution. This last property follows from the maximum principle worked out in Claim 2 of the proof below.

Theorem 5.5. *Let α, $\beta \in C^4([0,1])$ be such that $\alpha \leq \beta$, define $E = \{(t,u) \in [0,1] \times \mathbb{R} \mid \alpha(t) \leq u \leq \beta(t)\}$ and let $f : E \to \mathbb{R}$ be continuous.*

Assume that for all $(t,u) \in E$

$$\begin{array}{ll} \alpha^{(4)}(t) \leq f(t,u), & \beta^{(4)}(t) \geq f(t,u), \\ \alpha(0) \leq 0,\ \alpha(1) \leq 0, & \beta(0) \geq 0,\ \beta(1) \geq 0, \\ \alpha'(0) \leq 0,\ \alpha'(1) \geq 0, & \beta'(0) \geq 0,\ \beta'(1) \leq 0. \end{array}$$

Then the problem (5.9) *has at least one solution u such that $\alpha \leq u \leq \beta$.*

Proof. Consider the modified problem

$$\begin{aligned} u^{(4)} &= f(t, \gamma(t,u)), \\ u(0) &= 0,\ u(1) = 0, \\ u'(0) &= 0,\ u'(1) = 0, \end{aligned} \tag{5.10}$$

where $\gamma(t,u)$ is defined from (1.6).

Claim 1 – The problem (5.10) *has at least one solution.* If $\bar{f} \in C([0,1])$, the linear problem

$$\begin{aligned} u^{(4)} &= \bar{f}(t), \\ u(0) &= 0,\ u(1) = 0, \\ u'(0) &= 0,\ u'(1) = 0, \end{aligned}$$

has a unique solution

$$u(t) = \int_0^1 G(t,s)\bar{f}(s)\,ds,$$

where the Green's function $G(t,s)$, which can be computed explicitly, is continuous and nonnegative. We can then write (5.10) as an integral equation

$$u(t) = \int_0^1 G(t,s) f(s, \gamma(s, u(s)))\,ds$$

and prove existence of solutions from Schauder's Theorem.

Claim 2 – The solution u of (5.10) *is such that $\alpha \leq u \leq \beta$.* The function $w(t) = u(t) - \alpha(t)$ solves the problem

$$\begin{aligned} w^{(4)}(t) &= \bar{f}(t), \\ w(0) &= A \geq 0,\ w(1) = B \geq 0, \\ w'(0) &= C \geq 0,\ w'(1) = -D \leq 0, \end{aligned}$$

where $\bar{f}(t) := u^{(4)}(t) - \alpha^{(4)}(t) = f(t, \gamma(t, u(t))) - \alpha^{(4)}(t) \geq 0$. Its solution reads

$$w(t) = \int_0^1 G(t,s)\bar{f}(s)\,ds \\ +A(1-t)^2(1+2t) + Bt^2(1+2(1-t)) + Ct(1-t)^2 + Dt^2(1-t),$$

which is clearly positive. In a similar way, we prove that $\beta(t) - u(t) \geq 0$.

Conclusion – It follows from Claim 2 that the solution u of (5.10) solves (5.9). □

Remark If f is non-decreasing in u, the differential inequalities in Theorem 5.5 are satisfied provided

$$\beta^{(4)}(t) \geq f(t,\beta(t)) \quad \text{and} \quad \alpha^{(4)}(t) \leq f(t,\alpha(t)).$$

5.4 Bounded Solutions

We consider the second order differential equation

$$u'' = f(t,u,u'), \tag{5.11}$$

where f is a continuous function defined on $\mathbb{R}^3$. The following theorem gives sufficient conditions for the existence of a bounded solution $u \in \mathcal{C}^2(\mathbb{R})$ of (5.11).

Theorem 5.6. *Let α, $\beta \in \mathcal{C}^2(\mathbb{R})$ be such that $\alpha \leq \beta$, $E = \{(t,u,v) \in \mathbb{R}^3 \mid \alpha(t) \leq u \leq \beta(t)\}$ and $f : E \to \mathbb{R}$ be continuous.*

Assume that α and β are such that for all $t \in \mathbb{R}$

$$\alpha''(t) \geq f(t,\alpha(t),\alpha'(t)) \quad \text{and} \quad \beta''(t) \leq f(t,\beta(t),\beta'(t)).$$

Assume also that for any bounded interval I, there exists a positive continuous function $\varphi_I : \mathbb{R}^+ \to \mathbb{R}$ that satisfies

$$\int_0^\infty \frac{s\,ds}{\varphi_I(s)} = \infty$$

and for all $t \in I$, $(u,v) \in \mathbb{R}^2$ with $\alpha(t) \leq u \leq \beta(t)$,

$$|f(t,u,v)| \leq \varphi_I(|v|).$$

Then the equation (5.11) has at least one solution $u \in \mathcal{C}^2(\mathbb{R})$ such that for all $t \in \mathbb{R}$

$$\alpha(t) \leq u(t) \leq \beta(t).$$

It is clear that this theorem provides existence of a bounded solution of (5.11) if α and β are bounded.

Proof. Using Theorem 1.3 the problems

$$u'' = f(t, u, u'),$$
$$u(-n) = \alpha(-n), \; u(n) = \alpha(n),$$

have a solution u_n such that $\alpha \le u_n \le \beta$. From Proposition I-4.4 there exists $R_n > 0$ such that for all $i \ge n$, $\max_{t\in[-n,n]} |u_i'(t)| < R_n$. Hence, we can find $(u_n^1)_n$, which is a subsequence of $(u_n)_n$ that converges in $\mathcal{C}^1([-1, +1])$. Proceeding by induction for any $k \in \mathbb{N}$ we build $(u_n^k)_n$, which is a subsequence of $(u_n^{k-1})_n$ and converges in $\mathcal{C}^1([-k, +k])$. It follows that the diagonal sequence $(u_n^n)_n$ converges pointwise to a function $u : \mathbb{R} \to \mathbb{R}$ such that for any $k \in \mathbb{N}_0$ the convergence takes place in $\mathcal{C}^1([-k, k])$. Hence, u is a solution of (5.11) and $\alpha \le u \le \beta$. □

In the preceding theorem, the derivative u' is uniformly bounded provided the lower and upper solutions are bounded and φ_I is independent of I. It is enough to notice that, in such a case, the bound $R_n > 0$ can be chosen independent of n. The following theorem works out such a result using constant lower and upper solutions. Here the *a priori* bound on the derivative follows from an assumption on the vector field along lines $u' = \pm R$.

Theorem 5.7. *Let $f \in \mathcal{C}(\mathbb{R}^3)$ and assume*
(a) there exist $A \le B$ such that $f(t, A, 0) \le 0 \le f(t, B, 0)$ on $\mathbb{R}$;
(b) there exists $R > 0$ such that $f(t, u, R)f(t, u, -R) \ne 0$ on $\mathbb{R} \times [A, B]$.
Then there exists a solution $u \in \mathcal{C}^2(\mathbb{R})$ of (5.11) such that for all $t \in \mathbb{R}$

$$A \le u(t) \le B, \qquad |u'(t)| \le R.$$

Proof. Consider the modified equation

$$u'' = \bar{f}(t, u, u'), \tag{5.12}$$

where $\bar{f}(t, u, v) = f(t, u, \max(-R, \min(v, R)))$. If I is a bounded interval, the function $\bar{f}(t, u, v)$ is bounded on $D = I \times [A, B] \times \mathbb{R}$.

The constant functions $\alpha(t) = A$, $\beta(t) = B$ and $\varphi_I(v) = \max_D |\bar{f}(t, u, v)|$ satisfy the assumption of Theorem 5.6. Hence there exists a solution $u(t) \in [A, B]$ of (5.12).

Assume $f(t, u, R) > 0$ if $u \in [A, B]$ and suppose that for some t_0, $u'(t_0) > R$. From Lagrange's Theorem, there exists some $t \in [t_0, t_0 + (B - A)/R]$ such that $|u'(t)| \le R$. Let $t_1 = \max\{t \ge t_0 \mid u'(s) > R \text{ on } [t_0, t[\}$. We have then $t_1 \in [t_0, t_0 + (B - A)/R]$, $u'(t_1) = R$ and $u''(t_1) \le 0$. This implies $f(t, u(t_1), u'(t_1)) \le 0$ which contradicts the assumption. We obtain similar contradictions in the other cases. Hence, $|u'(t)| \le R$ and u solves (5.11). □

Example 5.2. Consider the differential equation

$$u'' = g(u) + |u'|^s + p(t), \tag{5.13}$$

where g and p are some continuous functions, p is lower bounded and $s > 0$. If there exist $A \leq B$ such that

$$g(A) + p(t) \leq 0 \leq g(B) + p(t)$$

then the equation (5.13) has a solution such that $A \leq u \leq B$. Notice that for $u \in [A, B]$ and $R > 0$ large enough, $g(u) + R^s + p(t) \geq \min_{[A,B]} g(u) + R^s + p(t) > 0$.

5.5 Ginzburg-Landau Model

Symmetric solutions of the Ginzburg-Landau model for superconductivity in a slab with a parallel external magnetic field are solutions of the boundary value problem

$$\begin{aligned} u'' &= c_1(u^2 + h^2v^2 - 1)u, & v'' &= c_2u^2v, \\ u'(0) &= 0,\ u'(1) = 0, & v(0) &= 0,\ v'(1) = 1. \end{aligned} \tag{5.14}$$

The theory of lower and upper solution can be used for such systems. More generally, we consider in this section the problem

$$\begin{aligned} u'' &= f(u, hv)u, & v'' &= g(u)v, \\ u'(0) &= 0,\ u'(1) = 0, & v(0) &= 0,\ v'(1) = 1. \end{aligned} \tag{5.15}$$

Theorem 5.8. *Assume $f : \mathbb{R} \times \mathbb{R}^+ \to \mathbb{R}$ and $g : \mathbb{R} \to \mathbb{R}^+$ are two continuous functions such that*
(a) for some $a > 0$ and all $v \geq 0$, $f(a, v) \geq 0$;
(b) $f(0, \cdot)$ is increasing;
(c) there exist $h_0 > 0$ and u_0 solution of

$$u'' = f(0, h_0t)u, \quad u'(0) = 0,\ u'(1) = 0,$$

with $u_0(t) > 0$ for any $t \in [0, 1]$.

Then, for every $h < h_0$, the problem (5.15) *has at least one solution (u, v) with $u > 0$ on $[0, 1]$.*

Proof. Part 1 – The operator

$$S : C([0, 1], \mathbb{R}) \to C([0, 1], \mathbb{R}^+),$$

where $S(u)$ is the solution of

$$v'' = g(u(t))v, \quad v(0) = 0,\ v'(1) = 1,$$

is well-defined and continuous. For any $u \in \mathcal{C}([0,1],\mathbb{R})$, we define $v(t;u)$ to be the solution of the Cauchy problem

$$v'' = g(u(t))v, \quad v(0) = 0, \; v'(0) = 1.$$

Notice that $v(\cdot;u) \in \mathcal{C}([0,1],\mathbb{R}^+)$ and $v'(1;u) \geq v'(0;u) = 1$. It follows that S is defined from

$$S(u)(t) = \frac{v(t;u)}{v'(1;u)}$$

and is continuous.

Part 2 – Construction of a lower solution α. Let $h \in [0, h_0[$. By continuous dependence, there exists $h_1 \in [h, h_0[$ such that the solution u_1 of

$$u'' = f(0, h_1 t)u, \quad u(0) = u_0(0), \; u'(0) = 0,$$

satisfies $u_1(t) > 0$ on $[0,1]$. Hence we have

$$\begin{aligned} u_1'(1)u_0(1) &= \int_0^1 \frac{d}{dt}(u_1' u_0 - u_1 u_0')(t)\, dt \\ &= \int_0^1 (f(0,h_1 t) - f(0,h_0 t))u_1(t)u_0(t)\, dt < 0, \end{aligned}$$

which proves $u_1'(1) < 0$.

From continuous dependence, we can find next $\alpha_0(t) > 0$ and $\epsilon > 0$ such that

$$\alpha_0''(t) = f(0, h_1 t + \epsilon)\alpha_0(t), \quad \|\alpha_0\|_\infty = 1, \; \alpha_0'(0) = 0, \; \alpha_0'(1) < 0.$$

Choose $\eta > 0$ such that for any $t \in [0,1]$

$$f(0, h_1 t + \epsilon) - f(0, h_1 t) \geq \eta$$

and $\delta \in]0, a]$ so that for any $|u| < \delta$ and $v \in [0, h_1]$,

$$|f(0,v) - f(u,v)| < \eta.$$

We define then $\alpha(t) = \delta \alpha_0(t)$.

Part 3 – Existence of a solution $u(t) \in [\alpha(t), a]$. Define

$$\gamma(t,u) = \max\{\alpha(t), \min\{u, a\}\}$$

and consider the modified problem

$$\begin{aligned} u'' - u &= [f(\gamma(t,u), hS(\gamma(\cdot,u))) - 1]\gamma(t,u), \\ u'(0) &= 0, \; u'(1) = 0. \end{aligned} \tag{5.16}$$

Using Green's function, we write (5.16) as a fixed point problem and existence of a solution u follows from Schauder's Theorem. If $\max_t u(t) = u(t_0) > a$ we have the contradiction

$$0 \geq u''(t_0) = f(a, hS(\gamma(\cdot, u))(t_0))a + u(t_0) - a > 0.$$

Suppose now that $\min_t(u(t) - \alpha(t)) = u(t_0) - \alpha(t_0) < 0$. This assumption leads to the contradiction

$$\begin{aligned}
0 &\leq u''(t_0) - \alpha''(t_0) \\
&\leq u(t_0) + [f(\alpha(t_0), hS(\gamma(\cdot, u))(t_0)) - 1]\alpha(t_0) - f(0, h_1 t_0 + \epsilon)\alpha(t_0) \\
&\leq u(t_0) - \alpha(t_0) + |f(\alpha(t_0), hS(\gamma(\cdot, u))(t_0)) - f(0, hS(\gamma(\cdot, u))(t_0))|\alpha(t_0) \\
&\qquad - [f(0, h_1 t_0 + \epsilon) - f(0, hS(\gamma(\cdot, u))(t_0))]\alpha(t_0) \\
&\leq u(t_0) - \alpha(t_0) + \eta\alpha(t_0) - [f(0, h_1 t_0 + \epsilon) - f(0, h_1 t_0)]\alpha(t_0) < 0.
\end{aligned}$$

As a conclusion $\gamma(t, u(t)) = u(t)$ and $(u, S(u))$ solves problem (5.15). □

Exercise 5.3. Prove that for c_2 large enough, there exists $0 < h^* < \infty$ such that

(i) for $h \in [0, h^*]$, the problem (5.14) has at least one solution (u, v) with $u > 0$ on $[0, 1]$;

(ii) for $h > h^*$, the problem (5.14) has no solution.

Hint : See [90].

6 PDE Problems

6.1 The Dirichlet Problem

The method of lower and upper solutions is easy to extend to partial differential equations. A natural framework for this method is the L^p-theory of elliptic equations. Consider for example the Dirichlet problem

$$\begin{aligned}
\Delta u + f(x, u) &= 0, \quad &&\text{in } \Omega, \\
u &= 0, &&\text{on } \partial\Omega,
\end{aligned} \tag{6.1}$$

where Ω is a bounded domain in $\mathbb{R}^N$ and $f : \Omega \times \mathbb{R} \to \mathbb{R}$.

To deal with such a problem, we need to recall basic facts on elliptic equations. A first one concerns the invertibility of the Laplacian operator.

Proposition 6.1. *Let Ω be a bounded domain in $\mathbb{R}^N$ with boundary $\partial\Omega$ of class $C^{1,1}$ and let $2 \leq p < \infty$. Then, there exists $\lambda_0 \in \mathbb{R}$ such that for all $\lambda > \lambda_0$ and $f \in L^p(\Omega)$, the problem*

$$\begin{aligned}
-\Delta u + \lambda u &= f, \quad &&\textit{in } \Omega, \\
u &= 0, &&\textit{on } \partial\Omega,
\end{aligned}$$

has a unique solution $u \in W^{2,p}(\Omega)$. *Further there exists a constant* $C > 0$ *such that*

$$\|u\|_{W^{2,p}} \leq C\|f\|_{L^p}.$$

Proof. See [302, Lemmas 3.21 and 3.22]. □

This result means that the operator

$$L : W_0^{2,p}(\Omega) \to L^p(\Omega), \tag{6.2}$$

defined by $Lu = -\Delta u + \lambda u$ has a continuous inverse. Another fundamental tool is the *strong maximum principle.*

Proposition 6.2. *Let* Ω *be a bounded domain in* $\mathbb{R}^N$ *with boundary* $\partial\Omega$ *of class* $C^{1,1}$, $p > N$ *and* $\lambda \geq 0$. *If* $u \in W^{2,p}(\Omega)$ *satisfies*

$$-\Delta u + \lambda u \leq 0, \qquad \text{in } \Omega,$$

then either u *does not have a positive maximum in* Ω *or* u *is constant.*

Proof. See [302, Theorem 3.27]. □

In this section, we consider nonlinearities $f : \Omega \times \mathbb{R}^k \to \mathbb{R}$, with $k = 1$ or $1 + N$. We shall say that f satisfies *Carathéodory conditions* if
(a) for a.e. $x \in \Omega$, the function $f(x, \cdot)$ is continuous,
(b) for all $v \in \mathbb{R}^k$, the function $f(\cdot, v)$ is measurable.
If further we have
(c) for all $r > 0$, there exists $h \in L^p(\Omega)$ such that for a.e. $x \in \Omega$ and all $v \in \mathbb{R}^k$ with $\|v\| \leq r$, $|f(x, v)| \leq h(x)$,
we say that f satisfies L^p-*Carathéodory conditions.*

We can now state the main result of the method which paraphrases Theorem 4.2.

Theorem 6.3. *Let* Ω *be a bounded domain in* $\mathbb{R}^N$ *with boundary* $\partial\Omega$ *of class* $C^{1,1}$, $p > N$ *and let* $f : \Omega \times \mathbb{R} \to \mathbb{R}$ *satisfy* L^p-*Carathéodory conditions. Consider* α *and* $\beta \in W^{2,p}(\Omega)$ *with* $\alpha \leq \beta$ *on* Ω *and suppose*

$$\begin{aligned} &\Delta\alpha(x) + f(x, \alpha(x)) \geq 0 \;\; \text{in } \Omega, \qquad && \alpha \leq 0 \;\; \text{on } \partial\Omega, \\ &\Delta\beta(x) + f(x, \beta(x)) \leq 0 \;\; \text{in } \Omega, \qquad && \beta \geq 0 \;\; \text{on } \partial\Omega. \end{aligned}$$

Then the problem (6.1) *has at least one solution* u *such that* $\alpha \leq u \leq \beta$ *on* Ω.

Proof. The proof is a remake of the proof of Theorem 4.2. We consider the modified problem

$$\begin{aligned} -\Delta u + \lambda u = f(x, \gamma(x,u)) + \lambda\gamma(x,u), &\quad \text{in } \Omega,\\ u(x) = 0, &\quad \text{on } \partial\Omega, \end{aligned} \tag{6.3}$$

where $\gamma(x,u) = \min\{\max\{\alpha(x), u\}, \beta(x)\}$, $\lambda > \lambda_0$ and λ_0 is defined in Proposition 6.1.

Claim 1 – The problem (6.3) *has at least one solution* $u \in W^{2,p}(\Omega)$. Define the operator $T : \mathcal{C}(\overline{\Omega}) \to \mathcal{C}(\overline{\Omega})$ given by

$$Tu = L^{-1}Nu,$$

where L^{-1} is the inverse of the operator $Lu = -\Delta u + \lambda u$, defined by (6.2), and $N : \mathcal{C}(\overline{\Omega}) \to L^p(\Omega)$ is defined from $(Nu)(x) = f(x, \gamma(x, u(x))) + \lambda\gamma(x, u(x))$. As $W^{2,p}(\Omega)$ is compactly embedded in $\mathcal{C}(\overline{\Omega})$, we can prove that T is completely continuous and bounded. Hence by Schauder's Fixed Point Theorem, T has a fixed point u which is a solution of (6.3) in $W^{2,p}(\Omega)$.

Claim 2 – Any solution u *of* (6.3) *satisfies* $\alpha \le u \le \beta$. Observe first that $\alpha(x) \le u(x) \le \beta(x)$ on $\partial\Omega$. Assume $\alpha - u$ has a positive maximum at some point $x_0 \in \Omega$. We can find Ω_0 regular enough and $x_1 \in \Omega_0$ such that

$$x_0 \in \Omega_0, \quad \alpha(x) \ge u(x) \text{ on } \Omega_0,$$

and

$$\alpha(x_1) - u(x_1) < \alpha(x_0) - u(x_0).$$

We compute then that on Ω_0,

$$-\Delta(\alpha - u) + \lambda(\alpha - u) \le f(x,\alpha) - f(x, \gamma(x,u)) + \lambda(\alpha - \gamma(x,u)) = 0$$

and deduce a contradiction from the strong maximum principle (Proposition 6.2). In a similar way, we prove $\beta - u$ does not have a negative minimum.

Conclusion – From Claim 2, the function u, solution of (6.3), solves (6.2). □

6.2 The Gradient Dependent Dirichlet Problem

Consider the problem

$$\begin{aligned} \Delta u + f(x, u, \nabla u) = 0, &\quad \text{in } \Omega,\\ u = 0, &\quad \text{on } \partial\Omega, \end{aligned} \tag{6.4}$$

where the nonlinearity f depends on ∇u. To deal with such a problem, we need *a priori* bounds on derivatives. These bounds can be obtained as in the ODE case using a Bernstein condition though the proof is somewhat more elaborate.

Proposition 6.4. *Let Ω be a bounded domain in $\mathbb{R}^N$ with boundary $\partial\Omega$ of class $C^{1,1}$, $p > N$ and α and $\beta \in W^{2,p}(\Omega)$ with $\alpha \leq \beta$. Assume also $K > 0$ and $h \in L^p(\Omega)$ are given. Then there exists $R > 0$ so that given any function $f : \Omega \times \mathbb{R} \times \mathbb{R}^N \to \mathbb{R}$ that satisfies L^p-Carathéodory conditions and is such that for a.e. $x \in \Omega$, all $u \in [\alpha(x), \beta(x)]$ and all $v \in \mathbb{R}^N$,*

$$|f(x,u,v)| \leq h(x) + K\|v\|^2, \tag{6.5}$$

and given any solution u of

$$\begin{aligned} \Delta u + f(x,u,\nabla u) &= 0, \quad && \text{in } \Omega, \\ u &= 0, && \text{on } \partial\Omega, \end{aligned}$$

satisfying $\alpha(x) \leq u(x) \leq \beta(x)$ on Ω, we have

$$\|\nabla u\|_\infty \leq R.$$

Proof. See [302, Lemma 5.10]. □

Theorem 6.5. *Let Ω be a bounded domain in $\mathbb{R}^N$ with boundary $\partial\Omega$ of class $C^{1,1}$, $p > N$ and let α and $\beta \in W^{2,p}(\Omega)$ with $\alpha \leq \beta$ on Ω. Assume $f : \Omega \times \mathbb{R} \times \mathbb{R}^N \to \mathbb{R}$ satisfies L^p-Carathéodory conditions and (6.5) for some $K > 0$ and $h \in L^p(\Omega)$. At last, suppose*

$$\begin{aligned} \Delta\alpha(x) + f(x,\alpha(x),\nabla\alpha(x)) &\geq 0 \text{ in } \Omega, \quad \alpha \leq 0 \text{ on } \partial\Omega, \\ \Delta\beta(x) + f(x,\beta(x),\nabla\beta(x)) &\leq 0 \text{ in } \Omega, \quad \beta \geq 0 \text{ on } \partial\Omega. \end{aligned}$$

Then the problem (6.4) has at least one solution u such that $\alpha \leq u \leq \beta$ on Ω.

Proof. Let λ_0 be given by Proposition 6.1, $\lambda > \lambda_0$ and $R > 0$ as defined in Proposition 6.4. Increasing R if necessary, we can suppose that $R > \max\{\|\nabla\alpha\|_\infty, \|\nabla\beta\|_\infty\}$. Let $N \in L^p(\Omega)$ be such that $N(x) \geq |f(x,u,v)|$ if $x \in \Omega$, $\alpha(x) \leq u \leq \beta(x)$ and $\|v\| \leq R$. Define then

$$\begin{aligned} \bar{f}(x,u,v) &= \min\{\max\{-N(x), f(x,\gamma(x,u),v)\}, N(x)\}, \\ \omega_1(x,\delta) &= \max_{\|v\|\leq\delta} |\bar{f}(x,\alpha(x),\nabla\alpha(x)+v) - \bar{f}(x,\alpha(x),\nabla\alpha(x))|, \\ \omega_2(x,\delta) &= \max_{\|v\|\leq\delta} |\bar{f}(x,\beta(x),\nabla\beta(x)+v) - \bar{f}(x,\beta(x),\nabla\beta(x))|, \end{aligned}$$

where $\gamma(x,u) = \min\{\max\{\alpha(x),u\},\beta(x)\}$. It is clear that ω_i are L^p-Carathéodory functions, non-decreasing in δ and such that $\omega_i(x,0) = 0$ and $|\omega_i(x,\delta)| \leq 2N(x)$.

Consider then the modified problem

$$\begin{aligned} -\Delta u + \lambda u &= \bar{f}(x,u,\nabla u) + \omega(x,u), && \text{in } \Omega, \\ u(x) &= 0, && \text{on } \partial\Omega, \end{aligned} \tag{6.6}$$

where

$$\begin{aligned} \omega(x,u) &= \lambda\beta(x) - \omega_2(x, u-\beta(x)), && \text{if } u > \beta(x), \\ &= \lambda u, && \text{if } \alpha(x) \leq u \leq \beta(x), \\ &= \lambda\alpha(x) + \omega_1(t, \alpha(x)-u), && \text{if } u < \alpha(x). \end{aligned}$$

Claim 1 – The problem (6.6) *has at least one solution* $u \in W^{2,p}(\Omega)$. This repeats the argument of the corresponding step in the proof of Theorem 6.3.

Claim 2 – Any solution u *of* (6.6) *satisfies* $\alpha \leq u \leq \beta$. Assume that $\alpha - u$ has a positive maximum at some point $x_0 \in \Omega$. In this case $\nabla\alpha(x_0) = \nabla u(x_0)$. Hence we can find Ω_0 regular enough and $x_1 \in \Omega_0$ such that

$$x_0 \in \Omega_0, \quad \|\nabla u(x) - \nabla\alpha(x)\| \leq \alpha(x) - u(x) \text{ on } \Omega_0,$$

and

$$\alpha(x_1) - u(x_1) < \alpha(x_0) - u(x_0).$$

We compute then that on Ω_0,

$$-\Delta(\alpha-u) + \lambda(\alpha-u) \leq \bar{f}(x,\alpha,\nabla\alpha) - \bar{f}(x,u,\nabla u) - \omega_1(x,\alpha(x)-u) \leq 0$$

and deduce a contradiction from Proposition 6.2. In a similar way, we prove that $\beta - u$ does not have a negative minimum.

Conclusion – Let u be a solution of (6.6) given by Claim 1. From Claim 2, $\alpha \leq u \leq \beta$ on Ω and u is a solution of

$$\begin{aligned} -\Delta u &= \bar{f}(x,u,\nabla u), && \text{in } \Omega, \\ u(x) &= 0, && \text{on } \partial\Omega. \end{aligned}$$

Notice now that for a.e. $x \in \Omega$, all $u \in [\alpha(x),\beta(x)]$, and all $v \in \mathbb{R}^N$,

$$|\bar{f}(x,u,v)| \leq h(x) + K\|v\|^2,$$

so that it follows from Proposition 6.4 that $\|\nabla u\|_\infty \leq R$ and u is a solution of (6.4). □

6.3 The General Problem

The method extends to the very general semilinear elliptic problem

$$\begin{aligned} Lu &= f(x,u,\nabla u), && \text{in } \Omega, \\ u(x) &= 0, && \text{on } \partial\Omega \setminus \Gamma, \\ Bu(x) &= \zeta(x), && \text{on } \Gamma, \end{aligned} \tag{6.7}$$

where Ω is a bounded domain in $\mathbb{R}^N$, Γ is an open and closed subset of the boundary $\partial\Omega$, L is a linear second order elliptic operator, B is a linear first order boundary operator, and ζ is a given function.

Theorem 6.6. *Let Ω be a bounded domain in $\mathbb{R}^N$ with boundary $\partial\Omega$ of class $\mathcal{C}^{1,1}$ and Γ an open and closed subset of $\partial\Omega$. Let $p > N$ and $L : W^{2,p}(\Omega) \to L^p(\Omega)$ be a strongly elliptic operator defined by*

$$Lu(x) = -\sum_{i,j=1}^{N} a_{i,j}(x) \frac{\partial^2 u}{\partial x_i \partial x_j}(x),$$

with $a_{i,j} \in \mathcal{C}(\overline{\Omega})$ such that for some $a > 0$, $\sum_{i,j=1}^{N} a_{i,j}(x)\xi_i\xi_j \geq a|\xi|^2$ for all $x \in \overline{\Omega}$. Let also $\zeta \in W^{1-\frac{1}{p},p}(\Gamma)$ and $B : W^{2,p}(\Omega) \to W^{1-\frac{1}{p},p}(\Gamma)$ be defined by

$$Bu(x) = \sum_{i=1}^{N} b_i(x) \frac{\partial u}{\partial x_i}(x) + b(x)u(x),$$

where b_i and $b \in \mathcal{C}^{0,1}(\Gamma)$ are such that, if $\nu(x)$ denotes the unit outward normal at $x \in \Gamma$, for all $x \in \Gamma$, $\sum_{i=1}^{N} b_i(x)\nu_i(x) \geq K > 0$.

Let α and $\beta \in W^{2,p}(\Omega)$ with $\alpha \leq \beta$ on Ω. Assume $f : \Omega \times \mathbb{R} \times \mathbb{R}^N \to \mathbb{R}$ satisfies L^p-Carathéodory conditions and (6.5) for some $K > 0$ and $h \in L^p(\Omega)$. At last, suppose

$$\begin{aligned} \Delta\alpha(x) + f(x,\alpha(x),\nabla\alpha(x)) &\geq 0, && \text{in } \Omega, \\ \alpha &\leq 0, && \text{on } \partial\Omega \setminus \Gamma, \\ B\alpha &\leq \zeta, && \text{on } \Gamma, \end{aligned}$$

$$\begin{aligned} \Delta\beta(x) + f(x,\beta(x),\nabla\beta(x)) &\leq 0, && \text{in } \Omega, \\ \beta &\geq 0, && \text{on } \partial\Omega \setminus \Gamma, \\ B\beta &\geq \zeta, && \text{on } \Gamma. \end{aligned}$$

Then the problem (6.4) has at least one solution u such that $\alpha \leq u \leq \beta$ in Ω.

Proof. See [93]. □

6.4 Weak Lower and Upper Solutions

The method of lower and upper solutions can be worked out for weak solutions. Let Ω be a bounded domain in $\mathbb{R}^N$ and $f : \Omega \times \mathbb{R} \to \mathbb{R}$. We consider the problem

Find $u \in H_0^1(\Omega)$ such that for any $v \in H_0^1(\Omega)$,

$$\int_\Omega [(\nabla u \mid \nabla v) - f(x,u)v]\,dx = 0. \tag{6.8}$$

A standard result using weak lower and upper solutions is as follows.

Theorem 6.7. *Let Ω be a bounded domain in $\mathbb{R}^N$ and let $f : \Omega \times \mathbb{R} \to \mathbb{R}$ satisfy Carathéodory conditions. Consider α and $\beta \in H^1(\Omega)$ with $\alpha \leq \beta$ on Ω, $\alpha^+ \in H_0^1(\Omega)$, $\beta^- \in H_0^1(\Omega)$ and suppose that for any $v \in H_0^1(\Omega)$, $v \geq 0$,*

$$\int_\Omega [(\nabla\alpha(x) \mid \nabla v(x)) - f(x,\alpha(x))v(x)]\,dx \leq 0, \tag{6.9}$$

$$\int_\Omega [(\nabla\beta(x) \mid \nabla v(x)) - f(x,\beta(x))v(x)]\,dx \geq 0. \tag{6.10}$$

Assume also there exists $h \in L^p(\Omega)$ with $p > 2N/(N+2)$ such that for a.e. $x \in \Omega$ and all $\alpha(x) \leq u \leq \beta(x)$, $|f(x,u)| \leq h(x)$.

Then the problem (6.8) has at least one solution u such that $\alpha \leq u \leq \beta$ on Ω.

Proof. We consider the modified problem

$$\int_\Omega [(\nabla u \mid \nabla v) - f(x,\gamma(x,u))v]\,dx = 0, \quad \forall v \in H_0^1(\Omega), \tag{6.11}$$

where $\gamma(x,u) = \min\{\max\{\alpha(x),u\},\beta(x)\}$.

Claim 1 – The problem (6.11) *has at least one solution.* Define $p' = p/(p-1)$ and the operator $T : L^{p'}(\Omega) \to L^{p'}(\Omega)$ where $Tu = w$ is the solution of

$$\int_\Omega [(\nabla w \mid \nabla v) - f(x,\gamma(x,u))v]\,dx = 0, \quad \forall v \in H_0^1(\Omega).$$

From Lax-Milgram theorem, such a solution exists, is unique, and $\|w\|_{H_0^1} \leq C\|h\|_{L^p}$. It follows that T is a bounded compact operator and the claim follows from Schauder's Theorem.

Claim 2 – Any solution u of (6.11) *satisfies* $\alpha \le u \le \beta$. Observe that, for all $v \in H_0^1(\Omega)$, $v \ge 0$,

$$\int_\Omega [(\nabla\alpha - \nabla u \mid \nabla v) - (f(x,\alpha) - f(x,\gamma(x,u)))v(x)]\,dx \le 0.$$

For $v = (\alpha - u)^+ \in H_0^1(\Omega)$, we obtain

$$\int_\Omega |\nabla(\alpha - u)^+|^2 \le 0$$

so that $(\alpha - u)^+ = 0$. In a similar way, we prove $\beta \ge u$.

Conclusion – From Claim 2, the function u, solution of (6.11), solves (6.8).
□

A function α (resp. β) that satisfies (6.9) (resp. (6.10)) is called a *weak lower solution* (resp. a *weak upper solution*) of (6.8). Such functions can also be used in case the nonlinearity is gradient dependent. For generalizations in the framework of Theorem 6.5, we refer to [38, 39].

Chapter III

Relation with Degree Theory

1 The Periodic Problem

It has been seen that the periodic problem

$$\begin{aligned} u'' &= f(t,u), \\ u(a) = u(b),\ & u'(a) = u'(b) \end{aligned} \tag{1.1}$$

is equivalent to the fixed point problem

$$u = Tu, \tag{1.2}$$

where $T : \mathcal{C}([a,b]) \to \mathcal{C}([a,b])$ is defined from

$$(Tu)(t) := \int_a^b G(t,s)[f(s,u(s)) - u(s)]\,ds \tag{1.3}$$

and $G(t,s)$ is the Green's function corresponding to (I-1.4). In this chapter, we associate the set

$$\Omega = \{u \in \mathcal{C}([a,b]) \mid \forall t \in [a,b],\ \alpha(t) < u(t) < \beta(t)\} \tag{1.4}$$

to the pair of lower and upper solutions (α, β) and wish to consider the degree

$$\deg(I - T, \Omega).$$

To this end we have to reinforce the concepts of lower and upper solutions so that the boundary of Ω does not contain solutions of (1.1). This modification amounts to requiring that the solutions cannot be tangent to the curves

$$u = \alpha(t) \quad \text{and} \quad u = \beta(t)$$

respectively from above or from below and motivates the definitions we introduce in the following section.

For problems with a nonlinearity f which depends on the derivative

$$\begin{aligned} u'' &= f(t,u,u'), \\ u(a) = u(b), &\ u'(a) = u'(b), \end{aligned} \tag{1.5}$$

the corresponding fixed point problem reads

$$u(t) = (Tu)(t) := \int_a^b G(t,s)[f(s,u(s),u'(s)) - u(s)]\, ds,$$

where $T : C^1([a,b]) \to C^1([a,b])$. Given lower and upper solutions α and β, and some $R > 0$, we shall compute the degree

$$\deg(I - T, \Omega),$$

where

$$\Omega = \{u \in C^1([a,b]) \mid \forall t \in [a,b],\ \alpha(t) < u(t) < \beta(t),\ |u'(t)| < R\}.$$

As above, this requirement will force us to reinforce the concepts of lower and upper solutions and furthermore to choose R to be an *a priori* bound on the derivative u'.

1.1 The Strict Lower and Upper Solutions

Definitions 1.1. *A C^2 or $W^{2,1}$-lower solution α of* (1.5) *(resp.* (1.1)*) is said to be* strict *if every solution u of* (1.5) *(resp.* (1.1)*) with $u \geq \alpha$ is such that $u(t) > \alpha(t)$ on $[a,b]$.*

Similarly, a C^2 or $W^{2,1}$-upper solution β of (1.5) *(resp.* (1.1)*) is said to be* strict *if every solution u of* (1.5) *(resp.* (1.1)*) with $u \leq \beta$ is such that $u(t) < \beta(t)$ on $[a,b]$.*

The classical way to obtain such a notion in the case of a continuous f and α or $\beta \in C^2([a,b])$ is described in the following proposition.

Proposition 1.1. *Let $f : [a,b] \times \mathbb{R}^2 \to \mathbb{R}$ be continuous and $\alpha \in C^2([a,b])$ be such that*
(a) for all $t \in [a,b]$, $\alpha''(t) > f(t,\alpha(t),\alpha'(t))$;
(b) $\alpha(a) = \alpha(b)$, $\alpha'(a) \geq \alpha'(b)$.

Then α is a strict C^2-lower solution of (1.5).

Proof. Let u be a solution of (1.5) such that $u \geq \alpha$ and assume by contradiction that

$$\min_{t\in[a,b]} (u(t) - \alpha(t)) = u(t_0) - \alpha(t_0) = 0.$$

We have $u'(t_0) - \alpha'(t_0) = 0$ which follows from assumption (b) in case $t_0 = a$ or b. Hence, we obtain the contradiction

$$0 \leq u''(t_0) - \alpha''(t_0) = f(t_0, \alpha(t_0), \alpha'(t_0)) - \alpha''(t_0) < 0. \qquad \square$$

Using the same argument we obtain the corresponding result for upper solutions.

Proposition 1.2. *Let $f : [a,b] \times \mathbb{R}^2 \to \mathbb{R}$ be continuous and $\beta \in C^2([a,b])$ be such that*
(a) for all $t \in [a,b]$, $\beta''(t) < f(t, \beta(t), \beta'(t))$;
(b) $\beta(a) = \beta(b)$, $\beta'(a) \leq \beta'(b)$.
Then β is a strict C^2-upper solution of (1.5).

If f is not continuous but L^p-Carathéodory, these last results no longer hold. In fact, even the stronger condition

$$\text{for a.e. } t \in [a,b], \quad \alpha''(t) \geq f(t, \alpha(t), \alpha'(t)) + 1 \tag{1.6}$$

does not prevent solutions u of (1.5) to be tangent to the curve $u = \alpha(t)$ from above. Such a situation occurs, for example, for the bounded function

$$\begin{aligned} f(t,u) &:= -1, & &\text{if } u \leq -1, \\ &:= -\frac{u^2 + \sin t}{1 + \sin t}, & &\text{if } -1 < u \leq \sin t, \\ &:= -\sin t, & &\text{if } \sin t < u, \end{aligned}$$

if we consider $\alpha(t) \equiv -1$, $u(t) \equiv \sin t$, $a = 0$ and $b = 2\pi$. This remark motivates the following proposition.

Proposition 1.3. *Let $f : [a,b] \times \mathbb{R}^2 \to \mathbb{R}$ be an L^1-Carathéodory function. Let $\alpha \in C([a,b])$ be such that $\alpha(a) = \alpha(b)$ and consider its periodic extension on $\mathbb{R}$ defined by $\alpha(t) = \alpha(t + b - a)$. Assume that α is not a solution of* (1.5) *and for any $t_0 \in \mathbb{R}$, either*
(a) $D_-\alpha(t_0) < D^+\alpha(t_0)$ or
(b) there exist an open interval I_0 and $\epsilon_0 > 0$ such that
$t_0 \in I_0$, $\alpha \in W^{2,1}(I_0)$ and
for a.e. $t \in I_0$, all $u \in [\alpha(t), \alpha(t)+\epsilon_0]$ and all $v \in [\alpha'(t)-\epsilon_0, \alpha'(t)+\epsilon_0]$,
$$\alpha''(t) \geq f(t,u,v).$$
Then α is a strict $W^{2,1}$-lower solution of (1.5).

Proof. The function α is a $W^{2,1}$-lower solution since clearly it satisfies Definition I-6.1. Let u be a solution of (1.5) such that $u \geq \alpha$. As α is not a solution, there exists t^* such that $u(t^*) > \alpha(t^*)$. Extend u and α by periodicity and assume by contradiction that

$$t_0 = \inf\{t > t^* \mid u(t) = \alpha(t)\}$$

exists. As $\alpha - u$ is maxima at t_0, we have $D_-\alpha(t_0) - u'(t_0) \geq D^+\alpha(t_0) - u'(t_0)$. Therefore, assumption (b) applies. We have that $u'(t_0) - \alpha'(t_0) = 0$ and there exist I_0 and $\epsilon_0 > 0$ according to (b). It follows that we can choose $t_1 \in I_0$ with $t_1 < t_0$ such that $u'(t_1) - \alpha'(t_1) < 0$ and for every $t \in]t_1, t_0[$

$$u(t) \leq \alpha(t) + \epsilon_0, \quad |u'(t) - \alpha'(t)| < \epsilon_0.$$

Hence, for almost every $t \in]t_1, t_0[$, we can write

$$\alpha''(t) \geq f(t, u(t), u'(t)),$$

which leads to the contradiction

$$0 < (u' - \alpha')(t_0) - (u' - \alpha')(t_1) = \int_{t_1}^{t_0} [f(t, u(t), u'(t)) - \alpha''(t)]\, dt \leq 0. \quad \square$$

In the same way we can prove the following result on strict upper solutions.

Proposition 1.4. *Let $f : [a, b] \times \mathbb{R}^2 \to \mathbb{R}$ be an L^1-Carathéodory function. Let $\beta \in C([a, b])$ be such that $\beta(a) = \beta(b)$ and consider its periodic extension on $\mathbb{R}$ defined by $\beta(t) = \beta(t + b - a)$. Assume that β is not a solution of* (1.5) *and for any $t_0 \in \mathbb{R}$, either*

(a) $D^-\beta(t_0) > D_+\beta(t_0)$ or

(b) there exist an open interval I_0 and $\epsilon_0 > 0$ such that $t_0 \in I_0$, $\beta \in W^{2,1}(I_0)$ and for a.e. $t \in I_0$, all $u \in [\beta(t) - \epsilon_0, \beta(t)]$ and all $v \in [\beta'(t) - \epsilon_0, \beta'(t) + \epsilon_0]$,

$$\beta''(t) \leq f(t, u, v).$$

Then β is a strict $W^{2,1}$-upper solution of (1.5).

Observe that in the continuous case, if α satisfies the conditions of Proposition 1.1, then it satisfies the conditions of Proposition 1.3. The converse, however, does not hold since Proposition 1.3 does not require strict inequalities.

To be able to say that a function α which satisfies (1.6) is a strict lower solution, we need some more regularity on f. In particular we have the following result.

Proposition 1.5. *Let $f : [a,b] \times \mathbb{R}^2 \to \mathbb{R}$ be an L^1-Carathéodory function that satisfies the assumption*
(A) for all $t_0 \in [a,b]$, $(u_0, v_0) \in \mathbb{R}^2$ and $\epsilon > 0$, there exists $\delta > 0$ such that

$$|t-t_0| < \delta,\ |u-u_0| < \delta,\ |v-v_0| < \delta \ \Rightarrow\ |f(t,u,v)-f(t,u_0,v_0)| < \epsilon.$$

Let $A > 0$ and $\alpha \in W^{2,1}(a,b)$ be such that

$$\begin{aligned} \alpha''(t) &\geq f(t,\alpha(t),\alpha'(t)) + A, \\ \alpha(a) &= \alpha(b),\ \alpha'(a) \geq \alpha'(b). \end{aligned}$$

Then α is a strict $W^{2,1}$-lower solution of (1.5).

Remark Notice that in this proposition, Condition (A) does not imply that f is continuous. For example, we can take $f(t,u,v) = g(u,v) + h(t)$ with g continuous and $h \in L^1(a,b)$ or $f(t,u,v) = g(u,v)h(t)$ with g continuous and $h \in L^\infty(a,b)$.

Proof. Let us deduce this proposition from Proposition 1.3. To this end, we will prove that for any $t_0 \in \mathbb{R}$, there exist an open interval I_0 and $\epsilon_0 > 0$ such that $t_0 \in I_0$ and for a.e. $t \in I_0$, for all u, v with $\alpha(t) \leq u \leq \alpha(t) + \epsilon_0$, $\alpha'(t) - \epsilon_0 \leq v \leq \alpha'(t) + \epsilon_0$, we have

$$\alpha''(t) \geq f(t,u,v).$$

Let $t_0 \in \mathbb{R}$ be fixed. Define $u_0 = \alpha(t_0)$, $v_0 = \alpha'(t_0)$ and $\epsilon = A/2$. We deduce then from Assumption (A) a $\delta > 0$ such that

$$\begin{aligned} &|t-t_0| < \delta,\ |u-\alpha(t_0)| < \delta,\ |v-\alpha'(t_0)| < \delta \\ &\qquad\qquad \Rightarrow\ |f(t,u,v) - f(t,\alpha(t_0),\alpha'(t_0))| < A/2. \end{aligned}$$

Let now $\eta < \delta$ be such that for any $t \in [t_0 - \eta, t_0 + \eta]$,

$$|\alpha(t) - \alpha(t_0)| < \delta/2, \quad |\alpha'(t) - \alpha'(t_0)| < \delta/2.$$

Choose $I_0 =]t_0 - \eta, t_0 + \eta[$ and $\epsilon_0 = \delta/2$. The result follows now since

$$\begin{aligned} \alpha''(t) - f(t,u,v) &= \alpha''(t) - f(t,\alpha(t),\alpha'(t)) \\ &\quad + f(t,\alpha(t),\alpha'(t)) - f(t,\alpha(t_0),\alpha'(t_0)) \\ &\quad + f(t,\alpha(t_0),\alpha'(t_0)) - f(t,u,v) \\ &\geq A - A/2 - A/2 = 0. \end{aligned}$$

□

In this last proposition, we can weaken the assumption $\alpha \in W^{2,1}(a,b)$ and allow angles as previously. Also, we can make Condition (A) one-sided and impose

(A') for all $t_0 \in [a,b]$, $(u_0, v_0) \in \mathbb{R}^2$, and $\epsilon > 0$, there exists $\delta > 0$ such that

$$\begin{aligned} &|t-t_0| < \delta,\ |u_1 - u_0| < \delta,\ u_2 \in [u_1, u_1 + \delta[, \\ &\qquad |v_1 - v_0| < \delta,\ |v_2 - v_0| < \delta \ \Rightarrow\ f(t,u_2,v_2) - f(t,u_1,v_1) < \epsilon. \end{aligned}$$

For strict upper solutions we have a similar result.

Proposition 1.6. *Let $f : [a,b] \times \mathbb{R}^2 \to \mathbb{R}$ be an L^1-Carathéodory function that satisfies Assumption (A) in Proposition 1.5. Let $B > 0$ and $\beta \in W^{2,1}(a,b)$ be such that*

$$\begin{aligned} \beta''(t) &\leq f(t, \beta(t), \beta'(t)) - B, \\ \beta(a) &= \beta(b), \;\; \beta'(a) \leq \beta'(b). \end{aligned}$$

Then β is a strict $W^{2,1}$-upper solution of (1.5).

Exercise 1.1. Generalize Propositions 1.5 and 1.6 to allow corners in α and β as in Propositions 1.3 and 1.4.

Another simple and interesting situation concerns the case where f satisfies a Lipschitz condition in v and a one-sided Lipschitz condition in u. Such assumptions are classical in studying monotone iterative methods.

Proposition 1.7. *Let $f : [a,b] \times \mathbb{R}^2 \to \mathbb{R}$ be an L^1-Carathéodory function such that for some k, $l \in L^1(a,b;\mathbb{R}^+)$,*

(a) for a.e. $t \in [a,b]$, all u_1, $u_2 \in \mathbb{R}$ with $u_1 \leq u_2$ and $v \in \mathbb{R}$,

$$f(t, u_2, v) - f(t, u_1, v) \leq k(t)\,(u_2 - u_1);$$

(b) for a.e. $t \in [a,b]$ and all u, v_1, $v_2 \in \mathbb{R}$,

$$|f(t, u, v_2) - f(t, u, v_1)| \leq l(t)|v_2 - v_1|.$$

Then every $W^{2,1}$-lower solution α (resp. every $W^{2,1}$-upper solution β) of (1.5) *which is not a solution is a strict $W^{2,1}$-lower solution (resp. a strict $W^{2,1}$-upper solution).*

Proof. Let u be a solution of (1.5) such that $u \geq \alpha$. As in the proof of Proposition 1.3, we find an open interval I_0, $t_0 \in I_0$ and $t_1 \in I_0$ with $t_1 < t_0$ such that $u(t_0) = \alpha(t_0)$, $u'(t_0) = \alpha'(t_0)$, $u'(t_1) - \alpha'(t_1) < 0$ and for a.e. $t \in I_0$

$$\alpha''(t) \geq f(t, \alpha(t), \alpha'(t)).$$

Let $w = u - \alpha$ and observe that, on $[t_1, t_0[$,

$$\begin{aligned} -w'' &+ l(t)\mathrm{sgn}(w'(t))\, w' + k(t)w \\ &\geq -f(t, u(t), u'(t)) + f(t, \alpha(t), \alpha'(t)) \\ &\qquad\qquad + l(t)|u'(t) - \alpha'(t)| + k(t)(u(t) - \alpha(t)) \\ &\geq 0. \end{aligned}$$

Defining

$$P(t) = -\int_{t_1}^{t} l(s)\operatorname{sgn}(w'(s))\,ds,$$

$$Q(t) = -\int_{t_1}^{t} e^{P(s)}k(s)\,ds, \qquad R(t) = -\int_{t_1}^{t} Q(s)e^{-P(s)}\,ds,$$

we compute

$$\begin{aligned}\frac{d}{dt}&[(w'(t)\;e^{P(t)} + Q(t)w(t))e^{R(t)}]\\ &= e^{R(t)}[e^{P(t)}(w''(t) - l(t)|w'(t)| - k(t)w(t)) - e^{-P(t)}Q^2(t)w(t)]\\ &\leq 0.\end{aligned}$$

Hence, we have the contradiction

$$0 \geq (w'(t)e^{P(t)} + Q(t)w(t))e^{R(t)}\Big|_{t_1}^{t_0} = -w'(t_1) > 0. \qquad \square$$

Exercise 1.2. Assume f satisfies the assumptions of Proposition 1.7 and α is a $W^{2,1}$-lower solution (resp. β is a $W^{2,1}$-upper solution) of (1.5) which is not a solution. Prove that any $W^{2,1}$-upper solution $\beta \geq \alpha$ (resp. $W^{2,1}$-lower solution $\alpha \leq \beta$) is such that $\beta > \alpha$.

Remark 1.1. In Proposition 1.7, we can assume (a) and (b) to hold only in a neighbourhood of $\{(t, \alpha(t), \alpha'(t)) \mid t \in [a,b] \text{ such that } \alpha'(t) \text{ exists}\}$ (resp. $\{(t, \beta(t), \beta'(t)) \mid t \in [a,b] \text{ such that } \beta'(t) \text{ exists}\}$). However this proposition does not hold without the Lipschitz conditions. Even a one-sided Hölder condition is not enough as shown by the following example

$$\begin{aligned} u'' &= 12|u|^{1/2},\\ u(-1) = u(1),&\;\; u'(-1) = u'(1). \end{aligned} \tag{1.7}$$

Here $\beta(t) = t^4$ is an upper solution which is not a solution of (1.7). On the other hand, $u(t) = 0$ is a solution such that $u(0) = \beta(0)$. Hence, β is not strict.

1.2 Existence and Multiplicity Results

Now we can prove the key result of this section.

Theorem 1.8. *Let α and $\beta \in C([a,b])$ be strict $W^{2,1}$-lower and upper solutions of problem (1.5) such that on $[a,b]$, $\alpha(t) < \beta(t)$. Define $A \subset [a,b]$ (resp. $B \subset [a,b]$) to be the set of points where α (resp. β) is derivable.*

Let E be defined by

$$E := \{(t,u,v) \in [a,b] \times \mathbb{R}^2 \mid \alpha(t) \leq u \leq \beta(t)\}, \tag{1.8}$$

and p, $q \in [1,\infty]$ *be such that* $\frac{1}{p}+\frac{1}{q}=1$. *Assume* $f: E \to \mathbb{R}$ *satisfies* L^p*-Carathéodory conditions and there exists* $N \in L^1(a,b)$, $N>0$ *such that for a.e.* $t \in A$ *(resp. for a.e.* $t \in B$*)*

$$f(t,\alpha(t),\alpha'(t)) \geq -N(t) \quad (\textit{resp. } f(t,\beta(t),\beta'(t)) \leq N(t)).$$

Assume moreover there exist $\varphi \in \mathcal{C}(\mathbb{R}^+,\mathbb{R}_0^+)$, $\psi \in L^p(a,b)$ *and* $R>0$ *such that*

$$\int_0^R \frac{s^{1/q}}{\varphi(s)}\,ds > \|\psi\|_{L^p}(\max_t \beta(t) - \min_t \alpha(t))^{1/q} \tag{1.9}$$

and that the function f *satisfies one of the one-sided Nagumo conditions (a), (b), (c), or (d) in Theorem I-6.1.*

Then

$$\deg(I-T,\Omega)=1, \tag{1.10}$$

where $T:\mathcal{C}^1([a,b]) \to \mathcal{C}^1([a,b])$ *is defined by*

$$(Tu)(t) := \int_a^b G(t,s)[f(s,u(s),u'(s)) - u(s)]\,ds, \tag{1.11}$$

$G(t,s)$ *is the Green's function corresponding to* (I-1.4) *and* Ω *is given by*

$$\Omega = \{u \in \mathcal{C}^1([a,b]) \mid \forall t \in [a,b],\ \alpha(t)<u(t)<\beta(t),\ |u'(t)|<R\}. \tag{1.12}$$

In particular, the problem (1.5) *has at least one solution* $u \in W^{2,p}(a,b)$ *such that for all* $t \in [a,b]$

$$\alpha(t)<u(t)<\beta(t).$$

Proof. Increasing N if necessary, we can assume $N(t) \geq |f(t,u,v)|$ if $t \in [a,b]$, $\alpha(t) \leq u \leq \beta(t)$ and $|v| \leq R$. Define then

$$\begin{aligned}
\bar f(t,u,v) &= \max\{\min\{f(t,\gamma(t,u),v),N(t)\},-N(t)\},\\
\omega_1(t,\delta) &= \chi_A(t)\max_{|v|\leq\delta} |\bar f(t,\alpha(t),\alpha'(t)+v) - \bar f(t,\alpha(t),\alpha'(t))|,\\
\omega_2(t,\delta) &= \chi_B(t)\max_{|v|\leq\delta} |\bar f(t,\beta(t),\beta'(t)+v) - \bar f(t,\beta(t),\beta'(t))|,
\end{aligned}$$

where γ is defined from (I-1.3), χ_A and χ_B are the characteristic functions of the sets A and B. It is clear that ω_i are L^1-Carathéodory functions, non-decreasing in δ such that $\omega_i(t,0)=0$ and $|\omega_i(t,\delta)| \leq 2N(t)$.

We consider now the modified problem

$$\begin{aligned}
u''-u &= \bar f(t,u,u') - \omega(t,u),\\
u(a) &= u(b),\ u'(a)=u'(b),
\end{aligned} \tag{1.13}$$

where

$$\begin{aligned}
\omega(t,u) &= \beta(t) - \omega_2(t,u-\beta(t)), && \text{if } u>\beta(t),\\
&= u, && \text{if } \alpha(t) \leq u \leq \beta(t),\\
&= \alpha(t) + \omega_1(t,\alpha(t)-u), && \text{if } u<\alpha(t).
\end{aligned}$$

This problem is equivalent to the fixed point problem

$$u = \bar{T}u,$$

where $\bar{T} : \mathcal{C}^1([a,b]) \to \mathcal{C}^1([a,b])$ is defined by

$$(\bar{T}u)(t) = \int_a^b G(t,s)[\bar{f}(s,u(s),u'(s)) - \omega(s,u(s))]ds.$$

Observe that $\bar{T}$ is completely continuous and there exists $\bar{R}$ large enough so that $\Omega \subset B(0,\bar{R})$ and $\bar{T}(\mathcal{C}^1([a,b])) \subset B(0,\bar{R})$. Hence we have, by the properties of the degree,

$$\deg(I - \bar{T}, B(0,\bar{R})) = 1.$$

We know that every fixed point u of $\bar{T}$ is a solution of (1.13). Arguing as in the proof of Theorem I-6.6, we see that $\alpha \le u \le \beta$ and $\|u'\|_\infty < R$. As α and β are strict, $\alpha < u < \beta$. Hence, every fixed point of $\bar{T}$ is in Ω and by the excision property we obtain

$$\deg(I - T, \Omega) = \deg(I - \bar{T}, \Omega) = \deg(I - \bar{T}, B(0,\bar{R})) = 1.$$

Existence of a solution u such that for all $t \in [a,b]$,

$$\alpha(t) < u(t) < \beta(t)$$

follows now from the properties of the degree. □

A generalization concerns the use of several lower and upper solutions.

Theorem 1.9. *Let $\alpha_i \in \mathcal{C}([a,b])$ $(i = 1,\dots,n)$ and $\beta_j \in \mathcal{C}([a,b])$ $(j = 1,\dots,m)$ be respectively $W^{2,1}$-lower and upper solutions of* (1.5). *Assume*

$$\alpha := \max_{1\le i\le n} \alpha_i \quad \text{and} \quad \beta := \min_{1\le j\le m} \beta_j$$

satisfy $\alpha(t) < \beta(t)$ on $[a,b]$ and are strict, i.e. any solution u of (1.5) *with $\alpha \le u \le \beta$ is such that $\alpha < u < \beta$. Define $A_i \subset [a,b]$ (resp. $B_j \subset [a,b]$) to be the set of points where α_i (resp. β_j) is derivable.*

Let $E = \{(t,u,v) \in [a,b] \times \mathbb{R}^2 \mid \min_i \alpha_i(t) \le u \le \max_j \beta_j(t)\}$ and p, $q \in [1,\infty]$ be such that $\frac{1}{p} + \frac{1}{q} = 1$. Assume $f : E \to \mathbb{R}$ satisfies L^p-Carathéodory conditions and there exists $N \in L^1(a,b)$, $N > 0$ such that for all i and a.e. $t \in A_i$ (resp. for all j and a.e. $t \in B_j$)

$$f(t,\alpha_i(t),\alpha_i'(t)) \ge -N(t) \quad (\textit{resp. } f(t,\beta_j(t),\beta_j'(t)) \le N(t)).$$

Assume moreover there exist $\varphi \in \mathcal{C}(\mathbb{R}^+,\mathbb{R}_0^+)$, $\psi \in L^p(a,b)$ and $R > 0$ that satisfy (1.9) *and that the function f satisfies one of the one-sided Nagumo conditions (a), (b), (c), or (d) in Theorem I-6.1.*

Then

$$\deg(I - T, \Omega) = 1,$$

where $T : \mathcal{C}^1([a,b]) \to \mathcal{C}^1([a,b])$ *is defined by* (1.11) *and* Ω *is given by* (1.12).

In particular, the problem (1.5) *has at least one solution* $u \in W^{2,p}(a,b)$ *such that for all* $t \in [a,b]$

$$\alpha(t) < u(t) < \beta(t).$$

Exercise 1.3. Prove Theorem 1.9 paraphrasing the argument of the previous proof and using the idea of Theorem I-6.10.

In case f is independent of u', this result reduces to

Theorem 1.10. *Let* $\alpha_i \in \mathcal{C}([a,b])$ $(i = 1, \ldots, n)$ *and* $\beta_j \in \mathcal{C}([a,b])$ $(j = 1, \ldots, m)$ *be respectively* $W^{2,1}$*-lower and upper solutions of* (1.1). *Assume*

$$\alpha := \max_{1 \le i \le n} \alpha_i \quad \text{and} \quad \beta := \min_{1 \le j \le m} \beta_j$$

satisfy $\alpha(t) < \beta(t)$ *on* $[a,b]$ *and are strict (see Theorem 1.9).*

Let $E := \{(t,u) \in [a,b] \times \mathbb{R} \mid \min_i \alpha_i(t) \le u \le \max_j \beta_j(t)\}$ *and assume* $f : E \to \mathbb{R}$ *satisfies* L^1*-Carathéodory conditions.*

Then

$$\deg(I - T, \Omega) = 1,$$

where $T : \mathcal{C}([a,b]) \to \mathcal{C}([a,b])$ *is defined by* (1.3) *and* Ω *is given by* (1.4).

In particular, the problem (1.1) *has at least one solution* $u \in W^{2,1}(a,b)$ *such that for all* $t \in [a,b]$

$$\alpha(t) < u(t) < \beta(t).$$

Observe that we can replace the Nagumo conditions by any condition so that an *a priori* bound in the space $\mathcal{C}([a,b])$ on solutions u of the corresponding modified problem implies an *a priori* bound on $\|u'\|_\infty$ which is the case for the Rayleigh and the Liénard equation.

Consider first the Rayleigh equation

$$\begin{aligned} &u'' + g(u') + h(t,u,u') = 0, \\ &u(a) = u(b), \ u'(a) = u'(b). \end{aligned} \tag{1.14}$$

Theorem 1.11. *Let* α *and* $\beta \in \mathcal{C}([a,b])$ *be strict* $W^{2,1}$*-lower and upper solutions of* (1.14) *such that on* $[a,b]$, $\alpha(t) < \beta(t)$. *Let* E *be defined by* (1.8), $g \in \mathcal{C}(\mathbb{R})$ *and* $h : E \to \mathbb{R}$ *be a Carathéodory function such that for some* $H \in L^2(a,b)$, *for a.e.* $t \in [a,b]$, *and for all* $(u,v) \in \mathbb{R}^2$ *with* $(t,u,v) \in E$,

$$|h(t,u,v)| \le H(t).$$

Then

$$\deg(I - T, \Omega) = 1,$$

where Ω *is defined by* (1.12) *(with* $R > 0$ *large enough),* $T : \mathcal{C}^1([a,b]) \to \mathcal{C}^1([a,b])$ *is defined by*

$$(Tu)(t) := -\int_a^b G(t,s)[g(u'(s)) + h(s,u(s),u'(s)) + C(s)u(s)]\, ds,$$

$G(t,s)$ *is the Green's function of*

$$\begin{gathered} u'' - C(t)u = f(t), \\ u(a) = u(b),\ u'(a) = u'(b), \end{gathered} \tag{1.15}$$

and $C \in L^1(a,b)$ *is chosen such that* $C(t) > |g(0)| + 1 + 3H(t)$ *on* $[a,b]$.

In particular, the problem (1.14) *has at least one solution* $u \in W^{2,2}(a,b)$ *such that for all* $t \in [a,b]$

$$\alpha(t) < u(t) < \beta(t).$$

Proof. The proof repeats the arguments of Theorem I-6.8. Using the notations therein, observing that α and β are strict, and denoting

$$\Omega_1 = \{u \in \mathcal{C}^1([a,b]) \mid \|u\|_\infty < \rho,\ \|u'\|_\infty < R\},$$

we have

$$\begin{aligned} \deg(I - T, \Omega) &= \deg(I - T_1, \Omega) = \deg(I - T_1, \Omega_1) \\ &= \deg(I - T_0, \Omega_1) = \deg(I - T_0, B(0, R_0)) = 1. \end{aligned}$$

This concludes the proof. □

The next result concerns the Liénard equation

$$\begin{gathered} u'' + g(u)u' + h(t,u) = 0, \\ u(a) = u(b),\ u'(a) = u'(b). \end{gathered} \tag{1.16}$$

Theorem 1.12. *Let* α *and* $\beta \in \mathcal{C}([a,b])$ *be strict* $W^{2,1}$*-lower and upper solutions of* (1.16) *such that on* $[a,b]$, $\alpha(t) < \beta(t)$. *Let* $E = \{(t,u) \in [a,b] \times \mathbb{R} \mid \alpha(t) \le u \le \beta(t)\}$, $g \in \mathcal{C}(\mathbb{R})$ *and* $h : E \to \mathbb{R}$ *be an* L^1*-Carathéodory function.*

Then

$$\deg(I - T, \Omega) = 1,$$

where Ω *is defined by* (1.12) *(with* $R > 0$ *large enough),* $T : \mathcal{C}^1([a,b]) \to \mathcal{C}^1([a,b])$ *is defined by*

$$(Tu)(t) := -\int_a^b G(t,s)[g(u(s))u'(s) + f(s,u(s)) + C(s)u(s)]\, ds,$$

$G(t,s)$ is the Green's function of (1.15) *and $C \in L^1(a,b)$ is chosen such that for a.e. $t \in [a,b]$, $|g(\alpha(t))| < C(t)$, $|g(\beta(t))| < C(t)$ and for every $(t,u) \in E$, $|f(t,u)| < C(t)$.*

In particular, the problem (1.16) *has at least one solution $u \in W^{2,1}(a,b)$ such that for all $t \in [a,b]$*

$$\alpha(t) < u(t) < \beta(t).$$

Proof. The proof repeats the arguments of Theorem 1.11 together with Theorem I-6.9. □

A first multiplicity result that we can deduce from Theorem 1.8 is obtained when we have two pairs of lower and upper solutions. In this case, we can prove existence of a third solution.

Theorem 1.13. (The Three Solutions Theorem) *Let α_1, β_1 and α_2, $\beta_2 \in C([a,b])$ be two pairs of $W^{2,1}$-lower and upper solutions of* (1.5) *such that on $[a,b]$*

$$\alpha_1(t) \le \beta_1(t),\ \alpha_1(t) \le \beta_2(t),\ \alpha_2(t) \le \beta_2(t)$$

and there exists $t_0 \in [a,b]$ with

$$\alpha_2(t_0) > \beta_1(t_0).$$

Assume further β_1 and α_2 are strict $W^{2,1}$-upper and lower solutions. Define $A_i \subset [a,b]$ (resp. $B_i \subset [a,b]$) to be the set of points where α_i (resp. β_i) is derivable.

Let E be defined by

$$E := \{(t,u,v) \in [a,b] \times \mathbb{R}^2 \mid \min_{i=1,2} \alpha_i(t) \le u \le \max_{i=1,2} \beta_i(t)\}, \tag{1.17}$$

and p, $q \in [1,\infty]$ be such that $\frac{1}{p} + \frac{1}{q} = 1$. Assume $f : E \to \mathbb{R}$ satisfies L^p-Carathéodory conditions and there exists $N \in L^1(a,b)$, $N > 0$ such that for $i = 1,2$ and a.e. $t \in A_i$ (resp. a.e. $t \in B_i$)

$$f(t,\alpha_i(t),\alpha_i'(t)) \ge -N(t) \quad (\text{resp. } f(t,\beta_i(t),\beta_i'(t)) \le N(t)).$$

Assume moreover there exist $\varphi \in C(\mathbb{R}^+, \mathbb{R}_0^+)$, $\psi \in L^p(a,b)$ and $R > 0$ such that (1.9) *holds and f satisfies one of the one-sided Nagumo conditions (a), (b), (c), or (d) in Theorem I-6.1, with $\alpha = \alpha_1$ and $\beta = \beta_2$.*

Then the problem (1.5) *has at least three solutions $u_1, u_2, u_3 \in W^{2,p}(a,b)$ such that for all $t \in [a,b]$*

$$\alpha_1(t) \le u_1(t) < \beta_1(t),\ \alpha_2(t) < u_2(t) \le \beta_2(t),\ u_1(t) \le u_3(t) \le u_2(t)$$

and there exist t_1, $t_2 \in [a,b]$ with

$$u_3(t_1) > \beta_1(t_1),\ u_3(t_2) < \alpha_2(t_2).$$

Notice that the conditions $u_3(t_1) > \beta_1(t_1)$ and $u_3(t_2) < \alpha_2(t_2)$ is a localization condition that implies that $u_3 \neq u_1$ and $u_3 \neq u_2$.

Proof. Define

$$\begin{aligned} g_i(t,u,v) &= f(t,\alpha_i(t),v) - \alpha_i(t) - \omega_{1i}(t,\alpha_i(t)-u), && \text{if } u < \alpha_i(t),\\ &= f(t,u,v) - u, && \text{if } u \geq \alpha_i(t),\end{aligned}$$

$$\begin{aligned} h_i(t,u,v) &= f(t,\beta_i(t),v) - \beta_i(t) + \omega_{2i}(t,u-\beta_i(t)), && \text{if } u > \beta_i(t),\\ &= f(t,u,v) - u, && \text{if } u \leq \beta_i(t),\end{aligned}$$

where

$$\begin{aligned} \omega_{1i}(t,\delta) &= \chi_{A_i}(t) \max_{|v|\leq\delta} |f(t,\alpha_i(t),\alpha_i'(t)+v) - f(t,\alpha_i(t),\alpha_i'(t))|,\\ \omega_{2i}(t,\delta) &= \chi_{B_i}(t) \max_{|v|\leq\delta} |f(t,\beta_i(t),\beta_i'(t)+v) - f(t,\beta_i(t),\beta_i'(t))|,\end{aligned}$$

χ_{A_i} and χ_{B_i} are the characteristic functions of the sets A_i and B_i.

We consider now the modified problem

$$\begin{aligned} u'' - u &= F(t,u,u'),\\ u(a) = u(b),&\ u'(a) = u'(b),\end{aligned} \tag{1.18}$$

where

$$\begin{aligned} F(t,u,v) &= g_1(t,u,v), && \text{if } u \leq \alpha_1(t),\\ &= f(t,u,v) - u, && \text{if } \alpha_1(t) < u < \beta_2(t),\\ &= h_2(t,u,v), && \text{if } \beta_2(t) \leq u.\end{aligned}$$

Let us choose k so that $\beta_1 \leq \beta_2 + k$ and $\alpha_1 - k \leq \alpha_2$, and define $T : \mathcal{C}^1([a,b]) \to \mathcal{C}^1([a,b])$ by

$$(Tu)(t) = \int_a^b G(t,s)F(s,u(s),u'(s))\,ds,$$

where $G(t,s)$ is the Green's function corresponding to (I-1.4).

Step 1 – Computation of $\deg(I-T,\Omega_{1,1})$, *where*

$$\Omega_{1,1} = \{u \in \mathcal{C}^1([a,b]) \mid \forall t \in [a,b], \alpha_1(t) - k < u(t) < \beta_1(t),\ |u'(t)| < R\}.$$

Define the alternative modified problem

$$\begin{aligned} u'' - u &= \bar{F}(t,u,u'),\\ u(a) = u(b),&\ u'(a) = u'(b),\end{aligned} \tag{1.19}$$

where

$$\begin{aligned} \bar{F}(t,u,u') &= g_1(t,u,v), && \text{if } u \leq \alpha_1(t),\\ &= f(t,u,v) - u, && \text{if } \alpha_1(t) < u < \min\{\beta_1(t),\beta_2(t)\},\\ &= \max\{h_1(t,u,v),h_2(t,u,v)\}, && \text{if } \min\{\beta_1(t),\beta_2(t)\} \leq u.\end{aligned}$$

Define next $\bar{T} : C^1([a,b]) \to C^1([a,b])$ by

$$(\bar{T}u)(t) = \int_a^b G(t,s)\bar{F}(s,u(s),u'(s))\,ds.$$

For any $\lambda \in [0,1]$, we consider then the homotopy $\mathcal{T}_\lambda = \lambda \bar{T} + (1-\lambda)T$.

Claim 1 : If $\lambda \in [0,1]$ *and* u *is a fixed point of* $\mathcal{T}_\lambda$, *we have* $\alpha_1 \le u \le \beta_2$. This result follows from the usual maximum principle argument as in Step 3 of the proof of Theorem I-6.6.

Claim 2 : If $\lambda \in [0,1]$ *and* $u \in \bar{\Omega}_{1,1}$ *is a fixed point of* $\mathcal{T}_\lambda$, *we have* $u < \beta_1$. Assume there exists $t_0 \in [a,b]$ such that $u(t_0) = \beta_1(t_0)$. We deduce from Claim 1 that $\alpha_1(t) \le u(t) \le \beta_2(t)$ for all $t \in [a,b]$. Hence we prove, as in Step 4 of the proof of Theorem I-6.6, that $\|u'\|_\infty < R$ so that u solves (1.5). As further β_1 is a strict upper solution, the claim follows.

Claim 3 : $\deg(I - T, \Omega_{1,1}) = 1$. It follows from the above claims that $\alpha_1 - k$ and β_1 are strict lower and upper solutions of (1.19) and we deduce from Theorem 1.8 and the properties of the degree that

$$\deg(I - T, \Omega_{1,1}) = \deg(I - \mathcal{T}_\lambda, \Omega_{1,1}) = \deg(I - \bar{T}, \Omega_{1,1}) = 1.$$

Step 2 – $\deg(I - T, \Omega_{2,2}) = 1$, *where*

$$\Omega_{2,2} = \{u \in C^1([a,b]) \mid \forall t \in [a,b], \alpha_2(t) < u(t) < \beta_2(t) + k,\ |u'(t)| < R\}.$$

The proof of this result parallels the proof of Step 1.

Step 3 – There exist three solutions $\bar{u}_i$ $(i = 1,2,3)$ *of* (1.18) *such that*

$$\alpha_1 - k < \bar{u}_1 < \beta_1,\ \alpha_2 < \bar{u}_2 < \beta_2 + k,\ \alpha_1 - k < \bar{u}_3 < \beta_2 + k$$

and there exist t_1, $t_2 \in [a,b]$ *with*

$$\bar{u}_3(t_1) > \beta_1(t_1),\ \bar{u}_3(t_2) < \alpha_2(t_2).$$

The two first solutions are obtained from the fact that

$$\deg(I - T, \Omega_{1,1}) = 1 \quad \text{and} \quad \deg(I - T, \Omega_{2,2}) = 1.$$

Define

$$\Omega_{1,2} = \{u \in C^1([a,b]) \mid \forall t \in [a,b],\ \alpha_1(t) - k < u(t) < \beta_2(t) + k,\ |u'(t)| < R\}.$$

We have

$$\begin{aligned} 1 &= \deg(I - T, \Omega_{1,2}) \\ &= \deg(I - T, \Omega_{1,1}) + \deg(I - T, \Omega_{2,2}) + \deg(I - T, \Omega_{1,2} \setminus (\bar{\Omega}_{1,1} \cup \bar{\Omega}_{2,2})) \end{aligned}$$

which implies

$$\deg(I - T, \Omega_{1,2} \setminus (\bar{\Omega}_{1,1} \cup \bar{\Omega}_{2,2})) = -1$$

and the existence of $\bar{u}_3 \in \Omega_{1,2} \setminus (\bar{\Omega}_{1,1} \cup \bar{\Omega}_{2,2})$ follows.

Step 4 – There exist solutions u_i $(i = 1, 2, 3)$ *of* (1.5) *such that*

$$\alpha_1 \leq u_1 < \beta_1,\ \alpha_2 < u_2 \leq \beta_2,\ u_1 \leq u_3 \leq u_2$$

and there exist $t_1,\ t_2 \in [a, b]$, *with*

$$u_3(t_1) > \beta_1(t_1),\ u_3(t_2) < \alpha_2(t_2).$$

We know that solutions u of (1.18) are such that

$$\alpha_1 \leq u \leq \beta_2 \quad \text{and} \quad \|u'\|_\infty \leq R,$$

i.e. they are solutions of (1.5). Next, from Theorem I-6.11, we know there exist extremal solutions u_{min} and u_{max} of (1.5) in $[\alpha_1, \beta_2]$. The claim follows then with $u_1 = u_{min}$, $u_2 = u_{max}$ and $u_3 = \bar{u}_3$. □

Observe that in this theorem $u_1 \leq \min\{\beta_1, \beta_2\}$ and $u_2 \geq \max\{\alpha_1, \alpha_2\}$.

Example 1.1. In Example I-1.4, we have proved that if $h \in C([0, 2\pi])$ is such that $\|\tilde{h}\|_{L^1} \leq 3$, $|\bar{h}| < \cos(\frac{\pi}{6}\|\tilde{h}\|_{L^1})$, the problem

$$\begin{aligned} u'' + \sin u = h(t),\\ u(0) = u(2\pi),\ u'(0) = u'(2\pi), \end{aligned} \tag{1.20}$$

has at least one solution. With Theorem 1.13, it is now easy to extend this example to functions $h \in L^1(0, 2\pi)$ and to complement it as follows. Let w to be the solution of

$$w'' = \tilde{h}(t), \quad w(0) = w(2\pi),\ w'(0) = w'(2\pi),\ \bar{w} = 0,$$

and

$$\begin{aligned} \alpha_1(t) = -\tfrac{3\pi}{2} + w(t), \quad \beta_1(t) = -\tfrac{\pi}{2} + w(t),\\ \alpha_2(t) = \tfrac{\pi}{2} + w(t), \quad \beta_2(t) = \tfrac{3\pi}{2} + w(t). \end{aligned}$$

Using Propositions 1.5, 1.6 and Theorem 1.13, we find three solutions of (1.20), i.e. $\alpha_1 < u_1 < \beta_1$, $\alpha_2 < u_2 < \beta_2$ and $\alpha_1 < u_3 < \beta_2$ with $u_3(t_1) \geq \beta_1(t_1)$ and $u_3(t_2) \leq \alpha_2(t_2)$ for some t_1 and $t_2 \in [0, 2\pi]$. Notice that u_1 might be $u_2 - 2\pi$ but $u_3 \neq u_1 \bmod 2\pi$. Hence, this problem has at least two geometrically different solutions.

Example 1.2. Consider the equation

$$(1 + e\cos t)u'' - (2e\sin t)u' + a\sin u = 4e\sin t \tag{1.21}$$

which describes the motions of a satellite in the plane of its orbit around its center of mass. Here u is twice the angle between the radius vector of the satellite and one of its principal axes of inertia lying in the plane of the orbit, t is the true anomaly, and a and e are parameters such that $|a| \leq 3$ and $0 < e < 1$.

Let us prove that for all a and $e \in]-1,1[$ with $|e| < |a|/4$, the equation (1.21) has at least two 2π-periodic solutions which do not differ by a multiple of 2π. To this end we apply, in case $a > 0$, Theorem 1.13 with

$$\alpha_1 = \tfrac{\pi}{2} < \beta_1 = \tfrac{3\pi}{2} < \alpha_2 = \tfrac{5\pi}{2} < \beta_2 = \tfrac{7\pi}{2}$$

and obtain three solutions

$$\alpha_1 < u_1 < \beta_1, \ \alpha_2 < u_2 < \beta_2 \ \text{ and } \ \alpha_1 < u_3 < \beta_2.$$

Notice that $u_3 \not\leq \beta_1$ and $\alpha_2 \not\leq u_3$ which implies $u_3 \neq u_1 \bmod 2\pi$. On the other hand, u_1 might be $u_2 - 2\pi$ so that we obtain only two geometrically different solutions of (1.21). A similar argument holds if $a < 0$.

The same conclusion holds for all a and $e \in \mathbb{R}$, with $|e| < e_0$, where $e_0 = 0.2982\ldots$ is the positive root of

$$9e^4 + (17 + \tfrac{41}{64})e^2 - (1 + \tfrac{41}{64}) = 0.$$

These conditions improve the preceding ones if a is small. To prove this result in case $a \geq 0$, let

$$z(t) = 2\int_0^t \left(\tfrac{(1-e^2)^{3/2}}{(1+e\cos s)^2} - 1\right) ds,$$

be a 2π-periodic solution of

$$(1 + e\cos t)z'' - (2e\sin t)z' = 4e\sin t.$$

If we can find a constant c such that on $[0, 2\pi]$

$$\alpha_1(t) = z(t) + c \in]0, \pi[,$$

this function is a strict lower solution as follows from Proposition 1.1. The result follows then from Theorem 1.13 with

$$\alpha_1 < \beta_1 = \alpha_1 + \pi < \alpha_2 = \alpha_1 + 2\pi < \beta_2 = \alpha_1 + 3\pi.$$

The extrema of α_1 are difficult to compute. However if we choose c so that $\bar{\alpha}_1 = \frac{1}{2\pi}\int_0^{2\pi} \alpha_1\, dt = \pi/2$, we can compute (using $|e| < e_0$ and Proposition A-4.1) that $\tilde{\alpha}_1 = \alpha_1 - \bar{\alpha}_1$ satisfies

$$\|\tilde{\alpha}_1\|_\infty^2 \leq \tfrac{\pi}{6}\int_0^{2\pi} \tilde{\alpha}_1'^2(t)\, dt = \tfrac{2\pi^2}{3}\left[\tfrac{2+3e^2}{(1-e^2)^{1/2}} - 2\right] < \tfrac{\pi^2}{4}.$$

The function $\alpha_1 = \tilde{\alpha}_1 + \pi/2$ works.

Example 1.3. Consider the problem

$$\begin{gathered} u'' + cu' - u^3 + 3u = p(t), \\ u(0) = u(2\pi),\ u'(0) = u'(2\pi), \end{gathered} \tag{1.22}$$

where $p \in L^\infty(0, 2\pi)$.

Claim 1 – For all $p \in L^\infty(0,2\pi)$, (1.22) has at least one solution. This result follows from Theorem I-6.9, where $\alpha_1 < -1$ and $\beta_2 > 1$ are constants such that $\alpha_1^3 - 3\alpha_1 < -\|p\|_\infty$ and $\beta_2^3 - 3\beta_2 > \|p\|_\infty$.

Claim 2 – For all $p \in L^\infty(0,2\pi)$ such that $\|p\|_\infty < 2$, (1.22) has at least three solutions. This result follows from Theorem 1.13, where α_1 and β_2 are chosen as in Claim 1, $\beta_1 = -1$ and $\alpha_2 = 1$. The functions β_1 and α_2 are strict upper and lower solutions as follows from Propositions 1.6 and 1.5.

Claim 3 – For all $p \in L^\infty(0,2\pi)$ such that $\|p\|_\infty = 2$, (1.22) has at least two solutions. Let α_i and β_i $(i = 1, 2)$ be chosen as in Claim 2. The existence of solutions $u_i \in [\alpha_i, \beta_i]$ follows then from Theorem I-6.9.

This result is best possible as follows from the case $c \neq 0$ and p constant. Here, we can give an exact count of the number of solutions. Multiplying the equation (1.22) by u' and integrating we obtain $\|u'\|_{L^2} = 0$. Hence, in this case, the problem admits only constant solutions. It is then easy to see that it has exactly
(a) three solutions for $p \in]-2, 2[$,
(b) two solutions for $p = -2$ or 2,
(c) one solution for $p \in]-\infty, -2[$ or $p \in]2, +\infty[$.
This situation is represented at figure 1.

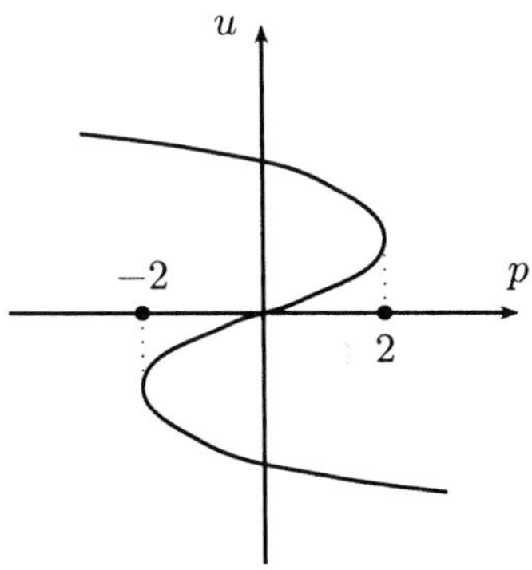

Fig. 1 : Diagram of the solutions of (1.22)

We can generalize this example and complement Exercice I-6.4.

Exercise 1.4. Consider the problem

$$\begin{aligned} u'' + g(u)\text{sgn}(u')|u'|^r + f(t,u) = s(t), \\ u(0) = u(2\pi),\ u'(0) = u'(2\pi), \end{aligned} \tag{1.23}$$

where f is an L^p-Carathéodory function that satisfies Condition (A) in Proposition 1.5, g is continuous, and $s \in L^\infty(0,2\pi)$. Assume $0 < r \le 2 - 1/p$,

$$\lim_{u\to+\infty} f(t,u) = -\infty, \qquad \lim_{u\to-\infty} f(t,u) = +\infty,$$

uniformly in t and there exist real numbers $v < u$ such that $f(t,v) < f(t,u)$.

Prove that for any $s \in L^\infty(0,2\pi)$ with $f(t,v) < s(t) < f(t,u)$, problem (1.23) has at least three solutions.

Hint : Use constant upper and lower solutions : $\alpha_1 < \beta_1 = v < \alpha_2 = u < \beta_2$.

Exercise 1.5. Adapt Theorem 1.13 to the Rayleigh and the Liénard equations (1.14) and (1.16).

Another way to obtain multiplicity results is to exhibit domains $\Omega_1 \supset \Omega$ such that

$$\deg(I - T, \Omega_1) = 0 \quad \text{and} \quad \deg(I - T, \Omega) = 1,$$

which is the basic idea of the following theorem.

Theorem 1.14. *Let $k > 0$ and α, $\beta \in C([a,b])$ be respectively a $W^{2,1}$-lower solution and a strict $W^{2,1}$-upper solution of the problem (1.5) with $\alpha \le \beta \le k$. Define $A \subset [a,b]$ (resp. $B \subset [a,b]$) to be the set of points where α (resp. β) is derivable.*

Let p, $q \in [1,\infty]$ with $\frac{1}{p} + \frac{1}{q} = 1$, and $E = \{(t,u,v) \in [a,b] \times \mathbb{R}^2 \mid \alpha(t) \le u\}$. Assume $f : E \to \mathbb{R}$ satisfies L^p-Carathéodory conditions and there exists $N \in L^1(a,b)$, $N > 0$ such that for a.e. $t \in A$ (resp. for a.e. $t \in B$)

$$f(t,\alpha(t),\alpha'(t)) \ge -N(t) \quad (\textit{resp. } f(t,\beta(t),\beta'(t)) \le N(t)).$$

Let $R > 0$, $\varphi \in C(\mathbb{R}^+, \mathbb{R}_0^+)$ and $\psi \in L^p(a,b)$ satisfy

$$\int_0^R \frac{s^{1/q}}{\varphi(s)}\, ds > \|\psi\|_{L^p}(k - \min_t \alpha(t))^{1/q},$$

and suppose that for a.e. $t \in [a,b]$ and all $(u,v) \in \mathbb{R}^2$ with $\alpha(t) \le u \le k$,

$$f(t,u,v) \le \psi(t)\varphi(|v|).$$

Assume at last that for every solution $u \in W^{2,1}(a,b)$ *of*

$$\begin{aligned} u'' &\leq f(t,u,u'), \\ u(a) = u(b), &\ u'(a) = u'(b), \end{aligned} \tag{1.24}$$

with $u \geq \alpha$ *we have* $u < k$ *on* $[a,b]$.

Then the problem (1.5) *has at least two solutions* $u_1,\ u_2 \in W^{2,p}(a,b)$ *such that for all* $t \in [a,b]$

$$\alpha(t) \leq u_1(t) < \beta(t), \quad u_1(t) \leq u_2(t) < k$$

and for some $t_1 \in [a,b]$,

$$u_2(t_1) > \beta(t_1).$$

Proof. Define

$$\bar{f}(t,u,v) = f(t,\gamma_1(t,u),v),$$

and

$$\omega_1(t,\delta) = \chi_A(t) \max_{|v|\leq\delta} |\bar{f}(t,\alpha(t),\alpha'(t)+v) - \bar{f}(t,\alpha(t),\alpha'(t))|,$$

where $\gamma_1(t,u) = \max\{\alpha(t),u\}$ and χ_A is the characteristic function of the set A.

We consider now the modified problem

$$\begin{aligned} u'' - u &= \bar{f}(t,u,u') - \omega(t,u) - s, \\ u(a) = u(b), &\ u'(a) = u'(b), \end{aligned} \tag{1.25}$$

where $s \geq 0$ and

$$\begin{aligned} \omega(t,u) &= \alpha(t) + \omega_1(t,\alpha(t)-u), && \text{if } u < \alpha(t), \\ &= u, && \text{if } u \geq \alpha(t). \end{aligned}$$

Solutions of (1.25) solve the fixed point problem

$$u(t) = (Tu)(t) + s,$$

where

$$(Tu)(t) = \int_a^b G(t,s)[\bar{f}(t,u,u') - \omega(t,u)]\,ds$$

and $G(t,s)$ is the Green's function corresponding to (I-1.4).

Claim 1 - Solutions u of (1.25), *with* $s \geq 0$, *are such that* $\alpha \leq u < k$. The claim $\alpha \leq u$ follows from the argument used in the proof of Theorem I-6.6. Hence u is a solution of (1.24) and the claim $u < k$ is an assumption.

Claim 2 - Solutions u of (1.25), *with* $s \geq 0$, *are such that* $\|u'\|_\infty < R$. From Claim 1, $\alpha \leq u < k$. Arguing then as in Theorem I-6.6, Claim 2 follows from Proposition I-4.8.

Claim 3 - $\deg(I - T, \Omega) = 1$, *where* $\Omega = \{u \in C^1([a, b]) \mid \forall t \in [a, b],\ \alpha(t) - 1 < u(t) < \beta(t),\ |u'(t)| < R\}$. Notice that $\alpha - 1$ is a lower solution of (1.25) and from Claim 1, this lower solution is strict. The claim follows then from Theorem 1.8.

Claim 4 - $\deg(I - T, \Omega_1) = 0$, *where* $\Omega_1 = \{u \in C^1([a, b]) \mid \forall t \in [a, b], \alpha(t) - 1 < u(t) < k,\ |u'(t)| < R\}$. Solutions u of (1.25) are such that

$$u = Tu + s.$$

As further they are in Ω_1, it is clear that $\|u - Tu\|_\infty = s$ is *a priori* bounded. Hence for $s = s_0$ large enough there is no solution of (1.25). It follows that

$$\deg(I - T, \Omega_1) = \deg(I - T - s_0, \Omega_1) = 0.$$

Conclusion - From Claim 3, there exists a solution $u_1 \in \Omega$. By Theorem I-6.11, we can choose u_1 to be the minimal solution in Ω. Next, by excision, we deduce from the previous claims that

$$\deg(I - T, \Omega_1 \setminus \bar{\Omega}) = -1,$$

and there exists a second solution $u_2 \in \Omega_1 \setminus \bar{\Omega}$. If $u_2 \not\geq u_1$, u_1 and u_2 are upper solution and we deduce from I-6.10 existence of a solution u_3 with $\alpha \leq u_3 \leq \min\{u_1, u_2\}$ which contradicts u_1 to be minimal. □

Notice that we cannot use a one-sided Nagumo condition such as

$$f(t, u, v) \geq -\psi(t)\varphi(|v|)$$

to obtain an *a priori* bound on the derivative of solutions of (1.25) which is uniform in $s \geq 0$ (see Claim 2).

Observe also that solutions of (1.24) are regular upper solutions of (1.5).

Exercise 1.6. Write and prove a result similar to the preceding theorem so that there exist solutions u_i with

$$\alpha < u_1 \leq \beta \quad \text{and} \quad -k < u_2 \leq u_1.$$

Exercise 1.7. Paraphrase Theorem 1.14 for the Rayleigh and the Liénard equations (1.14) and (1.16).

If f does not depend on the derivative u', Theorem 1.14 reduces to the following.

Theorem 1.15. *Let $k > 0$, α, $\beta \in \mathcal{C}([a,b])$ be respectively a $W^{2,1}$-lower solution and a strict $W^{2,1}$-upper solution of (1.1) such that $\alpha \leq \beta \leq k$. Let $E = \{(t,u) \in [a,b] \times \mathbb{R} \mid \alpha(t) \leq u\}$ and assume $f : E \to \mathbb{R}$ is an L^1-Carathéodory function.*

Suppose moreover that for all solutions $u \in W^{2,1}(a,b)$ of

$$\begin{aligned} u'' &\leq f(t,u), \\ u(a) = u(b),\ & u'(a) = u'(b), \end{aligned}$$

with $u \geq \alpha$ we have $u < k$ on $[a,b]$.

Then the problem (1.1) has at least two solutions u_1, $u_2 \in W^{2,1}(a,b)$ such that for all $t \in [a,b]$

$$\alpha(t) \leq u_1(t) < \beta(t), \quad u_1(t) \leq u_2(t) < k$$

and for some $t_1 \in [a,b]$,

$$u_2(t_1) > \beta(t_1).$$

Example 1.4. In Example I-1.2, we have proved that if $h \in \mathcal{C}([a,b])$ is such that for all $t \in [a,b]$, $-1/4 \leq h(t) \leq 0$, then $\alpha_1(t) = -1$ and $\alpha_2(t) = 0$ are lower solutions and $\beta_1(t) = -1/\sqrt{2}$ and $\beta_2(t) = 1/\sqrt{2}$ are upper solutions of

$$\begin{aligned} u'' + u^4 - u^2 &= h(t), \\ u(a) = u(b),\ & u'(a) = u'(b). \end{aligned} \tag{1.26}$$

If we reinforce the condition on h assuming $-1/4 < h(t) \leq 0$ for all $t \in [a,b]$, we can use Theorem 1.15 and prove the existence of a third solution u_3 such that for some $t_1 \in [a,b]$, $u_3(t_1) > 1/\sqrt{2}$. To this end, we apply this theorem with $\alpha = 0$ and $\beta = 1/\sqrt{2}$. The upper solution β is strict from Proposition 1.2. It remains to prove an *a priori* bound on the positive upper solutions of (1.26). Observe that $u^4 - u^2 \geq -\frac{1}{4}$ on $\mathbb{R}^+$. Hence, if u is an upper solutions of (1.26),

$$u'' \leq h(t) - (u^4 - u^2) \leq h(t) + \tfrac{1}{4} \leq \tfrac{1}{4}.$$

Moreover $\min_t u(t) = u(t_0) < 1$ as otherwise $u''(t) < 0$ on $[a,b]$ which contradicts the periodicity of u. Extending u and h by periodicity and as $u'(t_0) = 0$ we have, for all $t \in [t_0, t_0 + b - a[$,

$$u(t) = u(t_0) + \int_{t_0}^{t} u''(s)(t-s)\,ds \leq 1 + \tfrac{1}{8}(b-a)^2,$$

which proves the required *a priori* bound.

If we assume further $-1/4 < h(t) < 0$ for all $t \in [a,b]$, the lower solution $\alpha = 0$ is also strict as follows from Proposition 1.1 and we conclude, using

Theorem 1.13, that this problem has a fourth solution u_4 such that for all $t \in [a, b]$,

$$u_1 \le u_4 \le u_2.$$

Notice at last that if h is constant, the four solutions are easy to exhibit.

2 The Dirichlet Problem

2.1 The Strict Lower and Upper Solutions

Consider the Dirichlet problem

$$\begin{aligned} u'' &= f(t, u, u'), \\ u(a) &= 0, \ u(b) = 0, \end{aligned} \tag{2.1}$$

where f is an L^p-Carathéodory function. It has been seen that this problem is equivalent to the fixed point problem

$$u(t) = (Tu)(t) := \int_a^b G(t,s) f(s, u(s), u'(s))\, ds, \tag{2.2}$$

where $G(t, s)$ is the Green's function corresponding to

$$\begin{aligned} u'' &= f(t), \\ u(a) &= 0, \ \ u(b) = 0. \end{aligned} \tag{2.3}$$

In this chapter, we consider the degree of $I - T$ for an open set Ω of functions u that lie between the lower and upper solutions α and β. However, we want to allow α and β to satisfy the boundary conditions. In that case, the set

$$\Omega = \{u \in \mathcal{C}_0^1([a, b]) \mid \|u'\|_\infty < R, \forall t \in]a, b[, \ \alpha(t) < u(t) < \beta(t)\}$$

is not open in $\mathcal{C}_0^1([a, b])$. Similarly, if f does not depend on u', we cannot use the set

$$\Omega = \{u \in \mathcal{C}_0([a, b]) \mid \forall t \in]a, b[, \ \alpha(t) < u(t) < \beta(t)\},$$

which might not be open in $\mathcal{C}_0([a, b])$.

A way out is to impose some additional condition on the function at the boundary points $t = a$ and $t = b$. For example, the set

$$\begin{aligned} \{u \in \mathcal{C}_0^1([a, b]) \mid \|u'\|_\infty < R, \forall t \in]a, b[, \ \alpha(t) < u(t) < \beta(t), \\ D^+\alpha(a) < u'(a) < D_+\beta(a), \ D_-\alpha(b) > u'(b) > D^-\beta(b)\} \end{aligned}$$

is open in $C_0^1([a,b])$.

To work in a space of continuous function, we need to generalize the conditions $\alpha(t) < u(t)$ on $]a,b[$, $D^+\alpha(a) < u'(a)$ and $D^-\alpha(b) > u'(b)$ for such functions. Let us first introduce the following definition.

Definition 2.1. *Let u, $v \in C([a,b])$. We write $u \succ v$ or $v \prec u$ if*

$$\begin{array}{ll} u(t) > v(t), & \text{on }]a,b[, \\ D_+u(a) > D^+v(a), & \text{if } u(a) = v(a), \\ D^-u(b) < D_-v(b), & \text{if } u(b) = v(b), \end{array}$$

or, which is equivalent, if there exists $\epsilon > 0$ such that for any $t \in [a,b]$

$$u(t) - v(t) \geq \epsilon e(t),$$

where $e(t) := \sin(\pi \frac{t-a}{b-a})$ is the eigenfunction corresponding to the first eigenvalue of the problem

$$u'' + \lambda u = 0, \quad u(a) = 0, \ u(b) = 0.$$

The next step is to define the concept of strict lower and upper solutions.

Definitions 2.2. *A lower solution α of* (2.1) *is said* strict *if every solution u of* (2.1) *with $\alpha \leq u$ is such that $\alpha \prec u$.*

An upper solution β of (2.1) *is said* strict *if every solution u of* (2.1) *with $u \leq \beta$ is such that $u \prec \beta$.*

Observe that the notion of strict lower and upper solutions we have defined with the strict order $\prec$ is based on the same idea that the corresponding notion we have introduced for the periodic case. In this last case, a lower solution α is said strict if every solution u of the periodic problem with $\alpha \leq u$ is such that $\alpha < u$, i.e. if there exists $\epsilon > 0$ such that for any $t \in [a,b]$

$$u(t) - \alpha(t) \geq \epsilon \; 1.$$

The connection is obvious if we notice that $\sin(\pi \frac{t-a}{b-a})$ and 1 are the first eigenfunctions of the corresponding eigenvalue problems

$$u'' + \lambda u = 0, \quad u(a) = 0, \ u(b) = 0,$$

and

$$u'' + \lambda u = 0, \quad u(a) = u(b), \ u'(a) = u'(b).$$

More generally, a lower solution α is strict if for every solution u with $\alpha \leq u$, there exists $\epsilon > 0$ such that $\alpha(t) + \epsilon\varphi_1(t) \leq u(t)$, where φ_1 is the first eigenfunction of the corresponding eigenvalue problem.

Most of the results we present for the Dirichlet problem extend to the separated boundary conditions. Such generalizations are left to the reader as exercises.

A first result concerns lower and upper solutions which are $\mathcal{C}^2$.

Proposition 2.1. *Let $f : [a,b] \times \mathbb{R}^2 \to \mathbb{R}$ be continuous and $\alpha \in \mathcal{C}([a,b]) \cap \mathcal{C}^2(]a,b[)$ be such that*
(a) for all $t \in]a,b[$, $\alpha''(t) > f(t,\alpha(t),\alpha'(t))$;
(b) for $t_0 \in \{a,b\}$, either $\alpha(t_0) < 0$
or $\alpha(t_0) = 0$, $\alpha \in \mathcal{C}^2(]a,b[\cup \{t_0\})$, and $\alpha''(t_0) > f(t_0,\alpha(t_0),\alpha'(t_0))$.
Then α is a strict $\mathcal{C}^2$-lower solution of (2.1).

Proof. From the assumptions, α is a $\mathcal{C}^2$-lower solution of (2.1). Let then u be a solution of (2.1) such that $\alpha \le u$ and assume, by contradiction, that for any $\epsilon = 1/n$, there exists $t_n \in [a,b]$ such that

$$u(t_n) - \alpha(t_n) < \tfrac{1}{n} \sin(\pi \tfrac{t_n - a}{b-a}). \tag{2.4}$$

It follows there exists a subsequence of $(t_n)_n$ that converges to a point t_0 such that $u(t_0) = \alpha(t_0)$. If $t_0 \in]a,b[$, we have $u'(t_0) = \alpha'(t_0)$. On the other hand, if $t_0 = a$, we know that $\alpha \in \mathcal{C}^2([a,b[)$. Further, we deduce from (2.4) that

$$\frac{u(t_n) - u(a)}{t_n - a} \le \frac{\alpha(t_n) + \frac{1}{n}\sin(\pi\frac{t_n-a}{b-a}) - \alpha(a)}{t_n - a}.$$

This result implies that $u'(a) \le \alpha'(a)$. As further $u - \alpha$ is minimum at $t = a$, we also have $u'(a) \ge \alpha'(a)$. Hence, $u'(a) = \alpha'(a)$. A similar reasoning applies if $t_0 = b$ so that in all cases $u'(t_0) - \alpha'(t_0) = 0$. At last, we obtain the contradiction $0 \le u''(t_0) - \alpha''(t_0) = f(t_0,\alpha(t_0),\alpha'(t_0)) - \alpha''(t_0) < 0$. □

In a similar way, we can write the following.

Proposition 2.2. *Let $f : [a,b] \times \mathbb{R}^2 \to \mathbb{R}$ be continuous and $\beta \in \mathcal{C}([a,b]) \cap \mathcal{C}^2(]a,b[)$ be such that*
(a) for all $t \in]a,b[$, $\beta''(t) < f(t,\beta(t),\beta'(t))$;
(b) for $t_0 \in \{a,b\}$, either $\beta(t_0) > 0$
or $\beta(t_0) = 0$, $\beta \in \mathcal{C}^2(]a,b[\cup \{t_0\})$ and $\beta''(t_0) < f(t_0,\beta(t_0),\beta'(t_0))$.
Then β is a strict $\mathcal{C}^2$-upper solution of (2.1).

Notice that an upper solution β such that

$$\beta''(t) < f(t,\beta(t),\beta'(t)) \text{ on }]a,b[, \quad \beta(a) \ge 0, \quad \beta(b) > 0$$

is not necessarily strict. Consider for example the problem (2.1) defined on $[a,b] = [0,1]$ with

$$\begin{aligned} f(t,u,u') &= 0, && \text{if } u \le 0, \\ &= 7u/t^2, && \text{if } 0 < u \le t^3, \\ &= 7t, && \text{if } u > t^3. \end{aligned}$$

The function $u(t) = 0$ is a solution and $\beta(t) = t^3 \geq u(t)$ is an upper solution which is not strict (as $\beta(0) = u(0)$ and $\beta'(0) = u'(0)$) but satisfies $\beta''(t) < f(t, \beta(t), \beta'(t))$ on $]0, 1]$, $\beta(0) = 0$ and $\beta(1) > 0$.

In the Carathéodory case, we can use the following propositions.

Proposition 2.3. *Let $f : [a, b] \times \mathbb{R}^2 \to \mathbb{R}$ be an L^1-Carathéodory function. Assume that $\alpha \in C([a, b])$ is not a solution of (2.1) and that*

(a) for all $t_0 \in]a, b[$, either $D_-\alpha(t_0) < D^+\alpha(t_0)$ or there exist an open interval $I_0 \subset [a, b]$ and $\epsilon > 0$ such that $t_0 \in I_0$, $\alpha \in W^{2,1}(I_0)$ and for a.e. $t \in I_0$, all $u \in [\alpha(t), \alpha(t) + \epsilon \sin(\pi \frac{t-a}{b-a})]$ and all $v \in [\alpha'(t) - \epsilon, \alpha'(t) + \epsilon]$,

$$\alpha''(t) \geq f(t, u, v);$$

(b) either $\alpha(a) < 0$ or $\alpha(a) = 0$ and there exist $\epsilon > 0$ such that $\alpha \in W^{2,1}(a, a + \epsilon)$ and for a.e. $t \in [a, a + \epsilon]$, all $u \in [\alpha(t), \alpha(t) + \epsilon \sin(\pi \frac{t-a}{b-a})]$ and all $v \in [\alpha'(t) - \epsilon, \alpha'(t) + \epsilon]$,

$$\alpha''(t) \geq f(t, u, v);$$

(c) either $\alpha(b) < 0$ or $\alpha(b) = 0$ and there exist $\epsilon > 0$ such that $\alpha \in W^{2,1}(b - \epsilon, b)$ and for a.e. $t \in [b - \epsilon, b]$, all $u \in [\alpha(t), \alpha(t) + \epsilon \sin(\pi \frac{t-a}{b-a})]$ and all $v \in [\alpha'(t) - \epsilon, \alpha'(t) + \epsilon]$,

$$\alpha''(t) \geq f(t, u, v).$$

Then α is a strict $W^{2,1}$-lower solution of (2.1).

Proof. Notice first that α satisfies Definition II-2.1 and therefore is a $W^{2,1}$-lower solution.

Let u be a solution of (2.1) such that $u \geq \alpha$. Arguing by contradiction as in Proposition 2.1 there exists a sequence $(t_n)_n$ that satisfies (2.4) and that converges to a point t_0 such that $u(t_0) = \alpha(t_0)$ and $u'(t_0) = \alpha'(t_0)$.

As α is not a solution, we can find t^* such that $u(t^*) > \alpha(t^*)$. Assume $t_0 < t^*$ and define $t_1 = \max\{t < t^* \mid u(t) = \alpha(t)\}$. We prove then that $u(t_1) = \alpha(t_1)$, $u'(t_1) = \alpha'(t_1)$ and fix $\epsilon > 0$ from the assumptions. Next, for $t \geq t_1$ near enough t_1,

$$u(t) \in [\alpha(t), \alpha(t) + \epsilon \sin(\pi \tfrac{t-a}{b-a})] \quad \text{and} \quad u'(t) \in [\alpha'(t) - \epsilon, \alpha'(t) + \epsilon].$$

Hence, we compute

$$u'(t) - \alpha'(t) = \int_{t_1}^{t} [f(s, u(s), u'(s)) - \alpha''(s)]\, ds \leq 0$$

for $t \geq t_1$, near enough t_1 which contradicts the definition of t_1. A similar argument holds if $t_0 > t^*$. □

Strict $W^{2,1}$-upper solutions can be obtained from a similar proposition.

Proposition 2.4. *Let $f : [a,b] \times \mathbb{R}^2 \to \mathbb{R}$ be an L^1-Carathéodory function. Assume that $\beta \in C([a,b])$ is not a solution of (2.1) and that*

(a) for all $t_0 \in]a,b[$, either $D^-\beta(t_0) > D_+\beta(t_0)$ or there exist an open interval $I_0 \subset [a,b]$ and $\epsilon > 0$ such that $t_0 \in I_0$, $\beta \in W^{2,1}(I_0)$ and for a.e. $t \in I_0$, all $u \in [\beta(t) - \epsilon \sin(\pi \frac{t-a}{b-a}), \beta(t)]$ and all $v \in [\beta'(t) - \epsilon, \beta'(t) + \epsilon]$,

$$\beta''(t) \le f(t,u,v);$$

(b) either $\beta(a) > 0$ or $\beta(a) = 0$ and there exist $\epsilon > 0$ such that $\beta \in W^{2,1}(a, a+\epsilon)$ and for a.e. $t \in [a, a+\epsilon]$, all $u \in [\beta(t) - \epsilon \sin(\pi \frac{t-a}{b-a}), \beta(t)]$ and all $v \in [\beta'(t) - \epsilon, \beta'(t) + \epsilon]$,

$$\beta''(t) \le f(t,u,v);$$

(c) either $\beta(b) > 0$ or $\beta(b) = 0$ and there exist $\epsilon > 0$ such that $\beta \in W^{2,1}(b-\epsilon, b)$ and for a.e. $t \in [b-\epsilon, b]$, all $u \in [\beta(t) - \epsilon \sin(\pi \frac{t-a}{b-a}), \beta(t)]$ and all $v \in [\beta'(t) - \epsilon, \beta'(t) + \epsilon]$,

$$\beta''(t) \le f(t,u,v).$$

Then β is a strict $W^{2,1}$-upper solution of (2.1).

In these propositions the curves $u = \alpha(t)$ and $u = \beta(t)$ can have angles, provided their opening is from above for α and from below for β. Also, α and β can be zero at the end points a and b. In this case, the second alternative imposes some second order condition near these end points but restricted to some angular region above the curve $u = \alpha(t)$ and below the curve $u = \beta(t)$. Notice at last that the first alternative in (b) and (c), i.e. the strict inequalities on α and β at the end points, can be interpreted as a zero order condition and the first alternative in (a), the angular condition, as a first order one in order to prevent solutions to be tangent to α or β.

Observe also that, in the continuous case, if α satisfies the conditions of Proposition 2.1, it satisfies the conditions of Proposition 2.3. The converse, however, does not hold since Proposition 2.3 does not require strict inequalities.

As in the periodic case, we can give conditions on f to ensure that lower or upper solutions that satisfies conditions like

$$\alpha''(t) \ge f(t, \alpha(t), \alpha'(t)) + A$$

or

$$\beta''(t) \le f(t, \beta(t), \beta'(t)) - A,$$

with $A > 0$, are strict lower or upper solution.

Proposition 2.5. *Let $f : [a,b] \times \mathbb{R}^2 \to \mathbb{R}$ be an L^1-Carathéodory function that satisfies the assumption*
(B) for all $t_0 \in [a,b]$, $(u_0, v_0) \in \mathbb{R}^2$ and $\epsilon > 0$, there exists $\delta > 0$ such that

$$|t - t_0| < \delta, \ |u_1 - u_0| < \delta, \ u_2 \in [u_1, u_1 + \delta \sin(\pi \tfrac{t-a}{b-a})[,$$
$$|v_1 - v_0| < \delta, \ |v_2 - v_0| < \delta \ \Rightarrow \ f(t, u_2, v_2) - f(t, u_1, v_1) < \epsilon.$$

Let $A > 0$ and $\alpha \in W^{2,1}(a,b)$ be such that

$$\alpha''(t) \geq f(t, \alpha(t), \alpha'(t)) + A,$$
$$\alpha(a) \leq 0, \ \alpha(b) \leq 0.$$

Then α is a strict lower solution of (2.1).

Proof. The proof follows as for Proposition 1.5. □

Notice that Condition (B) holds if $f(t,u,v) = g(u,v) + h(t)$ with g continuous and $h \in L^1(a,b)$ or if $f(t,u,v) = g(u,v)h(t)$ with g continuous and $h \in L^\infty(a,b)$.

Proposition 2.6. *Let $f : [a,b] \times \mathbb{R}^2 \to \mathbb{R}$ be an L^1-Carathéodory function that satisfies Condition (B) of Proposition 2.5. Let $B > 0$ and $\beta \in W^{2,1}(a,b)$ be such that*

$$\beta''(t) \leq f(t, \beta(t), \beta'(t)) - B,$$
$$\beta(a) \geq 0, \ \beta(b) \geq 0.$$

Then β is a strict upper solutions of (2.1).

Notice that it is easy to generalize these results using lower and upper solutions with corners.

As for the periodic problem, we can study cases where f satisfies a Lipschitz condition in v and a one-sided Lipschitz condition in u.

Proposition 2.7. *Let $f : [a,b] \times \mathbb{R}^2 \to \mathbb{R}$ be an L^1-Carathéodory function such that for some k, $l \in L^1(a,b;\mathbb{R}^+)$,*

(a) for a.e. $t \in [a,b]$, all u_1, $u_2 \in \mathbb{R}$ with $u_1 \leq u_2$ and $v \in \mathbb{R}$,

$$f(t, u_2, v) - f(t, u_1, v) \leq k(t)\,(u_2 - u_1);$$

(b) for a.e. $t \in [a,b]$ and all u, v_1, $v_2 \in \mathbb{R}$,

$$|f(t, u, v_2) - f(t, u, v_1)| \leq l(t)|v_2 - v_1|.$$

Let α (resp. β) be a lower (resp. upper) solution of (2.1) *which is not a solution and assume*

(c) *either* $\alpha(a) < 0$ *(resp.* $\beta(a) > 0$*)*
or $\alpha(a) = 0$ *(resp.* $\beta(a) = 0$*) and there exists an interval* $I_0 = [a, c[\subset [a, b]$ *such that* $\alpha \in W^{2,1}(I_0)$ *(resp.* $\beta \in W^{2,1}(I_0)$*) and for a.e.* $t \in I_0$,

$$\alpha''(t) \geq f(t, \alpha(t), \alpha'(t)) \quad (\textit{resp. } \beta''(t) \leq f(t, \beta(t), \beta'(t)));$$

(d) *either* $\alpha(b) < 0$ *(resp.* $\beta(b) > 0$*)*
or $\alpha(b) = 0$ *(resp.* $\beta(b) = 0$*) and there exists an interval* $I_0 =]c, b] \subset [a, b]$ *such that* $\alpha \in W^{2,1}(I_0)$ *(resp.* $\beta \in W^{2,1}(I_0)$*) and for a.e.* $t \in I_0$,

$$\alpha''(t) \geq f(t, \alpha(t), \alpha'(t)) \quad (\textit{resp. } \beta''(t) \leq f(t, \beta(t), \beta'(t))).$$

Then α *(resp.* β*) is a strict lower solution (resp. a strict upper solution) of* (2.1).

Proof. Let u be a solution of (2.1) such that $u \geq \alpha$. We argue as in Proposition 2.1 to exhibit a point $t_0 \in [a, b]$ such that $u(t_0) = \alpha(t_0)$ and $u'(t_0) = \alpha'(t_0)$. Next, we conclude as in Proposition 1.7. □

Remark 2.1. As for periodic problem, we can see from the problem (1.7) that this proposition does not hold without the Lipschitz conditions on f. We can however assume (a) and (b) to hold only in a neighbourhood of $\{(t, \alpha(t), \alpha'(t)) \mid t \in [a, b]$ such that $\alpha'(t)$ exists$\}$ (resp. $\{(t, \beta(t), \beta'(t)) \mid t \in [a, b]$ such that $\beta'(t)$ exists$\}$).

Exercise 2.1. Assume f satisfies the assumptions of Proposition 2.7 and α, $\beta \geq \alpha$ are $W^{2,1}$-lower and upper solutions of (2.1). Assume further α or β is not a solution and satisfies conditions (c) and (d) of Proposition 2.7. Prove then that $\alpha \prec \beta$.

2.2 Existence and Multiplicity Results

The main result of this section relates strict lower and upper solutions with degree theory.

Theorem 2.8. *Let* α *and* $\beta \in C([a, b])$ *be strict lower and upper solutions of the problem* (2.1) *such that* $\alpha \prec \beta$. *Define* $A \subset [a, b]$ *(resp.* $B \subset [a, b]$*) to be the set of points where* α *(resp.* β*) is derivable. Let* E *be defined by* (1.8),

$$r = \max\{\frac{\beta(a) - \alpha(b)}{b - a}, \frac{\beta(b) - \alpha(a)}{b - a}\}$$

and let p, $q \in [1, \infty]$ *be such that* $\frac{1}{p} + \frac{1}{q} = 1$.

Assume $f : E \to \mathbb{R}$ satisfies L^p-Carathéodory conditions and there exists $N \in L^1(a,b)$, $N > 0$ such that for a.e. $t \in A$ (resp. for a.e. $t \in B$)

$$f(t,\alpha(t),\alpha'(t)) \geq -N(t) \quad (\textit{resp. } f(t,\beta(t),\beta'(t)) \leq N(t)).$$

Assume moreover there exist $\varphi \in C(\mathbb{R}^+, \mathbb{R}_0^+)$, $\psi \in L^p(a,b)$ and $R > r$ such that

$$\int_r^R \frac{s^{1/q}}{\varphi(s)}\, ds > \|\psi\|_{L^p}(\max_t \beta(t) - \min_t \alpha(t))^{1/q}$$

and that the function f satisfies for a.e. $t \in [a,b]$ and all (u,v) with $(t,u,v) \in E$,

$$|f(t,u,v)| \leq \psi(t)\varphi(|v|).$$

Then

$$\deg(I - T, \Omega) = 1,$$

where $T : C_0^1([a,b]) \to C_0^1([a,b])$ is defined by

$$(Tu)(t) := \int_a^b G(t,s) f(s,u(s),u'(s))\, ds,$$

with $G(t,s)$ the Green's function corresponding to (2.3) and

$$\Omega = \{u \in C_0^1([a,b]) \mid \alpha \prec u \prec \beta, \|u'\|_\infty < R\}. \tag{2.5}$$

In particular, the problem (2.1) has at least one solution $u \in W^{2,p}(a,b)$ such that

$$\alpha \prec u \prec \beta.$$

Proof. Increasing N if necessary, we can assume $N(t) \geq |f(t,u,v)|$ if $t \in [a,b]$, $\alpha(t) \leq u \leq \beta(t)$ and $|v| \leq R$. Define then

$$\begin{aligned}
\bar{f}(t,u,v) &= \max\{\min\{f(t,\gamma(t,u),v), N(t)\}, -N(t)\},\\
\omega_1(t,\delta) &= \chi_A(t) \max_{|v|\leq\delta} |\bar{f}(t,\alpha(t),\alpha'(t)+v) - \bar{f}(t,\alpha(t),\alpha'(t))|,\\
\omega_2(t,\delta) &= \chi_B(t) \max_{|v|\leq\delta} |\bar{f}(t,\beta(t),\beta'(t)+v) - \bar{f}(t,\beta(t),\beta'(t))|,
\end{aligned}$$

where γ is defined from (I-1.3), χ_A and χ_B are the characteristic functions of the sets A and B.

We consider now the modified problem

$$\begin{aligned}
u'' &= \bar{f}(t,u,u') - \omega(t,u),\\
u(a) &= 0, \ u(b) = 0,
\end{aligned} \tag{2.6}$$

where

$$\begin{aligned}
\omega(t,u) &= -\omega_2(t, u - \beta(t)), && \text{if } u > \beta(t),\\
&= 0, && \text{if } \alpha(t) \leq u \leq \beta(t),\\
&= \omega_1(t, \alpha(t) - u), && \text{if } u < \alpha(t).
\end{aligned}$$

This problem is equivalent to the fixed point problem

$$u = \bar{T}u,$$

where $\bar{T} : C_0^1([a,b]) \to C_0^1([a,b])$ is defined by

$$(\bar{T}u)(t) = \int_a^b G(t,s)[\bar{f}(s,u(s),u'(s)) - \omega(s,u(s))]\, ds.$$

Observe that $\bar{T}$ is completely continuous and we can find $\bar{R}$ large enough so that $\Omega \subset B(0,\bar{R})$ and $\bar{T}(C_0^1([a,b])) \subset B(0,\bar{R})$. Hence we have, by the properties of the degree,

$$\deg(I - \bar{T}, B(0,\bar{R})) = 1.$$

We know that every fixed point of $\bar{T}$ is a solution of (2.6) and arguing as in Step 3 and 4 of the proof of Theorem II-2.1 we prove $\alpha \le u \le \beta$ and $\|u'\|_\infty < R$. As α and β are strict, $\alpha \prec u \prec \beta$. Hence, every fixed point of $\bar{T}$ is in Ω and by the excision property we obtain

$$\deg(I - T, \Omega) = \deg(I - \bar{T}, \Omega) = \deg(I - \bar{T}, B(0,\bar{R})) = 1.$$

Existence of a solution u such that

$$\alpha \prec u \prec \beta$$

follows now from the properties of the degree. □

Exercise 2.2. Consider Theorem 2.8 with a one-sided Nagumo condition.

Exercise 2.3. Extend Theorem 2.8 to the separated boundary value problem

$$\begin{aligned} u'' &= f(t,u,u'), \\ a_1 u(a) - a_2 u'(a) &= A_0, \\ b_1 u(b) + b_2 u'(b) &= B_0, \end{aligned}$$

in case A_0, $B_0 \in \mathbb{R}$, a_1, $b_1 \in \mathbb{R}$, a_2, $b_2 \in \mathbb{R}^+$, $a_1^2 + a_2^2 > 0$ and $b_1^2 + b_2^2 > 0$.

Hint : See [31].

In the case where f does not depend on the derivative u', this result reduces to the following theorem.

Theorem 2.9. *Let α and $\beta \in \mathcal{C}([a,b])$ be strict $W^{2,1}$-lower and upper solutions of the problem*

$$\begin{aligned} u'' &= f(t,u), \\ u(a) &= 0,\, u(b) = 0, \end{aligned} \tag{2.7}$$

such that $\alpha \prec \beta$. Define $E := \{(t,u) \in [a,b] \times \mathbb{R} \mid \alpha(t) \le u \le \beta(t)\}$ and assume $f : E \to \mathbb{R}$ is an L^1-Carathéodory function.

Then, for $R > 0$ large enough,

$$\deg(I - T, \Omega) = 1,$$

where Ω is defined by (2.5), $T : \mathcal{C}_0^1([a,b]) \to \mathcal{C}_0^1([a,b])$ by

$$(Tu)(t) := \int_a^b G(t,s) f(s,u(s))\, ds$$

and $G(t,s)$ is the Green's function corresponding to the problem (2.3).

In particular, the problem (2.7) has at least one solution $u \in W^{2,1}(a,b)$ such that

$$\alpha \prec u \prec \beta.$$

In this theorem, it is essential to define T on $\mathcal{C}_0^1([a,b])$, since

$$\Omega^* := \{u \in \mathcal{C}_0([a,b]) \mid \alpha \prec u \prec \beta\}$$

is not open in the $\mathcal{C}$-topology.

As in Section II-4 we can also consider the case of an $\mathcal{A}$-Carathéodory function f. As already observed, in that case, we look for solutions in $W^{2,\mathcal{A}}(a,b)$ and it is not meaningful to define the corresponding fixed point problem on $\mathcal{C}_0^1([a,b])$. This remark forces us to make stronger assumptions on strict lower and upper solutions.

Theorem 2.10. *Let α and $\beta \in \mathcal{C}([a,b])$ be strict $W^{2,1}$-lower and upper solutions of the problem (2.7) such that*

$$\alpha(a) < 0 < \beta(a), \quad \alpha(b) < 0 < \beta(b) \quad \text{and} \quad \forall t \in \,]a,b[,\ \alpha(t) < \beta(t).$$

Define $E := \{(t,u) \in [a,b] \times \mathbb{R} \mid \alpha(t) \le u \le \beta(t)\}$ and assume $f : E \to \mathbb{R}$ is an $\mathcal{A}$-Carathéodory function.

Then

$$\deg(I - T, \Omega) = 1,$$

where

$$\Omega = \{u \in \mathcal{C}_0([a,b]) \mid \forall t \in [a,b],\ \alpha(t) < u(t) < \beta(t)\},$$

$T : C_0([a,b]) \to C_0([a,b])$ is defined by

$$(Tu)(t) := \int_a^b G(t,s) f(s,u(s))\, ds$$

and $G(t,s)$ is the Green's function corresponding to the problem (2.3).

In particular, the problem (2.7) *has at least one solution $u \in W^{2,\mathcal{A}}(a,b)$ such that for all $t \in [a,b]$,*

$$\alpha(t) < u(t) < \beta(t).$$

Exercise 2.4. Prove Theorem 2.10.

The simplest multiplicity result that we can deduce from Theorem 2.8 is obtained when we have two pairs of strict lower and upper solutions.

Theorem 2.11. (The Three Solutions Theorem) *Let α_1, β_1 and α_2, $\beta_2 \in C([a,b])$ be two pairs of $W^{2,1}$-lower and upper solutions of* (2.1) *such that on $[a,b]$*

$$\alpha_1(t) \le \beta_1(t),\ \alpha_1(t) \le \beta_2(t),\ \alpha_2(t) \le \beta_2(t)$$

and there exists $t_0 \in [a,b]$ with

$$\alpha_2(t_0) > \beta_1(t_0).$$

Assume further β_1 and α_2 are strict $W^{2,1}$-upper and lower solutions. Define $A_i \subset [a,b]$ (resp. $B_i \subset [a,b]$) to be the set of points where α_i (resp. β_i) is derivable.

Let E be defined by (1.17), *p, $q \in [1,\infty]$ be such that $\frac{1}{p} + \frac{1}{q} = 1$ and $r = \max\{\frac{\beta_2(a)-\alpha_1(b)}{b-a}, \frac{\beta_2(b)-\alpha_1(a)}{b-a}\}$. Suppose $f : E \to \mathbb{R}$ is an L^p-Carathéodory function and there exists $N \in L^1(a,b)$, $N > 0$ such that for $i = 1,2$ and a.e. $t \in A_i$ (resp. a.e. $t \in B_i$)*

$$f(t,\alpha_i(t),\alpha_i'(t)) \ge -N(t) \quad (\textit{resp. } f(t,\beta_i(t),\beta_i'(t)) \le N(t)).$$

Assume moreover there exist $\varphi \in C(\mathbb{R}^+, \mathbb{R}_0^+)$, $\psi \in L^p(a,b)$ and $R > r$ such that

$$\int_r^R \frac{s^{1/q}}{\varphi(s)}\, ds > \|\psi\|_{L^p} (\max_t \beta_2(t) - \min_t \alpha_1(t))^{1/q}$$

holds and for a.e. $t \in [a,b]$ and all (u,v) with $(t,u,v) \in E$,

$$|f(t,u,v)| \le \psi(t)\varphi(|v|).$$

Then the problem (2.1) *has at least three solutions $u_1, u_2, u_3 \in W^{2,p}(a,b)$ such that*

$$\alpha_1 \le u_1 \prec \beta_1,\ \alpha_2 \prec u_2 \le \beta_2,\ u_1 \le u_3 \le u_2$$

and there exist t_1, $t_2 \in [a,b]$ *with*

$$u_3(t_1) > \beta_1(t_1), \; u_3(t_2) < \alpha_2(t_2).$$

Exercise 2.5. Prove the preceding result adapting the argument of Theorem 1.13.

As in the periodic case, another way to obtain multiplicity result is to exhibit domains $\Omega_1 \supset \Omega$ such that

$$\deg(I - T, \Omega_1) = 0 \quad \text{and} \quad \deg(I - T, \Omega) = 1.$$

Such a result can be obtained if we assume that we have an *a priori* bound on all the upper solutions which satisfies the boundary conditions. Here we assume f does not depend on the derivative u'.

Theorem 2.12. *Let $k > 0$ and α, $\beta \in C([a,b])$ be respectively a $W^{2,1}$-lower solution and a strict $W^{2,1}$-upper solution of* (2.7) *with $\alpha \le \beta \le k$.*

Let $E = \{(t,u) \in [a,b] \times \mathbb{R} \mid \alpha(t) \le u\}$ and $f : E \to \mathbb{R}$ be an L^1-Carathéodory function.

Assume moreover that for all solutions $u \in W^{2,1}(a,b)$ of

$$u'' \le f(t,u),$$
$$u(a) = 0, \, u(b) = 0,$$

with $u \ge \alpha$ we have $u < k$ on $[a,b]$.

Then the problem (2.7) *has at least two solutions u_1 and $u_2 \in W^{2,1}(a,b)$ such that*

$$\alpha \le u_1 \prec \beta, \; u_1 \le u_2$$

and there exists $t_0 \in [a,b]$ with

$$u_2(t_0) > \beta(t_0).$$

Proof. Consider the modified problem

$$\begin{aligned} u'' &= f(t, \gamma_1(t,u)) - s, \\ u(a) &= 0, \; u(b) = 0, \end{aligned} \tag{2.8}$$

where $\gamma_1(t,u) = \max\{\alpha(t), u\}$. Solutions of (2.8) solve the fixed point problem

$$u(t) = (Tu)(t) - s\tfrac{(t-a)(b-t)}{2},$$

where

$$(Tu)(t) = \int_a^b G(t,s) f(t, \gamma_1(t,u))\, ds$$

and $G(t,s)$ is the Green's function corresponding to (II-4.2).

As in the proof of Theorem 1.14, we prove that solutions of (2.8) with $s \geq 0$ are such that $\alpha \leq u < k$. Hence $\|Tu - u\|_\infty = s\frac{(b-a)^2}{8}$ is *a priori* bounded, which implies that for $s = s_0$ large enough, (2.8) has no solution. Now we deduce from the equation that there exists some $R > 0$ so that solutions u of (2.8) with $0 \leq s \leq s_0$ satisfy $\|u'\|_\infty \leq R$. The rest of the proof follows as in Theorem 1.14. □

Again we can consider the case where f satisfies $\mathcal{A}$-Carathéodory conditions.

Theorem 2.13. *Let $k > 0$ and α, $\beta \in C([a,b])$ be respectively a $W^{2,1}$-lower solution and a strict $W^{2,1}$-upper solution of (2.7) with $\alpha \leq \beta \leq k$, $\beta(a) > 0$ and $\beta(b) > 0$.*

Let $E = \{(t,u) \in [a,b] \times \mathbb{R} \mid \alpha(t) \leq u\}$ and $f : E \to \mathbb{R}$ be an $\mathcal{A}$-Carathéodory function.

Assume moreover that for all solutions $u \in W^{2,1}(a,b)$ of

$$u'' \leq f(t,u),$$
$$u(a) = 0,\ u(b) = 0,$$

with $u \geq \alpha$ we have $u < k$ on $[a,b]$.

Then the problem (2.7) has at least two solutions u_1 and $u_2 \in W^{2,\mathcal{A}}(a,b)$ such that for all $t \in [a,b]$

$$\alpha(t) \leq u_1(t) < \beta(t),\ u_1(t) \leq u_2(t)$$

and there exists $t_0 \in [a,b]$ with

$$u_2(t_0) > \beta(t_0).$$

Exercise 2.6. Prove the above theorem.

3 Non Well-ordered Lower and Upper Solutions

3.1 Introduction

We noticed already in Chapter I that the method of lower and upper solutions depends strongly on the ordering $\alpha \leq \beta$. Consider for example the periodic problem

$$\begin{aligned} u'' + u &= \sin t, \\ u(0) = u(2\pi),\ u'(0) &= u'(2\pi). \end{aligned} \tag{3.1}$$

Clearly this problem has no solution although $\alpha(t) = 1$ is a lower solution and $\beta(t) = -1 \leq \alpha(t)$ is an upper solution. On the other hand, lower and upper solutions α, β satisfying the reversed ordering condition $\beta \le \alpha$ arise naturally in situations where the corresponding problem has a solution. As a very simple example, we can consider the linear problem

$$\begin{gathered} u'' + \tfrac{2}{3}u = \sin t, \\ u(0) = u(2\pi),\ u'(0) = u'(2\pi). \end{gathered}$$

The functions $\alpha = \frac{3}{2}$ and $\beta = -\frac{3}{2}$ are lower and upper solutions such that $\alpha \ge \beta$. Notice, however, that the unique solution $u(t) = -3\sin t$ does not lie between the lower and the upper solution.

More generally, we can consider the problem

$$\begin{gathered} u'' + \lambda u = \sin t, \\ u(0) = u(2\pi),\ u'(0) = u'(2\pi), \end{gathered} \tag{3.2}$$

where λ lies between the two first eigenvalues of the corresponding periodic problem, i.e. $0 < \lambda < 1$. It is easy to see that, in such a case, there exist constant lower and upper solutions of (3.2) which are in the reversed order. Further, the solution u of (3.2) lies between all constant upper and lower solution, $\beta \le u \le \alpha$, if and only if $0 < \lambda \le \frac{1}{2}$. We shall see in Section 3.5 that if we reinforce this condition, i.e. $\lambda \le \frac{1}{4}$, such a result holds true for any problem

$$\begin{gathered} u'' + \lambda u = f(t), \\ u(0) = u(2\pi),\ u'(0) = u'(2\pi), \end{gathered}$$

where f is continuous. More precisely, the solution lies between every pair of lower and upper solutions with $\alpha \ge \beta$. This limitation is optimal since we can find, for $\lambda > \frac{1}{4}$, forcings f and lower and upper solutions $\alpha \ge \beta$ such that the corresponding solution does not lie between β and α.

The use of the method of lower and upper solutions is strongly related to the interaction of the nonlinearity with the eigenvalues of the corresponding linear problem. Consider the Dirichlet problem

$$\begin{gathered} u'' + f(t,u) = 0, \\ u(a) = 0,\ u(b) = 0, \end{gathered} \tag{3.3}$$

together with lower and upper solutions α and β such that

$$\begin{aligned} \alpha''(t) + f(t,\alpha(t)) \ge 0, \quad \alpha(a) \le 0,\ \alpha(b) \le 0, \\ \beta''(t) + f(t,\beta(t)) \le 0, \quad \beta(a) \ge 0,\ \beta(b) \ge 0. \end{aligned}$$

Let us assume $\alpha \le \beta$ and

$$\frac{f(t,u) - f(t,v)}{u - v} \ge \lambda. \tag{3.4}$$

It follows that

$$-(\beta''(t)-\alpha''(t)) \geq f(t,\beta(t))-f(t,\alpha(t)) \geq \lambda(\beta(t)-\alpha(t))$$

and integrating twice by part, we obtain

$$\lambda \int_a^b (\beta(t)-\alpha(t)) \sin\left(\pi \tfrac{t-a}{b-a}\right) dt \leq \left(\tfrac{\pi}{b-a}\right)^2 \int_a^b (\beta(t)-\alpha(t)) \sin\left(\pi \tfrac{t-a}{b-a}\right) dt.$$

If furthermore $\lambda > \lambda_1$, where $\lambda_1 = (\frac{\pi}{b-a})^2$ is the first eigenvalue of the Dirichlet problem, then this inequality implies $\alpha = \beta$ which is then a solution of (3.3). This observation means that it is not possible to work the method of lower and upper solutions, with $\alpha \leq \beta$, if the nonlinearity satisfies (3.4) with $\lambda > \lambda_1$, i.e. if it is "above the first eigenvalue". This is a fundamental limitation of the method.

A dual conclusion holds if we assume α and β satisfy the reversed order $\alpha \geq \beta$ and

$$\frac{f(t,u)-f(t,v)}{u-v} \leq \lambda. \tag{3.5}$$

In this case, we deduce $\alpha(a) = \beta(a) = 0$, $\alpha(b) = \beta(b) = 0$,

$$-(\alpha''(t)-\beta''(t)) \leq f(t,\alpha(t))-f(t,\beta(t)) \leq \lambda(\alpha(t)-\beta(t))$$

and

$$\begin{aligned}\int_a^b (\alpha'-\beta')^2\,ds - \lambda \int_a^b (\alpha-\beta)^2\,ds \\ = -\int_a^b (\alpha-\beta)''(\alpha-\beta)\,ds - \lambda\int_a^b (\alpha-\beta)^2\,ds \leq 0.\end{aligned}$$

If $\lambda < \lambda_1$, the above inequality implies $\alpha = \beta$ and this function is a solution of (3.3). It follows now that it is not possible to work a method of lower and upper solutions, with $\beta \leq \alpha$, if the nonlinearity satisfies (3.5) with $\lambda < \lambda_1$, i.e. if it is "under the first eigenvalue".

It is easy to see that the same conclusions hold for other boundary value problems such as the periodic one

$$\begin{gathered}u'' + f(t,u) = 0,\\ u(0) = u(2\pi),\ u'(0) = u'(2\pi).\end{gathered}$$

Here the first eigenvalue is $\lambda_1 = 0$ and the method does not work with $\alpha \leq \beta$, if the nonlinearity satisfies (3.4) with $\lambda > 0$ or with $\beta \leq \alpha$, if the nonlinearity satisfies (3.5) with $\lambda < 0$. For the periodic problem we can complete this remark. If we assume the lower and upper solutions, α and β, to be not "well-ordered", i. e. $\alpha(t_0) > \beta(t_0)$ for some $t_0 \in [0, 2\pi]$, and

$$\frac{f(t,u)-f(t,v)}{u-v} \leq \tfrac{1}{4},$$

we can prove $\alpha \geq \beta$. Assume by contradiction $\alpha(t_1) < \beta(t_1)$ for some t_1. Extending $u := \alpha - \beta$ by periodicity, we define then

$$s_1 := \inf\{t \leq t_0 \mid u > 0 \text{ on } [t, t_0]\}$$

and

$$s_2 := \sup\{t \geq t_0 \mid u > 0 \text{ on } [t_0, t]\} < s_1 + 2\pi.$$

The function u satisfies the Dirichlet problem

$$u'' + q(t)u = \sigma(t), \quad u(s_1) = 0, \ u(s_2) = 0,$$

where $q(t) \leq \frac{1}{4}$ and $\sigma \geq 0$. It is easy to see that the first eigenvalue $\tilde{\lambda}$ of the problem

$$u'' + q(t)u + \lambda u = 0, \quad u(s_1) = 0, \ u(s_2) = 0,$$

is positive. It follows then from the maximum principle (Theorem A-5.2) that $u < 0$ on $]s_1, s_2[$ which is a contradiction.

The existence of lower an upper solutions α, β such that $\alpha \geq \beta$ is not sufficient to guarantee the solvability of this problem which is clear from Problem (3.1). The difficulty there comes from the interference of the nonlinearity with the second eigenvalue of the problem $\lambda_2 = 1$. In the theorems we present in this section, we will put assumptions so that the nonlinearity remains "below this second eigenvalue".

3.2 A Periodic Non-resonance Problem

Consider the problem

$$\begin{aligned} u'' + \lambda u &= f(t, u, u'), \\ B(u(a), u(b), u'(a), u'(b)) &= 0, \end{aligned}$$

where f is bounded and B represents the boundary conditions. This problem can be thought of as a perturbation of the linear eigenvalue problem

$$\begin{aligned} u'' + \lambda u &= 0, \\ B(u(a), u(b), u'(a), u'(b)) &= 0. \end{aligned}$$

In this section, we shall apply the method of lower and upper solutions to such problems when λ is the first eigenvalue.

Our first result concerns the periodic problem

$$\begin{aligned} u'' &= f(t, u, u'), \\ u(a) &= u(b), \ u'(a) = u'(b) \end{aligned} \tag{3.6}$$

with one-sided boundedness of the nonlinearity. Here the first eigenvalue is $\lambda_1 = 0$.

Theorem 3.1. *Let α and $\beta \in \mathcal{C}([a,b])$ be $W^{2,1}$-lower and upper solutions of (3.6) such that $\alpha \not\leq \beta$. Define $A \subset [a,b]$ (resp. $B \subset [a,b]$) to be the set of points where α (resp. β) is derivable.*

Assume $f : [a,b] \times \mathbb{R}^2 \to \mathbb{R}$ is an L^1-Carathéodory function and there exists $N \in L^1(a,b)$, $N > 0$ such that for a.e. $t \in A$ (resp. a.e. $t \in B$)

$$f(t, \alpha(t), \alpha'(t)) \geq -N(t) \quad (\textit{resp. } f(t, \beta(t), \beta'(t)) \leq N(t)).$$

Assume further that for some $h \in L^1(a,b)$ either

$$f(t,u,v) \leq h(t) \quad \text{on } [a,b] \times \mathbb{R}^2,$$

or

$$f(t,u,v) \geq h(t) \quad \text{on } [a,b] \times \mathbb{R}^2.$$

Then there exists a solution u of (3.6) in

$$\mathcal{S} := \{u \in \mathcal{C}^1([a,b]) \mid \exists t_1, t_2 \in [a,b],\ u(t_1) \geq \beta(t_1),\ u(t_2) \leq \alpha(t_2)\}. \quad (3.7)$$

Proof. For each $r \geq 2(b-a)^2$, we define

$$\begin{aligned} f_r(t,u,v) &= f(t,u,v), && \text{if } |u| < r, \\ &= (1 + r - |u|) f(t,u,v) + (|u| - r)\tfrac{u}{r}, && \text{if } r \leq |u| < r+1, \\ &= \tfrac{u}{r}, && \text{if } r + 1 \leq |u|, \end{aligned}$$

and consider the problem

$$\begin{gathered} u'' = f_r(t,u,u'), \\ u(a) = u(b),\ u'(a) = u'(b). \end{gathered} \quad (3.8)$$

Claim – There exists $k > 0$ such that for any $r \geq 2(b-a)^2$, solutions u of (3.8), which are in $\mathcal{S}$, are such that $\|u\|_\infty \leq k$. Consider the case $f(t,u,v) \leq h(t)$ on $[a,b] \times \mathbb{R}^2$. Let $u \in \mathcal{S}$ be a solution of (3.8) and let t_0, t_1 and $t_2 \in [a,b]$ be such that

$$u'(t_0) = 0, \quad u(t_1) \geq \beta(t_1) \geq -\|\beta\|_\infty \quad \text{and} \quad u(t_2) \leq \alpha(t_2) \leq \|\alpha\|_\infty.$$

Extending u by periodicity, we can write for $t \in [t_0, t_0 + b - a]$

$$u'(t) = -\int_t^{t_0+b-a} f_r(s, u(s), u'(s))\, ds \geq -\|h\|_{L^1} - \tfrac{\|u\|_\infty}{2(b-a)}.$$

It follows that for $t \in [t_1, t_1 + b - a]$

$$u(t) = u(t_1) + \int_{t_1}^t u'(s)\, ds \geq -\|\beta\|_\infty - \|h\|_{L^1}(b-a) - \tfrac{\|u\|_\infty}{2}$$

and for $t \in [t_2 - b + a, t_2]$

$$u(t) = u(t_2) - \int_t^{t_2} u'(s)\,ds \leq \|\alpha\|_\infty + \|h\|_{L^1}(b-a) + \tfrac{\|u\|_\infty}{2}.$$

Hence, we have

$$\|u\|_\infty \leq 2(\|\alpha\|_\infty + \|\beta\|_\infty + \|h\|_{L^1}(b-a)) =: k.$$

A similar argument holds if $f(t, u, v) \geq h(t)$.

Conclusion – Consider the problem (3.8), with $r > \max\{k, 2(b-a)^2\}$. It is easy to see that $\alpha_1 = -r - 2$ and $\beta_2 = r + 2$ are lower and upper solutions.

Assume β is not a strict upper solution. There exists then a solution u of (3.8) such that $u \leq \beta$ and for some $t_1 \in [a, b]$, $u(t_1) = \beta(t_1)$. As further $\alpha \not\leq \beta$, there exists $t_2 \in [a, b]$ such that $\alpha(t_2) > \beta(t_2)$. It follows that $\alpha(t_2) > u(t_2)$, $u \in \mathcal{S}$, and we deduce from the claim that $\|u\|_\infty \leq k$. Hence, u is a solution of (3.6) in $\mathcal{S}$.

We come to the same conclusion if α is not a strict lower solution.

Suppose now that $\beta_1 = \beta$ and $\alpha_2 = \alpha$ are strict upper and lower solutions. We deduce then from Theorem 1.13 the existence of three solutions of (3.8) one of them, u, being such that for some t_1, $t_2 \in [a, b]$, $u(t_1) > \beta(t_1)$ and $u(t_2) < \alpha(t_2)$. Hence, $u \in \mathcal{S}$ and from the claim $\|u\|_\infty < k$. This implies that u solves (3.6) and proves the theorem. □

Example 3.1. The problem

$$u'' = f(u), \quad u(-1) = u(1),\ u'(-1) = u'(1),$$

where

$$\begin{aligned} f(u) &= 12|u|^{1/2}, && \text{if } u \leq 4,\\ &= 28 - u, && \text{if } 4 < u \leq 28,\\ &= 0, && \text{if } 28 < u, \end{aligned}$$

is such that $\beta(t) = t^4$ and $\alpha(t) = 28 \geq \beta(t)$ are respectively upper and lower solutions. The solutions in the corresponding set $\mathcal{S}$ are the constant functions 0 and 28 but are not in the interior of $\mathcal{S}$.

A possible generalization concerns the use of several lower and upper solutions.

Exercise 3.1. Let $\alpha_i \in C([a, b])$ $(i = 1, \ldots, n)$ and $\beta_j \in C([a, b])$ $(j = 1, \ldots, m)$ be respectively $W^{2,1}$-lower and upper solutions of (3.6) such that

$$\alpha := \max_{1 \leq i \leq n} \alpha_i \not\leq \beta := \min_{1 \leq j \leq m} \beta_j.$$

Assume $f : [a,b] \times \mathbb{R}^2 \to \mathbb{R}$ is an Carathéodory function such that for some $h \in L^1(a,b)$ either

$$f(t,u,v) \leq h(t) \quad \text{on } [a,b] \times \mathbb{R}^2,$$

or

$$f(t,u,v) \geq h(t) \quad \text{on } [a,b] \times \mathbb{R}^2.$$

Prove there exists a solution $u \in \mathcal{S}$ of (3.6), where $\mathcal{S}$ is defined in (3.7).

In a similar way, we can consider the Liénard equation

$$\begin{aligned} &u'' + g(u)u' + f(t,u) = 0, \\ &u(a) = u(b), \ u'(a) = u'(b). \end{aligned} \tag{3.9}$$

Theorem 3.2. *Let α and $\beta \in C([a,b])$ be $W^{2,1}$-lower and upper solutions of (3.9) such that $\alpha \not\leq \beta$. Assume $g \in C(\mathbb{R})$ and $f : [a,b] \times \mathbb{R} \to \mathbb{R}$ is an L^1-Carathéodory function such that for some $h \in L^1(a,b)$ either*

$$f(t,u) \leq h(t) \quad \textit{on } [a,b] \times \mathbb{R},$$

or

$$f(t,u) \geq h(t) \quad \textit{on } [a,b] \times \mathbb{R}.$$

Then there exists a solution $u \in \mathcal{S}$ of (3.9), where $\mathcal{S}$ is defined in (3.7).

Proof. For each $r \geq 8(b-a)^2$, we define

$$\begin{aligned} f_r(t,u) &= f(t,u), & &\text{if } |u| < r, \\ &= (1 + r - |u|)f(t,u) - (|u| - r)\tfrac{u}{r}, & &\text{if } r \leq |u| < r+1, \\ &= -\tfrac{u}{r}, & &\text{if } r + 1 \leq |u|, \end{aligned}$$

and consider the problem

$$\begin{aligned} &u'' + g(u)u' + f_r(t,u) = 0, \\ &u(a) = u(b), \ u'(a) = u'(b). \end{aligned} \tag{3.10}$$

Claim – There exists $k > 0$ such that for any $r \geq 8(b-a)^2$, solutions u of (3.10), which are in $\mathcal{S}$, are such that $\|u\|_\infty \leq k$. Consider the case $f(t,u) \geq h(t)$ on $[a,b] \times \mathbb{R}$. Let $u \in \mathcal{S}$ be a solution of (3.10). Integrating (3.10), or multiplying this equation by u and integrating, we obtain respectively

$$\int_a^b f_r(t,u(t))\,dt = 0 \quad \text{and} \quad \int_a^b u'^2(t)\,dt = \int_a^b f_r(t,u(t))u(t)\,dt.$$

It follows that

$$\|u'\|_{L^2}^2 = \int_a^b f_r(t,u(t))(u(t)-\|u\|_\infty)\,dt \le 2(\|h\|_{L^1} + \tfrac{\|u\|_\infty}{r}(b-a))\|u\|_\infty,$$

i.e.

$$\|u'\|_{L^2} \le (2\|h\|_{L^1}\|u\|_\infty)^{1/2} + \tfrac{\|u\|_\infty}{2(b-a)^{1/2}}.$$

Define now t_1 and $t_2 \in [a,b]$ to be such that

$$u(t_1) \ge \beta(t_1) \ge -\|\beta\|_\infty \quad \text{and} \quad u(t_2) \le \alpha(t_2) \le \|\alpha\|_\infty.$$

Extending u by periodicity, we can write for $t \in [t_1, t_1+b-a]$

$$\begin{aligned} u(t) &= u(t_1) + \int_{t_1}^t u'(s)\,ds \ge -\|\beta\|_\infty - \|u'\|_{L^2}(b-a)^{1/2} \\ &\ge -\|\beta\|_\infty - (2(b-a)\|h\|_{L^1}\|u\|_\infty)^{1/2} - \tfrac{\|u\|_\infty}{2} \end{aligned}$$

and for $t \in [t_2-b+a, t_2]$

$$u(t) = u(t_2) - \int_t^{t_2} u'(s)\,ds \le \|\alpha\|_\infty + (2(b-a)\|h\|_{L^1}\|u\|_\infty)^{1/2} + \tfrac{\|u\|_\infty}{2}.$$

Hence, we obtain for some $k > 0$

$$\|u\|_\infty \le k.$$

A similar argument holds if $f(t,u) \le h(t)$.

Conclusion – The rest of the proof repeats the argument used in Theorem 3.1. □

3.3 Interaction with the First Fučík Curve

The reason of imposing in Theorem 3.1 a boundedness assumption, such as $f(t,u,v) \le h(t)$, is to make sure the nonlinearity does not interfere with the second eigenvalue of the corresponding linear problem. An alternative way to prevent such an interference is to assume some asymptotic control on the quotient $f(t,u,v)/u$ as $|u|$ goes to infinity. We can generalize further assuming different behaviours of this quotient as u goes to plus or minus infinity. The following theorem concerns such a generalization. To simplify, we consider the problem

$$\begin{gathered} u'' + f(t,u) = 0, \\ u(a) = u(b),\ u'(a) = u'(b). \end{gathered} \tag{3.11}$$

Theorem 3.3. *Let α and $\beta \in C([a,b])$ be $W^{2,1}$-lower and upper solutions of* (3.11) *such that $\alpha \not\leq \beta$.*

Let $f : [a,b] \times \mathbb{R} \to \mathbb{R}$ be an L^1-Carathéodory function such that for some functions $a_\pm \leq 0$, $b_\pm \geq 0$ in $L^1(a,b)$,

$$a_\pm(t) \leq \liminf_{u\to\pm\infty} \frac{f(t,u)}{u} \leq \limsup_{u\to\pm\infty} \frac{f(t,u)}{u} \leq b_\pm(t), \tag{3.12}$$

uniformly in $t \in [a,b]$.

Assume further that for any p, $q \in L^1(a,b)$, with $a_+ \leq p \leq b_+$ and $a_- \leq q \leq b_-$, the nontrivial solutions of

$$\begin{aligned} &u'' + p(t)u^+ - q(t)u^- = 0,\\ &u(a) = u(b),\ u'(a) = u'(b), \end{aligned} \tag{3.13}$$

where $u^+(t) = \max\{u(t),0\}$ and $u^-(t) = \max\{-u(t),0\}$, do not have zeros.

Then the problem (3.11) *has at least one solution $u \in \mathcal{S}$, where $\mathcal{S}$ is defined in* (3.7).

Proof. Step 1 – Claim : There exists $\epsilon > 0$ so that for any p, $q \in L^1(a,b)$, with $a_+ - \epsilon \leq p \leq b_+ + \epsilon$ and $a_- - \epsilon \leq q \leq b_- + \epsilon$, the nontrivial solutions of (3.13) *do not have zeros.* If the claim were wrong, there would exist sequences $(p_n)_n$, $(q_n)_n \subset L^1(a,b)$, $(t_n)_n \subset [a,b]$ and $(u_n)_n \subset W^{2,1}(a,b)$ so that $a_+ - 1/n \leq p_n \leq b_+ + 1/n$, $a_- - 1/n \leq q_n \leq b_- + 1/n$ and u_n is a solution of (3.13) (with $p = p_n$ and $q = q_n$) such that $\|u_n\|_{C^1} = 1$ and $u_n(t_n) = 0$. Going to subsequences we can assume, using the Dunford-Pettis Theorem (see [43]),

$$p_n \rightharpoonup p,\ q_n \rightharpoonup q \text{ in } L^1(a,b), \quad u_n \to u \text{ in } C^1([a,b]), \quad t_n \to t_0.$$

It follows that $a_+ \leq p \leq b_+$, $a_- \leq q \leq b_-$, u is a solution of (3.13) and $u(t_0) = 0$ which contradicts the assumption.

Step 2 – The modified problem. Let us choose $R > 0$ large enough so that

$$\begin{aligned} a_+ - \epsilon \leq g_+(t,u) &= \frac{f(t,u)}{u} \leq b_+ + \epsilon, \quad \text{for } u \geq R,\\ a_- - \epsilon \leq g_-(t,u) &= \frac{f(t,u)}{u} \leq b_- + \epsilon, \quad \text{for } u \leq -R \end{aligned}$$

and extend these functions on $[a,b] \times \mathbb{R}$ so that these inequalities remain valid. As f is L^1-Carathéodory, there exists $k \in L^1(a,b)$ such that

$$f(t,u) = g_+(t,u)u^+ - g_-(t,u)u^- + h(t,u)$$

and

$$|h(t,u)| \leq k(t).$$

Next, for each $r \geq 1$, we define

$$
\begin{aligned}
g_r^{\pm}(t,u) &= g_{\pm}(t,u), && \text{if } |u| < r,\\
&= (1+r-|u|)g_{\pm}(t,u), && \text{if } r \leq |u| < r+1,\\
&= 0, && \text{if } r+1 \leq |u|,
\end{aligned}
$$

$$
\begin{aligned}
h_r(t,u) &= h(t,u), && \text{if } |u| < r,\\
&= (1+r-|u|)h(t,u), && \text{if } r \leq |u| < r+1,\\
&= 0, && \text{if } r+1 \leq |u|
\end{aligned}
$$

and consider the modified problem

$$
\begin{gathered}
u'' + g_r^+(t,u)u^+ - g_r^-(t,u)u^- + h_r(t,u) = 0,\\
u(a) = u(b),\ u'(a) = u'(b).
\end{gathered}
\tag{3.14}
$$

Step 3 – Claim : There exists $k > 0$ such that, for any $r > k$, solutions u of (3.14), *which are in $\mathcal{S}$, are such that $\|u\|_\infty < k$.* Assume by contradiction, there exist sequences $(r_n)_n$ and $(u_n)_n \subset \mathcal{S}$, where $r_n \geq n$ and u_n is a solution of (3.14) with $r = r_n$ such that $\|u_n\|_\infty \geq n$. As $u_n \in \mathcal{S}$, there exist sequences $(t_{1n})_n$ and $(t_{2n})_n \subset [a,b]$ such that $u_n(t_{1n}) \geq \beta(t_{1n})$ and $u_n(t_{2n}) \leq \alpha(t_{2n})$.

Consider now the functions $v_n = u_n/\|u_n\|_\infty$ which solve the problems

$$
\begin{gathered}
v_n'' + g_{r_n}^+(t,u_n)v_n^+ - g_{r_n}^-(t,u_n)v_n^- + \tfrac{h_{r_n}(t,u_n)}{\|u_n\|_\infty} = 0,\\
v_n(a) = v_n(b),\ v_n'(a) = v_n'(b).
\end{gathered}
$$

Going to subsequence, we can assume as above

$$
g_{r_n}^+(\cdot,u_n) \rightharpoonup p,\ g_{r_n}^-(\cdot,u_n) \rightharpoonup q \ \text{ in } L^1(a,b), \qquad \tfrac{h_{r_n}(t,u_n)}{\|u_n\|_\infty} \to 0 \ \text{ in } L^1(a,b),
$$

$$
v_n \to v \ \text{ in } C^1([a,b]), \quad t_{1n} \to t_1, \quad t_{2n} \to t_2.
$$

It follows that v satisfies (3.13) and by assumption has no zeros. Hence, we come to a contradiction since $v(t_1) \geq 0$ and $v(t_2) \leq 0$ which implies v has a zero.

Conclusion – We deduce from Theorem 3.1 that problem (3.14) with $r > \max\{\|\alpha\|_\infty, \|\beta\|_\infty, k\}$ has a solution $u \in \mathcal{S}$ and conclude from Step 3. □

In this theorem we control asymptotically the nonlinearity using the functions $a_\pm$, $b_\pm$. Next we impose some admissibility condition on the box $[a_+,b_+] \times [a_-,b_-]$ which is to assume that for any $(p,q) \in [a_+,b_+] \times [a_-,b_-]$, the nontrivial solutions of problem (3.13) do not have zeros. Such a condition implies that the nonlinearity does not interfere with the second eigenvalue λ_2 of the periodic problem, i.e. $(\lambda_2,\lambda_2) \notin [a_+,b_+] \times [a_-,b_-]$.

This remark can be completed considering the second curve of the Fučík spectrum. The Fučík spectrum is the set $\mathcal{F}$ of points $(\mu, \nu) \in \mathbb{R}^2$ such that the problem

$$u'' + \mu u^+ - \nu u^- = 0,$$
$$u(a) = u(b), \;\; u'(a) = u'(b),$$

has nontrivial solutions. From explicit computations of the solution it is easy to see that

$$\mathcal{F} = \bigcup_{n=1}^{\infty} \mathcal{F}_n,$$

where

$$\mathcal{F}_1 = \{(\mu, 0) \mid \mu \in \mathbb{R}\} \cup \{(0, \nu) \mid \nu \in \mathbb{R}\}$$

and

$$\mathcal{F}_n = \{(\mu, \nu) \mid \tfrac{1}{\sqrt{\mu}} + \tfrac{1}{\sqrt{\nu}} = \tfrac{b-a}{\pi(n-1)}\}, \quad n = 2, 3, \ldots$$

The following proposition relates the admissibility of the box $[a_+, b_+] \times [a_-, b_-]$ with the Fučík spectrum.

Proposition 3.4. *Let $(\mu, \nu) \in \mathcal{F}_2$ and $p,\, q \in L^1(a, b)$. Assume that*

$$p(t) \leq \mu, \quad q(t) \leq \nu, \quad \textit{for a.e. } t \in [a, b]$$

and for some set $I \subset [a, b]$ of positive measure

$$p(t) < \mu, \quad q(t) < \nu, \quad \textit{for a.e. } t \in I.$$

Then the nontrivial solutions of problem (3.13) *have no zeros.*

Proof. Assume there exists a nontrivial solution u which has a zero. Extend u by periodicity and let t_0 and t_1 be consecutive zeros such that u is positive on $]t_0, t_1[$. Define $v(t) = \sin(\sqrt{\mu}(t - t_0))$ and compute

$$(uv' - vu')\Big|_{t_0}^{t_1} = \int_{t_0}^{t_1} (p(t) - \mu) u(t) v(t)\, dt.$$

If $t_1 - t_0 < \frac{\pi}{\sqrt{\mu}}$, we come to a contradiction

$$0 < -v(t_1) u'(t_1) \leq 0.$$

Hence $t_1 - t_0 \geq \frac{\pi}{\sqrt{\mu}}$ and we only have equality in case $p(t) = \mu$ on $[t_0, t_1]$. Similarly, we prove the distance between two consecutive zeros t_1 and t_2 with u negative on $]t_1, t_2[$ is such that $t_2 - t_1 \geq \frac{\pi}{\sqrt{\nu}}$ with equality if and only if $q(t) = \nu$ on $[t_1, t_2]$. It follows that

$$b - a \geq t_2 - t_0 \geq \frac{\pi}{\sqrt{\mu}} + \frac{\pi}{\sqrt{\nu}} = b - a.$$

These inequalities imply that $t_2 - t_0 = b - a$, $p(t) = \mu$ on $[t_0, t_1]$ and $q(t) = \nu$ on $[t_1, t_2]$ which contradicts the assumptions. □

Remark Assumption (3.12) can be replaced by the following:
For every $\varepsilon > 0$, there is $\gamma_\varepsilon \in L^1(a, b)$ such that, for a.e. $t \in [a, b]$ and every $u \in \mathbb{R}$,

$$f(t, u) = g_+(t, u)u^+ - g_-(t, u)u^- + h(t, u),$$

with $g_\pm$ and h, L^1-Carathéodory functions satisfying

$$a_\pm(t) - \epsilon \leq g_\pm(t, u) \leq b_\pm(t) + \epsilon$$

and

$$|h(t, u)| \leq \gamma_\varepsilon(t).$$

There is a vast literature concerning system which are asymptotic to positively homogeneous problem as in the above theorems. Most of these results can be adapted to a situation where there exist non well-ordered lower and upper solutions. We present here some extensions as exercises.

Exercise 3.2. *Let α and $\beta \in C([a, b])$ be $W^{2,1}$-lower and upper solutions of (3.11) such that $\alpha \not\leq \beta$.*

Let $f : [a, b] \times \mathbb{R} \to \mathbb{R}$ be an L^1-Carathéodory function such that for some functions $a_\pm \leq 0$, $b_- \geq 0$ in $L^1(a, b)$,

$$a_-(t) \leq \liminf_{u \to -\infty} \frac{f(t, u)}{u} \leq \limsup_{u \to -\infty} \frac{f(t, u)}{u} \leq b_-(t),$$
$$a_+(t) \leq \liminf_{u \to +\infty} \frac{f(t, u)}{u},$$

uniformly in $t \in [a, b]$.

Assume further that for any $q \in L^1(a, b)$, with $q \leq b_-$ and any $\bar{t} \in [a, b[$, the problem

$$u'' + q(t)u = 0,$$
$$u(\bar{t}) = 0, \; u(\bar{t} + b - a) = 0,$$

has only the trivial solution (where $q(t) = q(t - (b - a))$ for $t \in]b, \bar{t} + b - a]$).

Prove then that the problem (3.11) has at least one solution $u \in \mathcal{S}$, where $\mathcal{S}$ is defined by (3.7).

Hint : See [95].

Consider the problem

$$\begin{aligned} u'' + g(u) = h(t), \\ u(a) = u(b), \ u'(a) = u'(b). \end{aligned} \tag{3.15}$$

For such a problem, we can obtain similar result using conditions on the potential $G(u) = \int_0^u g(s)\,ds$. In this case lower bounds on $\frac{g(u)}{u}$ are not necessary anymore but we cannot prove the solution is in $\mathcal{S}$. The following problem concerns such a result.

Exercise 3.3. *Let α and $\beta \in C([a,b])$ be $W^{2,1}$-lower and upper solutions of (3.15). Let $h \in L^\infty(a,b)$, $g \in C(\mathbb{R})$ be such that for some $(\mu,\nu) \in \mathcal{F}_2$*

$$\begin{aligned} \limsup_{u\to+\infty} \frac{g(u)}{u} \le \mu \quad \textit{and} \quad \limsup_{u\to-\infty} \frac{g(u)}{u} \le \nu, \\ \liminf_{u\to+\infty} \frac{2G(u)}{u^2} < \mu \quad \textit{or} \quad \liminf_{u\to-\infty} \frac{2G(u)}{u^2} < \nu. \end{aligned}$$

Prove then that the problem (3.15) has at least one solution.

Hint : In case $\alpha > \beta$ see [147]. The proof without this condition can be built using the ideas in [79].

3.4 Multiplicity Results

Existence of several solutions are studied in Sections 1.2 and 2.2. Some additional results can be obtained using non well-ordered lower and upper solutions. The following result complements Theorem 3.3.

Theorem 3.5. *Let α_1 and $\alpha_2 \in C([a,b])$ be $W^{2,1}$-lower solutions of (3.11) and $\beta \in C([a,b])$ be a strict $W^{2,1}$-upper solution such that $\alpha_1 \le \beta$ and $\alpha_2 \not\le \beta$.*

Assume $f : [a,b] \times \mathbb{R} \to \mathbb{R}$ is an L^1-Carathéodory function such that for some function $a_+ \in L^1(a,b)$,

$$\liminf_{u\to+\infty} \frac{f(t,u)}{u} \ge a_+(t), \tag{3.16}$$

uniformly in $t \in [a,b]$.

Then the problem (3.11) has at least two solutions u_1 and u_2 such that

$$\alpha_1 \le u_1 < \beta, \quad u_2 \in \mathcal{S} \quad \textit{and} \quad u_1 \le u_2,$$

where $\mathcal{S}$ is defined in (3.7) with $\alpha = \max\{\alpha_1, \alpha_2\}$.

Proof. For any $r > \max\{\|\alpha_1\|_\infty, \|\alpha_2\|_\infty, \|\beta\|_\infty\}$, we consider the modified problem

$$\begin{gathered} u'' + f_r(t,u) = 0, \\ u(a) = u(b), \ u'(a) = u'(b), \end{gathered} \tag{3.17}$$

where

$$\begin{aligned} f_r(t,u) &= f(t,\alpha_1(t)) - u + \alpha_1(t), && \text{if } u \le \alpha_1(t), \\ &= f(t,u), && \text{if } \alpha_1(t) < u \le r, \\ &= (1+r-u)f(t,u), && \text{if } r < u \le r+1, \\ &= 0, && \text{if } r+1 < u. \end{aligned}$$

Claim 1 – Every solution of (3.17) *is such that* $u \ge \alpha_1$. This result follows from the usual maximum principle argument as it is used, for example, in the proof of Theorem I-3.1.

Claim 2 – There exists $k > 0$ *so that for any* $r > \max\{\|\alpha_1\|_\infty, \|\alpha_2\|_\infty, \|\beta\|_\infty\}$ *and any solution* $u \in \mathcal{S}$ *of* (3.17), *we have* $\|u\|_\infty < k$. As $u \in \mathcal{S}$, there exist t_0 and t_1 such that

$$u(t_0) = \min_{t\in[a,b]} u(t) \le u(t_1) \le \alpha(t_1) \le \|\alpha\|_\infty.$$

Further, we deduce from the asymptotic character of f that there exists $\hat{a}_+$ and $h \in L^1(a,b)$ such that for a.e. $t \in [a,b]$ and all $u \ge \alpha_1(t)$,

$$f_r(t,u) \ge -[\hat{a}_+(t)u + h(t)].$$

Hence, we have for $t \in [t_0, t_0 + b - a]$,

$$\begin{aligned} u(t) &= u(t_0) - \int_{t_0}^{t} f_r(s,u(s))(t-s)\,ds \\ &\le \|\alpha\|_\infty + \|h\|_{L^1}(b-a) + (b-a)\int_{t_0}^{t} \hat{a}_+(s)u(s)\,ds \end{aligned}$$

and the claim follows from Gronwall's Lemma.

Conclusion – Consider the problem (3.17) for

$$r > \max\{\|\alpha_1\|_\infty, \|\alpha_2\|_\infty, \|\beta\|_\infty, k\},$$

where k is given in Claim 2.

It follows from Theorem I-3.4 that there exists a solution u_1 of (3.17) which is minimal in $[\alpha_1, \beta]$. Hence, $\alpha_1(t) \le u_1(t) \le r$ which implies this solution solves also (3.11).

Next, we can apply Exercise 3.1 to obtain a solution $u_2 \in \mathcal{S}$ of (3.17). From Claim 1, $u_2 \ge \alpha_1$, and from Claim 2, $u_2 \le k < r$. Therefore u_2 is a solution of (3.11).

At last, notice that if $u_2 \not\geq u_1$, u_1 and u_2 are upper solutions of (3.17) and we deduce from Theorem I-3.2 the existence of a solution u_3 with $\alpha_1 \leq u_3 \leq \min\{u_1, u_2\}$ which contradicts u_1 to be minimal. □

Assumptions such as

(a) $f(t,u)$ is nonnegative or

(b) $f(t,0) \geq 0$ and $f(t,u_1) - f(t,u_2) \geq -M(u_1 - u_2)$, if $u_1 \geq u_2 \geq 0$,

imply the asymptotic behaviour (3.16) and we can choose $\alpha_1 = 0$. Hence, existence of nonnegative upper and lower solutions implies existence of two nonnegative solutions one of them being nontrivial.

Observe that the existence of the third solution of

$$u'' + u^4 - u^2 = h(t),$$
$$u(a) = u(b),\ u'(a) = u'(b),$$

if $-1/4 < h(t) \leq 0$ obtained in Example 1.4 can be deduced from Theorem 3.5 in a quite obvious way.

We can extend Theorem 3.5 to the Liénard equation (3.9).

Theorem 3.6. *Let α_1 and $\alpha_2 \in C([a,b])$ be $W^{2,1}$-lower solutions of* (3.9) *and $\beta \in C([a,b])$ be a strict $W^{2,1}$-upper solution such that $\alpha_1 \leq \beta$ and $\alpha_2 \not\leq \beta$.*

Assume $g \in C(\mathbb{R})$ and $f : [a,b] \times \mathbb{R} \to \mathbb{R}$ is an L^1-Carathéodory function such that for some function $h \in L^1(a,b)$,

$$\liminf_{u \to +\infty} f(t,u) \geq h(t),$$

uniformly in $t \in [a,b]$.

Then the problem (3.9) *has at least two solutions u_1 and u_2 such that*

$$\alpha_1 \leq u_1 < \beta, \quad u_2 \in \mathcal{S} \quad \text{and} \quad u_1 \leq u_2,$$

where $\mathcal{S}$ is defined in (3.7) *with $\alpha = \max\{\alpha_1, \alpha_2\}$.*

Proof. The proof is exactly the same as for Theorem 3.5 except Claim 2 which repeats the argument used to prove the claim in Theorem 3.2. □

Exercise 3.4. Consider the Liénard equation

$$\begin{aligned} u'' + f(u)u' + g(u) + h(t,u,u') = 0,\\ u(a) = u(b),\ u'(a) = u'(b), \end{aligned} \tag{3.18}$$

where f, $g : \mathbb{R} \to \mathbb{R}$ are continuous functions and $h : [a,b] \times \mathbb{R}^2 \to \mathbb{R}$ is continuous and bounded.

Prove that if $\inf_s |f(s)| > 0$, the existence of a lower and an upper solution implies the solvability of (3.18).

Hint : See [235].

3.5 Lower and Upper Solutions in the Reversed Order

In case $\alpha \geq \beta$ on $[a, b]$, we can ask whether the solution of (3.11) satisfies the "strong localization" $\beta \leq u \leq \alpha$ on $[a, b]$ as in the well ordered case. This is not always the case under the assumptions of Theorem 3.3 as shown by problem (3.2). But it is true if

$$\frac{f(t,u) - f(t,v)}{u - v} \leq \left(\frac{\pi}{b-a}\right)^2.$$

This analysis is the content of this section.

Theorem 3.7. *Let α and $\beta \in W^{2,1}(a, b)$ be lower and upper solutions of (3.11) such that $\alpha \geq \beta$, $E = \{(t, u) \in [a, b] \times \mathbb{R} \mid \beta(t) \leq u \leq \alpha(t)\}$ and $f : E \to \mathbb{R}$ be an L^1-Carathéodory function. Assume there exists $k \in]0, (\frac{\pi}{b-a})^2]$ such that for a.e. $t \in [a, b]$ and all $u \in [\beta(t), \alpha(t)]$,*

$$f(t, \alpha(t)) - k\alpha(t) \leq f(t, u) - ku \leq f(t, \beta(t)) - k\beta(t).$$

Then the problem (3.11) has at least one solution $u \in W^{2,1}(a, b)$ such that for all $t \in [a, b]$

$$\beta(t) \leq u(t) \leq \alpha(t).$$

Proof. We can assume that α and β are not solutions; otherwise, the result is trivial.

Let us consider the modified problem

$$\begin{gathered} u'' + \bar{f}(t, u) = 0, \\ u(a) = u(b), \; u'(a) = u'(b), \end{gathered} \tag{3.19}$$

where

$$\begin{aligned} \bar{f}(t, u) &= f(t, \alpha(t)) + k(u - \alpha(t)), && \text{if } u \geq \alpha(t), \\ &= f(t, u), && \text{if } \beta(t) \leq u \leq \alpha(t), \\ &= f(t, \beta(t)) + k(u - \beta(t)), && \text{if } u \leq \beta(t). \end{aligned}$$

By Theorem 3.3 and the Remark after Proposition 3.4, problem (3.19) has a solution $u \in \mathcal{S}$. It remains to prove the announced localization.

Let us prove $u \leq \alpha$. Observe that $v = \alpha - u$ satisfies

$$\begin{gathered} v'' + kv = (\alpha'' + \bar{f}(t, \alpha)) + (\bar{f}(t, u) - ku) - (\bar{f}(t, \alpha) - k\alpha) =: h(t), \\ v(a) - v(b) = 0, \; v'(a) - v'(b) \geq 0 \end{gathered}$$

where $h \in L^1(a, b)$ is such that $h \geq 0$, $h \not\equiv 0$. By Corollary A-6.3, $v \geq 0$, i.e. $u \leq \alpha$.

In the same way, we prove $u \geq \beta$. □

Remark Observe that the limitation on k in Theorem 3.7 is optimal if we want the strong localization. Consider the problem

$$u'' + ku = f(t),$$
$$u(a) = u(b), \ u'(a) = u'(b),$$

with $k > (\frac{\pi}{b-a})^2$.

Let c, d be such that

$$(\tfrac{\pi}{b-a})^2 < c^2 < k < d^2 \quad \text{and} \quad \tfrac{\pi}{c} + \tfrac{\pi}{d} = b - a,$$

and define

$$\begin{aligned} f(t) &:= \tfrac{k-c^2}{c} \sin c(t-a), && \text{if } t \in [a, a + \tfrac{\pi}{c}], \\ &:= \tfrac{k-d^2}{d} \sin d(t - \tfrac{\pi}{c} + \tfrac{\pi}{d} - a), && \text{if } t \in]a + \tfrac{\pi}{c}, b]. \end{aligned}$$

This problem has $\beta = 0$ and $\alpha = \frac{1}{k} \max\{\frac{k-c^2}{c}, \frac{d^2-k}{d}\}$ as upper and lower solutions satisfying the reversed ordering $\beta \le \alpha$. But its solution is given by

$$\begin{aligned} u(t) &:= \tfrac{1}{c} \sin c(t-a), && \text{if } t \in [a, a + \tfrac{\pi}{c}], \\ &:= \tfrac{1}{d} \sin d(t - \tfrac{\pi}{c} + \tfrac{\pi}{d} - a), && \text{if } t \in]a + \tfrac{\pi}{c}, b] \end{aligned}$$

which is not one-signed, i.e. it is not between the upper solution $\beta = 0$ and the lower solution.

Exercise 3.5. Prove that under the assumptions of Theorem 3.7, if for some $k \in]0, (\frac{\pi}{b-a})^2]$, for a.e. $t \in [a, b]$, and for all u, $v \in [\beta(t), \alpha(t)]$, $u \neq v$

$$\frac{f(t,u) - f(t,v)}{u - v} \le k,$$

then the solutions of (3.11) in $[\beta, \alpha]$ are ordered. Prove also there are solutions u_{min}, $u_{max} \in W^{2,1}(a, b)$ such that

$$\beta \le u_{min} \le u_{max} \le \alpha,$$

and any other solution u of (3.11) such that $\beta \le u \le \alpha$ satisfies

$$u_{min} \le u \le u_{max}.$$

3.6 The Dirichlet Problem

In this Section 3, we have worked out the periodic boundary value problem. These results can be extended to deal with other boundary conditions. We present here such an extension to the Dirichlet problem. The theorems we state are the counterpart of previously obtained results and their proof repeats previous arguments.

Our first concern is a non-resonance result for the problem

$$\begin{aligned} u'' + (\tfrac{\pi}{b-a})^2 u &= f(t, u, u'), \\ u(a) = 0, \ u(b) &= 0, \end{aligned} \tag{3.20}$$

where $(\frac{\pi}{b-a})^2$ is the first eigenvalue of the corresponding eigenvalue problem

$$\begin{aligned} u'' + \lambda u &= 0, \\ u(a) = 0, \ u(b) &= 0. \end{aligned} \tag{3.21}$$

Theorem 3.1 can then be modified as follows.

Theorem 3.8. *Assume α and $\beta \in C([a,b])$ are $W^{2,1}$-lower and upper solutions of (3.20) such that $\alpha \not\leq \beta$. Suppose $f : [a,b] \times \mathbb{R}^2 \to \mathbb{R}$ is a Carathéodory function such that for some $h \in L^1(a,b)$*

$$|f(t,u,v)| \leq h(t), \quad \text{on } [a,b] \times \mathbb{R}^2.$$

Then there exist a solution u of (3.20) in

$$\begin{aligned} \mathcal{S} := \operatorname{adh}_{C_0^1}\{u \in C_0^1([a,b]) \mid \\ \exists t_1, t_2 \in [a,b], \ u(t_1) > \beta(t_1), \ u(t_2) < \alpha(t_2)\}. \end{aligned} \tag{3.22}$$

Proof. For each $r \geq 1$, we define

$$\begin{aligned} f_r(t,u,v) &= f(t,u,v), && \text{if } |u| < r, \\ &= (1 + r - |u|) f(t,u,v) + (|u| - r)\tfrac{u}{r}, && \text{if } r \leq |u| < r+1, \\ &= \tfrac{u}{r}, && \text{if } r+1 \leq |u|, \end{aligned}$$

and consider the problem

$$\begin{aligned} u'' + (\tfrac{\pi}{b-a})^2 u &= f_r(t,u,u'), \\ u(a) = 0, \ u(b) &= 0. \end{aligned} \tag{3.23}$$

Claim 1 - For some $C > 0$, $\alpha(t) \prec C \sin(\pi\frac{t-a}{b-a})$ and $\beta(t) \succ -C\sin(\pi\frac{t-a}{b-a})$. Let $r > \max\{\|\alpha\|_\infty, \|\beta\|_\infty\}$. Define w to be the solution of

$$w'' + (\tfrac{\pi}{b-a})^2 w = \tfrac{w}{r}, \quad w(a) = r+2, \ w(b) = r+2,$$

$e(t) = \sin(\pi\frac{t-a}{b-a})$, choose $k > 0$ such that $ke + w \geq r+2$ on $[a,b]$ and define $\beta_2 = -\alpha_1 = ke + w$. Observe that β_2 and α_1 are upper and lower solutions of (3.23). Hence the claim follows from Remark II-2.1.

Claim 2 – There exists $k > 0$ such that for any $r > k$, solutions u of (3.23), *which are in $\mathcal{S}$, are such that $\|u\|_\infty < k$.* Assume by contradiction there exist sequences $(r_n)_n$ and $(u_n)_n \subset \mathcal{S}$, where $r_n \geq n$ and u_n is a solution of (3.23) with $r = r_n$, such that $\|u_n\|_\infty \geq n$. Let us write

$$u_n(t) = \tilde{u}_n(t) + \bar{u}_n e(t).$$

where $\int_a^b \tilde{u}_n(t)e(t)\,dt = 0$. We deduce from Proposition A-4.5 that for some $K > 0$

$$\begin{aligned}\|\tilde{u}_n\|_\infty &\leq K\int_a^b |\tilde{u}_n''(t) + (\tfrac{\pi}{b-a})^2\tilde{u}_n(t)|e(t)\,dt\\ &\leq K\int_a^b |f_{r_n}(t,u_n(t),u_n'(t))|e(t)\,dt \leq K\int_a^b (h(t) + \tfrac{|u_n(t)|}{r_n})e(t)\,dt.\end{aligned}$$

Hence, for some $C_1 > 0$ and n large enough, we can write

$$\|\tilde{u}_n\|_\infty \leq C_1(1 + \tfrac{|\bar{u}_n|}{r_n}). \tag{3.24}$$

Let t_0 be such that $\tilde{u}_n'(t_0) = 0$. It follows that

$$\begin{aligned}|\tilde{u}_n'(t)| &\leq |\int_{t_0}^t ((\tfrac{\pi}{b-a})^2\,|\tilde{u}_n(s)| + |f_{r_n}(s,u_n(s),u_n'(s))|)\,ds|\\ &\leq C_2(1 + \tfrac{|\bar{u}_n|}{r_n}),\end{aligned}$$

for some $C_2 > 0$.

Assume now that for some subsequence $\lim_{k\to\infty} \bar{u}_{n_k} = +\infty$. This assumption implies

$$u_{n_k}(t) \geq (\bar{u}_{n_k} - C_3\|\tilde{u}_{n_k}\|_{C^1})e(t) \geq (\bar{u}_{n_k} - C_4(1 + \tfrac{|\bar{u}_{n_k}|}{r_{n_k}}))e(t),$$

for some C_3 and $C_4 > 0$. Hence, for n_k large enough, $u_{n_k} \succ \alpha$ which contradicts $u_{n_k} \in \mathcal{S}$. In a similar way, we prove that no subsequence of $(\bar{u}_n)_n$ goes to $-\infty$. It follows that $\bar{u}_n$ and also, using (3.24), $\|u_n\|_\infty$ are bounded which contradicts the assumption.

Conclusion – Let $r > \max\{\|\alpha\|_\infty, \|\beta\|_\infty, k\}$. As in Claim 1, we define $\beta_2 = -\alpha_1 = ke + w$ and conclude as in the proof of Theorem 3.1. □

Next we consider the problem of interaction with Fučík spectrum. A result similar to Theorem 3.3 holds for the Dirichlet problem

$$\begin{aligned}u'' + f(t,u) = 0,\\ u(a) = 0,\ u(b) = 0.\end{aligned} \tag{3.25}$$

Theorem 3.9. *Let α, $\beta \in \mathcal{C}([a,b])$ be $W^{2,1}$-lower and upper solutions of* (3.25) *such that $\alpha \not\leq \beta$. Let $f : [a,b] \times \mathbb{R} \to \mathbb{R}$ be an L^1-Carathéodory function such that for some functions $a_\pm \leq (\frac{\pi}{b-a})^2$, $b_\pm \geq (\frac{\pi}{b-a})^2$ in $L^1(a,b)$,*

$$a_\pm(t) \leq \liminf_{u\to\pm\infty} \frac{f(t,u)}{u} \leq \limsup_{u\to\pm\infty} \frac{f(t,u)}{u} \leq b_\pm(t),$$

uniformly in $t \in [a,b]$.

Assume further that for any p, $q \in L^1(a,b)$, with $a_+ \leq p \leq b_+$ and $a_- \leq q \leq b_-$, the nontrivial solutions of

$$\begin{aligned} u'' + p(t)u^+ - q(t)u^- &= 0, \\ u(a) = 0, \ u(b) &= 0, \end{aligned} \tag{3.26}$$

where $u^+(t) = \max\{u(t), 0\}$ and $u^-(t) = \max\{-u(t), 0\}$, do not have interior zeros.

Then the problem (3.25) *has at least one solution $u \in \mathcal{S}$, where $\mathcal{S}$ is defined by* (3.22).

Exercise 3.6. Prove the above theorem.

The condition on p, q, can be related with the Fučík spectrum as in Proposition 3.4. Roughly speaking, the nontrivial solutions of the problem (3.26) do not have zeros if the box $[a_+, b_+] \times [a_-, b_-]$ lies under the second Fučík curve $\mathcal{F}_2$ of the Dirichlet problem.

At last, we can write multiplicity theorems for (3.25) which paraphrase results worked out in Section 3.4. We state here a situation with two lower solutions and a strict upper one.

Theorem 3.10. *Let α_1, $\alpha_2 \in \mathcal{C}([a,b])$ be $W^{2,1}$-lower solutions of* (3.25) *and $\beta \in \mathcal{C}([a,b])$ be a strict $W^{2,1}$-upper solution such that $\alpha_1 \leq \beta$ and $\alpha_2 \not\leq \beta$. Assume $f : [a,b] \times \mathbb{R} \to \mathbb{R}$ is an L^1-Carathéodory function such that for some function $a_+ \in L^1(a,b)$,* (3.16) *holds uniformly in $t \in [a,b]$.*

Then the problem (3.25) *has at least two solutions u_1 and u_2 such that*

$$\alpha_1 \leq u_1 \prec \beta, \quad u_2 \in \mathcal{S} \quad \text{and} \quad u_1 \leq u_2,$$

where $\mathcal{S}$ is defined in (3.22) *with $\alpha = \max\{\alpha_1, \alpha_2\}$.*

Exercise 3.7. Prove Theorem 3.10.

Hint : Adapt the proof of Theorem 3.5. In Claim 2, notice that $u \in \mathcal{S}$ implies the existence of sequences $(u_n)_n$ and $(t_n)_n$ such that $u_n \to u$ in $\mathcal{C}^1([a,b])$ and $u_n(t_n) < \alpha(t_n) \le Ce(t_n)$. Here $e(t) = \sin(\pi\frac{t-a}{b-a})$ and C is such that $\alpha \le Ce$ as in Remark II-2.1. We deduce then the existence of t_0 such that $u(t_0) \le Ce(t_0)$ and $u'(t_0) = Ce'(t_0)$ and compute

$$u(t) \le Ce(t_0) + Ce'(t_0)(t-t_0) - \int_{t_0}^{t} f_r(s,u(s))(t-s)\,ds.$$

The proof follows then as in Theorem 3.5.

In very much the same way, we can state a dual result with two upper solutions and a strict lower one.

Theorem 3.11. *Let $\alpha \in \mathcal{C}([a,b])$ be a strict $W^{2,1}$-lower solution of (3.25) and β_1, $\beta_2 \in \mathcal{C}([a,b])$ be $W^{2,1}$-upper solutions such that $\alpha \le \beta_2$ and $\alpha \not\le \beta_1$. Assume $f : [a,b] \times \mathbb{R} \to \mathbb{R}$ is an L^1-Carathéodory function such that for some function $a_- \in L^1(a,b)$,*

$$\liminf_{u\to-\infty} \frac{f(t,u)}{u} \ge a_-(t),$$

uniformly in $t \in [a,b]$.

Then the problem (3.25) has at least two solutions u_1 and u_2 such that

$$u_1 \in \mathcal{S}, \quad \alpha \le u_2 \prec \beta_2 \quad \text{and} \quad u_1 \le u_2,$$

where $\mathcal{S}$ is defined in (3.22) with $\beta = \min\{\beta_1, \beta_2\}$.

Exercise 3.8. Extend the results of this section to the separated boundary value problem

$$\begin{aligned} u'' &= f(t,u),\\ a_1u(a) - a_2u'(a) &= 0,\\ b_1u(b) + b_2u'(b) &= 0, \end{aligned}$$

in case a_1, $b_1 \in \mathbb{R}$, a_2, $b_2 \in \mathbb{R}^+$, $a_1^2 + a_2^2 > 0$ and $b_1^2 + b_2^2 > 0$.

Chapter IV

Variational Methods

1 The Minimization Method

Another approach in working with lower and upper solutions is to relate them with variational methods. Consider for example the problem

$$\begin{aligned} u'' + f(t,u) = 0, \\ u(a) = 0, \ u(b) = 0, \end{aligned} \tag{1.1}$$

where f is an L^1-Carathéodory function. It is well known that the related functional

$$\phi : H_0^1(a,b) \to \mathbb{R}, u \mapsto \int_a^b [\frac{u'^2(t)}{2} - F(t,u(t))]\,dt \tag{1.2}$$

with $F(t,u) = \int_0^u f(t,s)\,ds$ is of class C^1 and its critical points are the solutions of (1.1). A first link between the two methods is that existence of a well ordered pair of lower and upper solutions α and β implies the functional ϕ has a minimum on the convex set $[\alpha, \beta]$ which solves (1.1).

Theorem 1.1. *Let α and β be $W^{2,1}$-lower and upper solutions of* (1.1) *with $\alpha \leq \beta$ on $[a,b]$ and*

$$E = \{(t,u) \in [a,b] \times \mathbb{R} \mid \alpha(t) \leq u \leq \beta(t)\}. \tag{1.3}$$

Assume $f : E \to \mathbb{R}$ is an L^1-Carathéodory function.

Then the functional ϕ is minimum on $[\alpha, \beta]$, i.e. there exists u with $\alpha \leq u \leq \beta$ on $[a,b]$ so that

$$\phi(u) = \min_{\substack{v \in H_0^1(a,b) \\ \alpha \leq v \leq \beta}} \phi(v).$$

Further, u is a solution of (1.1).

Proof. Let us consider the modified problem

$$\begin{aligned} u'' + f(t,\gamma(t,u)) &= 0, \\ u(a) = 0, \ u(b) &= 0, \end{aligned} \tag{1.4}$$

where $\gamma(t,u) = \max\{\alpha(t), \min\{u, \beta(t)\}\}$. Define the functional

$$\overline{\phi} : H_0^1(a,b) \to \mathbb{R}, u \mapsto \int_a^b [\frac{u'^2(t)}{2} - \overline{F}(t,u(t))]\,dt,$$

where $\overline{F}(t,u) = \int_0^u f(t,\gamma(t,s))\,ds$. As $f(t,\gamma(t,u))$ is L^1-Carathéodory, $\overline{\phi}$ is of class $\mathcal{C}^1$ and its critical points are precisely the solutions of (1.4). Moreover $\overline{\phi}$ is weakly lower semi-continuous and coercive. Hence, $\overline{\phi}$ has a global minimum u (see Theorem A-2.2) which is a solution of (1.4). As in Theorem II-2.1, we prove that u satisfies $\alpha \le u \le \beta$, and hence is a solution of (1.1). Notice that on $\{u \in H_0^1(a,b) \mid \alpha \le u \le \beta\}$, the function $\bar{\phi}(u) - \phi(u)$ is constant. Hence both functions ϕ and $\bar{\phi}$ are minimized together between α and β so that the theorem follows. □

Example 1.1. Consider the problem

$$\begin{aligned} u'' + \lambda f(t,u) &= 0, \\ u(a) = 0, \ u(b) &= 0, \end{aligned} \tag{1.5}$$

where $f : [a,b] \times \mathbb{R} \to \mathbb{R}$ is an L^1-Carathéodory function that satisfies $f(t,0) = 0$, $f(t,R) \le 0$ for some $R > 0$ and there exists $\mu \in H_0^1(a,b)$, with $0 \le \mu(t) \le R$ on $[a,b]$ such that $\int_a^b F(t,\mu(t))\,dt > 0$, where $F(t,u) = \int_0^u f(t,s)\,ds$. Then, we can prove the existence of Λ such that for all $\lambda \ge \Lambda$, (1.5) has at least one nontrivial nonnegative solution.

We just have to observe that $\alpha = 0$ is a lower solution, $\beta = R$ is an upper solution and, for λ large enough, $\phi(\mu) = \int_a^b [\frac{\mu'^2(t)}{2} - \lambda F(t,\mu(t))]\,dt < 0$. Hence, for such values of λ, there exists $u \in [0,R]$ which solves (1.5) and minimizes ϕ on $[0,R]$, i.e.

$$\phi(u) = \min_{\substack{v \in H_0^1(a,b) \\ 0 \le v \le R}} \phi(v) \le \phi(\mu) < 0 = \phi(0).$$

This last inequality implies $u \not\equiv 0$.

The method applies to other boundary value problems such as the periodic problem

$$\begin{aligned} u'' + f(t,u) &= 0, \\ u(a) = u(b), \ u'(a) &= u'(b). \end{aligned} \tag{1.6}$$

Here, the associated functional reads

$$\phi : H^1_{per}(a,b) \to \mathbb{R}, u \mapsto \int_a^b [\frac{u'^2(t)}{2} - F(t,u(t))]\,dt, \tag{1.7}$$

with $F(t,u) = \int_0^u f(t,s)\,ds$ and $H^1_{per}(a,b) = \{u \in H^1(a,b) \mid u(a) = u(b)\}$.
We can also consider the Neumann problem

$$\begin{aligned} u'' + f(t,u) &= 0, \\ u'(a) = 0,\ u'(b) &= 0, \end{aligned} \tag{1.8}$$

with whom we associate the functional

$$\phi : H^1(a,b) \to \mathbb{R}, u \mapsto \int_a^b [\frac{u'^2(t)}{2} - F(t,u(t))]\,dt. \tag{1.9}$$

For both these problems, we can state the corresponding result.

Theorem 1.2. *Let α and β be $W^{2,1}$-lower and upper solutions of* (1.6) *(resp.* (1.8)*) with $\alpha \le \beta$ on $[a,b]$ and E be defined from* (1.3)*. Assume $f : E \to \mathbb{R}$ is an L^1-Carathéodory function.*

Then the functional ϕ defined by (1.7) *(resp.* (1.9)*) is minimum on $[\alpha,\beta]$, i.e. there exists u with $\alpha \le u \le \beta$ on $[a,b]$ so that*

$$\phi(u) = \min_{\substack{v \in H^1_{per}(a,b) \\ \alpha \le v \le \beta}} \phi(v) \quad (resp.\ \phi(u) = \min_{\substack{v \in H^1(a,b) \\ \alpha \le v \le \beta}} \phi(v)).$$

Further, u is a solution of (1.6) *(resp.* (1.8)*).*

Exercise 1.1. Prove Theorem 1.2.

As an application, consider the problem

$$\begin{aligned} u'' + \mu(t)g(u) + h(t) &= 0, \\ u(0) = 0,\ u(\pi) &= 0. \end{aligned} \tag{1.10}$$

Theorem 1.3. *Let μ, $h \in L^\infty(0,\pi)$ with $\mu_0 = \operatorname{essinf} \mu(t) > 0$ and $g : \mathbb{R} \to \mathbb{R}$ be a continuous function. Let us denote $G(u) = \int_0^u g(s)\,ds$. Assume that*

$$-\infty < \liminf_{u \to \pm\infty} \frac{G(u)}{u^2} \le 0 \quad and \quad \limsup_{u \to \pm\infty} \frac{G(u)}{u^2} = +\infty.$$

Then the problem (1.10) *has two infinite sequences of solutions $(u_n)_n$ and $(v_n)_n$ satisfying*

$$\ldots \le v_{n+1} \le v_n \le \ldots \le v_1 \le u_1 \le \ldots \le u_n \le u_{n+1} \le \ldots$$

and

$$\lim_{n \to \infty} (\max_t u_n(t)) = +\infty, \qquad \lim_{n \to \infty} (\min_t v_n(t)) = -\infty.$$

Proof. Step 1 – Claim: For every $M \geq 0$ there exists β, upper solution of (1.10), *with $\beta(t) \geq M$ on $[0, \pi]$.* First observe that if g is unbounded from below on $[0, +\infty[$, we have a sequence of constant upper solutions $\beta_n \to +\infty$. In the contrary, we can assume there exists $K \geq 0$ such that $g(u) \geq -K$ for $u \geq 0$.

Given $M > 0$, we can choose d so that

$$\|\mu\|_\infty(\frac{G(d)}{d^2} + \frac{K}{d}) + \frac{\|h\|_\infty}{d} \leq \frac{1}{8\pi^2} \quad \text{and} \quad d > 2M.$$

We define then β to be a solution of the Cauchy problem

$$\begin{aligned} u'' + \|\mu\|_\infty(g(u) + K) + \|h\|_\infty = 0, \\ u(0) = d, \ u'(0) = 0. \end{aligned} \tag{1.11}$$

Assume there exists $t_0 \in]0, \pi]$ such that $\beta(t) > M$ on $[0, t_0[$ and $\beta(t_0) = M$. Notice that on $[0, t_0]$, $\beta'(t) \leq 0$ and $\|\mu\|_\infty(G(\beta(t)) + K\beta(t)) + \|h\|_\infty\beta(t) \geq 0$. From the conservation of energy for (1.11), we have

$$\begin{aligned} \frac{\beta'^2(t)}{2} \leq \frac{\beta'^2(t)}{2} &+ \|\mu\|_\infty(G(\beta(t)) + K\beta(t)) + \|h\|_\infty\beta(t) \\ &= \|\mu\|_\infty(G(d) + Kd) + \|h\|_\infty d \leq \frac{d^2}{8\pi^2}, \end{aligned}$$

i.e.

$$0 \leq -\beta'(t) \leq \frac{d}{2\pi}.$$

It follows that for any $t \in [0, t_0]$,

$$d - \beta(t) \leq \frac{d}{2\pi}t_0 \leq \frac{d}{2},$$

which leads to the contradiction $\beta(t_0) \geq \frac{d}{2} > M$. Hence, the claim follows.

In a similar way, we prove the following.

Claim – For every $M \geq 0$ there exists a lower solution α of (1.10) *such that $\alpha(t) \leq -M$ on $[0, \pi]$.*

Step 2 – Claim: There exists a sequence of positive real numbers $(s_n)_n$ with $s_n \to +\infty$ and $z \succ 0$ such that $\phi(s_n z) \to -\infty$, where

$$\phi(u) = \int_0^\pi [\frac{u'^2(t)}{2} - \mu(t)G(u(t)) - h(t)u(t)]\, dt.$$

Define z to be a C^1-function such that $0 < z(t) \leq 1$ on $]0, \pi[$, $z(0) = 0$, $z(\pi) = 0$, $z'(0) > 0$, $z'(\pi) < 0$ and $z(t) = 1$ on $[\epsilon, \pi - \epsilon]$ for some $\epsilon > 0$. Choose $(s_n)_n$ a sequence of positive real numbers with $s_n \to +\infty$ and

$\frac{G(s_n)}{s_n^2} \to +\infty$. Recall that the assumptions imply $G(u) \geq -L(u^2+1)$, for some $L > 0$. We compute then

$$\begin{aligned}\phi(s_n z) &= \int_0^\pi [\frac{s_n^2 z'^2(t)}{2} - \mu(t)G(s_n z(t)) - h(t)s_n z(t)]\,dt \\ &= \int_{[0,\pi]\setminus[\epsilon,\pi-\epsilon]} \frac{s_n^2 z'^2(t)}{2}\,dt - G(s_n)\int_\epsilon^{\pi-\epsilon} \mu(t)\,dt \\ &\qquad - \int_{[0,\pi]\setminus[\epsilon,\pi-\epsilon]} \mu(t)G(s_n z(t))\,dt - s_n \int_0^\pi h(t)z(t)\,dt \\ &\leq s_n^2\|z'\|_\infty^2 \epsilon - G(s_n)\mu_0(\pi - 2\epsilon) + L(s_n^2+1)\|\mu\|_{L^1} + s_n\|h\|_{L^1}\|z\|_\infty.\end{aligned}$$

It follows that $\phi(s_n z) \to -\infty$.

Step 3 – Claim: There exists a sequence of negative real numbers $(t_n)_n$ with $t_n \to -\infty$ and $z \succ 0$ such that $\phi(t_n z) \to -\infty$. The argument is similar to Step 2.

Step 4 – Conclusion. By Step 1, we have α_1, β_1 lower and upper solutions of (1.10) with $\alpha_1 \leq \beta_1$. Hence, we obtain from Theorem 1.1 a solution u_1 of (1.10) such that $\alpha_1 \leq u_1 \leq \beta_1$.

From Step 2, we have z and s_1 such that $s_1 z \geq u_1$ and $\phi(s_1 z) < \phi(u_1)$. Moreover, Step 1 provides the existence of an upper solution β_2 with $u_1 \leq s_1 z \leq \beta_2$. Now, by Theorem 1.1, we have a solution u_2 of (1.10) satisfying

$$u_1 \leq u_2 \leq \beta_2$$

and

$$\phi(u_2) = \min_{\substack{v\in H_0^1(0,\pi)\\ u_1\leq v\leq \beta_2}} \phi(v) \leq \phi(s_1 z) < \phi(u_1).$$

It follows that $u_2 \neq u_1$.

Iterating this argument and reproducing it in the negative part, we prove the result. □

Remark 1.1. The condition on $G(u)/u^2$ cannot be replaced by analogous conditions on $g(u)/u$ as observed in [230].

Problem 1.2. Let $f : [0, +\infty[\to \mathbb{R}$ be a continuous function. Let us denote $F(u) = \int_0^u f(s)\,ds$. Assume that

$$\liminf_{u\to 0+} \frac{F(u)}{u^2} = 0 \quad \text{and} \quad \limsup_{u\to 0+} \frac{F(u)}{u^2} = +\infty.$$

Prove then that the problem

$$\begin{gathered} u'' + f(u) = 0, \\ u(0) = 0,\ u(\pi) = 0 \end{gathered}$$

has a sequence $(u_n)_n$ of positive solutions satisfying

$$\lim_{n\to\infty}(\max_t u_n(t)) = 0.$$

Hint : See [240].

As a next problem consider the following prescribed mean curvature equation

$$\begin{aligned}\left(\frac{u'}{\sqrt{1+u'^2}}\right)' &= \lambda f(t,u),\\ u(0) = 0,\ u(1) &= 0.\end{aligned} \tag{1.12}$$

This equation is equivalent to

$$u'' = \lambda(1+u'^2)^{3/2} f(t,u), \quad u(0) = 0,\ u(1) = 0.$$

As in Example 1.1, the following result provides a nontrivial nonnegative solution. Notice however that the equation does not satisfy a Nagumo condition which will force us to modify not only the dependence in u but also in the derivative u'.

Problem 1.3. Let $f : [0,1]\times\mathbb{R}\to\mathbb{R}$ be a continuous function such that $f(t,0) \leq 0$ and define $F(t,u) = \int_0^u f(t,s)\,ds$. Assume that for some $[a,b] \subset]0,1[$, $a \neq b$,

$$\lim_{u\to 0^+}[\max_{t\in[a,b]} \frac{F(t,u)}{u^2}] = -\infty.$$

Prove then that there exists a number $\lambda^* > 0$ such that the problem (1.12) has for each $\lambda \in]0,\lambda^*[$, a nontrivial nonnegative solution.

Hint : See [92] or [146]. Define

$$\begin{aligned} p(s) &= (1+s)^{-1/2}, && \text{if } 0 \leq s < 1,\\ &= \tfrac{1}{8\sqrt{2}}[(s-2)^2+7], && \text{if } 1 \leq s < 2,\\ &= \tfrac{7}{8\sqrt{2}}, && \text{if } 2 \leq s, \end{aligned}$$

$\alpha(t) = 0$ and $\beta(t) = t(1-t)$. Consider then the functional $\phi : H_0^1(0,1)\to\mathbb{R}$, defined by

$$\phi(u) = \int_0^1 [\tfrac{1}{2}P(u'^2) + \lambda\bar{F}(t,u(t))]\,dt,$$

where $P(v) = \int_0^v p(s)ds$, $\bar{F}(t,u) = \int_0^u f(t,\gamma(t,s))ds$ and as usually, $\gamma(t,u) = \max\{\alpha(t),\min\{u,\beta(t)\}\}$. Critical points of ϕ solve

$$(p(u'^2)u')' = \lambda f(t,\gamma(t,u)), \quad u(0) = 0,\ u(1) = 0,$$

which can also be written

$$u'' = \lambda(p(u'^2) + 2p'(u'^2)u'^2)^{-1} f(t, \gamma(t, u)), \quad u(0) = 0, \ u(1) = 0.$$

The proof follows then if we notice that for $\lambda > 0$ small enough, α and β are respectively lower and upper solutions and ϕ takes a negative value for some u with $\alpha \le u \le \beta$.

2 Local Minimum of the Functional

Theorem 1.1 defines the solution u of (1.1) as a minimizer of the functional ϕ, defined by (1.2), on the set $C = \{u \in H_0^1(a,b) \mid \alpha \le u \le \beta\}$, which does not imply that u is a local minimizer of that functional. Indeed, if α and β coincide at some point, the set C has empty interior and there is no evidence that the function u minimizes ϕ even locally. This observation raises difficulties for example in applications of the Mountain Pass Theorem (see Theorem A-2.3). In general, we can prove u is a local minimizer of the restricted functional

$$\hat{\phi} : C_0^1([a,b]) \to \mathbb{R}, u \mapsto \int_a^b [\frac{u'^2(t)}{2} - F(t, u(t))]\, dt. \tag{2.1}$$

The main result of this section proves that such a minimizer, minimizes locally ϕ on $H_0^1(a,b)$.

Theorem 2.1. *Let $f : [a,b] \times \mathbb{R} \to \mathbb{R}$ be an L^1-Carathéodory function. Assume $u_0 \in C_0^1([a,b])$ is a local minimizer of $\hat{\phi}$ defined by (2.1) i.e.*

$$\exists r > 0, \forall v \in C_0^1([a,b]), \quad \|v\|_{C^1} \le r \Rightarrow \hat{\phi}(u_0) \le \hat{\phi}(u_0 + v),$$

then u_0 is a local minimizer of ϕ defined by (1.2), i.e.

$$\exists r > 0, \forall v \in H_0^1(a,b), \quad \|v\|_{H_0^1} \le r \Rightarrow \phi(u_0) \le \phi(u_0 + v).$$

Remark 2.1. This result is somewhat surprising since a H_0^1-neighbourhood is much bigger than a C_0^1-neighbourhood. It is false in general, but here we can use the special structure of the functional.

Notice also that the above theorem still holds if $u_0 \in H_0^1(a,b)$.

Proof. Notice first that, as u_0 minimizes locally $\hat{\phi}$ in $C_0^1([a,b])$, we have

$$\int_a^b (u_0'\xi' - f(t, u_0)\xi)\, dt = 0,$$

for all $\xi \in C_0^1([a,b])$ and by density for all $\xi \in H_0^1(a,b)$.

Suppose the conclusion does not hold. Then

$$\forall r > 0,\ \exists v_r \in H_0^1(a,b), \quad \|v_r\|_{H_0^1} \le r \text{ and } \phi(u_0 + v_r) < \phi(u_0), \tag{2.2}$$

where $\|u\|^2_{H_0^1} = \int_a^b u'^2\,dt$. By a standard lower semi-continuity argument, we prove that the problem

$$\min_{u \in V} \phi(u),$$

where $V = \{u \in H_0^1(a,b) \mid \psi(u) = \int_a^b (u' - u_0')^2\,dt - r^2 \le 0\}$ has a solution $w_r \in H_0^1(a,b)$.

From Lagrange multiplier theory (see [69]), there exists $\mu_r \le 0$ so that

$$\int_a^b (w_r'\xi' - f(t, w_r)\xi)\,dt = d\phi(w_r)(\xi) = \frac{\mu_r}{2}\,d\psi(w_r)(\xi) = \mu_r \int_a^b (w_r' - u_0')\xi'\,dt,$$

for all $\xi \in H_0^1(a,b)$. It follows that

$$(1 - \mu_r)\int_a^b (w_r' - u_0')\xi'\,dt = \int_a^b (f(t, w_r) - f(t, u_0))\xi\,dt,$$

for all $\xi \in H_0^1(a,b)$, i.e. $w_r - u_0 \in W^{2,1}(a,b)$ and

$$-(1 - \mu_r)(w_r'' - u_0'') = f(t, w_r) - f(t, u_0).$$

Further, as $w_r \to u_0$ in $H_0^1(a,b)$ for $r \to 0$, the w_r are bounded in $\mathcal{C}([a,b])$ and there is a function $h \in L^1(a,b)$ such that for any $r \in [0,1]$, $|w_r''(t) - u_0''(t)| \le h(t)$ on $[a,b]$. We deduce now from Arzelà-Ascoli's Theorem that up to a subsequence, the w_r converge in $\mathcal{C}_0^1([a,b])$ as r goes to zero. Next as $w_r \to u_0$ in $H_0^1(a,b)$ we conclude $w_r \to u_0$ in $\mathcal{C}_0^1([a,b])$. As $\hat{\phi}(w_r) = \phi(w_r) < \hat{\phi}(u_0)$, this result proves that u_0 does not minimize $\hat{\phi}$ and concludes the proof. □

Theorem 2.2. *Let α and β be strict $W^{2,1}$-lower and upper solutions of (1.1) with $\alpha \prec \beta$ and E be defined by (1.3). Assume $f : E \to \mathbb{R}$ is an L^1-Carathéodory function.*

Then there exists a local minimizer u of the functional ϕ defined by (1.2). Further, this minimizer is a solution of (1.1) such that $\alpha \prec u \prec \beta$.

Proof. By Theorem 1.1, we have a solution u of (1.1) which is a minimizer of ϕ on $[\alpha, \beta]$. As the lower and upper solutions are strict, $\alpha \prec u \prec \beta$ and there is a $\mathcal{C}_0^1$-neighbourhood of u in $[\alpha, \beta]$. Hence u is a local minimizer of $\hat{\phi}$ defined from (2.1). By Theorem 2.1, u is also a local minimizer of ϕ. □

The following examples use the Palais-Smale Condition:

(PS) *If* $(u_n)_n \subset H_0^1(a,b)$ *is such that* $\phi(u_n)$ *is bounded and* $\nabla\phi(u_n) \stackrel{H_0^1}{\to} 0$ *then there exists some subsequence of* $(u_n)_n$ *which converges in* $H_0^1(a,b)$ *to a function* u *such that* $\nabla\phi(u) = 0$.

It is known, see Theorem A-2.1, that this condition is easy to verify if $\phi(u)$ is coercive or more generally if the Palais-Smale sequences $(u_n)_n$ are bounded in $H_0^1(a,b)$.

Example 2.1. The problem

$$u'' + u^3 = 0, \qquad u(0) = 0,\ u(\pi) = 0, \tag{2.3}$$

has a nontrivial nonnegative solution.

Proof. Critical points of the functional

$$\phi : H_0^1(0,\pi) \to \mathbb{R}, u \mapsto \int_0^\pi [\frac{u'^2(t)}{2} - \frac{(u^+)^4(t)}{4}]\, dt$$

are solutions of

$$u'' + (u^+)^3 = 0, \qquad u(0) = 0,\ u(\pi) = 0 \tag{2.4}$$

and, as these solutions are nonnegative, they are solutions of (2.3).

Observe then that ϕ has a mountain pass geometry. First, for $\epsilon > 0$ small enough $\alpha(t) = -\epsilon$ and $\beta(t) = \epsilon \sin t$ are strict lower and upper solution of (2.4). Hence by Theorem 2.2, the problem (2.4) has a solution u_0 which is a local minimizer of ϕ. Next, there exists $R > 0$ such that $\phi(R\sin t) < \phi(u_0)$.

Claim - ϕ satisfies the Palais-Smale Condition (PS). We compute

$$\tfrac{1}{4}\int_0^\pi u_n'^2\, dt = |\phi(u_n) - \tfrac{1}{4}d\phi(u_n)(u_n)| \le C(1 + \|u_n\|_{H_0^1}),$$

and hence, there exists $D > 0$ so that for all $n \in \mathbb{N}$, $\|u_n\|_{H_0^1} \le D$. This allows to conclude by Theorem A-2.1.

Conclusion - By the characterization of minimizers (Theorem A-2.4) either ϕ has two minimizers or we can apply the Mountain Pass Theorem (Theorem A-2.3). The conclusion follows in both cases.

Example 2.2. Consider the problem

$$u'' + g(u) = 0, \qquad u(0) = 0,\ u(\pi) = 0, \tag{2.5}$$

where $g \in \mathcal{C}([0, A])$ is such that $g(0) = 0$, $g(A) = 0$, and there exists $r > 0$ such that $g(u) \leq u$ on $[0, r]$. Assume there exists a strict lower solution α with $0 \prec \alpha \leq A$. Then the problem (2.5) has at least two solutions u_1, u_2, with $0 \lneq u_1 \lneq u_2 \leq A$ and $u_2 \succ \alpha$.

To prove this property we first extend g to the whole line

$$\begin{aligned} \bar{g}(u) &= g(u), \quad \text{if } 0 \leq u \leq A, \\ &= 0, \qquad \text{otherwise.} \end{aligned}$$

Define then

$$\bar{\phi} : H_0^1(0, \pi) \to \mathbb{R}, u \mapsto \int_0^\pi [\frac{u'^2(t)}{2} - \bar{G}(u(t))]\, dt,$$

where $\bar{G}(u) = \int_0^u \bar{g}(s)\, ds$. As in Example 2.1, it is easy to see that any nontrivial critical point u of $\bar{\phi}$ is a solution of (2.5) with $0 \leq u \leq A$.

Clearly $\bar{\phi}(u) \geq 0$ if $\|u\|_\infty \leq r$, i.e. if $\|u\|_{H^1}$ is small enough so that 0 is a local minimum of $\bar{\phi}$ in $H_0^1(0, \pi)$. Moreover, any $B > A$ is a strict upper solution so that by Theorem 2.2, $\bar{\phi}$ has a local minimizer $u_0 \in H_0^1(0, \pi)$ with $\alpha \prec u_0 \prec B$. This minimizer solves (2.5) so that $\alpha \prec u_0 \leq A$. Hence $\bar{\phi}$ has two distinct local minima. It satisfies the Palais-Smale Condition (PS) so that by a consequence of the Mountain Pass Theorem (Corollary A-2.5), $\bar{\phi}$ has another nontrivial critical point u_1. By Theorem II-2.6, there is a maximal solution u_2 of (2.5) in $[0, A]$. In particular $u_0 \leq u_2$ and $u_1 \leq u_2$ with u_0 and u_1 distinct. Therefore either u_0 or u_1 differs from u_2 and the result is proved.

The above results apply to other boundary value problems as seen in the following theorem.

Theorem 2.3. *Let α and β be strict $W^{2,1}$-lower and upper solutions of the periodic problem* (1.6) *(resp. of the Neumann problem* (1.8)*) with $\alpha \prec \beta$ and E be defined by* (1.3). *Assume $f : E \to \mathbb{R}$ is an L^1-Carathéodory function.*

Then there exists a local minimizer u of the functional ϕ defined by (1.7) *(resp. by* (1.9)*). Further, this minimizer is a solution of* (1.6) *(resp. of* (1.8)*) such that $\alpha \prec u \prec \beta$.*

Exercise 2.1. Prove the above result.

The following theorem applies the previous ideas to the parametric problem

$$\begin{aligned} u'' + \lambda f(t, u) &= 0, \\ u(a) = 0, \ u(b) &= 0. \end{aligned} \tag{2.6}$$

Theorem 2.4. *Let $f : [a,b] \times \mathbb{R} \to \mathbb{R}$ be an L^1-Carathéodory function. Assume*

(a) $f(t,0) \geq 0$ a.e. on $[a,b]$;

(b) there exist $B_1, \ldots, B_n$ such that $0 < B_1 < \ldots < B_n$ and for $i \in \{1, \ldots, n\}$,

$$f(t, B_i) \leq 0 \quad \text{a.e. on } [a,b];$$

(c) there exist $A_1, \ldots, A_{n-1}$ and $\nu_1, \ldots, \nu_{n-1} > 0$ such that

$$B_1 < A_1 < B_2 < \ldots < A_{n-1} < B_n$$

and for $i \in \{1, \ldots, n-1\}$ and all $(t,u) \in [a,b] \times [0, B_i[$

$$F(t, A_i) > F(t,u) + \nu_i,$$

where $F(t,u) = \int_0^u f(t,s)\,ds$;

(d) there exists $k \in L^1(a,b)$ such that for a.e. $t \in [a,b]$ and all $u \in [0, B_n]$, $f(t,u) + k(t)u$ is non-decreasing.

Then there exists $\bar{\lambda}$ such that for all $\lambda > \bar{\lambda}$, the problem (2.6) has at least $2n-1$ ordered solutions

$$0 \leq u_1 \prec u_2 \prec \ldots \prec u_{2n-1}.$$

Remark 2.2. (a) Roughly speaking the above hypotheses imply that for fixed t, the graph of $f(t,.)$ has n positive humps and n negative ones, each positive hump having greater area than the previous negative one.
(b) In case $F(t, A_i) > F(t,u)$ for all $(t,u) \in [a,b] \times [0, B_i[$ is not satisfied, the theorem is no more valid. Consider for example the problem

$$u'' + \lambda \cos u = 0, \quad u(a) = 0, \; u(b) = 0.$$

By elementary methods, it is easy to see that for $\lambda > 0$ this problem has a unique positive solution u with $\|u\|_\infty < \frac{\pi}{2}$.
(c) More qualitative properties of the set of solutions can be found in [98] and [202].

Proof. Define first

$$\begin{aligned} \overline{f}(t,u) &= f(t,u), \text{ if } u \geq 0, \\ &= f(t,0), \text{ if } u < 0 \end{aligned}$$

and

$$\overline{\phi} : H_0^1(a,b) \to \mathbb{R}, u \mapsto \int_a^b [\frac{u'^2(t)}{2} - \lambda \overline{F}(t, u(t))]\,dt,$$

where $\overline{F}(t,s) = \int_0^s \overline{f}(t,x)\,dx$. Notice that critical points of $\overline{\phi}$ are solutions of

$$\begin{aligned} u'' + \lambda \overline{f}(t,u) &= 0, \\ u(a) = 0, \; u(b) &= 0. \end{aligned} \tag{2.7}$$

Such solutions are nonnegative and therefore solve (2.6).

Let A_0 be a negative constant, and observe that it is a strict lower solution of (2.7). Next we deduce from assumptions (b), (d), and Proposition III-2.7 that for all $i \in \{1, \ldots, n\}$, B_i is a strict upper solution.

Step 1 - By Theorem 2.2, there exists a local minimizer u_1 of $\overline{\phi}$ in $H_0^1(a,b)$ which is a solution of (2.7) such that $A_0 \prec u_1 \prec B_1$. Solutions of (2.7) are nonnegative so that

$$\overline{\phi}(u_1) = \min_{\substack{v \in H_0^1(a,b) \\ 0 \le v \le B_1}} \overline{\phi}(v).$$

Step 2 - There exists $w \in H_0^1(a,b)$ and $\Lambda_1 > 0$ such that $u_1 \le w \le B_2$ and for $\lambda > \Lambda_1$, $\overline{\phi}(w) < \overline{\phi}(u_1)$. Let $h \in L^1(a,b)$ be such that for a.e. $t \in [a,b]$ and all $u \in [0, A_1]$, $|f(t,u)| \le h(t)$ and let $\delta > 0$ satisfy

$$2A_1 \int_{[a,b]\setminus[a+\delta,b-\delta]} h(t)\,dt < \nu_1(b-a).$$

Choose $w \in H_0^1(a,b)$ to be such that $w(t) = A_1$ on $[a+\delta, b-\delta]$ and $u_1 \le w \le A_1$. We compute, for λ large enough,

$$\begin{aligned}
\overline{\phi}(w) - \overline{\phi}(u_1) &= \int_a^b [\frac{w'^2(t)}{2} - \frac{u_1'^2(t)}{2}]\,dt - \lambda \int_a^b [F(t,A_1) - F(t,u_1(t))]\,dt \\
&\qquad + \lambda \int_{[a,b]\setminus[a+\delta,b-\delta]} [F(t,A_1) - F(t,w(t))]\,dt \\
&\le \int_a^b \frac{w'^2(t)}{2}\,dt - \lambda \left[\nu_1(b-a) - \int_{[a,b]\setminus[a+\delta,b-\delta]} h(t)A_1\,dt\right] \\
&\le \int_a^b \frac{w'^2(t)}{2}\,dt - \lambda\nu_1 \frac{(b-a)}{2} < 0.
\end{aligned}$$

Step 3 - By Step 2,

$$\min_{\substack{v \in H_0^1(a,b) \\ u_1 \le v \le B_2}} \overline{\phi}(v) < \overline{\phi}(u_1).$$

Hence, by Theorem 1.1, there exists u_3 solution of (2.7) such that $u_1 \lneqq u_3 \le B_2$ and

$$\overline{\phi}(u_3) = \min_{\substack{v \in H_0^1(a,b) \\ u_1 \le v \le B_2}} \overline{\phi}(v).$$

Arguing as in Proposition III-2.7 and using assumption (d), we prove $u_1 \prec u_3$. Hence using Theorem 2.1, we prove u_3 is a local minimizer of $\overline{\phi}$ in $H_0^1(a,b)$.

Step 4 - Consider now the modified function

$$\begin{aligned} \tilde{f}(t,u) &= \overline{f}(t,u_1(t)) = f(t,u_1(t)), && \text{if } u < u_1(t), \\ &= \overline{f}(t,u) = f(t,u), && \text{if } u_1(t) \le u \le u_3(t), \\ &= \overline{f}(t,u_3(t)) = f(t,u_3(t)), && \text{if } u_3(t) < u, \end{aligned}$$

and the corresponding functional

$$\tilde{\phi} : H_0^1(a,b) \to \mathbb{R}, u \mapsto \int_a^b [\frac{u'^2(t)}{2} - \lambda \tilde{F}(t,u(t))]\, dt,$$

where $\tilde{F}(t,s) = \int_0^s \tilde{f}(t,x)\, dx$. Let us prove that u_1 and u_3 are local minimum of $\tilde{\phi}$ in $H_0^1(a,b)$. To this end, observe that

$$\begin{aligned} \tilde{F}(t,u) &= \overline{f}(t,u_1(t))u, && \text{if } u < u_1(t), \\ &= \overline{f}(t,u_1(t))u_1(t) + \overline{F}(t,u) - \overline{F}(t,u_1(t)), && \text{if } u_1(t) \le u \le u_3(t), \\ &= \overline{f}(t,u_1(t))u_1(t) + \overline{F}(t,u_3(t)) \\ &\quad -\overline{F}(t,u_1(t)) + \overline{f}(t,u_3(t))(u-u_3(t)), && \text{if } u_3(t) < u. \end{aligned}$$

For u close to u_1 in C^1-norm, $u \le u_3$, and as u_1 is a solution of (2.7), we have

$$\begin{aligned} \tilde{\phi}(u) - \tilde{\phi}(u_1) &= \int_a^b [\frac{(u_1' + (u-u_1)^{+\prime})^2}{2} - \lambda \overline{F}(t, u_1 + (u-u_1)^+)]\, dt \\ &\quad - \int_a^b [\frac{u_1'^2(t)}{2} - \lambda \overline{F}(t,u_1)]\, dt + \int_a^b \frac{{(u-u_1)^{-}}'^2}{2}\, dt \\ &\quad - \int_a^b [u_1'(u-u_1)^{-\prime} - \lambda \overline{f}(t,u_1)(u-u_1)^-]\, dt \\ &= \overline{\phi}(u_1 + (u-u_1)^+) - \overline{\phi}(u_1) + \int_a^b \frac{{(u-u_1)^{-}}'^2}{2}\, dt \end{aligned}$$

and hence, by Theorem 2.1, u_1 is a local minimum of $\tilde{\phi}$ in $H_0^1(a,b)$. In the same way, we prove that u_3 is a local minimum of $\tilde{\phi}$. Notice that $\tilde{\phi}$ satisfies the Palais-Smale Condition so that by Corollary A-2.5, there exists u_2 critical point of $\tilde{\phi}$ and hence solution of

$$\begin{aligned} u'' + \lambda \tilde{f}(t,u) &= 0, \\ u(a) = 0,\ u(b) &= 0. \end{aligned}$$

As in Theorem II-2.1, we deduce from the definition of $\tilde{f}(t,u)$ that $u_1 \le u_2 \le u_3$. Using again assumption (d) we obtain $u_1 \prec u_2 \prec u_3$.

Conclusion - Iterating this process, we obtain the required solutions. □

We can now write a variational version of Amann's Three Solutions Theorem.

Exercise 2.2. Let α_1, β_1 and α_2, $\beta_2 \in \mathcal{C}([a,b])$ be two pairs of $W^{2,1}$-lower and upper solutions of (1.1) such that on $[a,b]$

$$\alpha_1(t) \le \beta_2(t),\ \alpha_1(t) \le \beta_1(t),\ \alpha_2(t) \le \beta_2(t)$$

and there exists $t_0 \in [a,b]$ with

$$\alpha_2(t_0) > \beta_1(t_0).$$

Assume further β_1 and α_2 are strict upper and lower solutions.

Let E be defined by (1.3) (with $\alpha = \min\{\alpha_1, \alpha_2\}$ and $\beta = \max\{\beta_1, \beta_2\}$) and suppose $f : E \to \mathbb{R}$ is an L^1-Carathéodory function.

Prove that the problem (1.1) has at least three solutions u_1, u_2, $u_3 \in W^{2,1}(a,b)$ such that

$$\alpha_1 \le u_1 \le \min\{\beta_1, \beta_2\},\ \max\{\alpha_1, \alpha_2\} \le u_2 \le \beta_2,\ u_1 \le u_3 \le u_2.$$

If moreover α_1 and β_2 are strict and $f(t,u) + k(t)u$ is non-decreasing in u for some $k \in L^1(a,b)$, prove that u_1 and u_2 are local minima of ϕ.

3 The Minimax Method

3.1 The Minus Gradient Flow

One of the main techniques in variational methods uses the deformation of paths or surfaces along the minus gradient (or pseudo-gradient) flow. In this section, we study the dynamical system associated with this flow.

We shall define the minus gradient flow using the following assumptions :

(H) Let $f : [a,b] \times \mathbb{R} \to \mathbb{R}, (t,u) \mapsto f(t,u)$ be an L^1-Carathéodory function, locally Lipschitz in u. Let also $m \in L^1(a,b)$ be such that $m > 0$ a.e. in $[a,b]$ and $f(t,u) + m(t)\,u$ is increasing in u.

Now, let us define on $H_0^1(a,b)$ the scalar product

$$(u,v)_{H_0^1} = \int_a^b [u'(t)v'(t) + m(t)u(t)v(t)]\,dt$$

and let

$$F(t,u) = \int_0^u f(t,s)\,ds.$$

It is then easy to see that the functional

$$\phi : H_0^1(a,b) \to \mathbb{R}, u \mapsto \int_a^b [\frac{u'^2(t)}{2} - F(t,u(t))]\,dt$$

is of class $\mathcal{C}^1$ and

$$\nabla\phi(u) = u - KN(u), \tag{3.1}$$

where

$$\begin{aligned} N : H_0^1(a,b) &\to L^1(a,b), u \mapsto f(t,u) + m(t)u, \\ K : L^1(a,b) &\to H_0^1(a,b), h \mapsto Kh \end{aligned} \tag{3.2}$$

and Kh is defined to be the unique solution of

$$-u'' + m(t)u = h(t), \quad u(a) = 0,\ u(b) = 0.$$

Let us notice at last that if Assumptions (H) are satisfied, the function

$$\nabla\phi : \mathcal{C}_0^1([a,b]) \to \mathcal{C}_0^1([a,b]),$$

defined from (3.1) is locally lipschitzian. Next, we define a $\mathcal{C}^1$-function $\psi_r : \mathbb{R} \to [0,1]$ such that $\psi_r(s) = 1$ if $s \geq r$ and $\psi_r(s) = 0$ if $s \leq r-1$.

We consider then the Cauchy problem

$$\begin{aligned} \tfrac{d}{dt}u = -\psi_r(\phi(u))\nabla\phi(u) &= -\psi_r(\phi(u))(u - KN(u)), \\ u(0) &= u_0, \end{aligned} \tag{3.3}$$

where $u_0 \in \mathcal{C}_0^1([a,b])$. From the theory of ordinary differential equations, we know that the solution $u(\,\cdot\,; u_0)$ of (3.3) exists, is unique, and is defined in the future on a maximal interval $[0, \omega(u_0)[$. We also know that for any $t \in [0, \omega(u_0)[$, the function $u(t;\,\cdot\,) : \mathcal{C}_0^1([a,b]) \to \mathcal{C}_0^1([a,b])$ is continuous. We call the *minus gradient flow* the local dynamical system defined on $\mathcal{C}_0^1([a,b])$ by $u(t;u_0)$.

We could have defined the minus gradient flow in $X = \mathcal{C}_0([a,b])$ or $H_0^1(a,b)$. However, such choices are not suitable in our context. We have to work with sets such as $\{u \in X \mid u \prec \beta\}$ and $\{u \in X \mid u \succ \alpha\}$, where α and β are lower and upper solutions that can satisfy the boundary conditions. With the $\mathcal{C}_0([a,b])$ or the $H_0^1(a,b)$-topology, these sets have in such a case empty interior which creates major difficulties.

A first result shows that the solutions of (3.3) are defined for all $t \geq 0$.

Proposition 3.1. *Let Assumptions (H) be satisfied and $u(t;u_0)$ be the minus gradient flow defined for some $r \in \mathbb{R}$. Then for any $u_0 \in \mathcal{C}_0^1([a,b])$ we have $\omega(u_0) = +\infty$.*

Proof. Notice that

$$\begin{aligned}\tfrac{d}{dt}\phi(u(t;u_0)) &= (\nabla\phi(u(t;u_0)), \tfrac{d}{dt}u(t;u_0))_{H_0^1}\\ &= -\psi_r(\phi(u(t;u_0)))\|\nabla\phi(u(t;u_0))\|^2_{H_0^1},\end{aligned}$$

which implies that for all $t \in [0, \omega(u_0)[$

$$\phi(u(t;u_0)) \le \phi(u_0). \tag{3.4}$$

Observe also that $\phi(u(t;u_0)) \ge \min\{r-1, \phi(u_0)\} =: C$. At last we have for any $0 \le t_1 < t_2 < \omega(u_0)$

$$\begin{aligned}\|u(t_2;u_0) - u(t_1;u_0)\|_{H_0^1} &\le \int_{t_1}^{t_2} \psi_r(\phi(u(s;u_0)))\|\nabla\phi(u(s;u_0))\|_{H_0^1}\, ds\\ &\le [\int_{t_1}^{t_2} \psi_r(\phi(u(s;u_0)))\|\nabla\phi(u(s;u_0))\|^2_{H_0^1}\, ds]^{\frac12}[\int_{t_1}^{t_2}\psi_r(\phi(u(s;u_0)))\, ds]^{\frac12}\\ &\le [-\int_{t_1}^{t_2} \tfrac{d}{ds}\phi(u(s;u_0))\, ds]^{\frac12}\sqrt{t_2-t_1} \le [\phi(u_0) - C]^{\frac12}\sqrt{t_2-t_1}.\end{aligned}$$

Hence, if $\omega(u_0) < +\infty$, there exists $u^* \in H_0^1(a,b)$ such that $u(t;u_0) \overset{H_0^1}{\to} u^*$ as $t \to \omega(u_0)$. It follows that the function $u(\,\cdot\,;u_0) : [0, \omega(u_0)] \to \mathcal{C}([a,b])$, where $u(\omega(u_0);u_0) = u^*$, is continuous and

$$KNu(\,\cdot\,;u_0) \in \mathcal{C}([0,\omega(u_0)], \mathcal{C}_0^1([a,b])).$$

Let $a(t) = \psi_r(\phi(u(t;u_0)))$. For all $t \in [0, \omega(u_0)]$, we can write

$$u(t;u_0) = u_0 e^{-\int_0^t a(r)\,dr} + \int_0^t e^{-\int_s^t a(r)\,dr} a(s)KNu(s;u_0)\, ds \in \mathcal{C}_0^1([a,b]).$$

Hence, $u(\,\cdot\,;u_0) : [0,\omega(u_0)] \to \mathcal{C}_0^1([a,b])$ is continuous, which implies

$$u(t;u_0) \overset{\mathcal{C}_0^1}{\to} u^* \;\text{ as }\; t \to \omega(u_0),$$

and contradicts the maximality of $\omega(u_0)$. □

3.2 Invariant Sets

An important property of the cones

$$C_\alpha = \{u \in \mathcal{C}_0^1([a,b]) \mid u \succ \alpha\} \tag{3.5}$$

and

$$C^\beta = \{u \in \mathcal{C}_0^1([a,b]) \mid u \prec \beta\}, \tag{3.6}$$

which are associated to lower and upper solutions α and β of (1.1) is that they are positively invariant. To make this precise, let us introduce the following definitions.

Definition 3.1. *Let $u(t;u_0)$ be the minus gradient flow defined for some $r \in \mathbb{R}$. A nonempty set $M \subset C_0^1([a,b])$ is called a* positively invariant set *if*

$$\forall u_0 \in M, \ \forall t \geq 0, \quad u(t;u_0) \in M.$$

As a first example, notice that (3.4) implies that the set

$$\phi^c = \{u \in C_0^1([a,b]) \mid \phi(u) < c\}$$

is positively invariant. Also, unions and intersections of positively invariant sets are positively invariant.

To investigate the cones C_α and C^β we need the following lemma.

Lemma 3.2. *Let Assumptions (H) be satisfied. Assume $\alpha \in W^{2,1}(a,b)$ is a lower solution of* (1.1).

Then for all $u \geq \alpha$, we have $KNu \geq \alpha$, where K and N are defined by (3.2).

Proof. Let $u \geq \alpha$, set $w = KNu - \alpha$ and observe that w satisfies

$$\begin{aligned} -w'' + m(t)w &= f(t,u(t)) + m(t)u(t) - (-\alpha''(t) + m(t)\alpha(t)) \\ &\geq f(t,u(t)) + m(t)u(t) - (f(t,\alpha(t)) + m(t)\alpha(t)) \geq 0, \\ &\qquad w(a) \geq 0, \ w(b) \geq 0. \end{aligned}$$

It follows from the maximum principle that $w \geq 0$. □

Proposition 3.3. *Let Assumptions (H) be satisfied and $u(t;u_0)$ be the minus gradient flow defined for some $r \in \mathbb{R}$.*

Assume $\alpha \in W^{2,1}(a,b)$ is a lower solution of (1.1). *Then the set C_α defined from* (3.5) *is positively invariant.*

Similarly, if $\beta \in W^{2,1}(a,b)$ is an upper solution of (1.1), *the set C^β defined from* (3.6) *is positively invariant.*

Proof. If the claim is wrong, we can find $u_0 \in C_\alpha$ and $t_1 > 0$ so that for all $t \in [0,t_1[$, $u(t;u_0) \in C_\alpha$ and $u(t_1;u_0) \in \partial C_\alpha$.

Let $w(t) = u(t;u_0) - \alpha$, define $a(t) = \psi_r(\phi(u(t;u_0)))$ and observe that for all $t \in [0,t_1[$

$$\tfrac{d}{dt}w(t) = \tfrac{d}{dt}u(t;u_0) = -a(t)(u(t,u_0) - KN(u(t;u_0))) = -a(t)w(t) + h(t),$$

where from Lemma 3.2, $h(t) = a(t)(KN(u(t;u_0)) - \alpha) \geq 0$ for $t \in [0, t_1[$. As $w(0) = u_0 - \alpha \succ 0$ we have

$$w(t_1) = w(0)e^{-\int_0^{t_1} a(r)\,dr} + \int_0^{t_1} e^{-\int_s^{t_1} a(r)\,dr} h(s)\,ds \succ 0,$$

which contradicts $u(t_1; u_0) \in \partial C_\alpha$. □

3.3 Non Well-ordered Lower and Upper Solutions

The first result of this section provides Palais-Smale type sequences from non well-ordered lower and upper solutions. As usual, this proposition gives a solution of (1.1) with the help of the Palais-Smale condition.

Proposition 3.4. *Let Assumptions* (H) *be satisfied. Suppose* α *and* $\beta \in W^{2,1}(a,b)$ *are lower and upper solutions of* (1.1) *and* $\alpha \not\leq \beta$. *Define* C_α *and* C^β *from* (3.5) *and* (3.6),

$$\begin{aligned} \Gamma &= \{\gamma \in \mathcal{C}([0,1], \mathcal{C}_0^1([a,b])) \mid \gamma(0) \in C^\beta,\ \gamma(1) \in C_\alpha\}, \\ T_\gamma &= \{s \in [0,1] \mid \gamma(s) \in \mathcal{C}_0^1([a,b]) \setminus (C^\beta \cup C_\alpha)\} \end{aligned} \tag{3.7}$$

and assume

$$c := \inf_{\gamma \in \Gamma} \max_{s \in T_\gamma} \phi(\gamma(s)) \in \mathbb{R},$$

where $\phi(u)$ *is defined from* (1.2). *At last, let* $u(t;u_0)$ *be the minus gradient flow defined with* $r = c - 1$.

Then, for any $\delta \in]0,1[$ *there exist* $u_0 \in \mathcal{C}_0^1([a,b])$ *such that*

$$\forall t > 0, \quad u(t;u_0) \in \phi^{-1}([c-\delta, c+\delta]) \setminus (C^\beta \cup C_\alpha)$$

and there exists an increasing unbounded sequence $(t_n)_n \subset \mathbb{R}^+$ *such that*

$$\nabla\phi(u(t_n;u_0)) \overset{H_0^1}{\to} 0 \quad \textit{as} \quad n \to \infty.$$

Proof. Let us fix $\delta \in]0,1[$ and define $E = \phi^{c-\delta} \cup C_\alpha \cup C^\beta$. Observe that E is positively invariant. Define $A(E) = \{u_0 \in \mathcal{C}_0^1([a,b]) \mid \exists t \geq 0,\ u(t;u_0) \in E\}$. Obviously, this set is open and positively invariant. Consider a path $\gamma \in \Gamma$ so that for all $s \in T_\gamma$, $\phi(\gamma(s)) \leq c + \delta$.

Claim : There exists $u_0 \in \gamma(T_\gamma) \setminus A(E)$. Assume by contradiction that for every $s \in T_\gamma$, $\gamma(s) \in A(E)$, i.e. that for every $s \in [0,1]$, $\gamma(s) \in A(E)$.

Let us prove first that in such a case there exists $T \geq 0$ such that for all $s \in [0,1]$, $u(T;\gamma(s)) \in E$. For any $s \in [0,1]$, we can find $t_s \geq 0$

such that $u(t_s;\gamma(s)) \in E$. Assume next that for every $n \in \mathbb{N}$, there exists $s_n \in [0,1]$ such that $u(n,\gamma(s_n)) \notin E$. Going to a subsequence, we can assume $s_n \to s^* \in [0,1]$ and using the contradiction assumption, there exists $t_{s^*} \geq 0$ such that $u(t_{s^*};\gamma(s^*)) \in E$. As E is open, for all n large enough $u(t_{s^*};\gamma(s_n)) \in E$ which leads to a contradiction as E is positively invariant and $u(n;\gamma(s_n)) \notin E$.

Notice now that $\bar{\gamma} = u(T;\gamma(\cdot))$ is in Γ and such that $\phi(\bar{\gamma}(\cdot)) \leq c - \delta$ on $T_{\bar{\gamma}}$ which contradicts the definition of c.

Conclusion – By construction of $A(E)$ and as $\phi(u(\cdot;u_0))$ is non-increasing, we have for all $t > 0$, $u(t;u_0) \in \phi^{-1}([c-\delta, c+\delta]) \setminus (C^\beta \cup C_\alpha)$. Hence, u satisfies

$$\frac{d}{dt}u = -\nabla\phi(u)$$

and there exists an increasing unbounded sequence $(t_n)_n$ which verifies

$$\frac{d}{dt}\phi(u(t_n;u_0)) = -\|\nabla\phi(u(t_n;u_0))\|^2_{H^1_0} \to 0. \qquad \square$$

In order to obtain existence of solutions of (1.1), we need to prove that the sequence $(u(t_n;u_0))_n$ converges towards such a solution, which holds true in case we assume the Palais-Smale Condition (PS).

Theorem 3.5. *Let Assumptions (H) be satisfied. Suppose α and $\beta \in W^{2,1}(a,b)$ are lower and upper solutions of (1.1) and $\alpha \not\leq \beta$. Define Γ and T_γ from (3.7) and assume*

$$c := \inf_{\gamma\in\Gamma}\max_{s\in T_\gamma}\phi(\gamma(s)) \in \mathbb{R},$$

where $\phi(u)$ is defined from (1.2). Assume at last the Palais-Smale Condition (PS) is satisfied.

Then there exists $v \in C^1_0([a,b]) \setminus (C^\beta \cup C_\alpha)$ solution of (1.1) such that $\phi(v) = c$.

Proof. Let $u(t;u_0)$ be the minus gradient flow defined with $r = c - 1$.

Part 1 – For any $k \in \mathbb{N}$, there exists $v_k \in H^1_0(a,b)$ such that

$$c - \tfrac{1}{k} \leq \phi(v_k) \leq c + \tfrac{1}{k} \quad \text{and} \quad \nabla\phi(v_k) = 0.$$

Let us fix $k \in \mathbb{N}$. From Proposition 3.4, there exist $u_k \in C^1_0([a,b])$ such that

$$\forall t > 0, \quad u(t;u_k) \in \phi^{-1}([c - \tfrac{1}{k}, c + \tfrac{1}{k}]) \setminus (C^\beta \cup C_\alpha).$$

and there exists an increasing unbounded sequence $(t_n)_n \subset \mathbb{R}^+$ such that

$$\nabla\phi(u(t_n;u_k)) \overset{H^1_0}{\to} 0 \quad \text{as} \quad n \to \infty.$$

Notice that $\phi(u(t;u_k)) > c-1$ so that $u(t;u_k)$ solves

$$\tfrac{d}{dt}u = -\nabla\phi(u), \quad u(0) = u_k.$$

Using the Palais-Smale Condition (PS), there exists a subsequence that we still write $(u(t_n;u_k))_n$ and $v_k \in H_0^1(a,b)$ such that

$$u(t_n;u_k) \overset{H_0^1}{\to} v_k \quad \text{as } n \to \infty,$$
$$c - \tfrac{1}{k} \le \phi(v_k) \le c + \tfrac{1}{k} \text{ and } \nabla\phi(v_k) = 0.$$

Part 2 – $v_k \in \mathcal{C}_0^1([a,b]) \setminus (C^\beta \cup C_\alpha)$.

Claim 1 : There exists $R > 0$ *such that for all* $s \in [0, +\infty[$,

$$\|u(s;u_k)\|_{H_0^1} \ge R \text{ implies } \|\nabla\phi(u(s;u_k))\|_{H_0^1} \ge \tfrac{1}{R}.$$

If not, there exists a sequence $(s_m)_m \subset [0, +\infty[$ such that

$$\|u(s_m;u_k)\|_{H_0^1} \ge m \tag{3.8}$$

and $\|\nabla\phi(u(s_m;u_k))\|_{H_0^1} < \frac{1}{m}$. As $\phi(u(t;u_k))$ is bounded, by Palais-Smale Condition, there exists a subsequence $(s_{m_j})_j$ so that $u(s_{m_j};u_k)$ converges in $H_0^1(a,b)$ which contradicts (3.8).

Claim 2 : $\|u(t;u_k)\|_{H_0^1}$ *is bounded on* $\mathbb{R}^+$. Assume that for some $t > 0$, $\|u(t;u_k)\|_{H_0^1} > R_0 = \max\{\|u_k\|_{H_0^1}, R\}$. Then there exists $t_1 \in [0,t]$ so that $\|u(t_1;u_k)\|_{H_0^1} = R_0$ and for any $s \in [t_1,t]$, $\|u(s;u_k)\|_{H_0^1} \ge R_0 \ge R$. It follows that

$$|\phi(u(t;u_k)) - \phi(u(t_1;u_k))| = \int_{t_1}^t \|\nabla\phi(u(s;u_k))\|_{H_0^1}^2 \, ds \ge \tfrac{1}{R^2}(t-t_1).$$

On the other hand, we have

$$\begin{aligned}
\|u(t;u_k) - u(t_1;u_k)\|_{H_0^1} &= \|\int_{t_1}^t \nabla\phi(u(s;u_k))\, ds\|_{H_0^1} \\
&\le [\int_{t_1}^t \|\nabla\phi(u(s;u_k))\|_{H_0^1}^2 \, ds]^{\frac{1}{2}} (t-t_1)^{\frac{1}{2}} \\
&\le |\phi(u(t;u_k)) - \phi(u(t_1;u_k))|^{\frac{1}{2}} (t-t_1)^{\frac{1}{2}} \\
&\le |\phi(u(t;u_k)) - \phi(u(t_1;u_k))| \, R.
\end{aligned}$$

As $\phi(u(t;u_k))$ is bounded, the claim follows.

Claim 3 : $v_k \in C_0^1([a,b]) \setminus (C^\beta \cup C_\alpha)$. To prove this claim let us show that for some subsequence $u(t_n; u_k) \overset{C_0^1}{\to} v_k$. Consider the sequence $(w_n)_n \subset C_0^1([a,b])$, defined by

$$w_n(r) = \int_0^{t_n} e^{-(t_n-s)}(KNu(s;u_k))(r)\,ds,$$

with K and N defined from (3.2). As $\|u(t;u_k)\|_{H_0^1}$ is bounded, there exists $h \in L^1(a,b)$ so that

$$\begin{aligned}|w_n''(r)| &= |\int_0^{t_n} e^{-(t_n-s)}[f(\cdot,u(s;u_k)) + m(\cdot)(u(s;u_k) - KNu(s;u_k))](r)\,ds| \\ &\le \int_0^{t_n} e^{-(t_n-s)}h(r)\,ds \le h(r).\end{aligned}$$

Using Arzelà-Ascoli Theorem, we can find a subsequence $(w_{n_i})_i$ converging in $C_0^1([a,b])$. The same holds true for

$$u(t_n;u_k) = u_k e^{-t_n} + \int_0^{t_n} e^{-(t_n-s)}KNu(s;u_k)\,ds,$$

i.e.

$$u(t_{n_i};u_k) \overset{C_0^1}{\to} \hat{v}_k$$

for some $\hat{v}_k$ and as $u(t_n;u_k) \overset{H_0^1}{\to} v_k$ the claim follows.

Conclusion – From Palais-Smale Condition, a subsequence of $(v_k)_k$ converges in $H_0^1(a,b)$ to some function v. As $v_k = KNv_k$, the convergence also holds in $C_0^1([a,b])$ so that

$$v \in C_0^1([a,b]) \setminus (C^\beta \cup C_\alpha), \quad \phi(v) = c \quad \text{and} \quad \nabla\phi(v) = 0. \qquad \square$$

Theorem 3.6. *Let Assumptions (H) be satisfied. Suppose the functions* α_1, $\alpha_2 \in W^{2,1}(a,b)$ *are lower solutions of* (1.1) *and* β_1, $\beta_2 \in W^{2,1}(a,b)$ *are upper solutions of* (1.1), *none of them being solution. Assume they satisfy*

$$\alpha_1 \le \beta_1, \quad \alpha_2 \le \beta_2, \qquad \alpha_1 \not\le \beta_2, \quad \alpha_2 \not\le \beta_1.$$

Assume at last the Palais-Smale Condition (PS) is satisfied.

Then the problem (1.1) *has at least 7 solutions such that*
$\alpha_1 \prec u_1 \prec \beta_1$, $\alpha_2 \prec u_2 \prec \beta_2$,
$\alpha_1 \prec u_3$, $u_3 \not\prec \beta_1$, $u_3 \not\succ \alpha_2$, $\alpha_2 \prec u_4$, $u_4 \not\prec \beta_2$, $u_4 \not\succ \alpha_1$,
$u_5 \prec \beta_2$, $u_5 \not\prec \beta_1$, $u_5 \not\succ \alpha_2$, $u_6 \prec \beta_1$, $u_6 \not\prec \beta_2$, $u_6 \not\succ \alpha_1$,
u_7 *is not comparable with* α_i *and* β_i *for* $i = 1, 2$.

Proof. Notice that from Proposition III-2.7 and Assumptions (H), the lower and upper solutions α_i and β_i are strict. The two first solutions u_1 and u_2 are then obtained by Theorem 2.2, while the solutions u_i for $i = 3, 4, 5, 6$ are obtained repeating the argument of Theorem 3.5 with, for example, for u_3,

$$\Gamma = \{\gamma \in \mathcal{C}([0,1], C_{\alpha_1}) \mid \gamma(0) \in C^{\beta_1},\ \gamma(1) \in C_{\alpha_2}\}.$$

For that which concerns the last solution, let us define

$$\begin{aligned}\Sigma = \{\sigma \in \mathcal{C}([0,1]^2, \mathcal{C}_0^1([a,b])) \mid \\ \sigma(r,0) \in C^{\beta_2},\ \sigma(r,1) \in C_{\alpha_1},\ \sigma(0,s) \in C^{\beta_1},\ \sigma(1,s) \in C_{\alpha_2}\},\end{aligned}$$

$$T_\sigma = \{(r,s) \in [0,1]^2 \mid \sigma(r,s) \in \mathcal{C}_0^1([a,b]) \setminus (C^{\beta_1} \cup C_{\alpha_1} \cup C^{\beta_2} \cup C_{\alpha_2})\},$$

$$c = \inf_{\sigma\in\Sigma}\big(\max_{(r,s)\in T_\sigma} \phi(\sigma(r,s))\big),$$

and observe that $c \in \mathbb{R}$.

Let $u(t;u_0)$ be the minus gradient flow defined with $r = c - 1$. We will prove that for any $\delta \in]0,1[$, there exists $u_0 \in \mathcal{C}_0^1([a,b])$ such that

$$\forall t \in [0,+\infty[, \quad u(t,u_0) \in \phi^{-1}([c-\delta, c+\delta]) \setminus (C^{\beta_1} \cup C_{\alpha_1} \cup C^{\beta_2} \cup C_{\alpha_2}).$$

Fix $\delta \in]0,1[$, let $\sigma \in \Sigma$ be such that

$$c \le \max_{(r,s)\in T_\sigma} \phi(\sigma(r,s)) \le c + \delta$$

and define $E = \phi^{c-\delta} \cup (C^{\beta_1} \cup C_{\alpha_1} \cup C^{\beta_2} \cup C_{\alpha_2})$. Observe that E is an open positively invariant set. Define $A(E) = \{u_0 \in \mathcal{C}_0^1([a,b]) \mid \exists t \ge 0,\ u(t;u_0) \in E\}$.

Let us prove there exists $u_0 \in \sigma(T_\sigma) \setminus A(E)$. Otherwise, as in Proposition 3.4, we prove there exists T so that for all $(r,s) \in [0,1]^2$, $u(T;\sigma(r,s)) \in E$. Notice that $\bar\sigma(r,s) = u(T;\sigma(r,s)) \in \Sigma$, and

$$\max_{T_{\bar\sigma}} \phi(\bar\sigma(r,s)) < c - \delta,$$

which contradicts the definition of c and proves the claim.

Hence, by construction of $A(E)$ and as $\phi(u(\cdot;u_0))$ is non-increasing, for all $t \in [0,+\infty[$,

$$u(t;u_0) \in \phi^{-1}([c-\delta, c+\delta]) \setminus (C^{\beta_1} \cup C_{\alpha_1} \cup C^{\beta_2} \cup C_{\alpha_2}).$$

As the Palais-Smale Condition holds, we conclude repeating the proof of Theorem 3.5. □

3.4 A Four Solutions Theorem

This section deals with a problem (1.1) which has the trivial solution $u = 0$. We consider assumptions which imply existence of lower and upper solutions α_i, β_i so that

$$\alpha_1 \leq \beta_1 \leq 0 \leq \alpha_2 \leq \beta_2.$$

The Three Solutions Theorem (Theorem III-2.11) provides three solutions, two one-sign solutions $u_1 \in [\alpha_1, \beta_1]$ and $u_2 \in [\alpha_2, \beta_2]$ and a third one that can be the zero solution. The difficulty is to obtain a third nontrivial solution. Here such a result is obtained assuming the slope $\frac{f(t,u)}{u}$ crosses the two first eigenvalues.

Theorem 3.7. *Let Assumptions (H) be satisfied and assume:*

(i) there exist $\lambda > \lambda_2 = \frac{4\pi^2}{(b-a)^2}$ and $\delta > 0$ such that for a.e. $t \in [a, b]$ and all $u \in [-\delta, \delta]$,

$$\frac{f(t,u)}{u} \geq \lambda;$$

(ii) there exist $\mu < \lambda_1 = \frac{\pi^2}{(b-a)^2}$ and $R > 0$ such that for a.e. $t \in [a, b]$ and all $u \in \mathbb{R}$ with $|u| \geq R$,

$$\frac{f(t,u)}{u} \leq \mu.$$

Then the problem (1.1) *has at least three nontrivial solutions u_i such that $u_1 \prec 0$, $u_2 \succ 0$ and u_3 changes sign.*

Proof. Claim – There exists $\alpha_1 \prec 0$ which is a strict lower solution of (1.1). Let $h \in L^1(a, b)$ be such that $h \geq 0$ and for a.e. $t \in [a, b]$ and all $u \leq 0$,

$$f(t,u) > \mu u - h(t).$$

Define then α_1 to be the solution of

$$u'' + \mu u = h(t), \quad u(a) = 0, \; u(b) = 0.$$

As $\mu < \lambda_1$ and $h \geq 0$ we have from the maximum principle $\alpha_1 \prec 0$ and

$$\alpha_1''(t) + f(t, \alpha_1(t)) > \alpha_1''(t) + \mu\alpha_1(t) - h(t) = 0, \quad \alpha_1(a) = 0, \; \alpha_1(b) = 0,$$

i.e. $\alpha_1 \prec 0$ is a lower solution which is not a solution. Using Assumptions (*H*), it follows from Proposition III-2.7 that α_1 is a strict lower solution.

Claim – There exists $\beta_2 \succ 0$ which is a strict upper solution of (1.1). We construct β_2 as we did for α_1.

The modified problem. Consider the modified problem

$$\begin{aligned} u'' + \bar{f}(t,u) = 0, \\ u(a) = 0, \; u(b) = 0, \end{aligned} \tag{3.9}$$

where

$$\begin{aligned} \bar{f}(t,u) &= f(t,\alpha_1(t)), && \text{if } u < \alpha_1(t), \\ &= f(t,u), && \text{if } \alpha_1(t) \le u \le \beta_2(t), \\ &= f(t,\beta_2(t)), && \text{if } u > \beta_2(t), \end{aligned}$$

and the corresponding functional

$$\bar{\phi}(u) = \int_a^b [\frac{u'^2(t)}{2} - \bar{F}(t,u(t))]\,dt,$$

with $\bar{F}(t,u) = \displaystyle\int_0^u \bar{f}(t,s)\,ds$. As usual (see Theorem II-2.1) it is easy to see that every solution u of (3.9) satisfies $\alpha_1 \le u \le \beta_2$ and is a solution of (1.1).

Existence of the solutions u_1 and u_2. Define $\varphi_1(t) = \sin(\pi\frac{t-a}{b-a})$ and fix $\epsilon > 0$ small enough so that $\epsilon < \min\{\delta/4, \lambda - \lambda_2\}$, $-4\epsilon\varphi_1 \succ \alpha_1$ and $4\epsilon\varphi_1 \prec \beta_2$. It is easy to see that $\beta_1 = -\epsilon\varphi_1$ and $\alpha_2 = \epsilon\varphi_1$ are respectively upper and lower solutions of (1.1) but are not solutions since

$$\begin{aligned} -\beta_1''(t) = -\epsilon\lambda_1\varphi_1(t) > f(t,-\epsilon\varphi_1(t)) = f(t,\beta_1(t)), \\ -\alpha_2''(t) = \epsilon\lambda_1\varphi_1(t) < f(t,\epsilon\varphi_1(t)) = f(t,\alpha_2(t)). \end{aligned}$$

Using Assumptions (H) and Proposition III-2.7, these are strict upper and lower solutions. Theorem 1.1 applies then with $\alpha = \alpha_i$ and $\beta = \beta_i$, which implies the existence of solutions $u_1 \in C^{\beta_1}$ and $u_2 \in C_{\alpha_2}$.

Existence of a third nontrivial solution. Observe that as $\bar{\phi}$ is coercive, it satisfies the Palais-Smale Condition and we can apply Theorem 3.5 with $\alpha = \alpha_2$ and $\beta = \beta_1$. This observation proves the existence of a solution $u_3 \in C_0^1([a,b]) \setminus (C^{\beta_1} \cup C_{\alpha_2})$, i.e. $u_3 \ne u_1$ and $u_3 \ne u_2$. The main problem is to see that u_3 is not the trivial solution. To this aim, we prove that $c = \bar{\phi}(u_3) < 0 = \bar{\phi}(0)$.

Define Γ from (3.7), with $\alpha = \alpha_2$, $\beta = \beta_1$, and $\gamma \in \Gamma$ by

$$\gamma(s) = 2\epsilon[(2s-1)\varphi_1 + (1 - |2s-1|)\varphi_2],$$

where $\varphi_2(t) = \sin(2\pi\frac{t-a}{b-a})$. Observe that

$$\begin{gathered} \gamma(0) = -2\epsilon\varphi_1 \prec \beta_1, \quad \gamma(1) = 2\epsilon\varphi_1 \succ \alpha_2, \\ \alpha_1 \prec -4\varepsilon\varphi_1 \le \gamma(s) \le 4\varepsilon\varphi_1 \prec \beta_2 \quad \text{for all } s \in [0,1]. \end{gathered}$$

Moreover,

$$\begin{aligned}\bar{\phi}(\gamma(s)) &= \int_a^b [2\epsilon^2[(2s-1)^2\varphi_1'^2(t) + (1-|2s-1|)^2\varphi_2'^2(t)] \\ &\qquad -\bar{F}(t, 2\epsilon[(2s-1)\varphi_1(t) + (1-|2s-1|)\varphi_2(t)])]\, dt \\ &\le \int_a^b [2\epsilon^2[(2s-1)^2\lambda_1\varphi_1^2(t) + (1-|2s-1|)^2\lambda_2\varphi_2^2(t)] \\ &\qquad -\lambda 2\epsilon^2[(2s-1)\varphi_1(t) + (1-|2s-1|)\varphi_2(t)]^2]\, dt \\ &\le \epsilon^2(b-a)[(2s-1)^2(\lambda_1-\lambda) + (1-|2s-1|)^2(\lambda_2-\lambda)] \\ &\le \epsilon^2(b-a)[(2s-1)^2 + (1-|2s-1|)^2](\lambda_2-\lambda) \\ &\le -\tfrac{\epsilon^3}{2}(b-a).\end{aligned}$$

Hence $c \le -\frac{\epsilon^3}{2}(b-a) < 0$, which implies the third solution u_3 is nontrivial.

Claim – The function u_3 changes sign. Assume $u_3 \ge 0$ and define $\eta = \max\{\tau \ge 0 \mid u_3 - \tau\varphi_1 \ge 0\}$. Observe first that

$$\begin{aligned}-u_3'' + m(t)u_3 = f(t,u_3) + m(t)u_3 \ge f(t,0) = 0,\\ u_3(a) = 0,\ u_3(b) = 0.\end{aligned}$$

As $u_3 \ne 0$, we deduce from the maximum principle that $u_3 \succ 0$ which implies that $\eta > 0$. Let us assume now that $\eta < \delta$. We can find then $t_0 \in [a,b]$ such that $u_3(t_0) - \eta\varphi_1(t_0) = 0$, $u_3'(t_0) - \eta\varphi_1'(t_0) = 0$ and for t close enough to t_0

$$\begin{aligned}-(t-t_0)(u_3-\eta\varphi_1)'(t) &= (t-t_0)\int_{t_0}^t (f(s,u_3(s)) - \lambda_1\eta\varphi_1(s))\, ds \\ &\ge (\lambda-\lambda_1)(t-t_0)\eta\int_{t_0}^t \varphi_1(s)\, ds > 0,\end{aligned}$$

which contradicts the minimality of $u_3 - \eta\varphi_1$ for $t = t_0$. It follows that $u_3 \ge \delta\varphi_1 \succ \alpha_2$ which contradicts the localization of u_3.

We prove in a similar way that u_3 cannot be negative. Therefore u_3 changes sign. □

An additional solution can be obtained combining variational method and degree theory. Here we impose that the slope $\frac{f(t,u)}{u}$ lies between two consecutive eigenvalues for small values of u.

Theorem 3.8. *Assume that $f \in C^1([a,b]\times\mathbb{R})$ satisfies Assumptions (H) together with*

(i) there exist p, q, $k \geq 2$ $(k \in \mathbb{N})$ *and* $\delta > 0$ *such that for a.e.* $t \in [a, b]$ *and all* $u \in [-\delta, \delta]$,

$$\lambda_k := \frac{k^2\pi^2}{(b-a)^2} < p \leq \frac{f(t,u)}{u} \leq q < \frac{(k+1)^2\pi^2}{(b-a)^2} =: \lambda_{k+1};$$

(ii) there exist $\mu < \lambda_1 = \frac{\pi^2}{(b-a)^2}$ *and* $R > 0$ *such that for a.e.* $t \in [a, b]$ *and all* $u \in \mathbb{R}$ *with* $|u| \geq R$,

$$\frac{f(t,u)}{u} \leq \mu.$$

Then the problem (1.1) *has at least four nontrivial solutions* u_i *such that* $u_1 \prec 0$, $u_2 \succ 0$ *and* u_3, u_4 *change sign.*

Proof. Part 1 – Existence of solutions $u_1 \prec 0 \prec u_2$. As in the proof of Theorem 3.7, we choose strict lower solutions α_i and strict upper solutions β_i such that

$$\alpha_1 \prec -\delta\varphi_1 \prec \beta_1 = -\varepsilon\varphi_1 \quad \text{and} \quad \alpha_2 = \varepsilon\varphi_1 \prec \delta\varphi_1 \prec \beta_2,$$

where $\varepsilon \in]0, \delta[$. From Theorem II-2.6, the problem (1.1) has two solutions $u_1 \prec \beta_1$ and $u_2 \succ \alpha_2$ such that u_1 is the maximum solution in $[\alpha_1, \beta_1]$ and u_2 is the minimum solution in $[\alpha_2, \beta_2]$. Moreover we can prove as in the proof of Theorem 3.7 that

$$u_1 \prec -\delta\varphi_1 \quad \text{and} \quad u_2 \succ \delta\varphi_1.$$

Part 2 – Existence of a changing sign solution u_3. Consider now the modified problem

$$\begin{aligned} u'' + \bar{f}(t,u) &= 0, \\ u(a) = 0, \ u(b) &= 0, \end{aligned} \tag{3.10}$$

where

$$\begin{aligned} \bar{f}(t,u) &= f(t, u_1(t)), && \text{if } u < u_1(t), \\ &= f(t,u), && \text{if } u_1(t) \leq u < u_2(t), \\ &= f(t, u_2(t)), && \text{if } u_2(t) \leq u, \end{aligned}$$

and the corresponding functional

$$\bar{\phi}(u) = \int_a^b [\frac{u'^2}{2} - \bar{F}(t, u(t))]\, dt,$$

with $\bar{F}(t, u(t)) = \int_0^u \bar{f}(t,s)\, ds$. As usual (see Theorem II-2.1), it is easy to see that every solution of (3.10) satisfies $u_1 \leq u \leq u_2$ and is a solution of (1.1). As in the proof of Theorem 3.7, we see that the problem (3.10) and hence (1.1) has a third solution $u_3 \neq 0$, which changes sign and is such that

$$\bar\phi(u_3) = \inf_{\gamma\in\Gamma} \max_{s\in T_\gamma} \bar\phi(\gamma(s)) < 0,$$

where Γ and T_γ are defined from (3.7). Notice also that $\bar\phi$ satisfies the Palais-Smale Condition.

Part 3 – Existence of the fourth nontrivial solution u_4. Assume by contradiction that the only solutions of (3.10) are u_1, u_2, u_3 and 0. As $u_1 - 1$ and α_2 are strict lower solutions of (3.10) and β_1 and $u_2 + 1$ are strict upper solutions of (3.10), we have, by Theorem III-2.8,

$$\deg(I - K\bar N, \Omega_1) = 1 \quad \text{and} \quad \deg(I - K\bar N, \Omega_2) = 1$$

where, for R large enough,

$$\begin{aligned} \Omega_1 &= \{u \in C_0^1([a,b]) \mid u_1 - 1 \prec u \prec \beta_1,\ \|u'\|_\infty < R\},\\ \Omega_2 &= \{u \in C_0^1([a,b]) \mid \alpha_2 \prec u \prec u_2 + 1,\ \|u'\|_\infty < R\}, \end{aligned}$$

and K, $\bar N$ are defined from

$$\bar N : C_0^1([a,b]) \to L^1(a,b), u \mapsto \bar f(t,u) + m(t)u,$$
$$K : L^1(a,b) \to C_0^1([a,b]), h \mapsto Kh$$

where Kh is the unique solution of

$$-u'' + m(t)u = h(t), \quad u(a) = 0,\ u(b) = 0$$

and m is defined in Assumption (H).

Suppose $\bar\phi(u_1) \geq \bar\phi(u_2)$, and a similar argument holds if $\bar\phi(u_2) > \bar\phi(u_1)$. By Theorem A-2.4, there exists $\gamma > 0$ such that

$$\inf\{\bar\phi(u) \mid \|u - u_1\|_{H_0^1} = \gamma\} > \bar\phi(u_1).$$

Next, by Theorem A-2.3, there is a critical point of mountain pass type. As $k \geq 2$, it cannot be 0. Hence this point is u_3 and from Corollary A-2.7 and Theorem A-1.7, there exists $\varepsilon > 0$ such that

$$\deg(I - K\bar N, B(u_3, \varepsilon)) = -1.$$

Moreover, as $K\bar N(C_0^1([a,b])) \subset B(0, \bar R)$ for some $\bar R > 0$,

$$\deg(I - K\bar N, B(0, \bar R)) = 1.$$

Let us prove next that, for r small enough,

$$|\deg(I - K\bar N, B(0, r))| = 1.$$

Consider the homotopy

$$\begin{array}{c} u'' + s\bar{f}(t,u) + (1-s)\frac{p+q}{2}u = 0, \\ u(a) = 0, \ u(b) = 0. \end{array} \tag{3.11}$$

Notice that for a.e. $t \in [a,b]$ and all $u \in [\max(-\delta, u_1(t)), \min(u_2(t), \delta)]$

$$\lambda_k < p \leq s\frac{\bar{f}(t,u)}{u} + (1-s)\frac{p+q}{2} \leq q < \lambda_{k+1}.$$

Any solution u of (3.11) is such that

$$u'' + A(t)u = 0, \quad u(a) = 0, \ u(b) = 0,$$

with $A(t) := s\frac{\bar{f}(t,u(t))}{u(t)} + (1-s)\frac{p+q}{2}$. We can find $r > 0$ small enough such that if $u \in \partial B(0,r)$ we have $A(t) \in [p,q] \subset]\lambda_k, \lambda_{k+1}[$ for $t \in [a,b]$. By eigenvalue comparison, we conclude that $u \equiv 0$. Hence, using Theorem A-1.6, we can write

$$|\deg(I - K\bar{N}, B(0,r))| = |\deg(I - K(\tfrac{p+q}{2}I), B(0,r))| = 1.$$

We come now to the contradiction

$$\begin{aligned} &\deg(I - K\bar{N}, B(0,\bar{R})) \\ &\quad = \deg(I - K\bar{N}, \Omega_1) + \deg(I - K\bar{N}, \Omega_2) \\ &\qquad + \deg(I - K\bar{N}, B(u_3, \varepsilon)) + \deg(I - K\bar{N}, B(0,r)) \\ &\quad = 2 - 1 \pm 1 \neq 1, \end{aligned}$$

which proves existence of an additional nontrivial solution u_4 of (3.10). Recall that such a solution lies in $[u_1, u_2]$ and from the definition of u_1 and u_2, $u_4 \notin C^{\beta_1} \cup C_{\alpha_2}$. Arguing as in Theorem 3.7, we prove then that this solution changes sign. □

3.5 A Dual Application

The result we consider in this section is somewhat dual to Theorem 3.7. Here we assume the slope $\frac{f(t,u)}{u}$ crosses the two first eigenvalues the other way round. It is lower than the first eigenvalue for small values of $|u|$ and between two consecutive ones for large $|u|$.

Theorem 3.9. *Let Assumptions (H) be satisfied together with*

(i) there exist $\mu < \lambda_1 = \frac{\pi^2}{(b-a)^2}$ and $\delta > 0$ such that for a.e. $t \in [a,b]$ and all $u \in \mathbb{R}$ with $|u| \leq \delta$,

$$\frac{f(t,u)}{u} \leq \mu;$$

(ii) there exist p, q, $k \geq 2$ ($k \in \mathbb{N}$) and $R > 0$ such that for a.e. $t \in [a,b]$ and all $u \in \mathbb{R}$ with $|u| \geq R$,

$$\lambda_k = \frac{k^2\pi^2}{(b-a)^2} < p \leq \frac{f(t,u)}{u} \leq q < \frac{(k+1)^2\pi^2}{(b-a)^2} = \lambda_{k+1}.$$

Then the problem (1.1) *has at least three nontrivial solutions u_i such that $u_1 \prec 0$, $u_2 \succ 0$ and u_3 changes sign.*

Proof. As in the proof of Theorem 3.7 we introduce α, α_2, strict lower solutions and β_1, β strict upper solutions of (1.1) such that

$$\beta_1 \prec \alpha \prec 0 \prec \beta \prec \alpha_2.$$

Part 1 – Existence of a solution $u_1 \prec 0$ of (1.1). Define the sequence of modified problems

$$\begin{aligned} u'' + f_\rho(t,u) &= 0, \\ u(a) = 0, \ u(b) &= 0, \end{aligned} \tag{3.12}$$

where

$$\begin{aligned} f_\rho(t,u) &= f(t,u), & &\text{if } -\rho \leq u < 0, \\ &= (1+\rho+u)f(t,-\rho), & &\text{if } -\rho-1 \leq u < -\rho, \\ &= 0, & &\text{if } u < -\rho-1 \text{ or } 0 \leq u. \end{aligned}$$

Notice first that solutions of (3.12) are such that $u \leq 0$. Next arguing as in Theorem III-3.3 there exists $K > 0$ such that for all $\rho > 0$ and all $u \in W^{2,1}(a,b)$, solution of (3.12) with $u \notin C^{\beta_1} \cup C_\alpha$, we have $\|u\|_{C^1} < K$.

Let $\rho > K$ be fixed and consider the modified functional

$$\phi_\rho(u) = \int_a^b [\frac{u'^2(t)}{2} - F_\rho(t,u(t))]dt,$$

with $F_\rho(t,u) = \displaystyle\int_0^u f_\rho(t,s)\,ds$. Since ϕ_ρ is coercive on $H_0^1(a,b)$, the Palais-Smale Condition is satisfied and Theorem 3.5 with $\alpha = \alpha$ and $\beta = \beta_1$ produces then a solution u_1 of (3.12) which is such that $u_1 \notin C^{\beta_1} \cup C_\alpha$. Hence $-\rho < -K < u_1 \leq 0$ and u_1 is a solution of (1.1). Also, as $u_1 \notin C_\alpha$, we have $u_1 \neq 0$ and using the maximum principle $u_1 \prec 0$.

Part 2 – Existence of a solution $u_2 \succ 0$ of (1.1). The argument is similar to the proof of Part 1.

Part 3 – Existence of a solution u_3 of (1.1) *which changes sign.* Let $\varphi_1(t) = \sin(\pi\frac{t-a}{b-a})$, $\varphi_2(t) = \sin(2\pi\frac{t-a}{b-a})$ and define $\phi(u) = \int_a^b [\frac{u'^2(t)}{2} - F(t,u(t))]dt$, with $F(t,u) = \int_0^u f(t,s)\,ds$.

Claim 1 : $\phi(R((2r-1)\varphi_1+(1-|2r-1|)\varphi_2)) \to -\infty$ *as* $|R| \to \infty$, *uniformly in* $r \in [0,1]$. Let $h \in L^1(a,b)$, $h(t) \geq 0$, be such that

$$f(t,u) \geq pu - h(t), \text{ if } u \geq 0,$$
$$f(t,u) \leq pu + h(t), \text{ if } u \leq 0.$$

It follows that

$$F(t,u) \geq \tfrac{p}{2}u^2 - h(t)|u|,$$

$$\phi(u) \leq \tfrac{1}{2}\int_a^b [u'^2(t) - p(t)u^2(t)]\,dt + \int_a^b h(t)|u(t)|\,dt.$$

Hence, we compute

$$\begin{aligned}
&\phi(R((2r-1)\varphi_1 + (1-|2r-1|)\varphi_2)) \\
&\qquad \leq \tfrac{R^2}{2}\int_a^b [((2r-1)^2\varphi_1'^2(t) + (1-|2r-1|)^2\varphi_2'^2(t)) \\
&\qquad\qquad\qquad -p((2r-1)^2\varphi_1^2(t) + (1-|2r-1|)^2\varphi_2^2(t))]\,dt \\
&\qquad\quad +R\int_a^b h(t)|(2r-1)\varphi_1(t) + (1-|2r-1|)\varphi_2(t)|\,dt \\
&\qquad \leq -(p-\lambda_2)\tfrac{R^2}{2}\int_a^b ((2r-1)^2\varphi_1^2(t) + (1-|2r-1|)^2\varphi_2^2(t))\,dt \\
&\qquad\quad +R\int_a^b h(t)|(2r-1)\varphi_1(t) + (1-|2r-1|)\varphi_2(t)|\,dt
\end{aligned}$$

and the claim follows.

Notations – Introduce the following notations :

$$\begin{aligned}
&m = \inf_{C^\beta \cap C_\alpha} \phi(u) > -\infty, \\
&\Sigma = \{\sigma \in \mathcal{C}([0,1]^2, \mathcal{C}_0^1([a,b])) \mid \\
&\qquad\qquad \sigma(r,0) = 0,\ \sigma(0,s) \in C_\alpha,\ \sigma(1,s) \in C^\beta,\ \phi(\sigma(r,1)) < m-1\}, \\
&T_\sigma = \{(r,s) \in [0,1]^2 \mid \sigma(r,s) \in \mathcal{C}_0^1([a,b]) \setminus (C^\beta \cup C_\alpha)\}, \\
&c = \inf_{\sigma\in\Sigma}\Big(\max_{(r,s)\in T_\sigma} \phi(\sigma(r,s))\Big).
\end{aligned}$$

Observe first that if R is large enough,

$$\sigma(r,s) = Rs((2r-1)\varphi_1 + (1-|2r-1|)\varphi_2)$$

defines a surface element in Σ. Also, it is easy to see that $c \geq m$.

Claim 2 : Let $u(t; u_0)$ be the minus gradient flow defined with $r = m - 1$. Then, for any $\delta \in]0, 1[$, there exists $u_0 \in \mathcal{C}_0^1([a, b])$ such that

$$\forall t \in [0, +\infty[, \quad u(t, u_0) \in \phi^{-1}([c - \delta, c + \delta]) \setminus (C^\beta \cup C_\alpha).$$

Fix $\delta \in]0, 1[$, let $\sigma \in \Sigma$ be such that

$$c \leq \max_{(r,s) \in T_\sigma} \phi(\sigma(r, s)) \leq c + \delta$$

and define $E = \phi^{c-\delta} \cup C^\beta \cup C_\alpha$. Observe that E is an open positively invariant set. Define $A(E) = \{u_0 \in \mathcal{C}_0^1([a, b]) \mid \exists t \geq 0, \ u(t; u_0) \in E\}$.

Let us prove there exists $u_0 \in \sigma(T_\sigma) \setminus A(E)$. Otherwise, as in Proposition 3.4, we prove there exists T so that for all $(r, s) \in [0, 1]^2$, $u(T; \sigma(r, s)) \in E$. Notice that $\bar{\sigma}(r, s) = u(T; \sigma(r, s)) \in \Sigma$ and

$$\max_{T_{\bar{\sigma}}} \phi(\bar{\sigma}(r, s)) < c - \delta,$$

which contradicts the definition of c and proves the claim.

Hence, by construction of $A(E)$ and as $\phi(u(\cdot; u_0))$ is non-increasing, for all $t \in [0, +\infty[$,

$$u(t; u_0) \in \phi^{-1}([c - \delta, c + \delta]) \setminus (C_\alpha \cup C^\beta).$$

Claim 3 : The Palais-Smale Condition (PS) holds. Let us decompose the space $H_0^1(a, b)$ into orthogonal components $H_0^1(a, b) = H^- \oplus H^+$, where $H^- = \text{span}\{\varphi_1, \ldots, \varphi_k\}$, $H^+ = \overline{\text{span}}\{\varphi_{k+1}, \ldots\}$ and $\varphi_k(t) = \sin(k\pi \frac{t-a}{b-a})$. For $u \in H_0^1(a, b)$, we write $u = u_- + u_+$, where $u_- \in H^-$ and $u_+ \in H^+$. Let $\bar{\lambda} = \frac{\lambda_k + \lambda_{k+1}}{2}$ and consider

$$\begin{aligned}
(\nabla\phi(u), u_+ - u_-)_{H_0^1} &= \int_a^b [(u'_+ + u'_-)(u'_+ - u'_-) - f(s, u)(u_+ - u_-)] \, ds \\
&= \int_a^b [((u'_+)^2 - (u'_-)^2) - \bar{\lambda}((u_+)^2 - (u_-)^2)] \, ds \\
&\qquad - \int_a^b (f(s, u) - \bar{\lambda} u)(u_+ - u_-) \, ds.
\end{aligned}$$

We deduce from the assumptions that

$$\int_a^b (f(s, u) - \bar{\lambda} u)(u_+ - u_-) \, ds \leq \delta \|u\|_{L^2}^2 + C_0 \|u\|_{L^2} \leq \delta \|u\|_{L^2}^2 + C_1 \|u\|_{H_0^1}$$

for some $\delta < \frac{\lambda_{k+1} - \lambda_k}{2}$ and some C_0, $C_1 > 0$. On the other hand, we have

$$\int_a^b [((u'_+)^2 - (u'_-)^2) - \bar{\lambda}((u_+)^2 - (u_-)^2)] \, ds \geq \delta \|u\|_{L^2}^2 + \varepsilon \|u\|_{H_0^1}^2,$$

for some $\varepsilon > 0$. It follows then

$$(\nabla\phi(u), u_+ - u_-) \geq \varepsilon\|u\|^2_{H^1_0} - C_1\|u\|_{H^1_0}. \tag{3.13}$$

Let now $(u_n)_n$ be such that $\phi(u_n)$ is bounded and $\nabla\phi(u_n) \stackrel{H^1_0}{\to} 0$. We have

$$(\nabla\phi(u_n), (u_n)_+ - (u_n)_-) \leq C_2\|u_n\|_{H^1_0}.$$

We deduce from (3.13) that $\|u_n\|_{H^1_0}$ is bounded and the Palais-Smale Condition follows from Theorem A-2.1.

Conclusion – For every $\delta \in]0,1[$, we have proved the existence of u_0 such that for all $t \in [0,+\infty[$,

$$u(t;u_0) \in \phi^{-1}([c-\delta, c+\delta]) \setminus (C_\alpha \cup C^\beta).$$

Hence, there exists an increasing unbounded sequence $(t_n)_n$ so that

$$\frac{d}{dt}\phi(u(t_n;u_0)) = -\|\nabla\phi(u(t_n;u_0))\|^2_{H^1_0} \to 0.$$

As the Palais-Smale Condition holds, we prove as in Theorem 3.5 the existence of $u_3 \in C^1_0([a,b]) \setminus (C_\alpha \cup C^\beta)$ solution of (1.1) such that $\phi(u_3) = c$. We deduce then from the localization $u_3 \in C^1_0([a,b]) \setminus (C_\alpha \cup C^\beta)$ that u_3 changes sign. □

3.6 A Seven Solutions Theorem

Combining the results of the previous sections we obtain the existence of at least seven solutions in case the slope $\frac{f(t,u)}{u}$ is bigger than the second eigenvalue for small and large values of $|u|$ and there exist lower and upper solutions.

Theorem 3.10. *Let $f \in C^1([a,b] \times \mathbb{R})$ satisfy Assumptions (H) together with*

(i) there exist μ, λ, $k \geq 2$ $(k \in \mathbb{N})$ and $\delta > 0$ such that for a.e. $t \in [a,b]$ and all $u \in [-\delta, \delta]$,

$$\lambda_k = \frac{k^2\pi^2}{(b-a)^2} < \mu \leq \frac{f(t,u)}{u} \leq \lambda < \frac{(k+1)^2\pi^2}{(b-a)^2} = \lambda_{k+1};$$

(ii) there exist p, q, $l \geq 2$ $(l \in \mathbb{N})$ and $R > 0$ such that for a.e. $t \in [a,b]$ and all $u \in \mathbb{R}$ with $|u| \geq R$,

$$\lambda_l = \frac{l^2\pi^2}{(b-a)^2} < p \leq \frac{f(t,u)}{u} \leq q < \frac{(l+1)^2\pi^2}{(b-a)^2} = \lambda_{l+1};$$

(iii) there exist α and β lower and upper solutions of (1.1) *which are not solutions of* (1.1) *such that $\alpha \prec 0 \prec \beta$.*

Then the problem (1.1) *has at least seven nontrivial solutions such that two of them are negative, two are positive, and three change sign.*

Proof. We now observe that arguing as in Theorem 3.8, we obtain four solutions in $C_\alpha \cap C^\beta$, one negative, one positive and two changing sign, and using the arguments of Theorem 3.9 we prove the existence of three solutions outside $C_\alpha \cap C^\beta$, one negative, one positive, and one changing sign. □

3.7 Crossing of the Second Eigenvalue

Using the arguments of the preceding sections, we can also consider the case where the nonlinearity crosses only the second eigenvalue and prove the existence of a solution which changes sign. This use of the lower and upper solution method in a case of interaction only with the second eigenvalue is the main interest of this section.

Theorem 3.11. *Let Assumptions (H) be satisfied together with*
(i) there exist μ, λ and $\delta > 0$ such that for a.e. $t \in [a,b]$ and all $u \in [-\delta, \delta]$,

$$\lambda_1 = \frac{\pi^2}{(b-a)^2} < \mu \leq \frac{f(t,u)}{u} \leq \lambda < \frac{4\pi^2}{(b-a)^2} = \lambda_2;$$

(ii) there exist p, q, $k \geq 2$ $(k \in \mathbb{N})$ and $R > 0$ such that for a.e. $t \in [a,b]$ and all $u \in \mathbb{R}$ with $|u| \geq R$,

$$\lambda_k = \frac{k^2\pi^2}{(b-a)^2} < p \leq \frac{f(t,u)}{u} \leq q < \frac{(k+1)^2\pi^2}{(b-a)^2} = \lambda_{k+1}.$$

Then the problem (1.1) *has at least one nontrivial changing sign solution.*

Proof. Let $\varphi_1(t) = \sin(\pi\frac{t-a}{b-a})$, $\varphi_2(t) = \sin(2\pi\frac{t-a}{b-a})$ and define $\phi(u) = \int_a^b [\frac{u'^2(t)}{2} - F(t,u(t))]dt$, where $F(t,u) = \int_0^u f(t,s)\,ds$.

Claim 1 – For every $\rho \in [-\delta,\delta] \setminus \{0\}$ we have $\phi(\rho\varphi_1) < 0$. Notice that for a.e. $t \in [a,b]$ and all $u \in [-\delta,\delta]$,

$$F(t,u) \geq \tfrac{\mu}{2}u^2,$$

which implies that for $\rho \in [\delta, \delta] \setminus \{0\}$,

$$\begin{aligned}\phi(\rho\varphi_1) &= \tfrac{1}{2}\int_a^b \rho^2 \varphi_1'^2(t)\,dt - \int_a^b F(t, \rho\varphi_1(t))\,dt \\ &\le -\tfrac{\mu-\lambda_1}{2}\rho^2 \int_a^b \varphi_1^2(t)\,dt < 0.\end{aligned}$$

Claim 2 – Let $\mathcal{E} = \{u \in C_0^1([a,b]) \mid \int_a^b u(t)\varphi_1(t)\,dt = 0\}$, *then there exist an* $\eta > 0$ *and an* $\varepsilon > 0$ *so that for all* $u \in \mathcal{E}$ *with* $\|u\|_{H_0^1} = \eta$, *we have* $\phi(u) \ge \varepsilon$. Observe that for a.e. $t \in [a,b]$ and all $u \in [-\delta, \delta]$,

$$F(t,u) \le \tfrac{\lambda}{2}u^2.$$

Choose $\eta > 0$ so that $\|u\|_{H_0^1} := [\int_a^b |u'(t)|^2\,dt]^{1/2} = \eta$ implies $\|u\|_\infty \le \delta$. We can now pick $\varepsilon \in]0, \frac{\eta}{2}(1 - \frac{\lambda}{\lambda_2})]$ so that for any $u \in \mathcal{E}$ with $\|u\|_{H_0^1} = \eta$ we compute

$$\begin{aligned}\phi(u) &= \tfrac{1}{2}\int_a^b u'^2(t)\,dt - \int_a^b F(t,u(t))\,dt \\ &\ge \tfrac{1}{2}\int_a^b u'^2(t)\,dt - \tfrac{\lambda}{2}\int_a^b u^2(t)\,dt \ge \tfrac{1}{2}(1 - \tfrac{\lambda}{\lambda_2})\int_a^b u'^2(t)\,dt \ge \varepsilon.\end{aligned}$$

Claim 3 – $\phi(R((2r-1)\varphi_1 + (1-|2r-1|)\varphi_2)) \to -\infty$ *as* $|R| \to \infty$, *uniformly in* $r \in [0,1]$. The proof is similar to the one in Theorem 3.9.

Notations – Introduce the following quantities :

$\bar{\delta} = \min(\delta, \eta\frac{\sqrt{b-a}}{\pi})$,

$\bar{R} > \sqrt{\frac{5(b-a)}{2}}\,\frac{\eta}{\pi}$ such that $\max_{r\in[0,1]} \phi(\bar{R}((2r-1)\varphi_1 + (1-|2r-1|)\varphi_2)) < -1$,

$\alpha(t) = \frac{\bar{\delta}}{2}\varphi_1(t)$ and $\beta(t) = -\frac{\bar{\delta}}{2}\varphi_1(t)$,

$\Sigma = \{\sigma \in \mathcal{C}([0,1]^2, C_0^1([a,b])) \mid \sigma(r,0) = \bar{\delta}(2r-1)\varphi_1,\ \sigma(0,s) \in C^\beta,$
$\sigma(1,s) \in C_\alpha,\ \sigma(r,1) = \bar{R}((2r-1)\varphi_1 + (1-|2r-1|)\varphi_2)\}$,

$T_\sigma = \{(r,s) \in [0,1]^2 \mid \sigma(r,s) \in C_0^1([a,b]) \setminus (C^\beta \cup C_\alpha)\}$,

$c = \inf_{\sigma\in\Sigma}\big(\max_{(r,s)\in T_\sigma} \phi(\sigma(r,s))\big)$.

As in Theorem 3.7, we prove that α and β are strict lower and upper solutions. Observe next that

$$\sigma(r,s) = \bar{R}s((2r-1)\varphi_1 + (1-|2r-1|)\varphi_2) + \bar{\delta}(1-s)(2r-1)\varphi_1$$

defines a surface element in Σ.

Using a Generalized Intermediate Value Theorem or a degree argument, it is easy to see that given $\sigma \in \Sigma$, the function $(\int_a^b \sigma(r,s)\varphi_1\,dt, \|\sigma(r,s)\|_{H_0^1} - \eta)$ has a zero $(r_0, s_0) \in [0,1]^2$, i.e. any surface $\sigma \in \Sigma$ intersects the set $\{u \in \mathcal{E} \mid \|u\|_{H_0^1} = \eta\}$. It follows then from Claim 2 that for such a σ we have $\max_{(r,s)\in T_\sigma} \phi(\sigma(r,s)) \geq \varepsilon$, which implies $c \geq \varepsilon > 0$.

Claim 4 – Let $u(t;u_0)$ be the minus gradient flow defined by

$$\begin{aligned} \tfrac{d}{dt}u &= -\tilde{\psi}(\phi(u))\nabla\phi(u), \\ u(0) &= u_0, \end{aligned}$$

where $u_0 \in C_0^1([a,b])$ and $\tilde{\psi} \in C^1(\mathbb{R},[0,1])$ is such that $\tilde{\psi}(s) = 1$ if $s \geq \varepsilon/2$ and $\tilde{\psi}(s) = 0$ if $s \leq 0$. Then for any $\delta \in]0,\varepsilon[$, there exists a $u_0 \in C_0^1([a,b])$ such that

$$\forall t \in [0,+\infty[, \quad u(t;u_0) \in \phi^{-1}([c-\delta, c+\delta]) \setminus (C^\beta \cup C_\alpha).$$

Fix $\delta \in]0,\varepsilon[$, let $\sigma \in \Sigma$ be such that

$$c \leq \max_{(r,s)\in T_\sigma} \phi(\sigma(r,s)) \leq c + \delta$$

and define $E = \phi^{c-\delta} \cup C^\beta \cup C_\alpha$. Observe that E is an open positively invariant set. Define $A(E) = \{u_0 \in C_0^1([a,b]) \mid \exists t \geq 0,\ u(t;u_0) \in E\}$. We prove as in Theorem 3.9 there exists $u_0 \in \sigma(T_\sigma) \setminus A(E)$ such that

$$\forall t \in [0,+\infty[, \quad u(t;u_0) \in \phi^{-1}([c-\delta, c+\delta]) \setminus (C^\beta \cup C_\alpha).$$

Claim 5 : The Palais-Smale Condition (PS) holds. The argument is the same as in Theorem 3.9.

Conclusion – For every $\delta \in]0,\varepsilon[$, we have proved the existence of u_0 such that for all $t \in [0,+\infty[$,

$$u(t,u_0) \in \phi^{-1}([c-\delta, c+\delta]) \setminus (C_\alpha \cup C^\beta).$$

Hence, there exists an increasing unbounded sequence $(t_n)_n$ so that

$$\frac{d}{dt}\phi(u(t_n;u_0)) = -\|\nabla\phi(u(t_n;u_0))\|_{H_0^1}^2 \to 0.$$

As the Palais-Smale Condition holds, we prove as in Theorem 3.5 the existence of $v \in C_0^1([a,b]) \setminus (C_\alpha \cup C^\beta)$ solution of (1.1) such that $\phi(v) = c$. As $c > 0$ we know that $v \not\equiv 0$, and as in Theorem 3.7, we prove that v changes sign. □

We can also consider the case where f crosses the second eigenvalue the other way on.

Theorem 3.12. *Let Assumptions (H) be satisfied and assume :*

(i) there exist $\lambda > \lambda_2 = \frac{4\pi^2}{(b-a)^2}$ and $\delta > 0$ such that for a.e. $t \in [a,b]$ and all $u \in [-\delta, \delta]$,

$$\frac{f(t,u)}{u} \geq \lambda;$$

(ii) there exist p, q and $R > 0$ such that for a.e. $t \in [a,b]$ and all $u \in \mathbb{R}$ with $|u| \geq R$,

$$\lambda_1 = \frac{\pi^2}{(b-a)^2} < p \leq \frac{f(t,u)}{u} \leq q < \lambda_2 = \frac{4\pi^2}{(b-a)^2}.$$

Then the problem (1.1) *has at least one nontrivial solution u which changes sign.*

Exercise 3.1. Prove Theorem 3.12.

4 A Reduction Method

The main difficulty in using the lower and upper solutions method is to find these functions in concrete situations. Consider a variational problem associated to a functional $\phi : H \to \mathbb{R}$, where H is a Hilbert space. The idea of this section is to consider a foliation described by an equation $\Gamma(u) = \xi$, where $\Gamma : H \to \mathbb{R}$ and ξ is a real parameter. Assume $\phi(u)$ has a minimum on each of the sets $\Gamma^{-1}(\xi)$. We can then define the *reduction function*

$$\varphi(\xi) := \min_{\Gamma(u)=\xi} \phi(u).$$

In this section, we show how this function can help to find lower and upper solutions so that we can apply the previous theory.

4.1 Lower and Upper Solutions in a Weak Sense

The lower and upper solutions we obtain using the reduction have to be considered in a weak sense which is related to the notion of weak solutions i.e. of the derivative of the functional.

Definition 4.1. *A function* $\alpha \in H^1(a,b)$ *is a* weak lower solution of (1.1) *if* $\alpha(a) \le 0$, $\alpha(b) \le 0$ *and for every* $v \in H_0^1(a,b)$, $v \ge 0$,

$$\int_a^b [\alpha'(t)v'(t) - f(t,\alpha(t))v(t)]\,dt \le 0.$$

In a similar way, a function $\beta \in H^1(a,b)$ *is a* weak upper solution of (1.1) *if* $\beta(a) \ge 0$, $\beta(b) \ge 0$ *and for every* $v \in H_0^1(a,b)$, $v \ge 0$,

$$\int_a^b [\beta'(t)v'(t) - f(t,\beta(t))v(t)]\,dt \ge 0.$$

Definition of strict weak lower and upper solutions can be made as in Definition III-2.2.

Remark 4.1. Notice that if $\alpha \in W^{2,1}(a,b)$ is a $W^{2,1}$-lower solution of (1.1), it is a weak lower solution. Similarly a $W^{2,1}$-upper solution of (1.1), which is in $W^{2,1}(a,b)$, is a weak upper solution.

Remark 4.2. If $\phi(u)$ is defined from (1.2), the function $\alpha \in H_0^1(a,b)$ is a weak lower solution of (1.1) if for every $v \in H_0^1(a,b)$, $v \ge 0$,

$$(\nabla\phi(\alpha), v)_{H_0^1} \le 0.$$

Similarly the function $\beta \in H_0^1(a,b)$ is a weak upper solution of (1.1) if for every $v \in H_0^1(a,b)$, $v \ge 0$,

$$(\nabla\phi(\beta), v)_{H_0^1} \ge 0.$$

This remark turns out to be a key to prove existence of such lower and upper solutions.

Theorem 4.1. *Assume* $f : [a,b] \times \mathbb{R} \to \mathbb{R}$ *is* L^1*-Carathéodory and there exist* α, $\beta \in H^1(a,b)$ *weak lower and upper solutions of* (1.1) *with* $\alpha \le \beta$. *Then the problem* (1.1) *has at least one solution* $u \in W^{2,1}(a,b)$ *such that for all* $t \in [a,b]$,

$$\alpha(t) \le u(t) \le \beta(t).$$

If moreover α *and* β *are strict, then for* $R > 0$ *large enough,* $\deg(I - T, \Omega) = 1$, *where* T *and* Ω *are defined by*

$$T : C_0^1([a,b]) \to C_0^1([a,b]) : u \mapsto \int_a^b G(t,s)f(s,u(s))\,ds$$

with $G(t,s)$ the Green's function of

$$\begin{gathered} -u'' = h(t), \\ u(a) = 0, \; u(b) = 0 \end{gathered}$$

and

$$\Omega = \{u \in C_0^1([a,b]) \mid \alpha \prec u \prec \beta, \|u'\|_\infty < R\}.$$

Proof. Consider the modified problem

$$\begin{gathered} u'' + f(t, \gamma(t,u)) = 0, \\ u(a) = 0, \; u(b) = 0, \end{gathered} \tag{4.1}$$

where $\gamma(t,u) = \max\{\alpha(t), \min\{u, \beta(t)\}\}$. As $f(t, \gamma(t,u))$ is bounded, this problem has a solution.

Let us prove that every solution u of (4.1) is such that $\alpha \le u \le \beta$. Assume on the contrary that $\max_t(\alpha - u) > 0$. Observe that $(\alpha-u)^+ \in H_0^1(a,b)$, $(\alpha - u)^+ \gneqq 0$ and we have the contradiction

$$\begin{aligned} 0 &= \int_a^b [u'((\alpha-u)^+)' - f(t,\alpha)(\alpha-u)^+]\,dt \\ &\le -\int_a^b (\alpha-u)'((\alpha-u)^+)'\,dt = -\int_a^b ((\alpha-u)^+)'^2\,dt < 0. \end{aligned}$$

Hence $u \ge \alpha$. In a similar way, we prove $u \le \beta$.

The degree result is proved as in Theorem III-2.8. □

In a similar way we can consider for example the periodic problem (1.6).

Definition 4.2. *A function $\alpha \in H^1_{per}(a,b)$ is a* weak lower solution *of (1.6) if for every $v \in H^1_{per}(a,b)$, $v \ge 0$,*

$$\int_a^b [\alpha'(t)v'(t) - f(t,\alpha(t))v(t)]\,dt \le 0.$$

In a similar way, a function $\beta \in H^1_{per}(a,b)$ is a weak upper solution *of (1.6) if for every $v \in H^1_{per}(a,b)$, $v \ge 0$,*

$$\int_a^b [\beta'(t)v'(t) - f(t,\beta(t))v(t)]\,dt \ge 0.$$

Definition of strict weak lower and upper solutions can be made as in Definition III-1.1.

Remark 4.3. Notice that if $\alpha \in W^{2,1}(a,b)$ is a $W^{2,1}$-lower solution of (1.6), it is a weak lower solution. Similarly a $W^{2,1}$-upper solution of (1.6), which is in $W^{2,1}(a,b)$, is a weak upper solution.

Remark 4.4. If $\phi(u)$ is defined from (1.7), the function $\alpha \in H^1_{per}(a,b)$ is a weak lower solution of (1.6) if for every $v \in H^1_{per}(a,b)$, $v \geq 0$,

$$(\nabla\phi(\alpha), v)_{H^1} \leq 0.$$

Similarly the function $\beta \in H^1_{per}(a,b)$ is a weak upper solution of (1.6) if for every $v \in H^1_{per}(a,b)$, $v \geq 0$,

$$(\nabla\phi(\beta), v)_{H^1} \geq 0.$$

Theorem 4.2. *Assume $f : [a,b] \times \mathbb{R} \to \mathbb{R}$ is L^1-Carathéodory and there exist α, $\beta \in H^1_{per}(a,b)$ weak lower and upper solutions of* (1.6) *with $\alpha \leq \beta$. Then the problem* (1.6) *has at least one solution $u \in W^{2,1}(a,b)$ such that for all $t \in [a,b]$,*

$$\alpha(t) \leq u(t) \leq \beta(t).$$

If moreover α and β are strict, then $\deg(I - T, \Omega) = 1$, *where T and Ω are defined by*

$$T : C([a,b]) \to C([a,b]) : u \mapsto \int_a^b G(t,s)[f(s,u(s)) + u(s)]\, ds$$

with $G(t,s)$ the Green's function of

$$-u'' + u = h(t),$$
$$u(a) = u(b),\ u'(a) = u'(b)$$

and

$$\Omega = \{u \in C([a,b]) \mid \alpha(t) < u(t) < \beta(t) \text{ on } [a,b]\}.$$

Proof. The first part of the proof follows as in Theorem I-1.1 together with Theorem 4.1 and the second part follows as in Theorem III-1.8. □

4.2 Foliations and Reductions

In this section we present the framework of the reduction method in some abstract Hilbert space.

Definition 4.3. *Given a Hilbert space H (with scalar product $(\cdot,\cdot)$ and norm $\|\cdot\|$), we say that*

$$\Gamma : H \to \mathbb{R}$$

is a foliation of H *if it is weakly continuous, surjective, and convex. The sets $\Gamma^{-1}(\xi)$, where $\xi \in \mathbb{R}$, are called the* leaves *of the foliation.*

Let us first recall some well known facts of standard convex analysis.

For a convex, continuous function Γ a notion of *subgradient* $\partial\Gamma(u)$ can be defined (see for instance [248]). The subgradient can be equivalently characterized by means of the right Gateaux derivative

$$D_r\Gamma(u)(v) := \lim_{t\to 0^+} \frac{\Gamma(u+tv)-\Gamma(u)}{t}$$

which exists for all $v \in H$ due to the convexity of Γ. Precisely, it is well known that $w \in \partial\Gamma(u)$ if and only if

$$(w,v) \le D_r\Gamma(u)(v), \qquad \forall v \in H. \tag{4.2}$$

With this notion we can state a necessary condition for solutions of a constraint minimization problem.

Proposition 4.3. *Let H be a Hilbert space, $\phi : H \to \mathbb{R}$ be of class C^1, $\Gamma : H \to \mathbb{R}$ be a foliation and $\xi \in \mathbb{R}$. Assume u is a solution of*

$$\min_{\Gamma(u)=\xi} \phi(u).$$

Then there exist $\lambda \in \mathbb{R}$ and $w \in \partial\Gamma(u)$ such that

$$\nabla\phi(u) = \lambda w.$$

Proof. See, for instance, [69, Theorem 6.1.1]. □

The number λ is said to be a *Lagrange multiplier of* u, and the set of all the multipliers will be denoted by Λ_u.

Some compatibility condition between Γ and ϕ has to be fulfilled in order to prove that the reduction function $\varphi(\xi) = \min_{\Gamma(u)=\xi} \phi(u)$ is well defined.

Definition 4.4. *We say that a foliation $\Gamma : H \to \mathbb{R}$ is* admissible *for $\phi : H \to \mathbb{R}$ if, for any two constants C_1, C_2, there exists C_3 such that*

$$\phi(u) \le C_1, \quad |\Gamma(u)| \le C_2 \qquad \textit{implies} \qquad \|u\| \le C_3.$$

Proposition 4.4. *Let H be a Hilbert space, $\phi : H \to \mathbb{R}$ be a C^1, weakly lower semicontinuous function and let $\Gamma : H \to \mathbb{R}$ be a foliation admissible for ϕ.*

Then the reduction function

$$\varphi(\xi) = \min_{\Gamma(u)=\xi} \phi(u)$$

is well defined, continuous, and satisfies

$$\varphi(\Gamma(u)) \le \phi(u), \qquad \forall u \in H. \tag{4.3}$$

Proof. To see that φ is well defined, for any given $\xi \in \mathbb{R}$, take a sequence $(u_n)_n$ such that

$$\Gamma(u_n) = \xi, \qquad \lim_{n\to\infty} \phi(u_n) = \inf_{\Gamma(v)=\xi} \phi(v).$$

Due to the admissibility condition on Γ, $(u_n)_n$ is bounded in H. Denote again by $(u_n)_n$ a subsequence which is weakly convergent and let u be its weak limit. Since Γ is weakly continuous, $\Gamma(u) = \xi$ and, since ϕ is weakly lower semicontinuous,

$$\phi(u) \le \liminf_{n\to\infty} \phi(u_n) = \inf_{\Gamma(v)=\xi} \phi(v).$$

Hence the infimum is a minimum and $\varphi(\xi)$ is well defined. The formula (4.3) follows trivially.

Under the same general setting, φ is also lower semicontinuous. Otherwise there exists $\epsilon > 0$ and $\xi_n \to \xi$ such that

$$\varphi(\xi_n) \le \varphi(\xi) - \epsilon.$$

Then, taking $u_n \in H$ such that

$$\phi(u_n) = \min_{\Gamma(u)=\xi_n} \phi(u) = \varphi(\xi_n),$$

the admissibility condition says that $(u_n)_n$ is bounded in H. If u is the weak limit of a convergent subsequence, which will be denoted again by $(u_n)_n$, once more $\Gamma(u) = \xi$ and we have the contradiction

$$\varphi(\xi) \le \phi(u) \le \liminf_{n\to\infty} \phi(u_n) = \liminf_{n\to\infty} \varphi(\xi_n) \le \varphi(\xi) - \epsilon.$$

To prove the upper semicontinuity of φ, we need the full force of our hypotheses on Γ. For a given $\xi \in \mathbb{R}$, choose $u \in H$ such that

$$\phi(u) = \min_{\Gamma(v)=\xi} \phi(v) = \varphi(\xi).$$

First observe that, as Γ is surjective, $0 \notin \partial\Gamma(u)$ which means that for some $v \in H$

$$D_r\Gamma(u)(v) < 0.$$

Also if $w \in \partial\Gamma(u)$, we have $w \neq 0$ and

$$0 < \|w\|^2 \le D_r\Gamma(u)(w).$$

We use w to prove that φ is upper semicontinuous at ξ from the right. By contradiction assume that for some $\epsilon > 0$ and $\xi_n \to \xi$, $\xi_n \geq \xi$, we have

$$\varphi(\xi_n) \geq \varphi(\xi) + \epsilon.$$

Since $D_r\Gamma(u)(w) > 0$, we can choose $t_n \to 0^+$ such that $\Gamma(u + t_n w) = \xi_n$. Then we should have

$$\varphi(\xi) + \epsilon \leq \varphi(\xi_n) \leq \phi(u + t_n w) \quad \text{and} \quad \phi(u + t_n w) \to \phi(u) = \varphi(\xi),$$

which is a contradiction. The same argument, where w is replaced by v, leads to the upper semicontinuity from the left. □

The set of the constrained minimizers on the leaves of the associated foliation will be denoted by

$$M_\xi := \{u \in H \mid \Gamma(u) = \xi,\ \phi(u) = \varphi(\xi)\}.$$

Going back to the problems (1.1) and (1.6), remember that a lower solution (resp. an upper solution) is such that for all $v \geq 0$

$$(\nabla\phi(u), v) \leq 0 \quad (\text{resp. } (\nabla\phi(u), v) \geq 0).$$

On the other hand, any solution u of the corresponding constraint problem

$$\min_{\Gamma(u)=\xi} \phi(u)$$

is such that for some $\lambda \in \mathbb{R}$ and $w \in \partial\Gamma(u)$,

$$\nabla\phi(u) = \lambda w.$$

Hence, if $\lambda(w, v)$ is one-signed for all $v \geq 0$, the corresponding solution u turns out to be a lower or an upper solution.

A first step in this direction requires that the Hilbert space H is endowed with a partial ordering and that Γ is monotone.

Definition 4.5. *Let H be a Hilbert space with partial ordering $\leq$. We say that Γ is an* increasing *function on H when*

$$u \leq v \quad \textit{implies} \quad \Gamma(u) \leq \Gamma(v).$$

Lemma 4.5. *Let H be a Hilbert space with partial ordering $\leq$ and $\Gamma : H \to \mathbb{R}$ be convex and increasing. Then for any $w \in \partial\Gamma(u)$ and any $v \in H$ with $v \geq 0$ we have*

$$(w, v) \geq 0.$$

Proof. For all $v \geq 0$, we can write $(w, v) \geq \Gamma(u) - \Gamma(u - v) \geq 0$. □

Consider a situation where φ has a maximizer, i.e. we have $\gamma_1 < \xi_0 < \gamma_2$ such that $\max\{\varphi(\gamma_1), \varphi(\gamma_2)\} < \varphi(\xi_0)$. We can deduce the existence of two points $\xi_1 \in]\gamma_1, \xi_0[$ and $\xi_2 \in]\xi_0, \gamma_2[$ such that

$$D^+\varphi(\xi_1) > 0, \qquad D_-\varphi(\xi_2) < 0 \tag{4.4}$$

(see for instance [220]). The question is now how to connect the sign of the Dini derivative of φ at a given ξ to the sign of the Lagrange multipliers of a given $u \in M_\xi$. The next statement provides a satisfactory answer.

Proposition 4.6. *Let H be a Hilbert space, $\phi : H \to \mathbb{R}$ be a C^1, weakly lower semicontinuous function and let $\Gamma : H \to \mathbb{R}$ be a foliation admissible for ϕ. Let also $\xi \in \mathbb{R}$.*

Then

(a) $D^+\varphi(\xi) \geq 0$ (resp. $D^+\varphi(\xi) > 0$) implies that the Lagrange multipliers of any $u \in M_\xi$ are nonnegative (resp. positive);

(b) $D_-\varphi(\xi) \leq 0$ (resp. $D_-\varphi(\xi) < 0$) implies that the Lagrange multipliers of any $u \in M_\xi$ are nonpositive (resp. negative).

Proof. Let $\xi \in \mathbb{R}$, $u \in M_\xi$, $\lambda \in \Lambda_u$ and $w \in \partial\Gamma(u)$ such that $\nabla\phi(u) = \lambda w$. By definition of φ we know that

$$\frac{\varphi(\Gamma(u + tv)) - \varphi(\Gamma(u))}{t} \leq \frac{\phi(u + tv) - \phi(u)}{t}$$

for all $t > 0$ and all $v \in H$. The idea is to pass to the limit when $t \to 0$, for some appropriate choice of v.

To prove (a), choose $v = w$ and note that by (4.2) (with $v = w$), the function $\Gamma(u + tw)$ is strictly increasing for $t \in \mathbb{R}^+$. We compute then

$$\limsup_{t \to 0^+} \frac{\varphi(\Gamma(u + tw)) - \varphi(\Gamma(u))}{t} = D^+\varphi(\Gamma(u))\ [D_r\Gamma(u)(w)] \geq D^+\varphi(\xi)\, \|w\|^2 \geq 0.$$

Since we have also

$$\limsup_{t \to 0^+} \frac{\varphi(\Gamma(u + tw)) - \varphi(\Gamma(u))}{t} \leq \lim_{t \to 0} \frac{\phi(u + tw) - \phi(u)}{t} = (\nabla\phi(u), w) = \lambda\|w\|^2,$$

we conclude that $\lambda \geq 0$, the strict inequality arising as soon as $D^+\varphi(\xi) > 0$.

To prove (b), let us choose a $v \in H$ such that $D_r\Gamma(u)(v) < 0$. Recall that this is possible as $0 \notin \partial\Gamma(u)$. It follows then that $\Gamma(u+tv)$ is a strictly decreasing function for $t \in \mathbb{R}^+$ small and

$$\limsup_{t\to 0^+} \frac{\varphi(\Gamma(u+tv)) - \varphi(\Gamma(u))}{t} = D_-\varphi(\xi)[D_r\Gamma(u)(v)] \geq 0.$$

Also,

$$\limsup_{t\to 0^+} \frac{\varphi(\Gamma(u+tv)) - \varphi(\Gamma(u))}{t} \leq \lim_{t\to 0} \frac{\phi(u+tv) - \phi(u)}{t} = (\nabla\phi(u), v) = \lambda(w, v).$$

As $(w, v) \leq D_r\Gamma(u)(v) < 0$, we conclude that $\lambda \leq 0$, the strict inequality arising as soon as $D_-\varphi(\xi) < 0$. □

Remark 4.5. It is not difficult to realize that everything works unchanged when the convex foliation Γ is replaced by a smooth function, which is moreover a regular constraint, which means that $\nabla\Gamma(u) \neq 0$ holds for all $u \in H$. In that case, we can even sharpen the previous result, by proving that

$$D^+\varphi(\xi) \leq \lambda \leq D_-\varphi(\xi)$$

holds for the (unique) Lagrange multiplier λ of any $u \in M_\xi$.

As a consequence, Proposition 4.6 and Lemma 4.5 trivially imply the following statement.

Proposition 4.7. *Let H be a Hilbert space with partial ordering, $\phi : H \to \mathbb{R}$ be a C^1, weakly lower semicontinuous function and let $\Gamma : H \to \mathbb{R}$ be an increasing foliation admissible for ϕ. Let further $\xi \in \mathbb{R}$, $u \in M_\xi$.*

Then

$$\begin{aligned} D^+\varphi(\xi) \geq 0 \quad &\textit{implies} \quad (\nabla\phi(u), v) \geq 0 \textit{ for all } v \geq 0, \\ D_-\varphi(\xi) \leq 0 \quad &\textit{implies} \quad (\nabla\phi(u), v) \leq 0 \textit{ for all } v \geq 0. \end{aligned}$$

Hence, if φ is not strictly monotone, we have u_1, $u_2 \in H$ such that for all $v \geq 0$,

$$(\nabla\phi(u_1), v) \geq 0 \quad \text{and} \quad (\nabla\phi(u_2), v) \leq 0.$$

The main limitation in applying these results is that we have no information on the ordering of the lower and upper solutions. In fact, in case we have a maximum of the reduction φ, we have $\xi_1 < \xi_2$ satisfying

$$D^+\varphi(\xi_1) \geq 0, \qquad D_-\varphi(\xi_2) \leq 0$$

(see for instance [220]) and $u_1 \in M_{\xi_1}$ and $u_2 \in M_{\xi_2}$. This means that u_1, u_2 are respectively upper and lower solutions of the corresponding problem. As the foliation is increasing, we see that $u_2 \not\leq u_1$. Nevertheless we can apply for example Theorem III-3.3 or Exercise III-3.2 in the periodic case or Theorem III-3.9 for Dirichlet boundary conditions to obtain existence results. This is what is done in the next sections.

4.3 Application to the Periodic Problem

In this section, we use the previous ideas to deal with a nonlinearity which is non-resonant with the first Fučík curve. The basic condition (4.5) is a variant of the so-called Ahmad, Lazer and Paul condition (see [4]).

Theorem 4.8. *Let $f : [a,b] \times \mathbb{R} \to \mathbb{R}$ be an L^1-Carathéodory function such that*

$$\limsup_{u \to \pm\infty} \int_a^b F(t,u)\,dt = +\infty, \tag{4.5}$$

where $F(t,u) = \int_0^u f(t,s)\,ds$ and for some $a_\pm \leq 0$, $b_\pm \geq 0$ in $L^1(a,b)$,

$$a_\pm(t) \leq \liminf_{u \to \pm\infty} \frac{f(t,u)}{u} \leq \limsup_{u \to \pm\infty} \frac{f(t,u)}{u} \leq b_\pm(t),$$

uniformly in t. Assume moreover there exist $\mu \geq \nu > 0$ with

$$\frac{1}{\sqrt{\mu}} + \frac{1}{\sqrt{\nu}} = \frac{b-a}{\pi}$$

and a set $I \subset [a,b]$ of positive measure such that

$$\begin{aligned} &b_+(t) \leq \mu,\ b_-(t) \leq \nu, && \text{for a.e. } t \in [a,b], \\ &b_+(t) < \mu,\ b_-(t) < \nu, && \text{for a.e. } t \in I. \end{aligned}$$

Then the periodic problem (1.6) has at least one solution $u \in W^{2,1}(a,b)$.

Proof. We just have to prove that

$$\phi(u) = \int_a^b [\frac{u'^2}{2} - F(t,u)]\,dt \quad \text{and} \quad \Gamma(u) = \frac{1}{b-a}\int_a^b [\mu u^+ - \nu u^-]\,dt$$

defined on $H^1_{per}(a,b)$ verify the conditions of Proposition 4.7 and that φ, defined in Proposition 4.4, is not monotone. In that case, we apply Proposition 4.7, Theorem III-3.3 (with the modification given by Theorem 4.2), and Proposition III-3.4 to conclude.

It is well known that $\phi : H^1_{per}(a,b) \to \mathbb{R}$ is C^1 and weakly lower semicontinuous and it is obvious that Γ is weakly continuous, convex, and increasing. To see that Γ is surjective, we just have to observe that if $\xi \geq 0$, $\Gamma(\xi/\mu) = \xi$ and if $\xi < 0$, $\Gamma(\xi/\nu) = \xi$. Let us prove that Γ is admissible.

If it is not the case, there exists a sequence $(u_n)_n$ such that $|\Gamma(u_n)| \leq C_2$, $\phi(u_n) \leq C_1$ and $\|u_n\|_{H^1_{per}} \to \infty$. First observe that this result implies $\|u_n\|_\infty \to \infty$. If not, we have for some D_1,

$$C_1 \geq \phi(u_n) = \int_a^b [\frac{u_n'^2}{2} - F(t, u_n)]\, dt \geq \int_a^b \frac{u_n'^2}{2}\, dt - D_1$$

and $\|u_n\|_{H^1_{per}}$ is bounded which contradicts the assumption. For each n, set $v_n = u_n/\|u_n\|_\infty$. Clearly v_n satisfies

$$|\frac{1}{b-a}\int_a^b [\mu v_n^+ - \nu v_n^-]\, dt| \leq \frac{C_2}{\|u_n\|_\infty}, \tag{4.6}$$

and, for every $\epsilon > 0$, there exists $\gamma_\epsilon \in L^1(a, b)$ such that

$$\begin{aligned}\frac{C_1}{\|u_n\|_\infty^2} &\geq \int_a^b [\frac{v_n'^2}{2} - \frac{F(t, u_n)}{\|u_n\|_\infty^2}]\, dt \\ &\geq \int_a^b [\frac{v_n'^2}{2} - (b_+\frac{v_n^{+2}}{2} + b_-\frac{v_n^{-2}}{2} + \epsilon\frac{v_n^2}{2} + v_n\frac{\gamma_\epsilon}{\|u_n\|_\infty})]\, dt.\end{aligned} \tag{4.7}$$

In particular, for some $D_2 > 0$ and every n, $\|v_n\|_{H^1_{per}} \leq D_2$. Up to a subsequence, $(v_n)_n$ converges weakly in H^1_{per} and strongly in $\mathcal{C}([a, b])$ to some function v with $\|v\|_\infty = 1$. As ϵ is arbitrary, passing to the limit in (4.6) and (4.7), we have

$$\int_a^b [\mu v^+ - \nu v^-]\, dt = 0,$$

$$\int_a^b [\frac{v'^2}{2} - (\mu\frac{v^{+2}}{2} + \nu\frac{v^{-2}}{2})]\, dt \leq \int_a^b [\frac{v'^2}{2} - (b_+\frac{v^{+2}}{2} + b_-\frac{v^{-2}}{2})]\, dt \leq 0.$$

By the variational characterization of the first Fučík curve given in [78, 120],

$$\int_a^b [\frac{v'^2}{2} - (\mu\frac{v^{+2}}{2} + \nu\frac{v^{-2}}{2})]\, dt = \int_a^b [\frac{v'^2}{2} - (b_+\frac{v^{+2}}{2} + b_-\frac{v^{-2}}{2})]\, dt = 0$$

and using [78, Lemma 4.3], we know that v satisfies

$$\begin{gathered} v'' + [\mu v^+ - \nu v^-] = 0, \\ v(a) = v(b),\ v'(a) = v'(b). \end{gathered}$$

Hence either $v \equiv 0$ or $v \neq 0$ a.e. on $[a, b]$. The second possibility is excluded as then

$$0 \geq \int_a^b [\frac{v'^2}{2} - (b_+\frac{v^{+2}}{2} + b_-\frac{v^{-2}}{2})]\, dt > \int_a^b [\frac{v'^2}{2} - (\mu\frac{v^{+2}}{2} + \nu\frac{v^{-2}}{2})]\, dt = 0.$$

We conclude that $v \equiv 0$ which contradicts $\|v\|_\infty = 1$.

As

$$\liminf_{\xi\to+\infty} \varphi(\xi) \le -\limsup_{\xi\to+\infty} \int_a^b F(t,\xi/\mu)\,dt = -\infty,$$
$$\liminf_{\xi\to-\infty} \varphi(\xi) \le -\limsup_{\xi\to-\infty} \int_a^b F(t,\xi/\nu)\,dt = -\infty,$$

we conclude that φ has a maximum. The result follows. □

Our second result concerns the case f is one-sided superlinear.

Theorem 4.9. *Let $f : [a,b]\times\mathbb{R} \to \mathbb{R}$ be an L^1-Carathéodory function such that*

$$\limsup_{u\to\pm\infty} \int_a^b F(t,u)\,dt = +\infty,$$

where $F(t,u) = \int_0^u f(t,s)\,ds$ and for some $a_\pm \le 0$, $b_- \ge 0$ in $L^1(a,b)$,

$$a_-(t) \le \liminf_{u\to-\infty} \frac{f(t,u)}{u} \le \limsup_{u\to-\infty} \frac{f(t,u)}{u} \le b_-(t),$$
$$a_+(t) \le \liminf_{u\to+\infty} \frac{f(t,u)}{u},$$

uniformly in t. Assume moreover $b_- \le (\frac{\pi}{b-a})^2$ a.e. on $[a,b]$ and $b_- < (\frac{\pi}{b-a})^2$ on a subset of positive measure of $[a,b]$. Then the problem (1.6) has at least one solution $u \in W^{2,1}(a,b)$.

Proof. Again we just have to prove that

$$\phi(u) = \int_a^b [\frac{u'^2}{2} - F(t,u)]\,dt \qquad \text{and} \qquad \Gamma(u) = \max_t u(t)$$

defined on $H^1_{per}(a,b)$ verify the conditions of Proposition 4.7, and that φ, defined in Proposition 4.4, is not monotone. In that case, we apply Proposition 4.7 and Exercise III-3.2 (with the modification given by Theorem 4.2) to conclude. As in the previous case, the only nontrivial assumption is the admissibility.

If it is not the case, there exists a sequence $(u_n)_n$ such that $|\Gamma(u_n)| \le C_2$, $\phi(u_n) \le C_1$ and $\|u_n\|_{H^1_{per}} \to \infty$. Arguing as in the proof of Theorem 4.8, we see also that $\|u_n^-\|_\infty \to +\infty$.

For each n, set $v_n = \frac{u_n}{\|u_n^-\|_\infty}$. Clearly v_n satisfies

$$|\max_t v_n(t)| \le \frac{C_2}{\|u_n^-\|_\infty}, \tag{4.8}$$

and for every $\epsilon > 0$, there exists $\gamma_\epsilon \in L^1(a,b)$ such that

$$\begin{aligned}\frac{C_1}{\|u_n^-\|_\infty^2} &\geq \int_a^b [\frac{v_n'^2}{2} - \frac{F(t,u_n)}{\|u_n^-\|_\infty^2}]\,dt \\ &\geq \int_a^b [\frac{v_n'^2}{2} - b_-\frac{v_n^{-2}}{2} - \epsilon\frac{v_n^2}{2} + v_n\frac{\gamma_\epsilon}{\|u_n^-\|_\infty}]\,dt.\end{aligned} \tag{4.9}$$

In particular, there exists D_2 such that for every n, $\|v_n\|_{H^1_{per}} \leq D_2$. Hence, up to a subsequence, $(v_n)_n$ converges weakly in H^1_{per} and strongly in $C([a,b])$ to some function v with $\|v^-\|_\infty = 1$. Passing to the limit in (4.8) and (4.9) and as ϵ is arbitrary, we have

$$\max_t v(t) = 0 \quad \text{and} \quad \int_a^b [\frac{v'^2}{2} - (\frac{\pi}{b-a})^2\frac{v^2}{2}]\,dt \leq \int_a^b [\frac{v'^2}{2} - b_-\frac{v^2}{2}]\,dt \leq 0.$$

As $(\frac{\pi}{b-a})^2$ is the first eigenvalue of the Dirichlet problem on an interval of lenght $b-a$ and for some $t_1 \in [a,b]$, $v \in H_0^1(t_1, t_1+b-a)$, we have

$$\begin{aligned}0 &\leq \int_{t_1}^{t_1+b-a} [\frac{v'^2}{2} - (\frac{\pi}{b-a})^2\frac{v^2}{2}]\,dt \\ &= \int_a^b [\frac{v'^2}{2} - (\frac{\pi}{b-a})^2\frac{v^2}{2}]\,dt \leq \int_a^b [\frac{v'^2}{2} - b_-\frac{v^2}{2}]\,dt \leq 0.\end{aligned}$$

Hence $v(t) = -\sin(\frac{\pi}{b-a}(t-t_1))$ and the above equality contradicts $b_- < (\frac{\pi}{b-a})^2$ on a subset of positive measure of $[a,b]$, which proves the admissibility and concludes the proof. □

4.4 Application to the Dirichlet Problem

Using Theorem III-3.9 and Theorem 4.1, we can obtain the equivalent of Theorem 4.8 for the Dirichlet problem, which we leave as an exercise. Instead we concentrate on other types of results in this section. These results have also their counterpart in the periodic case.

Our first result concerns a strong resonance with the first eigenvalue.

Theorem 4.10. *Let $h \in L^1(0,\pi)$ with $\int_0^\pi h(t)\sin t\,dt = 0$, $f : [0,\pi] \times \mathbb{R} \to \mathbb{R}$ be an L^1-Carathéodory function, define $F(t,u) = \int_0^u f(t,s)\,ds$ and assume*

$$\lim_{|u|\to\infty} f(t,u) = 0 \quad \text{and} \quad \lim_{|u|\to\infty} F(t,u) = 0,$$

uniformly in $t \in [0, \pi]$. *Then the problem*

$$u'' + u + f(t, u) = h(t),\\ u(0) = 0,\ u(\pi) = 0,$$

has at least one solution.

Proof. We apply Theorem 4.1 or Theorem III-3.9 (with the modification given by Theorem 4.1) and Proposition 4.7 with

$$\phi : H_0^1(0,\pi) \to \mathbb{R}, u \to \int_0^\pi [\frac{u'^2(t)}{2} - \frac{u^2(t)}{2} - F(t, u(t)) + h(t)u(t)]\, dt$$

and

$$\Gamma : H_0^1(0,\pi) \to \mathbb{R}, u \to \int_0^\pi u(t) \sin t\, dt.$$

It is easy to see that ϕ is C^1 and weakly lower semicontinuous and that Γ is weakly continuous, convex, increasing and surjective. Moreover the admissibility of Γ can be proved as in Theorem 4.8. It remains to prove that φ, defined by Proposition 4.4, is not strictly monotone to apply Proposition 4.7.

Let w_∞ be the unique solution of

$$w'' + w = h(t), \quad w(0) = 0,\ w(\pi) = 0, \quad \int_0^\pi w(t) \sin t\, dt = 0,$$

and $c_\infty = \int_0^\pi [\frac{w_\infty'^2(t)}{2} - \frac{w_\infty^2(t)}{2} + h(t) w_\infty(t)]\, dt$. We shall prove that

$$\lim_{|\xi| \to \infty} \varphi(\xi) = c_\infty.$$

Let $(\xi_n)_n$ be an unbounded sequence and $u_{\xi_n} \in M_{\xi_n}$. We write $u_{\xi_n} = \xi_n \sin t + w_{\xi_n}$ with $\int_0^\pi w_{\xi_n}(t) \sin t\, dt = 0$. Observe that w_{ξ_n} solves

$$w_{\xi_n}'' + w_{\xi_n} + f(t, \xi_n \sin t + w_{\xi_n}) = h(t) + \mu_n \sin t,\\ w_{\xi_n}(0) = 0,\ w_{\xi_n}(\pi) = 0, \tag{4.10}$$

for some $\mu_n \in \mathbb{R}$. Multiplying (4.10) by w_{ξ_n} and by $\sin t$ and integrating, we have

$$\int_0^\pi [\frac{w_{\xi_n}'^2(t)}{2} - \frac{w_{\xi_n}^2(t)}{2} - f(t, \xi_n \sin t + w_{\xi_n}(t)) w_{\xi_n}(t) + h(t) w_{\xi_n}(t)]\, dt = 0,$$
$$\mu_n = \frac{2}{\pi} \int_0^\pi f(t, \xi_n \sin t + w_{\xi_n}(t)) \sin t\, dt.$$

We deduce then from the above equalities that $(w_{\xi_n})_n \subset H_0^1(0,\pi)$ and $(\mu_n)_n$ are bounded. As w_{ξ_n} is a solution of (4.10), these results imply the existence of a function $k \in L^1(0,\pi)$ such that for all n, $|w''_{\xi_n}(t)| \le k(t)$ on $[0,\pi]$. Hence, up to a subsequence, $w_{\xi_n} \to w$ in $H_0^1(0,\pi)$. Passing to the limit in (4.10) we see that $w = w_\infty$. Moreover

$$\varphi(\xi_n) = \phi(u_{\xi_n}) = \int_0^\pi [\frac{w'^2_{\xi_n}(t)}{2} - \frac{w^2_{\xi_n}(t)}{2} - F(t, \xi_n \sin t + w_{\xi_n}(t)) + h(t) w_{\xi_n}(t)] dt$$

and passing to the limit, we observe that

$$\lim_{n\to\infty} \varphi(\xi_n) = \int_0^\pi [\frac{w'^2_\infty(t)}{2} - \frac{w^2_\infty(t)}{2} + h(t) w_\infty(t)]\, dt = c_\infty.$$

As the sequence $(\xi_n)_n$ is arbitrary, we conclude that

$$\lim_{|\xi|\to\infty} \varphi(\xi) = c_\infty,$$

and φ is not strictly monotone. □

Our second application concerns oscillating potentials.

Theorem 4.11. *Let $f : [0,\pi] \times \mathbb{R} \to \mathbb{R}$ be an L^1-Carathéodory function such that, for a.e. $t \in [0,\pi]$ and all $u \in \mathbb{R}$,*

$$|f(t,u)| \le h(t),$$

for some $h \in L^1(0,\pi)$. Suppose moreover that

$$\liminf_{|\xi|\to\infty} \int_0^\pi F(t, \xi \sin t)\, dt = -\infty \quad \textit{and} \quad \limsup_{|\xi|\to\infty} \int_0^\pi F(t, \xi \sin t)\, dt = +\infty.$$

where $F(t,u) = \int_0^u f(t,s)\, ds$. Then the problem

$$u'' + u + f(t,u) = 0,$$
$$u(0) = 0, \ u(\pi) = 0,$$

has infinitely many solutions.

Proof. We apply Proposition 4.7 with

$$\phi : H_0^1(0,\pi) \to \mathbb{R}, u \to \int_0^\pi [\frac{u'^2(t)}{2} - \frac{u^2(t)}{2} - F(t, u(t))]\, dt$$

and

$$\Gamma(u) = \int_0^\pi u(t) \sin t\, dt.$$

It is easy to see that ϕ is C^1 and weakly lower semicontinuous, and that Γ is weakly continuous, convex, increasing, and surjective. Moreover the admissibility of Γ can be proved as in Theorem 4.8. It remains to study the reduction function φ defined by Proposition 4.4.

For every ξ and $u_\xi \in M_\xi$, denote $u_\xi = \xi \sin t + w_\xi$ with $\int_0^\pi w_\xi(t) \sin t \, dt = 0$. As in the proof of Theorem 4.10, we prove that $\|w_\xi\|_{H_0^1}$ and, using (4.10), $\|w_\xi\|_{C^1}$ are bounded independently of ξ. Hence we have

$$\begin{aligned}\varphi(\xi) = \phi(u_\xi) &= \int_0^\pi [\frac{w_\xi'^2(t)}{2} - \frac{w_\xi^2(t)}{2} - F(t, \xi \sin t + w_\xi(t))] \, dt \\ &= -\int_0^\pi F(t, \xi \sin t) \, dt + O(1),\end{aligned}$$

as $|\xi| \to \infty$ and

$$\liminf_{|\xi|\to\infty} \varphi(\xi) = -\infty \quad \text{and} \quad \limsup_{|\xi|\to\infty} \varphi(\xi) = +\infty.$$

These results imply that we have two unbounded sequences $(\xi_i)_i$ and $(\nu_i)_i$ with $D^+\varphi(\xi_i) \geq 0$ and $D_-\varphi(\nu_i) \leq 0$. By Proposition 4.7, we have an infinite number of lower and upper solutions $\alpha_i = \nu_i \sin t + w_{\nu_i}$ and $\beta_i = \xi_i \sin t + w_{\xi_i}$. These functions can be ordered $\alpha_i \prec \beta_i \prec \alpha_{i+1}$ because $\|w_\mu\|_{C^1}$ is bounded independently of μ. We conclude by applying Theorem 4.1. $\square$

Chapter V

Monotone Iterative Methods

1 An Abstract Formulation

Let Z be a Banach space. An *order cone* $K \subset Z$ is a closed set such that

for all u and $v \in K$, $u + v \in K$,
for all $t \in \mathbb{R}^+$ and $u \in K$, $tu \in K$,
if $u \in K$ and $-u \in K$ then $u = 0$.

Such a cone K induces an *order* on Z:

$$u \leq v \quad \text{if and only if} \quad v - u \in K.$$

We write equivalently $u \leq v$ or $v \geq u$. The cone is said to be *normal* if there exists $c > 0$ such that $0 \leq u \leq v$ implies $\|u\| \leq c\|v\|$.

The following theorem gives conditions for an increasing sequence $(\alpha_n)_n$ to converge to a fixed point of T.

Theorem 1.1. *Let $X \subset Z$ be continuously included Banach spaces so that Z has a normal order cone. Let α and $\beta \in X$, $\alpha \leq \beta$,*

$$\mathcal{E} = \{u \in X \mid \alpha \leq u \leq \beta\} \tag{1.1}$$

and let $T : \mathcal{E} \to X$ be completely continuous in X. Assume that the sequence $(\alpha_n)_n$ defined by

$$\alpha_0 = \alpha, \quad \alpha_n = T\alpha_{n-1} \tag{1.2}$$

is bounded in X and for all $n \in \mathbb{N}$

$$\alpha_n \leq \alpha_{n+1} \leq \beta.$$

Then the sequence $(\alpha_n)_n$ converges monotonically in X to a fixed point u of T such that

$$\alpha \leq u \leq \beta.$$

Proof. Claim – The sequence $(\alpha_n)_n$ *converges in* X. The sequence $(\alpha_n)_n$ is increasing and included in $\mathcal{E}$. As the set $A = \{\alpha_n \mid n \in \mathbb{N}\}$ is bounded in X, $T(A)$ is relatively compact in X. Hence, any sequence $(\alpha_{n_k})_k \subset (\alpha_n)_n$ has a converging subsequence in X and therefore in Z. As the order cone is normal and the sequence is monotone, the sequence itself converges in Z, i.e. there exists $u \in Z$ so that

$$\alpha \leq u \leq \beta \text{ and } \alpha_n \xrightarrow{Z} u.$$

It follows that all such subsequence converging in X have the same limit u, which implies

$$\alpha_n \xrightarrow{X} u.$$

Next, we deduce from the continuity of T that u is a fixed point of T. □

A similar result holds to prove the convergence of the sequence $(\beta_n)_n$.

Theorem 1.2. *Let* $X \subset Z$ *be continuously included Banach spaces so that* Z *has a normal order cone. Let* α *and* $\beta \in X$, $\alpha \leq \beta$, $\mathcal{E}$ *be defined by* (1.1) *and* $T : \mathcal{E} \to X$ *be completely continuous in* X. *Assume the sequence* $(\beta_n)_n$ *defined by*

$$\beta_0 = \beta, \quad \beta_n = T\beta_{n-1}, \tag{1.3}$$

is bounded in X *and for all* $n \in \mathbb{N}$

$$\beta_n \geq \beta_{n+1} \geq \alpha.$$

Then the sequence $(\beta_n)_n$ *converges monotonically in* X *to a fixed point* v *of* T *such that*

$$\alpha \leq v \leq \beta.$$

As a corollary we can write the following result which deals with maps T that are *monotone increasing*, i.e. $u \leq v$ implies $Tu \leq Tv$.

Theorem 1.3. *Let* $X \subset Z$ *be continuously included Banach spaces so that* Z *has a normal order cone. Let* α *and* $\beta \in X$, $\alpha \leq \beta$, $\mathcal{E}$ *be defined by* (1.1) *and let* $T : \mathcal{E} \to X$ *be continuous and monotone increasing. Assume* $T(\mathcal{E})$ *is relatively compact in* X *and*

$$\alpha \leq T\alpha \quad \text{and} \quad T\beta \leq \beta.$$

Then, the sequence $(\alpha_n)_n$ *and* $(\beta_n)_n$ *defined by* (1.2) *and* (1.3) *converge monotonically in* X *to fixed points* u_{min} *and* u_{max} *of* T *such that*

$$\alpha \leq u_{min} \leq u_{max} \leq \beta.$$

Further, any fixed point $u \in \mathcal{E}$ *of* T *verifies*

$$u_{min} \leq u \leq u_{max}.$$

Proof. Claim 1 – The sequence $(\alpha_n)_n$ *converges in* X *to a fixed point* u_{min} *of* T *such that* $\alpha \le u_{min} \le \beta$. As T is monotone increasing, we prove by recurrence that for any $n \in \mathbb{N}$, $\alpha_n \le \alpha_{n+1} \le \beta$. Hence, $(\alpha_n)_n \subset \mathcal{E}$ and since $T(\mathcal{E})$ is relatively compact in X the sequence $(\alpha_n)_n$ is bounded in X. The claim follows now from Theorem 1.1.

Claim 2 – The sequence $(\beta_n)_n$ *converges in* X *to a fixed point* u_{max} *of* T *such that* $u_{min} \le u_{max} \le \beta$. As for Claim 1 we prove existence of a fixed point u_{max} such that $\alpha \le u_{max} \le \beta$. Next, as T is monotone increasing and $\alpha \le \beta$ we deduce by recurrence that $\alpha_n \le \beta_n$. At last, going to the limit we obtain $u_{min} \le u_{max}$.

Claim 3 – Any fixed point $u \in \mathcal{E}$ *of* T *verifies* $u_{min} \le u \le u_{max}$. Since $\alpha \le u \le \beta$, we deduce by recurrence $\alpha_n = T\alpha_{n-1} \le Tu = u \le T\beta_{n-1} = \beta_n$. The claim follows now going to the limit. □

Remark 1.1. In Theorem 1.1, we cannot prove the fixed point u is minimal since we only have some control on $T(u)$ for $u \in \{\alpha_n \mid n \in \mathbb{N}\}$. Consider for example any continuous function $T : [-1/2, 1/2] \to \mathbb{R}$ such that

$$T(-\tfrac{1}{2n}) = -\tfrac{1}{2(n+1)}, \quad T(-\tfrac{1}{2n+1}) = -\tfrac{1}{2n+1}.$$

The assumptions of the theorem are satisfied for $\alpha = -1/2$ and $\beta = 1/2$. The sequence $(\alpha_n)_n = (-\frac{1}{2n})_n$ converges to the fixed point $u = 0$. However $u = 0$ is not minimal since the points $u_n = -\frac{1}{2n+1} \in [-\frac{1}{2}, \frac{1}{2}]$ are fixed points of T.

2 Well-ordered Lower and Upper Solutions

2.1 The Periodic Problem

Consider the periodic boundary value problem

$$\begin{gathered} u'' = f(t,u), \\ u(a) = u(b), \ u'(a) = u'(b), \end{gathered} \tag{2.1}$$

where f is a continuous function.

Our aim is to build an approximation scheme, easy to compute, that converges to solutions of (2.1). To this end, given continuous functions α and β, and $M > 0$, we consider the sequences $(\alpha_n)_n$ and $(\beta_n)_n$ defined by

$$\begin{gathered} \alpha_0 = \alpha, \\ \alpha_n'' - M\alpha_n = f(t,\alpha_{n-1}) - M\alpha_{n-1}, \\ \alpha_n(a) = \alpha_n(b), \ \alpha_n'(a) = \alpha_n'(b) \end{gathered} \tag{2.2}$$

and

$$\begin{aligned} &\beta_0 = \beta, \\ &\beta_n'' - M\beta_n = f(t, \beta_{n-1}) - M\beta_{n-1}, \\ &\beta_n(a) = \beta_n(b), \ \beta_n'(a) = \beta_n'(b). \end{aligned} \tag{2.3}$$

The approximations α_n and β_n are "easy to compute", in the sense that for every n, the problems (2.2) and (2.3) are linear and have unique solutions which read explicitely

$$\alpha_n(t) = \int_a^b G(t,s)(f(s,\alpha_{n-1}(s)) - M\alpha_{n-1}(s))\,ds,$$
$$\beta_n(t) = \int_a^b G(t,s)(f(s,\beta_{n-1}(s)) - M\beta_{n-1}(s))\,ds,$$

where $G(t,s)$ is the Green function of the problem

$$\begin{aligned} u'' - Mu = f(t), \\ u(a) = u(b), \ u'(a) = u'(b). \end{aligned} \tag{2.4}$$

Clearly this method does not avoid numerical difficulties such as those related to stiff systems.

The next theorem proves the convergence of the α_n and β_n.

Theorem 2.1. *Let α and $\beta \in C^2([a,b])$, $\alpha \le \beta$ and*

$$E := \{(t,u) \in [a,b] \times \mathbb{R} \mid \alpha(t) \le u \le \beta(t)\}. \tag{2.5}$$

Assume $f : E \to \mathbb{R}$ is a continuous function, there exists $M > 0$ such that for all (t, u_1), $(t, u_2) \in E$,

$$u_1 \le u_2 \text{ implies } f(t,u_2) - f(t,u_1) \le M(u_2 - u_1)$$

and for all $t \in [a,b]$

$$\alpha''(t) \ge f(t,\alpha(t)), \quad \alpha(a) = \alpha(b), \ \alpha'(a) \ge \alpha'(b),$$
$$\beta''(t) \le f(t,\beta(t)), \quad \beta(a) = \beta(b), \ \beta'(a) \le \beta'(b).$$

Then the sequences $(\alpha_n)_n$ and $(\beta_n)_n$ defined by (2.2) and (2.3) converge monotonically in $C^1([a,b])$ to solutions u_{min} and u_{max} of (2.1) such that

$$\alpha \le u_{min} \le u_{max} \le \beta.$$

Further, any solution u of (2.1) with graph in E verifies

$$u_{min} \le u \le u_{max}.$$

Proof. Let $X = C^1([a,b])$, $Z = C([a,b])$, $K = \{u \in Z \mid u(t) \ge 0 \text{ on } [a,b]\}$ be the ordered cone in Z and $\mathcal{E}$ be defined from (1.1). Define the operator $T : \mathcal{E} \to X$ by

$$Tu(t) = \int_a^b G(t,s)(f(s,u(s)) - Mu(s))\,ds,$$

where $G(t,s)$ is the Green function of (2.4). This operator is continuous in X and monotone increasing. Further, $T(\mathcal{E})$ is relatively compact in X, $\alpha \leq T\alpha$ and $\beta \geq T\beta$. The proof follows now from Theorem 1.3. □

Remark Notice that the assumption α and $\beta \in \mathcal{C}^2([a,b])$ is not restrictive. If these functions are only $\mathcal{C}^2$-lower and upper solutions, the first iterates α_1 and β_1 satisfy the assumptions of the theorem and are such that $\alpha \leq \alpha_1 \leq \beta_1 \leq \beta$.

Remark Recall that existence of the minimal and maximal solutions u_{min} and u_{max} follows from Theorem I-2.4.

Next, we consider a derivative dependent problem

$$\begin{aligned} u'' &= f(t,u,u'), \\ u(a) &= u(b), \; u'(a) = u'(b). \end{aligned} \tag{2.6}$$

As above, given α, $\beta \in \mathcal{C}^1([a,b])$ and $L > 0$, we consider the approximation schemes

$$\begin{aligned} \alpha_0 &= \alpha, \\ \alpha_n'' - L\alpha_n &= f(t,\alpha_{n-1},\alpha_{n-1}') - L\alpha_{n-1}, \\ \alpha_n(a) &= \alpha_n(b), \; \alpha_n'(a) = \alpha_n'(b) \end{aligned} \tag{2.7}$$

and

$$\begin{aligned} \beta_0 &= \beta, \\ \beta_n'' - L\beta_n &= f(t,\beta_{n-1},\beta_{n-1}') - L\beta_{n-1}, \\ \beta_n(a) &= \beta_n(b), \; \beta_n'(a) = \beta_n'(b). \end{aligned} \tag{2.8}$$

Such problems lead to a major difficulty. A straightforward application of the previous ideas would be to assume that for any u_1, u_2, v_1 and v_2,

$$u_1 \leq u_2 \text{ implies } f(t,u_2,v_2) - f(t,u_1,v_1) \leq L(u_2 - u_1).$$

This would mean that f does not depend on derivatives.

The following result turns out the difficulty.

Theorem 2.2. *Let α and $\beta \in \mathcal{C}^2([a,b])$, $\alpha \leq \beta$ and*

$$E := \{(t,u,v) \in [a,b] \times \mathbb{R}^2 \mid \alpha(t) \leq u \leq \beta(t)\}. \tag{2.9}$$

Assume $f : E \to \mathbb{R}$ is a continuous function, there exists $M \geq 0$ such that for all (t,u_1,v), $(t,u_2,v) \in E$,

$$u_1 \leq u_2 \quad \textit{implies} \quad f(t,u_2,v) - f(t,u_1,v) \leq M(u_2 - u_1); \tag{2.10}$$

there exists $N \geq 0$ such that for all (t,u,v_1), $(t,u,v_2) \in E$,

$$|f(t,u,v_2) - f(t,u,v_1)| \leq N|v_2 - v_1|; \tag{2.11}$$

and for all $t \in [a,b]$

$$\begin{aligned} \alpha''(t) &\geq f(t,\alpha(t),\alpha'(t)), \quad \alpha(a)=\alpha(b),\ \alpha'(a)\geq\alpha'(b),\\ \beta''(t) &\leq f(t,\beta(t),\beta'(t)), \quad \beta(a)=\beta(b),\ \beta'(a)\leq\beta'(b). \end{aligned}$$

At last, let $L > 0$ *be such that*

$$L \geq M + \frac{N^2}{2} + \frac{N}{2}\sqrt{N^2+4M} \tag{2.12}$$

and for all $t \in [a,b]$

$$f(t,\alpha(t),\alpha'(t)) - f(t,\beta(t),\beta'(t)) + L(\beta(t)-\alpha(t)) \geq 0. \tag{2.13}$$

Then, the sequences $(\alpha_n)_n$ *and* $(\beta_n)_n$ *defined by* (2.7) *and* (2.8) *converge monotonically in* $\mathcal{C}^1([a,b])$ *to solutions* u *and* v *of* (2.6) *such that*

$$\alpha \leq u \leq v \leq \beta.$$

Remarks (a) The function $w = \beta - \alpha \geq 0$ satisfies

$$-w'' + N|w'| + (M+1)w = h(t) \geq 0, \quad w(a) = w(b),\ w'(b) \geq w'(a).$$

Thus, using the maximum principle (Theorem A-5.3 with $p(t) = -N\text{sign}w'$ and $q(t) = -(M+1)$) we can prove that if $\alpha \neq \beta$, our assumptions imply $\alpha < \beta$ on $[a,b]$. Also if u is a solution of (2.6) such that $\alpha \lneqq u \lneqq \beta$, we have $\alpha < u < \beta$.

(b) It is clear from Remark (a) that the assumptions on L are satisfied if L is large enough so that the theorem applies for any values of M and N which satisfy the assumptions (2.10) and (2.11).

(c) The conditions on L are immediately satisfied with $L = M$ if the function f does not depend on the derivative u' (i.e. $N = 0$).

(d) If α (resp. β) is a solution of (2.6), we have $\alpha_n = \alpha$ (resp. $\beta_n = \beta$) for all $n \in \mathbb{N}$.

Proof of Theorem 2.2 : The proof uses Theorems 1.1 and 1.2 with $X = \mathcal{C}^1([a,b])$, $Z = \mathcal{C}([a,b])$ and $K = \{u \in Z \mid u(t) \geq 0 \text{ on } [a,b]\}$ as the ordered cone in Z. Let $\mathcal{E}$ be defined from (1.1). The operator $T : \mathcal{E} \to X$, defined by

$$Tu(t) = \int_a^b G(t,s)(f(s,u(s),u'(s)) - Lu(s))\,ds,$$

where $G(t,s)$ is the Green function of (2.4) with $M = L$, is completely continuous in X. With these notations, the approximation schemes (2.7) and (2.8) are equivalent to (1.2) and (1.3).

A : *Claim – Let* $L > 0$ *satisfy* (2.12). *Then the functions* α_n *defined recursively by* (2.7) *are such that for all* $n \in \mathbb{N}$,
(a) α_n *is a lower solution, i.e.*

$$\begin{gathered}\alpha_n''(t) \geq f(t, \alpha_n(t), \alpha_n'(t)),\\ \alpha_n(a) = \alpha_n(b), \ \alpha_n'(a) \geq \alpha_n'(b),\end{gathered} \tag{2.14}$$

(b) $\alpha_{n+1} \geq \alpha_n$.
The proof is by recurrence.

Initial step : $n = 0$. The condition (2.14) for $n = 0$ is an assumption. Next, $w = \alpha_1 - \alpha_0$ is a solution of

$$\begin{gathered}-w'' + Lw = \alpha_0''(t) - f(t, \alpha_0(t), \alpha_0'(t)) \geq 0,\\ w(a) = w(b), \ w'(a) \leq w'(b).\end{gathered}$$

Hence, we deduce (b) from the maximum principle (Theorem A-5.3).

Recurrence step – 1st part : assume (a) and (b) hold for some n and let us prove that

$$\begin{gathered}\alpha_{n+1}''(t) \geq f(t, \alpha_{n+1}(t), \alpha_{n+1}'(t)),\\ \alpha_{n+1}(a) = \alpha_{n+1}(b), \ \alpha_{n+1}'(a) \geq \alpha_{n+1}'(b).\end{gathered}$$

Let $w = \alpha_{n+1} - \alpha_n$. We have

$$\begin{aligned}-\alpha_{n+1}'' &+ f(t, \alpha_{n+1}, \alpha_{n+1}')\\ &= f(t, \alpha_{n+1}, \alpha_{n+1}') - f(t, \alpha_n, \alpha_n') - L(\alpha_{n+1} - \alpha_n)\\ &\leq M(\alpha_{n+1} - \alpha_n) + N|\alpha_{n+1}' - \alpha_n'| - L(\alpha_{n+1} - \alpha_n)\\ &= (M - L)w + N|w'|.\end{aligned}$$

On the other hand, w satisfies

$$-w'' + Lw = h(t), \quad w(a) = w(b), \ w'(b) - w'(a) = A, \tag{2.15}$$

with $h(t) := \alpha_n''(t) - f(t, \alpha_n(t), \alpha_n'(t)) \geq 0$ and $A \geq 0$. Its solution w reads

$$\begin{aligned}w(t) = k \Bigg[&\int_a^t h(s) \cosh \sqrt{L}(\tfrac{b-a}{2} + s - t)\, ds\\ &+ \int_t^b h(s) \cosh \sqrt{L}(\tfrac{b-a}{2} + t - s)\, ds + A \cosh \sqrt{L}(t - \tfrac{a+b}{2}) \Bigg],\end{aligned}$$

where

$$k = (2\sqrt{L} \sinh \sqrt{L}\tfrac{b-a}{2})^{-1}.$$

Hence, to prove α_{n+1} is a lower solution, we only have to verify

$$\begin{aligned}\int_a^t \Big[&(M - L) \cosh \sqrt{L}(\tfrac{b-a}{2} + s - t)\\ &+ N\sqrt{L}|\sinh \sqrt{L}(\tfrac{b-a}{2} + s - t)| \Big] h(s)\, ds \leq 0,\end{aligned}$$

$$\int_t^b \Big[(M-L)\cosh\sqrt{L}(\tfrac{b-a}{2}+t-s) \\ +N\sqrt{L}|\sinh\sqrt{L}(\tfrac{b-a}{2}+t-s)|\Big]\, h(s)\, ds \le 0$$

and

$$(M-L)\cosh\sqrt{L}(t-\tfrac{a+b}{2}) + N\sqrt{L}|\sinh\sqrt{L}(t-\tfrac{a+b}{2})| \le 0.$$

Since h is nonpositive and

$$(M-L)\cosh x + N\sqrt{L}|\sinh x| \le (M-L+N\sqrt{L})|\sinh x|$$

for all $x \in \mathbb{R}$, we obtain $(M-L)w + N|w'| \le 0$ if $M - L + N\sqrt{L} \le 0$, which follows from (2.12).

Recurrence step – 2d part : assume (a) and (b) hold for some n and let us prove that $\alpha_{n+2} \ge \alpha_{n+1}$. The function $w = \alpha_{n+2} - \alpha_{n+1}$ satisfies (2.15), where

$$h(t) := \alpha''_{n+1}(t) - f(t, \alpha_{n+1}(t), \alpha'_{n+1}(t)) \quad \text{and} \quad A = 0.$$

From the previous step $h(t) \ge 0$ and the claim follows from the maximum principle (Theorem A-5.3).

B : *Claim – Let* $L > 0$ *satisfy* (2.12). *Then the functions* β_n *defined recursively by* (2.8) *are such that for all* $n \in \mathbb{N}$,
(a) β_n *is an upper solution, i.e.*

$$\beta''_n(t) \le f(t, \beta_n(t), \beta'_n(t)),$$
$$\beta_n(a) = \beta_n(b), \ \ \beta'_n(a) \le \beta'_n(b),$$

(b) $\beta_{n+1} \le \beta_n$.
The proof of this claim parallels the proof of Claim A.

C : *Claim* – $\alpha_n \le \beta_n$. Define, for all $i \in \mathbb{N}$, $w_i = \beta_i - \alpha_i$ and

$$h_i(t) := f(t, \alpha_i(t), \alpha'_i(t)) - f(t, \beta_i(t), \beta'_i(t)) + L(\beta_i(t) - \alpha_i(t)).$$

The proof of the claim is by recurrence.
Initial step : $\alpha_1 \le \beta_1$. The function w_1 is a solution of (2.15) with $h = h_0 \ge 0$ and $A = 0$. Using the maximum principle (Theorem A-5.3), we deduce that $w_1 \ge 0$, i.e. $\alpha_1 \le \beta_1$.

Recurrence step : Let $n \ge 2$. *If* $h_{n-2} \ge 0$ *and* $\alpha_{n-1} \le \beta_{n-1}$, *then* $h_{n-1} \ge 0$ *and* $\alpha_n \le \beta_n$. First, let us prove that for all $t \in [a, b]$, the function h_{n-1} is nonnegative. Indeed, we have

$$\begin{aligned} h_{n-1} &= f(\cdot, \alpha_{n-1}, \alpha'_{n-1}) - f(\cdot, \beta_{n-1}, \beta'_{n-1}) + L(\beta_{n-1} - \alpha_{n-1}) \\ &\ge -M(\beta_{n-1} - \alpha_{n-1}) - N|\beta'_{n-1} - \alpha'_{n-1}| + L(\beta_{n-1} - \alpha_{n-1}) \\ &= (L-M)w_{n-1} - N|w'_{n-1}|. \end{aligned}$$

Notice that w_{n-1} is a solution of (2.15) with $h(t) = h_{n-2}(t) \geq 0$ and $A = 0$. Hence, we can proceed as in the proof of Claim A to show that $h_{n-1} \geq 0$. It follows then from the maximum principle (Theorem A-5.3) that w_n is nonnegative, i.e. $\alpha_n \leq \beta_n$.

D : *Claim* – *There exists $R > 0$ such that any solution u of*

$$u'' \geq f(t, u, u'), \quad u(a) = u(b), \ u'(a) = u'(b),$$

with $\alpha \leq u \leq \beta$ satisfies $\|u'\|_\infty < R$. We deduce from the assumptions that

$$u'' = f(t, u, u') + h(t),$$

where $h(t) \geq 0$,

$$f(t, u, u') + h(t) \geq -\max_F |f(t, u, 0)| - N|u'| \tag{2.16}$$

and $F = \{(t, u) \mid t \in [a, b], \alpha(t) \leq u \leq \beta(t)\}$. The proof follows now using Proposition I-4.5.

E : *Claim* – *There exists $R > 0$ such that any solution u of*

$$u'' \leq f(t, u, u'), \quad u(a) = u(b), \ u'(a) = u'(b),$$

with $\alpha \leq u \leq \beta$ satisfies $\|u'\|_\infty < R$. The proof repeats the argument of Claim D.

F : *Conclusion* – We deduce from Theorems 1.1 and 1.2 that the sequences $(\alpha_n)_n$ and $(\beta_n)_n$ converge monotonically in $C^1([a, b])$ to functions u and v which are solutions of (2.6) such that $\alpha \leq u \leq \beta$ and $\alpha \leq v \leq \beta$. Further, as $\alpha_n \leq \beta_n$ for any n, we have $u \leq v$. □

Remark 2.1. In Claim D or E, the *a priori* bound on $\|u'\|_\infty$ has to be independent of h. We cannot deduce it from an usual Nagumo condition, the control we have on $f(t, u, u') + h(t)$ being basically one-sided (see (2.16)), which is why we used a one-sided Nagumo condition.

Convergence to minimal and maximal solutions are easy to prove in case L is large enough.

Corollary 2.3. *Let the assumptions of Theorem 2.2 be satisfied.*

Then, for L large enough, the sequences $(\alpha_n)_n$ and $(\beta_n)_n$ defined by (2.7) and (2.8) converge monotonically in $C^1([a, b])$ to solutions u_{min} and u_{max} of (2.6) such that

$$\alpha \leq u_{min} \leq u_{max} \leq \beta.$$

Further, any solution u of (2.6), such that $\alpha \leq u \leq \beta$, verifies

$$u_{min} \leq u \leq u_{max}.$$

Proof. Existence of extremal solutions u_{min} and u_{max} follows from Theorem I-5.6. If α is not a solution, we deduce from the above Remark (a) that $u_{min} > \alpha$. Hence, choosing $L > 0$ large enough so that (2.12) is satisfied and

$$f(t, \alpha(t), \alpha'(t)) - f(t, u_{min}(t), u'_{min}(t)) + L(u_{min}(t) - \alpha(t)) \geq 0$$

on $[a, b]$, we can apply Theorem 2.2 with $\beta = u_{min}$ and obtain that $u = \lim_{n\to\infty} \alpha_n$ is such that $\alpha \leq u \leq u_{min}$, whence $u = u_{min}$.

Similarly, we prove $\lim_{n\to\infty} \beta_n = u_{max}$. □

Remark Observe that the problem

$$\begin{gathered} u'' - u + (u')^2 = \sin t, \\ u(0) = u(2\pi), \; u'(0) = u'(2\pi), \end{gathered}$$

cannot be worked out from Theorem 2.2 as it does not satisfy (2.11). However, we can proceed as follows. Notice first that such a problem satisfies a Nagumo condition. Next, we know that lower and upper solutions, α and $\beta \in [-1, 1]$, of problems that satisfy this Nagumo condition have *a priori* bounded derivatives: $\|\alpha'\|_\infty$ and $\|\beta'\|_\infty \leq R$. We can modify then the equation for $|u'| \geq R$ so that the same *a priori* bound on the derivatives can be obtained for the modified problem together with (2.11). It follows then that the approximations defined from (2.7) and (2.8) are the corresponding approximations for the modified problem so that convergence follows from Theorem 2.2.

The following theorem uses an approximation scheme which, from a computational point of view, is more involved since it uses piecewise linear equations. On the other hand, the analysis is simplified since the coefficients in the approximation scheme are the Lipschitz constants on the nonlinearity. Notice also that in this theorem we use different approximation schemes for the α_n and for the β_n.

Theorem 2.4. *Let α and $\beta \in C^2([a, b])$, $\alpha \leq \beta$ and E be defined by (2.9). Assume $f : E \to \mathbb{R}$ is a continuous function, there exists $M > 0$ such that for all (t, u_1, v), $(t, u_2, v) \in E$,*

$$u_1 \leq u_2 \quad \textit{implies} \quad f(t, u_2, v) - f(t, u_1, v) \leq M(u_2 - u_1);$$

there exists $N \geq 0$ such that for all (t, u, v_1), $(t, u, v_2) \in E$,

$$|f(t, u, v_2) - f(t, u, v_1)| \leq N|v_2 - v_1|;$$

and for all $t \in [a, b]$

$$\begin{aligned} &\alpha''(t) \geq f(t, \alpha(t), \alpha'(t)), \quad \alpha(a) = \alpha(b), \; \alpha'(a) \geq \alpha'(b), \\ &\beta''(t) \leq f(t, \beta(t), \beta'(t)), \quad \beta(a) = \beta(b), \; \beta'(a) \leq \beta'(b). \end{aligned}$$

Let $\alpha_0 = \alpha$ and $\beta_0 = \beta$. Then the problems

$$\begin{gathered}\alpha_n'' - N|\alpha_n' - \alpha_{n-1}'| - M\alpha_n = f(t, \alpha_{n-1}, \alpha_{n-1}') - M\alpha_{n-1},\\ \alpha_n(a) = \alpha_n(b),\ \alpha_n'(a) = \alpha_n'(b)\end{gathered}$$

and

$$\begin{gathered}\beta_n'' + N|\beta_n' - \beta_{n-1}'| - M\beta_n = f(t, \beta_{n-1}, \beta_{n-1}') - M\beta_{n-1},\\ \beta_n(a) = \beta_n(b),\ \beta_n'(a) = \beta_n'(b),\end{gathered}$$

define sequences $(\alpha_n)_n$ and $(\beta_n)_n$ that converge monotonically in $\mathcal{C}^1([a,b])$ to solutions u_{min} and u_{max} of (2.6) such that

$$\alpha \leq u_{min} \leq u_{max} \leq \beta.$$

Furthermore, any solution u of (2.6), such that $\alpha \leq u \leq \beta$, verifies

$$u_{min} \leq u \leq u_{max}.$$

Proof. Let $X = \mathcal{C}^1([a,b])$, $Z = \mathcal{C}([a,b])$, $K = \{u \in Z \mid u(t) \geq 0 \text{ on } [a,b]\}$ be the ordered cone in Z and $\mathcal{E}$ be defined from (1.1).

A : *Definition of a completely continuous operator T. Claim – For any $u \in X$ the problem*

$$\begin{gathered}v'' - N|v' - u'(t)| - Mv = f(t, u(t), u'(t)) - Mu(t),\\ v(a) = v(b),\ v'(a) = v'(b),\end{gathered} \tag{2.17}$$

has a unique solution v. Notice that if $k > 0$ is large enough, $-k$ and k are well-ordered lower and upper solutions of (2.17). Existence of a solution of this problem follows then from Theorem I-5.3 with $\varphi(s) = K(s+1)$ and $K > 0$ large enough.

Let v_1 and v_2 be two solutions of (2.17). Define $w = v_2 - v_1$ and assume that for some $t_0 \in [a,b]$

$$w(t_0) = \max_{t \in [a,b]} w(t) > 0.$$

We compute $w'(t_0) = v_2'(t_0) - v_1'(t_0) = 0$ and obtain the contradiction $w''(t_0) = Mw(t_0) + N(|v_2'(t_0) - u'(t_0)| - |v_1'(t_0) - u'(t_0)|) > 0$. Hence $w \leq 0$. Similarly we prove $w \geq 0$, which implies $w = 0$, and the solution of (2.17) is unique.

Now, we can define the operator

$$T : \mathcal{E} \to X, u \mapsto Tu,$$

where $\mathcal{E}$ is defined from (1.1) and Tu is the solution of (2.17). This operator is completely continuous.

B : *Claim – For all* $n \in \mathbb{N}$, $\alpha_{n+1} \geq \alpha_n$.

Initial step : $w = \alpha_1 - \alpha_0 = T\alpha - \alpha \geq 0$. Notice that w solves the problem

$$\begin{gathered} w'' - N|w'| - Mw = f(t, \alpha(t), \alpha'(t)) - \alpha''(t) \leq 0, \\ w(b) - w(a) = 0, \ w'(b) - w'(a) = \alpha'(a) - \alpha'(b) \geq 0. \end{gathered}$$

The claim follows from the maximum principle argument used in Claim A.

Recurrence step – Assume $\alpha_{n+1} - \alpha_n \geq 0$ *and prove* $w = \alpha_{n+2} - \alpha_{n+1} \geq 0$. The function w satisfies

$$\begin{aligned} w'' &- N|w'| - Mw \\ &= f(t, \alpha_{n+1}, \alpha'_{n+1}) - f(t, \alpha_n, \alpha'_n) - M(\alpha_{n+1} - \alpha_n) - N|\alpha'_{n+1} - \alpha'_n| \\ &\leq 0 \end{aligned}$$

and

$$w(a) = w(b), \ w'(a) = w'(b).$$

The claim follows from the same maximum principle argument.

C : *Claim –* $\alpha_n \leq \beta$. The proof repeats the argument of the previous claims.

D : *Claim – The sequence* $(\alpha_n)_n$ *is bounded in* X. The functions α_n solve problems

$$u'' + N|u'| - Mu = g_n(t), \quad u(a) = u(b), \ u'(a) = u'(b),$$

where

$$\begin{aligned} g_n = {} & N(|\alpha'_n| + |\alpha'_n - \alpha'_{n-1}| - |\alpha'_{n-1}|) \\ & + N|\alpha'_{n-1}| + f(t, \alpha_{n-1}, \alpha'_{n-1}) - f(t, \alpha_{n-1}, 0) \\ & + f(t, \alpha_{n-1}, 0) - M\alpha_{n-1} \\ \geq {} & f(t, \alpha_{n-1}, 0) - M\alpha_{n-1} \geq f(t, \beta, 0) - M\beta. \end{aligned}$$

As the g_n are uniformly lower bounded, the boundedness of the α_n follows now from Proposition I-4.5.

E : *Conclusion –* The convergence of the sequence $(\alpha_n)_n$ to a solution u_{min} of (2.6) follows from Theorem 1.1. To prove u_{min} is a minimum solution, let u be any solution of (2.6) such that $\alpha \leq u \leq \beta$. Notice that, if we choose $\beta = u$ as an upper solution, the previous claims still hold and Theorem 1.1 applies. Hence $u_{min} = \lim_{n \to \infty} \alpha_n \leq u$.

Similarly, we prove the convergence of the sequence $(\beta_n)_n$ to a maximal solution using Theorem 1.2. □

2.2 The Neumann Problem

The Neumann problem

$$\begin{gathered} u'' = f(t, u), \\ u'(a) = 0, \ u'(b) = 0, \end{gathered} \tag{2.18}$$

where f is a continuous function, can be investigated along the lines of the periodic problem.

Theorem 2.5. *Let α and $\beta \in C^2([a,b])$, $\alpha \leq \beta$ and let E be defined from* (2.5). *Assume $f : E \to \mathbb{R}$ is a continuous function, there exists $M > 0$ such that for all (t, u_1), $(t, u_2) \in E$,*

$$u_1 \leq u_2 \text{ implies } f(t,u_2) - f(t,u_1) \leq M(u_2 - u_1)$$

and for all $t \in [a,b]$

$$\begin{aligned} \alpha''(t) \geq f(t,\alpha(t)), \quad & \alpha'(a) \geq 0, \ \alpha'(b) \leq 0, \\ \beta''(t) \leq f(t,\beta(t)), \quad & \beta'(a) \leq 0, \ \beta'(b) \geq 0. \end{aligned}$$

Then the sequences $(\alpha_n)_n$ and $(\beta_n)_n$ defined by

$$\alpha_0 = \alpha,$$

$$\begin{aligned} \alpha_n'' - M\alpha_n &= f(t,\alpha_{n-1}) - M\alpha_{n-1}, \\ \alpha_n'(a) &= 0, \ \alpha_n'(b) = 0, \end{aligned}$$

and

$$\beta_0 = \beta,$$

$$\begin{aligned} \beta_n'' - M\beta_n &= f(t,\beta_{n-1}) - M\beta_{n-1}, \\ \beta_n'(a) &= 0, \ \beta_n'(b) = 0, \end{aligned}$$

converge monotonically in $C^1([a,b])$ to solutions u_{min} and u_{max} of (2.18) *such that*

$$\alpha \leq u_{min} \leq u_{max} \leq \beta.$$

Further, any solution u of (2.18) *with graph in E verifies*

$$u_{min} \leq u \leq u_{max}.$$

Exercise 2.1. Prove the above theorem.

Hint : See the proof of Theorem 2.1.

To deal with derivative dependent equations

$$\begin{aligned} u'' &= f(t,u,u'), \\ u'(a) &= 0, \ u'(b) = 0, \end{aligned} \tag{2.19}$$

we can use the ideas of Theorem 2.2.

Theorem 2.6. *Let α and $\beta \in C^2([a,b])$, $\alpha \leq \beta$ and let E be defined from* (2.9). *Assume $f : E \to \mathbb{R}$ is a continuous function, there exists $M \geq 0$ such that for all (t, u_1, v), $(t, u_2, v) \in E$,*

$$u_1 \leq u_2 \quad \text{implies} \quad f(t,u_2,v) - f(t,u_1,v) \leq M(u_2 - u_1);$$

there exists $N \geq 0$ *such that for all* (t, u, v_1), $(t, u, v_2) \in E$,

$$|f(t, u, v_2) - f(t, u, v_1)| \leq N|v_2 - v_1|;$$

and for all $t \in [a, b]$

$$\alpha''(t) \geq f(t, \alpha(t), \alpha'(t)), \quad \alpha'(a) \geq 0, \ \alpha'(b) \leq 0,$$
$$\beta''(t) \leq f(t, \beta(t), \beta'(t)), \quad \beta'(a) \leq 0, \ \beta'(b) \geq 0.$$

Then, if L *is large enough, the sequences* $(\alpha_n)_n$ *and* $(\beta_n)_n$ *defined by*

$$\alpha_0 = \alpha,$$
$$\alpha_n'' - L\alpha_n = f(t, \alpha_{n-1}, \alpha_{n-1}') - L\alpha_{n-1},$$
$$\alpha_n'(a) = 0, \ \alpha_n'(b) = 0$$

and

$$\beta_0 = \beta,$$
$$\beta_n'' - L\beta_n = f(t, \beta_{n-1}, \beta_{n-1}') - L\beta_{n-1},$$
$$\beta_n'(a) = 0, \ \beta_n'(b) = 0$$

converge monotonically in $C^1([a, b])$ *to solutions* u_{min} *and* u_{max} *of* (2.19) *such that*

$$\alpha \leq u_{min} \leq u_{max} \leq \beta.$$

Further, any solution u *of* (2.19), *such that* $\alpha \leq u \leq \beta$, *verifies*

$$u_{min} \leq u \leq u_{max}.$$

Exercise 2.2. Prove the above theorem.

Hint : See the proofs of Theorem 2.2 and Corollary 2.3, or [65].

We can investigate problem (2.19) along the lines of Theorem 2.4.

Theorem 2.7. *Let* α *and* $\beta \in C^2([a, b])$, $\alpha \leq \beta$ *and let* E *be defined from* (2.9). *Assume* $f : E \to \mathbb{R}$ *is a continuous function, there exists* $M > 0$ *such that for all* (t, u_1, v), $(t, u_2, v) \in E$,

$$u_1 \leq u_2 \quad \text{implies} \quad f(t, u_2, v) - f(t, u_1, v) \leq M(u_2 - u_1);$$

there exists $N \geq 0$ *such that for all* (t, u, v_1), $(t, u, v_2) \in E$,

$$|f(t, u, v_2) - f(t, u, v_1)| \leq N|v_2 - v_1|;$$

and for all $t \in [a, b]$

$$\alpha''(t) \geq f(t, \alpha(t), \alpha'(t)), \quad \alpha'(a) \geq 0, \ \alpha'(b) \leq 0,$$
$$\beta''(t) \leq f(t, \beta(t), \beta'(t)), \quad \beta'(a) \leq 0, \ \beta'(b) \geq 0.$$

Let $\alpha_0 = \alpha$ *and* $\beta_0 = \beta$. *Then the problems*

$$\alpha_n'' - N|\alpha_n' - \alpha_{n-1}'| - M\alpha_n = f(t, \alpha_{n-1}, \alpha_{n-1}') - M\alpha_{n-1},$$
$$\alpha_n'(a) = 0,\ \alpha_n'(b) = 0$$

and

$$\beta_n'' + N|\beta_n' - \beta_{n-1}'| - M\beta_n = f(t, \beta_{n-1}, \beta_{n-1}') - M\beta_{n-1},$$
$$\beta_n'(a) = 0,\ \beta_n'(b) = 0$$

define sequences $(\alpha_n)_n$ *and* $(\beta_n)_n$ *that converge monotonically in* $C^1([a,b])$ *to solutions* u_{min} *and* u_{max} *of* (2.19) *such that*

$$\alpha \le u_{min} \le u_{max} \le \beta.$$

Further, any solution u *of* (2.19), *such that* $\alpha \le u \le \beta$, *verifies*

$$u_{min} \le u \le u_{max}.$$

Exercise 2.3. Prove the above theorem.

Hint : See the proof of Theorem 2.4 or [234].

2.3 The Dirichlet Problem

Consider the Dirichlet problem

$$\begin{aligned} u'' &= f(t,u), \\ u(a) &= 0,\ u(b) = 0, \end{aligned} \tag{2.20}$$

where f is a continuous function.

Theorem 2.8. *Let* α *and* $\beta \in C^2([a,b])$, $\alpha \le \beta$ *and let* E *be defined from* (2.5). *Assume* $f : E \to \mathbb{R}$ *is a continuous function, there exists* $M \ge 0$ *such that for all* (t, u_1), $(t, u_2) \in E$,

$$u_1 \le u_2 \text{ implies } f(t,u_2) - f(t,u_1) \le M(u_2 - u_1),$$

and for all $t \in [a,b]$

$$\alpha''(t) \ge f(t,\alpha(t)),\quad \alpha(a) \le 0,\ \alpha(b) \le 0,$$
$$\beta''(t) \le f(t,\beta(t)),\quad \beta(a) \ge 0,\ \beta(b) \ge 0.$$

Then the sequences $(\alpha_n)_n$ *and* $(\beta_n)_n$ *defined by*

$$\alpha_0 = \alpha,$$

$$\alpha_n'' - M\alpha_n = f(t,\alpha_{n-1}) - M\alpha_{n-1},$$
$$\alpha_n(a) = 0,\ \alpha_n(b) = 0$$

and

$$\beta_0 = \beta,$$
$$\beta_n'' - M\beta_n = f(t, \beta_{n-1}) - M\beta_{n-1},$$
$$\beta_n(a) = 0, \ \beta_n(b) = 0$$

converge monotonically in $C^1([a,b])$ *to solutions* u_{min} *and* u_{max} *of* (2.20) *such that*

$$\alpha \le u_{min} \le u_{max} \le \beta.$$

Further, any solution u *of* (2.20) *with graph in* E *verifies*

$$u_{min} \le u \le u_{max}.$$

Exercise 2.4. Prove the above theorem.

Hint : See the proof of Theorem 2.1.

In case of derivative dependent equations

$$\begin{aligned} u'' &= f(t, u, u'), \\ u(a) &= 0, \ u(b) = 0, \end{aligned} \tag{2.21}$$

approximation schemes similar to (2.7) and (2.8) do not work. Indeed if we try to repeat the argument of Theorem 2.2, we have to prove *a priori* bounds on the derivatives of lower and upper solutions (see Parts D and E). As noted in Remark 2.1, it implies we use one-sided Nagumo condition which, for the Dirichlet problem, imposes the lower and upper solutions to satisfy the boundary conditions. We can think this condition is not very restrictive since the first iterates α_1 and β_1 already satisfy such conditions. However L must also verify (2.13) and this might not be the case even for large values of L. We have then to consider alternative approximation schemes.

Theorem 2.9. *Let* α *and* $\beta \in C^2([a,b])$, $\alpha \le \beta$ *and let* E *be defined from* (2.9). *Assume* $f : E \to \mathbb{R}$ *is a continuous function, there exists* $M \ge 0$ *such that for all* (t, u_1, v), $(t, u_2, v) \in E$,

$$u_1 \le u_2 \quad \textit{implies} \quad f(t, u_2, v) - f(t, u_1, v) \le M(u_2 - u_1);$$

there exists $N \ge 0$ *such that for all* (t, u, v_1), $(t, u, v_2) \in E$,

$$|f(t, u, v_2) - f(t, u, v_1)| \le N|v_2 - v_1|;$$

and for all $t \in [a,b]$

$$\alpha''(t) \ge f(t, \alpha(t), \alpha'(t)), \quad \alpha(a) = 0, \ \alpha(b) = 0,$$
$$\beta''(t) \le f(t, \beta(t), \beta'(t)), \quad \beta(a) = 0, \ \beta(b) = 0.$$

At last, let $K_0 \in \mathcal{C}([a,b])$ *be such that* $K_0(a) > 0$ *and for all* $t \in [a,b]$, $K_0(t) = -K_0(b+a-t)$.

Then, for L large enough, the sequences $(\alpha_n)_n$ *and* $(\beta_n)_n$ *defined by* $\alpha_0 = \alpha$, $\beta_0 = \beta$,

$$\alpha''_{n+1} - \sqrt[3]{L}\, K_0(t)\alpha'_{n+1} - L\alpha_{n+1} = f(t, \alpha_n, \alpha'_n) - \sqrt[3]{L}\, K_0(t)\alpha'_n - L\alpha_n,$$
$$\alpha_{n+1}(a) = 0, \ \alpha_{n+1}(b) = 0$$

and

$$\beta''_{n+1} - \sqrt[3]{L}\, K_0(t)\beta'_{n+1} - L\beta_{n+1} = f(t, \beta_n, \beta'_n) - \sqrt[3]{L}\, K_0(t)\beta'_n - L\beta_n,$$
$$\beta_{n+1}(a) = 0, \ \beta_{n+1}(b) = 0$$

converge monotonically in $C^1([a,b])$ *to solutions* u_{min} *and* u_{max} *of* (2.21) *such that*

$$\alpha \le u_{min} \le u_{max} \le \beta.$$

Further, any solution u of (2.21) *with graph in E verifies*

$$u_{min} \le u \le u_{max}.$$

Exercise 2.5. Prove the above theorem.

Hint : See the proofs of Theorem 2.2 and Corollary 2.3 or [63].

2.4 Bounded Solutions

In this section we consider bounded solutions of the differential equation

$$u'' + cu' = f(t, u). \tag{2.22}$$

To this end, we define the set $\mathcal{B}C(\mathbb{R}) = \{u \in C(\mathbb{R}) \mid u \in L^\infty(\mathbb{R})\}$.

Theorem 2.10. *Let* α *and* $\beta \in C^2(\mathbb{R}) \cap L^\infty(\mathbb{R})$, $\alpha \le \beta$ *and let* $E := \{(t,u) \in \mathbb{R}^2 \mid \alpha(t) \le u \le \beta(t)\}$. *Assume* $c \in \mathbb{R}$, $f : E \to \mathbb{R}$ *is a continuous bounded function and there exists* $M > 0$ *such that for all* (t, u_1), $(t, u_2) \in E$,

$$u_1 \le u_2 \Rightarrow f(t, u_2) - f(t, u_1) \le M(u_2 - u_1).$$

Assume further that for all $t \in \mathbb{R}$

$$\alpha''(t) + c\alpha'(t) \ge f(t, \alpha(t)),$$
$$\beta''(t) + c\beta'(t) \le f(t, \beta(t)).$$

Let $\alpha_0 = \alpha$ *and* $\beta_0 = \beta$. *Then the equations*

$$\alpha''_{n+1} + c\alpha'_{n+1} - M\alpha_{n+1} = f(t, \alpha_n) - M\alpha_n,$$
$$\beta''_{n+1} + c\beta'_{n+1} - M\beta_{n+1} = f(t, \beta_n) - M\beta_n$$

define sequences $(\alpha_n)_n$ *and* $(\beta_n)_n \subset \mathcal{BC}(\mathbb{R})$ *that converge monotonically and uniformly on bounded intervals of* $\mathbb{R}$ *to solutions* u_{min} *and* u_{max} *of* (2.22) *such that*

$$\alpha \le u_{min} \le u_{max} \le \beta.$$

Further, any solution u *of* (2.22) *with graph in* E *verifies*

$$u_{min} \le u \le u_{max}.$$

Proof. First recall that given $p \in \mathcal{BC}(\mathbb{R})$, the problem

$$y'' + cy' - \lambda y = p(t)$$

with $\lambda > 0$ has a unique solution in $\mathcal{BC}(\mathbb{R})$ given by

$$y(t) = \int_{-\infty}^{+\infty} G(t,s)p(s)\,ds,$$

where

$$G(t,s) = -\tfrac{1}{2\nu}\exp(\tfrac{c}{2}(s-t))\exp(-\nu|t-s|),$$

with $\nu = \sqrt{\lambda + \frac{c^2}{4}}$. This result implies the $(\alpha_n)_n$ and $(\beta_n)_n$ are uniquely defined.

Claim 1 – If α_n *is such that for all* $t \in \mathbb{R}$

$$\begin{gathered} \alpha_n''(t) + c\alpha_n'(t) \ge f(t,\alpha_n(t)), \\ \alpha_n(t) \le \beta(t), \end{gathered}$$

then the function $\alpha_{n+1} \in \mathcal{BC}(\mathbb{R})$ *defined by*

$$\alpha_{n+1}'' + c\alpha_{n+1}' - M\alpha_{n+1} = f(t,\alpha_n) - M\alpha_n,$$

satisfies for all $t \in \mathbb{R}$,

$$\begin{gathered} \alpha_{n+1}''(t) + c\alpha_{n+1}'(t) \ge f(t,\alpha_{n+1}(t)), \\ \alpha_n(t) \le \alpha_{n+1}(t) \le \beta(t). \end{gathered}$$

It is enough to observe that α_n and β satisfy for all $t \in \mathbb{R}$

$$\begin{gathered} \alpha_n''(t) + c\alpha_n'(t) - M\alpha_n(t) \ge f(t,\alpha_n(t)) - M\alpha_n(t), \\ \beta''(t) + c\beta'(t) - M\beta(t) \le f(t,\beta(t)) - M\beta(t) \le f(t,\alpha_n(t)) - M\alpha_n(t). \end{gathered}$$

Hence, by Theorem II-5.6, the solution α_{n+1} of

$$u'' + cu' - Mu = f(t,\alpha_n) - M\alpha_n$$

is such that $\alpha_n \leq \alpha_{n+1} \leq \beta$ on $\mathbb{R}$ and then satisfies also for all $t \in \mathbb{R}$

$$\begin{aligned}\alpha''_{n+1}(t) + c\alpha'_{n+1}(t) - M\alpha_{n+1}(t) &= f(t, \alpha_n(t)) - M\alpha_n(t) \\ &\geq f(t, \alpha_{n+1}(t)) - M\alpha_{n+1}(t).\end{aligned}$$

Claim 2 – If β_n is such that for all $t \in \mathbb{R}$

$$\begin{gathered}\beta''_n(t) + c\beta'_n(t) \leq f(t, \beta_n(t)), \\ \alpha(t) \leq \beta_n(t)\end{gathered}$$

then the function β_{n+1} defined by

$$\beta''_{n+1} + c\beta'_{n+1} - M\beta_{n+1} = f(t, \beta_n) - M\beta_n,$$

satisfies for all $t \in \mathbb{R}$,

$$\begin{gathered}\beta''_{n+1}(t) + c\beta'_{n+1}(t) \leq f(t, \beta_{n+1}(t)), \\ \alpha(t) \leq \beta_{n+1}(t) \leq \beta_n(t).\end{gathered}$$

The proof of this claim is similar to the proof of Claim 1.

Conclusion – For every bounded interval $I \subset \mathbb{R}$, we deduce from Proposition I-4.4, the existence of K such that for all n, $\|\alpha_n\|_{C^1(I)} \leq K$ and $\|\beta_n\|_{C^1(I)} \leq K$. Hence, we deduce from the Arzelà-Ascoli Theorem the convergence to solutions u_{min} and u_{max}. As usually if u is a solution such that $\alpha \leq u \leq \beta$, we can take u as a lower solution and prove that $u_{max} = \lim_{n\to\infty} \beta_n \geq u$. Similarly we have $u_{min} \leq u$. □

3 Lower and Upper Solutions in Reversed Order

The monotone approximation method can be used in case the lower and upper solutions are in the reversed order $\beta \leq \alpha$. This method works for any boundary value problem such that a uniform anti-maximum principle holds. This is the case for the periodic and the Neumann problems. We can also work bounded solutions on $\mathbb{R}$. However the method does not apply to the Dirichlet problem since, in this case, we only have a non-uniform anti-maximum principle.

3.1 The Periodic Problem

Consider the periodic boundary value problem

$$\begin{gathered}u'' = f(t, u), \\ u(a) = u(b), \; u'(a) = u'(b),\end{gathered} \tag{3.1}$$

where f is a continuous function.

In Section 2, we have built an approximation scheme for solutions of (3.1) based on the maximum principle. Here, we consider a similar approach based on the anti-maximum principle. Given continuous functions α and β, and $M > 0$, we consider the sequences $(\alpha_n)_n$ and $(\beta_n)_n$ defined by

$$\begin{gathered} \alpha_0 = \alpha, \\ \alpha_n'' + M\alpha_n = f(t, \alpha_{n-1}) + M\alpha_{n-1}, \\ \alpha_n(a) = \alpha_n(b),\ \alpha_n'(a) = \alpha_n'(b) \end{gathered} \tag{3.2}$$

and

$$\begin{gathered} \beta_0 = \beta, \\ \beta_n'' + M\beta_n = f(t, \beta_{n-1}) + M\beta_{n-1}, \\ \beta_n(a) = \beta_n(b),\ \beta_n'(a) = \beta_n'(b). \end{gathered} \tag{3.3}$$

If M is not an eigenvalue of the periodic problem, i.e. $M \neq (\frac{2n\pi}{b-a})^2$ with $n \in \mathbb{N}$, the functions α_n and β_n, solutions of (3.2) and (3.3) can be written explicitly

$$\begin{aligned} \alpha_n(t) &= \int_a^b G(t,s)(f(s, \alpha_{n-1}(s)) + M\alpha_{n-1}(s))\, ds, \\ \beta_n(t) &= \int_a^b G(t,s)(f(s, \beta_{n-1}(s)) + M\beta_{n-1}(s))\, ds, \end{aligned}$$

where $G(t,s)$ is the Green function of the problem

$$\begin{gathered} u'' + Mu = f(t), \\ u(a) = u(b),\ u'(a) = u'(b). \end{gathered} \tag{3.4}$$

The next theorem indicates a framework to obtain convergence of the α_n and β_n to extremal solutions of (3.1).

Theorem 3.1. *Let α and $\beta \in C^2([a,b])$, $\beta \le \alpha$ and*

$$E := \{(t,u) \in [a,b] \times \mathbb{R} \mid \beta(t) \le u \le \alpha(t)\}. \tag{3.5}$$

Assume $f : E \to \mathbb{R}$ is a continuous function, there exists $M \in]0, (\frac{\pi}{b-a})^2]$ such that for all (t,u_1), $(t,u_2) \in E$,

$$u_1 \le u_2 \text{ implies } f(t,u_2) - f(t,u_1) \ge -M(u_2 - u_1)$$

and for all $t \in [a,b]$

$$\begin{aligned} \alpha''(t) &\ge f(t, \alpha(t)), \quad \alpha(a) = \alpha(b),\ \alpha'(a) \ge \alpha'(b), \\ \beta''(t) &\le f(t, \beta(t)), \quad \beta(a) = \beta(b),\ \beta'(a) \le \beta'(b). \end{aligned}$$

Then the sequences $(\alpha_n)_n$ *and* $(\beta_n)_n$ *defined by* (3.2) *and* (3.3) *converge monotonically in* $\mathcal{C}^1([a,b])$ *to solutions* u_{max} *and* u_{min} *of* (3.1) *such that*

$$\beta \leq u_{min} \leq u_{max} \leq \alpha.$$

Further, any solution u *of* (3.1) *with graph in* E *verifies*

$$u_{min} \leq u \leq u_{max}.$$

Proof. Let $X = \mathcal{C}^1([a,b])$, $Z = \mathcal{C}([a,b])$, $K = \{u \in Z \mid u(t) \geq 0 \text{ on } [a,b]\}$ be the ordered cone in Z and

$$\mathcal{E} = \{u \in X \mid \beta \leq u \leq \alpha\}. \tag{3.6}$$

The operator $T : \mathcal{E} \to X$, defined by

$$Tu(t) = \int_a^b G(t,s)(f(s,u(s)) + Mu(s))\, ds,$$

where $G(t,s)$ is the Green function of (3.4), is continuous in X and monotone increasing (see Corollary A-6.3). Further, $T(\mathcal{E})$ is relatively compact in X, $\beta \leq T\beta$ and $\alpha \geq T\alpha$. The proof follows now from Theorem 1.3, where α and β have to be interchanged. □

Next, we consider the derivative dependent problem

$$\begin{gathered} u'' = f(t,u,u'), \\ u(a) = u(b), \ u'(a) = u'(b). \end{gathered} \tag{3.7}$$

As above, given α, $\beta \in \mathcal{C}^1([a,b])$ and $L > 0$, we consider the approximation schemes

$$\begin{gathered} \alpha_0 = \alpha, \\ \alpha_n'' + L\alpha_n = f(t,\alpha_{n-1},\alpha_{n-1}') + L\alpha_{n-1}, \\ \alpha_n(a) = \alpha_n(b), \ \alpha_n'(a) = \alpha_n'(b) \end{gathered} \tag{3.8}$$

and

$$\begin{gathered} \beta_0 = \beta, \\ \beta_n'' + L\beta_n = f(t,\beta_{n-1},\beta_{n-1}') + L\beta_{n-1}, \\ \beta_n(a) = \beta_n(b), \ \beta_n'(a) = \beta_n'(b). \end{gathered} \tag{3.9}$$

The following result paraphrases Theorem 2.2 in a case where the anti-maximum principle applies.

Theorem 3.2. *Let* α *and* $\beta \in \mathcal{C}^2([a,b])$, $\beta \leq \alpha$ *and*

$$E := \{(t,u,v) \in [a,b] \times \mathbb{R}^2 \mid \beta(t) \leq u \leq \alpha(t)\}. \tag{3.10}$$

Assume $f : E \to \mathbb{R}$ is a continuous function, there exists $M \in]0, (\frac{\pi}{b-a})^2]$ such that for all (t, u_1, v), $(t, u_2, v) \in E$,

$$u_1 \leq u_2 \quad \text{implies} \quad f(t, u_2, v) - f(t, u_1, v) \geq -M(u_2 - u_1);$$

there exists $N \geq 0$ such that for all (t, u, v_1), $(t, u, v_2) \in E$,

$$|f(t, u, v_2) - f(t, u, v_1)| \leq N|v_2 - v_1|;$$

and for all $t \in [a, b]$

$$\begin{array}{ll} \alpha''(t) \geq f(t, \alpha(t), \alpha'(t)), & \alpha(a) = \alpha(b), \ \alpha'(a) \geq \alpha'(b), \\ \beta''(t) \leq f(t, \beta(t), \beta'(t)), & \beta(a) = \beta(b), \ \beta'(a) \leq \beta'(b). \end{array}$$

At last, let $L \in [M, (\frac{\pi}{b-a})^2]$ be such that

$$(L - M) \cos \sqrt{L}(\tfrac{b-a}{2}) - N\sqrt{L} \sin \sqrt{L}(\tfrac{b-a}{2}) \geq 0 \tag{3.11}$$

and

$$f(t, \alpha(t), \alpha'(t)) - f(t, \beta(t), \beta'(t)) + L(\alpha(t) - \beta(t)) \geq 0.$$

Then, the sequences $(\alpha_n)_n$ and $(\beta_n)_n$ defined by (3.8) and (3.9) converge monotonically in $C^1([a, b])$ to solutions u and v of (3.7) such that

$$\beta \leq v \leq u \leq \alpha.$$

Remark In case f is independent of u', we have $N = 0$ and we can choose $L = M$ so that Theorem 3.2 reduces to Theorem 3.1.

Proof. The proof uses Theorems 1.1 and 1.2 with $X = C^1([a, b])$, $Z = C([a, b])$ and $K = \{u \in Z \mid u(t) \geq 0 \text{ on } [a, b]\}$ as the ordered cone in Z. Let $\mathcal{E}$ be defined from (3.6). The operator $T : \mathcal{E} \to X$, defined by

$$Tu(t) = \int_a^b G(t, s)(f(s, u(s), u'(s)) + Lu(s)) \, ds,$$

where $G(t, s)$ is the Green function of (3.4) with M replaced by L, is completely continuous in X. With these notations, the approximation schemes (3.8) and (3.9) are equivalent to (1.3) and (1.2).

A : *Claim* – *Let $L \in [M, (\frac{\pi}{b-a})^2]$ satisfy (3.11). Then the functions α_n defined recursively by (3.8) are such that for all $n \in \mathbb{N}$,*
(a) α_n is a lower solution, i.e.

$$\begin{array}{c} \alpha_n''(t) \geq f(t, \alpha_n(t), \alpha_n'(t)), \\ \alpha_n(a) = \alpha_n(b), \ \alpha_n'(a) \geq \alpha_n'(b), \end{array} \tag{3.12}$$

(b) $\alpha_{n+1} \leq \alpha_n$.

The proof is by recurrence.

Initial step : $n = 0$. The condition (3.12) for $n = 0$ is an assumption. Next, $w = \alpha_0 - \alpha_1$ is a solution of

$$\begin{gathered} w'' + Lw = \alpha_0''(t) - f(t, \alpha_0(t), \alpha_0'(t)) \geq 0, \\ w(a) = w(b), \ w'(a) \geq w'(b). \end{gathered}$$

Hence, we deduce (b) from the anti-maximum principle (Corollary A-6.3).

Recurrence step – 1st part : assume (a) and (b) hold for some n and let us prove that

$$\begin{gathered} \alpha_{n+1}''(t) \geq f(t, \alpha_{n+1}(t), \alpha_{n+1}'(t)), \\ \alpha_{n+1}(a) = \alpha_{n+1}(b), \ \alpha_{n+1}'(a) \geq \alpha_{n+1}'(b). \end{gathered}$$

Let $w = \alpha_n - \alpha_{n+1} \geq 0$. We have

$$\begin{aligned} -\alpha_{n+1}'' &+ f(t, \alpha_{n+1}, \alpha_{n+1}') \\ &= -f(t, \alpha_n, \alpha_n') + f(t, \alpha_{n+1}, \alpha_{n+1}') - L(\alpha_n - \alpha_{n+1}) \\ &\leq M(\alpha_n - \alpha_{n+1}) + N|\alpha_{n+1}' - \alpha_n'| - L(\alpha_n - \alpha_{n+1}) \\ &= (M - L)w + N|w'|. \end{aligned}$$

On the other hand, w satisfies

$$w'' + Lw = h(t), \quad w(a) = w(b), \ w'(a) - w'(b) = C, \tag{3.13}$$

with $h(t) := \alpha_n''(t) - f(t, \alpha_n(t), \alpha_n'(t)) \geq 0$ and $C \geq 0$. Observe that

$$\begin{aligned} w(t) = &\frac{1}{2\sqrt{L}\sin\sqrt{L}(\frac{b-a}{2})}\Big[C\cos\sqrt{L}(\tfrac{a+b}{2} - t) \\ &+ \int_a^t h(s)\cos\sqrt{L}(\tfrac{b-a}{2} + s - t)\,ds + \int_t^b h(s)\cos\sqrt{L}(\tfrac{b-a}{2} + t - s)\,ds\Big]. \end{aligned}$$

Hence, using (3.11) and denoting $D = 2\sqrt{L}\sin\sqrt{L}(\frac{b-a}{2})$, we compute

$$\begin{aligned} (M &-L)w(t) + N|w'(t)| \\ &\leq \frac{1}{D}\Big[C\,[(M - L)\cos\sqrt{L}(\tfrac{a+b}{2} - t) + N\sqrt{L}|\sin\sqrt{L}(\tfrac{a+b}{2} - t)|] \\ &\quad + \int_a^t h(s)[(M - L)\cos\sqrt{L}(\tfrac{b-a}{2} + s - t) \\ &\qquad\qquad + N\sqrt{L}|\sin\sqrt{L}(\tfrac{b-a}{2} + s - t)|]\,ds \\ &\quad + \int_t^b h(s)[(M - L)\cos\sqrt{L}(\tfrac{b-a}{2} + t - s) \\ &\qquad\qquad + N\sqrt{L}|\sin\sqrt{L}(\tfrac{b-a}{2} + t - s)|]\,ds\Big] \\ &\leq 0. \end{aligned}$$

Hence α_{n+1} is a lower solution.

Recurrence step – 2d part : assume (a) and (b) hold for some n and let us prove that $\alpha_{n+2} \le \alpha_{n+1}$. The function $w = \alpha_{n+1} - \alpha_{n+2}$ satisfies (3.13), where

$$h(t) := \alpha''_{n+1}(t) - f(t, \alpha_{n+1}(t), \alpha'_{n+1}(t)) \quad \text{and} \quad C = 0.$$

From the previous step $h(t) \ge 0$ and the claim follows from the anti-maximum principle (Corollary A-6.3).

B : *Claim – Let* $L \in [M, (\frac{\pi}{b-a})^2]$ *satisfy* (3.11). *Then the functions* β_n *defined recursively by* (3.9) *are such that for all* $n \in \mathbb{N}$,
(a) β_n *is an upper solution, i.e.*

$$\begin{gathered}\beta''_n(t) \le f(t, \beta_n(t), \beta'_n(t)),\\ \beta_n(a) = \beta_n(b),\ \beta'_n(a) \le \beta'_n(b),\end{gathered}$$

(b) $\beta_{n+1} \ge \beta_n$.
The proof of this claim parallels the proof of Claim A.

C : *Claim –* $\alpha_n \ge \beta_n$. Define, for all $i \in \mathbb{N}$, $w_i = \alpha_i - \beta_i$ and

$$h_i(t) := f(t, \alpha_i(t), \alpha'_i(t)) - f(t, \beta_i(t), \beta'_i(t)) + L(\alpha_i(t) - \beta_i(t)).$$

The proof of the claim is by recurrence.
Initial step : $\alpha_1 \ge \beta_1$. The function w_1 is a solution of (3.13) with $h = h_0 \ge 0$ and $C = 0$. Using the anti-maximum principle (Corollary A-6.3), we deduce that $w_1 \ge 0$, i.e. $\alpha_1 \ge \beta_1$.
Recurrence step : Let $n \ge 2$. *If* $h_{n-2} \ge 0$ *and* $\alpha_{n-1} \ge \beta_{n-1}$, *then* $h_{n-1} \ge 0$ *and* $\alpha_n \ge \beta_n$. First, let us prove that for all $t \in [a, b]$, the function h_{n-1} is nonnegative. Indeed, we have

$$\begin{aligned}h_{n-1} &= f(t, \alpha_{n-1}, \alpha'_{n-1}) - f(t, \beta_{n-1}, \beta'_{n-1}) + L(\alpha_{n-1} - \beta_{n-1})\\ &\ge -M(\alpha_{n-1} - \beta_{n-1}) - N|\alpha'_{n-1} - \beta'_{n-1}| + L(\alpha_{n-1} - \beta_{n-1})\\ &= (L - M)w_{n-1} - N|w'_{n-1}|.\end{aligned}$$

Recall that w_{n-1} is a solution of (3.13) with $h(t) = h_{n-2}(t) \ge 0$ and $C = 0$. Hence, we can proceed as in the proof of Claim A to show that $h_{n-1} \ge 0$. It follows then from the anti-maximum principle (Corollary A-6.3) that w_n is nonnegative, i.e. $\alpha_n \ge \beta_n$.

D : *Claim – There exists* $R > 0$ *such that any solution* u *of*

$$u'' \ge f(t, u, u'), \quad u(a) = u(b),\ u'(a) = u'(b),$$

with $\beta \le u \le \alpha$ *satisfies* $\|u'\|_\infty < R$. We deduce from the assumptions that

$$u'' = f(t, u, u') + h(t),$$

where $h(t) \geq 0$, $f(t,u,u') + h(t) \geq -\max_F |f(t,u,0)| - N|u'|$ and $F = \{(t,u) \mid t \in [a,b], \beta(t) \leq u \leq \alpha(t)\}$. The proof follows now using Proposition I-4.5.

E : *Claim* – *There exists $R > 0$ such that any solution u of*

$$u'' \leq f(t,u,u'), \quad u(a) = u(b),\ u'(a) = u'(b),$$

with $\beta \leq u \leq \alpha$ satisfies $\|u'\|_\infty < R$. The proof repeats the argument of Claim D.

F : *Conclusion* – We deduce from Theorems 1.2 and 1.1, where α and β have to be interchanged, that the sequences $(\alpha_n)_n$ and $(\beta_n)_n$ converge monotonically in $C^1([a,b])$ to functions u and v such that $\beta \leq v \leq \alpha$ and $\beta \leq u \leq \alpha$. Further Claim C implies $v \leq u$. □

Example 3.1. Consider the problem

$$\begin{aligned} &u'' - k \arctan u' + cu^3 = \sin t, \\ &u(0) = u(2\pi),\ u'(0) = u'(2\pi), \end{aligned}$$

where $c > 0$ and $k > 0$. This problem has $\alpha = c^{-1/3}$ and $\beta = -c^{-1/3}$ as lower and upper solutions. To satisfy the assumptions of Theorem 3.2, let $M = 3c^{1/3}$ and $N = k$, choose $L \in]3c^{1/3}, 1/4[$ and assume

$$c < 12^{-3} \quad \text{and} \quad k \leq \tfrac{L-3c^{1/3}}{\sqrt{L}} \operatorname{cotan} \sqrt{L}\pi.$$

As in Section 2, we can modify the analysis in case we agree to compute the approximations α_n and β_n from piecewise linear problems. The following result adapts Theorem 2.4 to lower and upper solutions in reversed order.

Theorem 3.3. *Let α and $\beta \in C^2([a,b])$, $\beta \leq \alpha$ and let E be defined by (3.10). Assume $f : E \to \mathbb{R}$ is a continuous function, there exists $M > 0$ such that for all (t,u_1,v), $(t,u_2,v) \in E$,*

$$u_1 \leq u_2 \quad \textit{implies} \quad f(t,u_2,v) - f(t,u_1,v) \geq -M(u_2 - u_1),$$

there exists $N \geq 0$ such that for all (t,u,v_1), $(t,u,v_2) \in E$,

$$|f(t,u,v_2) - f(t,u,v_1)| \leq N|v_2 - v_1|$$

and for all $t \in [a,b]$

$$\begin{aligned} &\alpha''(t) \geq f(t,\alpha(t),\alpha'(t)), \quad \alpha(a) = \alpha(b),\ \alpha'(a) \geq \alpha'(b), \\ &\beta''(t) \leq f(t,\beta(t),\beta'(t)), \quad \beta(a) = \beta(b),\ \beta'(a) \leq \beta'(b). \end{aligned}$$

Assume also $b - a \leq 2\theta(M, N/2)$, *where*

$$\theta(M,\tilde{N}) = \begin{cases} \frac{1}{\sqrt{\tilde{N}^2 - M}} \operatorname{arctanh} \frac{\sqrt{\tilde{N}^2 - M}}{\tilde{N}}, & \text{if } 0 < M < \tilde{N}^2, \\ \frac{1}{\tilde{N}}, & \text{if } M = \tilde{N}^2, \\ \frac{1}{\sqrt{M - \tilde{N}^2}} \left(\frac{\pi}{2} - \arctan \frac{\tilde{N}}{\sqrt{M - \tilde{N}^2}} \right), & \text{if } 0 \leq \tilde{N}^2 < M. \end{cases} \tag{3.14}$$

Let $\alpha_0 = \alpha$ *and* $\beta_0 = \beta$. *Then the problems*

$$\begin{aligned} \alpha_n'' - N|\alpha_n' - \alpha_{n-1}'| + M\alpha_n = f(t, \alpha_{n-1}, \alpha_{n-1}') + M\alpha_{n-1}, \\ \alpha_n(a) = \alpha_n(b),\ \alpha_n'(a) = \alpha_n'(b) \end{aligned}$$

and

$$\begin{aligned} \beta_n'' + N|\beta_n' - \beta_{n-1}'| + M\beta_n = f(t, \beta_{n-1}, \beta_{n-1}') + M\beta_{n-1}, \\ \beta_n(a) = \beta_n(b),\ \beta_n'(a) = \beta_n'(b), \end{aligned}$$

define sequences $(\alpha_n)_n$ *and* $(\beta_n)_n$ *that converge monotonically in* $C^1([a,b])$ *to solutions* u_{max} *and* u_{min} *of* (3.7) *such that*

$$\beta \leq u_{min} \leq u_{max} \leq \alpha.$$

Further, any solution u *of* (3.7), *such that* $\beta \leq u \leq \alpha$, *verifies*

$$u_{min} \leq u \leq u_{max}.$$

Proof. Let $X = C^1([a,b])$, $Z = C([a,b])$, $K = \{u \in Z \mid u(t) \geq 0 \text{ on } [a,b]\}$ be the ordered cone in Z and $\mathcal{E}$ be defined from (1.1).

A : *Definition of a completely continuous operator* T. *Claim 1 – For any* $u \in X$ *the problem*

$$\begin{aligned} v'' - N|v' - u'(t)| + Mv = f(t, u(t), u'(t)) + Mu(t), \\ v(a) = v(b),\ v'(a) = v'(b), \end{aligned} \tag{3.15}$$

has a solution v. Consider the problem

$$v'' - s|v' - u'(t)| + Mv = \sigma(t), \quad v(a) = v(b),\ v'(a) = v'(b), \tag{3.16}$$

with $s \in [0, N]$ as an homotopy parameter and $\sigma \in C([a,b])$. Assume that solutions of (3.16) are not *a priori* bounded in X. Hence, there exists sequences $(s_n)_n \subset [0, N]$ and $(v_n)_n \subset X$ such that v_n solves (3.16) with $s = s_n$ and $\lim_{n\to\infty} \|v_n\|_{C^1} = +\infty$. We define then $\overline{v}_n = v_n/\|v_n\|_{C^1}$ and going to subsequences we can assume

$$\overline{v}_n \xrightarrow{C^1} \overline{v}, \quad \overline{v}_n'' \xrightarrow{L^2} v'' \quad \text{and} \quad s_n \to s.$$

Going to the limit in (3.16), we obtain that $\overline{v}$ solves

$$\overline{v}'' - s|\overline{v}'| + M\overline{v} = 0, \quad \overline{v}(a) = \overline{v}(b), \ \overline{v}'(a) = \overline{v}'(b).$$

If $M \le s^2/4$, Theorem A-6.2 implies $\bar{v} = 0$. On the other hand, if $M > s^2/4$, we have $b - a \le 2\theta(M, N/2) \le 2\theta(M, s/2) \le 2\chi(M, -s/2)$ (see (A-6.3) and (A-6.4)) and the conclusion $\overline{v} = 0$ follows from the same theorem. Hence, in all cases, this contradicts $\lim_{C^1} \overline{v}_n = \overline{v}$. The proof of the claim follows now from the *a priori* bounds and classical arguments in degree theory.

Claim 2 – The solution of problem (3.15) *is unique* Let v_1 and v_2 be two solutions of (3.15). Define $w = v_2 - v_1$ and compute

$$w'' + N|w'| + Mw = N(|v_1' - v_2'| + |v_2' - u'| - |v_1' - u'|) \ge 0.$$

Hence, we deduce from the anti-maximum principle (Theorem A-6.2) that $w \ge 0$. Similarly we prove $w \le 0$, which implies $w = 0$, and the solution of (3.15) is unique.

Now, we can define the operator

$$T : \mathcal{E} \to X, u \mapsto Tu,$$

where $\mathcal{E}$ is defined from (1.1) and Tu is the solution of (3.15). This operator is completely continuous.

B : *Claim – For all* $n \in \mathbb{N}$, $\alpha_n \ge \alpha_{n+1}$.

Initial step : $w = \alpha_0 - \alpha_1 = \alpha - T\alpha \ge 0$. Notice that w solves the problem

$$w'' + N|w'| + Mw = \alpha''(t) - f(t, \alpha(t), \alpha'(t)) \ge 0,$$
$$w(a) - w(b) = 0, \ w'(a) - w'(b) = \alpha'(a) - \alpha'(b) \ge 0.$$

The claim follows from the anti-maximum principle (Theorem A-6.2).

Recurrence step – Assume $\alpha_n - \alpha_{n+1} \ge 0$ *and prove* $w = \alpha_{n+1} - \alpha_{n+2} \ge 0$. The function w satisfies

$$\begin{aligned} w'' &+ N|w'| + Mw \\ &= f(t, \alpha_n, \alpha_n') - f(t, \alpha_{n+1}, \alpha_{n+1}') + N|\alpha_{n+1}' - \alpha_n'| + M(\alpha_n - \alpha_{n+1}) \\ &\ge 0 \end{aligned}$$

and the claim follows from the anti-maximum principle (Theorem A-6.2).

C : *Claim –* $\alpha_n \ge \beta$. The proof repeats the argument of the previous claims.

D : *Claim – The sequence* $(\alpha_n)_n$ *is bounded in* X. The functions α_n, $n \geq 1$, solve the problems

$$u'' + N|u'| + Mu = g_n(t), \quad u(a) = u(b), \; u'(a) = u'(b),$$

where

$$\begin{aligned} g_n = & \, N(|\alpha_n'| + |\alpha_n' - \alpha_{n-1}'| - |\alpha_{n-1}'|) \\ & + N|\alpha_{n-1}'| + f(t, \alpha_{n-1}, \alpha_{n-1}') - f(t, \alpha_{n-1}, 0) \\ & + f(t, \alpha_{n-1}, 0) + M\alpha_{n-1} \\ \geq & \, f(t, \alpha_{n-1}, 0) + M\alpha_{n-1} \geq f(t, \beta, 0) + M\beta. \end{aligned}$$

As the g_n are uniformly lower bounded, the boundedness of the α_n follows now from Proposition I-4.5.

E : *Conclusion* – The convergence of the sequence $(\alpha_n)_n$ to a solution u_{max} of (3.7) follows from Theorem 1.2, where α and β have to be interchanged. To prove u_{max} is the maximum solution, let u be any solution of (3.7) such that $\beta \leq u \leq \alpha$. We apply then Theorem 1.2 with the α and β of this theorem replaced respectively by u and α and obtain $u_{max} = \lim_{n\to\infty} \alpha_n \geq u$.

Similarly, we prove the convergence of the sequence $(\beta_n)_n$ to a mimimal solution using Theorem 1.1. □

3.2 The Neumann Problem

The Neumann problem

$$\begin{gathered} u'' = f(t, u), \\ u'(a) = 0, \; u'(b) = 0, \end{gathered} \tag{3.17}$$

where f is a continuous function, can be investigated along the lines of the periodic problem.

Theorem 3.4. *Let* α *and* $\beta \in C^2([a, b])$, $\beta \leq \alpha$ *and let* E *be defined from* (3.5). *Assume* $f : E \to \mathbb{R}$ *is a continuous function, there exists* $M \in]0, (\frac{\pi}{2(b-a)})^2]$ *such that for all* (t, u_1), $(t, u_2) \in E$,

$$u_1 \leq u_2 \text{ implies } f(t, u_2) - f(t, u_1) \geq -M(u_2 - u_1)$$

and for all $t \in [a, b]$

$$\begin{aligned} \alpha''(t) \geq f(t, \alpha(t)), \quad \alpha'(a) \geq 0, \; \alpha'(b) \leq 0, \\ \beta''(t) \leq f(t, \beta(t)), \quad \beta'(a) \leq 0, \; \beta'(b) \geq 0. \end{aligned}$$

Then the sequences $(\alpha_n)_n$ *and* $(\beta_n)_n$ *defined by*

$$\alpha_0 = \alpha,$$

$$\begin{gathered} \alpha_n'' + M\alpha_n = f(t, \alpha_{n-1}) + M\alpha_{n-1}, \\ \alpha_n'(a) = 0, \; \alpha_n'(b) = 0, \end{gathered}$$

and

$$\beta_0 = \beta,$$
$$\beta_n'' + M\beta_n = f(t, \beta_{n-1}) + M\beta_{n-1},$$
$$\beta_n'(a) = 0, \ \beta_n'(b) = 0,$$

converge monotonically in $C^1([a,b])$ *to solutions* u_{max} *and* u_{min} *of* (3.17) *such that*

$$\beta \leq u_{min} \leq u_{max} \leq \alpha.$$

Further, any solution u *of* (3.17) *with graph in* E *verifies*

$$u_{min} \leq u \leq u_{max}.$$

Exercise 3.1. Deduce the above theorem from Theorem 1.3.

To deal with derivative dependent equations

$$\begin{aligned} u'' &= f(t, u, u'), \\ u'(a) &= 0, \ u'(b) = 0, \end{aligned} \tag{3.18}$$

we can use the ideas of Theorem 3.2.

Theorem 3.5. *Let* α *and* $\beta \in C^2([a,b])$, $\beta \leq \alpha$ *and let* E *be defined from* (3.10). *Assume* $f : E \to \mathbb{R}$ *is a continuous function, there exists* $M \in]0, (\frac{\pi}{2(b-a)})^2]$ *such that for all* (t, u_1, v), $(t, u_2, v) \in E$,

$$u_1 \leq u_2 \quad \textit{implies} \quad f(t, u_2, v) - f(t, u_1, v) \geq -M(u_2 - u_1);$$

there exists $N \geq 0$ *such that for all* (t, u, v_1), $(t, u, v_2) \in E$,

$$|f(t, u, v_2) - f(t, u, v_1)| \leq N|v_2 - v_1|;$$

and for all $t \in [a,b]$

$$\begin{aligned} \alpha''(t) &\geq f(t, \alpha(t), \alpha'(t)), \quad \alpha'(a) \geq 0, \ \alpha'(b) \leq 0, \\ \beta''(t) &\leq f(t, \beta(t), \beta'(t)), \quad \beta'(a) \leq 0, \ \beta'(b) \geq 0. \end{aligned}$$

At last, let $L \in [M, (\frac{\pi}{2(b-a)})^2]$ *be such that*

$$(L - M)\cos\sqrt{L}(b-a) - N\sqrt{L}\sin\sqrt{L}(b-a) \geq 0$$

and

$$f(t, \alpha(t), \alpha'(t)) - f(t, \beta(t), \beta'(t)) + L(\alpha(t) - \beta(t)) \geq 0.$$

Then, the sequences $(\alpha_n)_n$ *and* $(\beta_n)_n$ *defined by*

$$\alpha_0 = \alpha,$$

$$\alpha_n'' + L\alpha_n = f(t, \alpha_{n-1}, \alpha_{n-1}') + L\alpha_{n-1},$$
$$\alpha_n'(a) = 0, \ \alpha_n'(b) = 0,$$

and

$$\beta_0 = \beta,$$

$$\beta_n'' + L\beta_n = f(t, \beta_{n-1}, \beta_{n-1}') + L\beta_{n-1},$$
$$\beta_n'(a) = 0, \ \beta_n'(b) = 0,$$

converge monotonically in $C^1([a,b])$ *to solutions* u *and* v *of* (3.18) *such that*

$$\beta \leq v \leq u \leq \alpha.$$

Exercise 3.2. Prove the above theorem along the lines of the proof of Theorem 3.2.

Hint : See [65].

We can investigate problem (3.18) using an analog of Theorem 3.3.

Theorem 3.6. *Let* α *and* $\beta \in C^2([a,b])$, $\beta \leq \alpha$ *and let* E *be defined from* (3.10). *Assume* $f : E \to \mathbb{R}$ *is a continuous function, there exists* $M > 0$ *such that for all* (t, u_1, v), $(t, u_2, v) \in E$,

$$u_1 \leq u_2 \quad \textit{implies} \quad f(t, u_2, v) - f(t, u_1, v) \geq -M(u_2 - u_1);$$

there exists $N \geq 0$ *such that for all* (t, u, v_1), $(t, u, v_2) \in E$,

$$|f(t, u, v_2) - f(t, u, v_1)| \leq N|v_2 - v_1|;$$

and for all $t \in [a,b]$

$$\alpha''(t) \geq f(t, \alpha(t), \alpha'(t)), \quad \alpha'(a) \geq 0, \ \alpha'(b) \leq 0,$$
$$\beta''(t) \leq f(t, \beta(t), \beta'(t)), \quad \beta'(a) \leq 0, \ \beta'(b) \geq 0.$$

Assume also $b - a \leq \theta(M, N/2)$, *where* θ *is defined from* (3.14).

Let $\alpha_0 = \alpha$ *and* $\beta_0 = \beta$. *Then the problems*

$$\alpha_n'' - N|\alpha_n' - \alpha_{n-1}'| + M\alpha_n = f(t, \alpha_{n-1}, \alpha_{n-1}') + M\alpha_{n-1},$$
$$\alpha_n'(a) = 0, \ \alpha_n'(b) = 0,$$

and

$$\beta_n'' + N|\beta_n' - \beta_{n-1}'| + M\beta_n = f(t, \beta_{n-1}, \beta_{n-1}') + M\beta_{n-1},$$
$$\beta_n'(a) = 0, \ \beta_n'(b) = 0,$$

define sequences $(\alpha_n)_n$ *and* $(\beta_n)_n$ *that converge monotonically in* $C^1([a,b])$ *to solutions* u_{max} *and* u_{min} *of* (3.18) *such that*

$$\beta \leq u_{min} \leq u_{max} \leq \alpha.$$

Further, any solution u of (3.18), *such that* $\beta \le u \le \alpha$ *verifies*

$$u_{min} \le u \le u_{max}.$$

Exercise 3.3. Prove the above theorem along the lines of the proof of Theorem 3.3.

Hint : See [51].

3.3 Bounded Solutions

Existence of bounded solutions of (2.22) can be deduced from lower and upper solutions in the reversed order. The following result parallels Theorem 2.10. Here we use $\mathcal{B}C^2(\mathbb{R}) = \{u \in C^2(\mathbb{R}) \mid u, u', u'' \in L^\infty(\mathbb{R})\}$.

Theorem 3.7. *Let α and $\beta \in \mathcal{B}C^2(\mathbb{R})$, $\alpha \ge \beta$ and let $E := \{(t,u) \in \mathbb{R}^2 \mid \beta(t) \le u \le \alpha(t)\}$. Assume $c > 0$, $f : E \to \mathbb{R}$ is a continuous bounded function such that for all (t, u_1), $(t, u_2) \in E$,*

$$u_1 \le u_2 \Rightarrow f(t,u_2) - f(t,u_1) \ge -\tfrac{c^2}{4}(u_2 - u_1).$$

Assume further that for all $t \in \mathbb{R}$

$$\begin{aligned} \alpha''(t) + c\alpha'(t) &\ge f(t,\alpha(t)),\\ \beta''(t) + c\beta'(t) &\le f(t,\beta(t)). \end{aligned}$$

Let $\alpha_0 = \alpha$ and $\beta_0 = \beta$. Then the equations

$$\begin{aligned} \alpha''_{n+1} + c\alpha'_{n+1} + \tfrac{c^2}{4}\alpha_{n+1} &= f(t,\alpha_n) + \tfrac{c^2}{4}\alpha_n,\\ \beta''_{n+1} + c\beta'_{n+1} + \tfrac{c^2}{4}\beta_{n+1} &= f(t,\beta_n) + \tfrac{c^2}{4}\beta_n, \end{aligned}$$

define sequences $(\alpha_n)_n$ and $(\beta_n)_n \subset \mathcal{B}C(\mathbb{R})$ that converge monotonically and uniformly on bounded intervals of $\mathbb{R}$ to solutions u_{max} and u_{min} of (2.22) *such that*

$$\beta \le u_{min} \le u_{max} \le \alpha.$$

Further, any solution u of (2.22), *such that* $\beta \le u \le \alpha$ *verifies*

$$u_{min} \le u \le u_{max}.$$

Proof. First observe that given $p \in \mathcal{B}C(\mathbb{R})$, the unique solution in $\mathcal{B}C(\mathbb{R})$ of

$$y'' + cy' + \tfrac{c^2}{4}y = p(t)$$

is given by

$$y(t) = \int_{-\infty}^{+\infty} G(t,s)p(s)\,ds,$$

where

$$G(t,s) = (t-s)\exp(-\tfrac{c}{2}(t-s)), \quad \text{if } s \le t,$$
$$= 0, \qquad\qquad\qquad\qquad \text{if } s > t.$$

Claim – If α_n is such that for all $t \in \mathbb{R}$

$$\alpha_n''(t) + c\alpha_n'(t) \ge f(t, \alpha_n(t)),$$
$$\alpha_n(t) \ge \beta(t)$$

then the function α_{n+1} defined by

$$\alpha_{n+1}'' + c\alpha_{n+1}' + \tfrac{c^2}{4}\alpha_{n+1} = f(t, \alpha_n) + \tfrac{c^2}{4}\alpha_n,$$

satisfies for all $t \in \mathbb{R}$,

$$\alpha_{n+1}''(t) + c\alpha_{n+1}'(t) \ge f(t, \alpha_{n+1}(t)),$$
$$\alpha_n(t) \ge \alpha_{n+1}(t) \ge \beta(t).$$

It is enough to observe that $w = \alpha_n - \alpha_{n+1}$ solves

$$w'' + cw' + \tfrac{c^2}{4}w = \alpha_n'' + c\alpha_n' - f(t, \alpha_n).$$

As $\alpha_n'' + c\alpha_n' - f(t, \alpha_n(t)) \ge 0$ we deduce from the form of the solution that $w \ge 0$, i.e. $\alpha_n \ge \alpha_{n+1}$ and also, for all $t \in \mathbb{R}$,

$$\alpha_{n+1}''(t) + c\alpha_{n+1}'(t) = f(t, \alpha_n(t)) + \tfrac{c^2}{4}(\alpha_n(t) - \alpha_{n+1}(t)) \ge f(t, \alpha_{n+1}(t)).$$

The inequality $\beta \le \alpha_{n+1}$ can be proved in the same way.

The rest of the proof follows as in Theorem 2.10. □

4 A Mixed Approximation Scheme

In the previous section, we considered sequences $(\alpha_n)_n$ and $(\beta_n)_n$ which converge to solutions of a boundary value problem such as (2.1). The reader might have noticed that computing α_n and β_n can be a difficult problem. In this section, we give a method of approximations which is simple but provides only bounds on the solutions. We also give assumptions which imply these bounds to be equal, in which case they are solutions.

Consider a Dirichlet problem

$$u'' = f(t, u, u), \quad u(a) = 0, \; u(b) = 0. \tag{4.1}$$

To simplify the argument, we assume here f is independent of the derivative u', which is not essential and could be developed as in the previous sections.

We shall use the following definition.

Definition 4.1. *Functions α and $\beta \in \mathcal{C}([a,b])$ are* coupled lower and upper quasi-solutions *of* (4.1) *if*

(a) for any $t \in [a,b]$, $\alpha(t) \le \beta(t)$;

(b) for any $t_0 \in]a,b[$, either $D_-\alpha(t_0) < D^+\alpha(t_0)$, or there exists an open interval $I_0 \subset]a,b[$ such that $t_0 \in I_0$, $\alpha \in W^{2,1}(I_0)$, and for a.e. $t \in I_0$

$$\alpha''(t)) \ge f(t, \alpha(t), \beta(t));$$

(c) for any $t_0 \in]a,b[$, either $D^-\beta(t_0) > D_+\beta(t_0)$, or there exists an open interval $I_0 \subset]a,b[$ such that $t_0 \in I_0$, $\beta \in W^{2,1}(I_0)$, and for a.e. $t \in I_0$

$$\beta''(t) \le f(t, \beta(t), \alpha(t));$$

(d) $\alpha(a) \le 0 \le \beta(a)$, $\alpha(b) \le 0 \le \beta(b)$.

Consider the following auxiliary problem

$$\begin{aligned} u'' &= f(t,u,v), \quad u(a) = 0, \ u(b) = 0, \\ v'' &= f(t,v,u), \quad v(a) = 0, \ v(b) = 0. \end{aligned} \tag{4.2}$$

Proposition 4.1. *Let α_0, $\beta_0 \in \mathcal{C}([a,b])$,*

$$E := \{(t,u,v) \mid t \in [a,b], \ u, v \in [\alpha_0(t), \beta_0(t)]\} \tag{4.3}$$

and $f : E \to \mathbb{R}$ be an L^1-Carathéodory function such that $f(t,u,v)$ is non-increasing in u and non-decreasing in v. Assume α_0 and β_0 are coupled lower and upper quasi-solutions of (4.1).

Then, the sequences $(\alpha_n)_n$ and $(\beta_n)_n$, defined for $n \ge 1$ by

$$\begin{aligned} \alpha_n'' &= f(t, \alpha_{n-1}, \beta_{n-1}), \quad \alpha_n(a) = 0, \ \alpha_n(b) = 0, \\ \beta_n'' &= f(t, \beta_{n-1}, \alpha_{n-1}), \quad \beta_n(a) = 0, \ \beta_n(b) = 0, \end{aligned} \tag{4.4}$$

converge monotonically in $\mathcal{C}^1([a,b])$ to functions u_{min} and u_{max}. The pair (u_{min}, u_{max}) is a solution of (4.2) such that

$$\alpha_0 \le u_{min} \le u_{max} \le \beta_0.$$

Moreover, any solution (u,v) of (4.2) with $\alpha_0 \le u \le \beta_0$, $\alpha_0 \le v \le \beta_0$ is such that

$$u_{min} \le u \le u_{max}, \quad u_{min} \le v \le u_{max}.$$

Proof. Let $X = C^1([a,b])\times C^1([a,b])$, $Z = C([a,b])\times C([a,b])$, $K = \{(u,v) \in Z \mid u \geq 0, v \leq 0\}$ and $\mathcal{E} = \{(u,v) \in X \mid \alpha_0 \leq u \leq \beta_0,\ \alpha_0 \leq v \leq \beta_0\}$. We define $T : \mathcal{E} \to X, (u,v) \mapsto T(u,v)$, where $T(u,v)$ is the solution (x,y) of

$$\begin{aligned} x'' &= f(t,u,v), \quad & x(a) = 0,\ x(b) = 0,\\ y'' &= f(t,v,u), \quad & y(a) = 0,\ y(b) = 0 \end{aligned}$$

and verify T is continuous, monotone increasing, $T(\mathcal{E})$ is relatively compact in X and

$$(\alpha_0,\beta_0) \leq T(\alpha_0,\beta_0), \quad (\beta_0,\alpha_0) \geq T(\beta_0,\alpha_0).$$

Theorem 1.3 applies with $\alpha = (\alpha_0,\beta_0)$ and $\beta = (\beta_0,\alpha_0)$, and the claims follow. □

Notice that if u is a solution of the given problem (4.1), then (u,u) is a solution of the auxiliary problem (4.2), thus u_{min} and u_{max} are bounds on solutions of (4.1).

If the bounds u_{min} and u_{max} given in Proposition 4.1 are equal, u_{min} is a solution of the initial problem (4.1), and this theorem provides an approximation scheme to a solution of (4.1). In particular, this will be the case if solutions of the auxiliary problem (4.2) are unique. Unfortunately, these solutions are in general not unique as follows from the problem

$$\begin{aligned} u'' + se^u &= 0, \quad & u(0) = u(1) = 0,\\ v'' + se^v &= 0, \quad & v(0) = v(1) = 0 \end{aligned}$$

and Theorem VIII-3.6, if we choose $s > 0$ small enough. The next proposition proves, under appropriate assumptions, the uniqueness of solutions of the auxiliary problem (4.2). Hence, it also proves convergence of the sequences defined in Proposition 4.1 to the unique solution of the given problem (4.1).

Theorem 4.2. *Suppose the assumptions of Proposition 4.1 hold. Assume moreover*

(i) there exists $\epsilon > 0$ *such that* $\alpha_0 \geq \epsilon\beta_0$*;*
(ii) for every $s \in [\epsilon, 1[$*, almost every* $t \in [a,b]$ *and every* $u,\ v \in [\alpha_0,\beta_0]$ *with* $sv \leq u \leq v$*,*

$$sf(t,\tfrac{u}{s},sv) > f(t,u,v).$$

Then, the functions u_{min} *and* u_{max} *defined in Proposition 4.1 are equal, i.e. are solutions of* (4.1).

Proof. From assumption (i), we deduce

$$\epsilon u_{max} \leq u_{min} \leq u_{max}.$$

Let $s_0 = \sup\{s \mid su_{max} \leq u_{min}\}$. It is obvious that $s_0 \in [\epsilon, 1]$ and that $s_0 u_{max} \leq u_{min}$. From the definition of s_0, we deduce the existence of $t_0 \in [a, b]$ such that

$$u_{min}(t_0) - s_0 u_{max}(t_0) = 0, \quad u'_{min}(t_0) - s_0 u'_{max}(t_0) = 0.$$

If not we have

$$\begin{aligned} u_{min}(t) - s_0 u_{max}(t) > 0, \quad & \text{for } t \in]a, b[, \\ u'_{min}(a) - s_0 u'_{max}(a) > 0, \quad & u'_{min}(b) - s_0 u'_{max}(b) < 0. \end{aligned}$$

Hence there exists $\epsilon > 0$ so that $u_{min}(t) - s_0 u_{max}(t) \geq \epsilon u_{max}(t)$ on $[a, b]$. This contradicts the definition of s_0.

If $t_0 \neq b$, we also have $t_1 > t_0$ such that $u'_{min}(t_1) - s_0 u'_{max}(t_1) \geq 0$. Assume now that $s_0 < 1$. Hence, we can write

$$\begin{aligned} s_0 u''_{max} &= s_0 f(t, u_{max}, u_{min}) \geq s_0 f(t, \tfrac{1}{s_0} u_{min}, s_0 u_{max}) \\ &> f(t, u_{min}, u_{max}) = u''_{min}, \end{aligned}$$

which leads to the contradiction

$$0 \leq (u'_{min} - s_0 u'_{max})\Big|_{t_0}^{t_1} = \int_{t_0}^{t_1} (u''_{min}(t) - s_0 u''_{max}(t))\, dt < 0.$$

A similar argument holds if $t_0 = b$. Hence $s_0 = 1$ and $u_{max} = u_{min}$. □

Conditions of Theorem 4.2 can be checked on any pair of lower and upper quasi-solutions. In some cases, it is useful to use some iterate (α_n, β_n) from the sequence defined in (4.4).

Theorem 4.3. *Let α_0, $\beta_0 \in C([a, b])$ and E be defined from (4.3). Suppose $f : E \to \mathbb{R}$ is an L^1-Carathéodory function such that for some $K > -(\frac{\pi}{b-a})^2$, $f(t, u, v) - Kv$ is non-increasing in u, non-decreasing in v and for almost every t and all u, v with $\alpha_0(t) \leq u < v \leq \beta_0(t)$,*

$$[f(t, u, v) - f(t, v, u)](v - u) < (\tfrac{\pi^2}{(b-a)^2} + 2K)(v - u)^2.$$

Assume α_0 and β_0 are coupled lower and upper quasi-solutions of (4.1).

Then, both sequences $(\alpha_n)_n$ and $(\beta_n)_n$, defined for $n \geq 1$ by

$$\begin{aligned} \alpha''_n - K\alpha_n &= f(t, \alpha_{n-1}, \beta_{n-1}) - K\beta_{n-1}, \ \alpha_n(a) = 0, \ \alpha_n(b) = 0, \\ \beta''_n - K\beta_n &= f(t, \beta_{n-1}, \alpha_{n-1}) - K\alpha_{n-1}, \ \beta_n(a) = 0, \ \beta_n(b) = 0, \end{aligned}$$

converge to the same solution u of (4.1).

Proof. As in Proposition 4.1, we can prove that the sequences $(\alpha_n)_n$ and $(\beta_n)_n$ converge respectively to functions $u = u_{min}$ and $v = u_{max} \geq u$ solutions of

$$\begin{aligned} u'' - Ku &= f(t,u,v) - Kv, \quad u(a) = 0,\ u(b) = 0, \\ v'' - Kv &= f(t,v,u) - Ku, \quad v(a) = 0,\ v(b) = 0. \end{aligned}$$

If $u \neq v$, we compute

$$\begin{aligned} &\int_a^b [(v'' - u'') - K(v-u)](u-v)\,dt \\ &\qquad = \int_a^b [f(t,v,u) - f(t,u,v) - K(u-v)](u-v)\,dt \\ &\qquad < \left(\tfrac{\pi^2}{(b-a)^2} + K\right)\int_a^b (u-v)^2\,dt \end{aligned}$$

and also

$$\begin{aligned} -&\int_a^b [(v'' - u'') - K(v-u)](v-u)\,dt \\ &\qquad = \int_a^b [(v'-u')^2 + K(v-u)^2]\,dt \\ &\qquad \geq \left(\tfrac{\pi^2}{(b-a)^2} + K\right)\int_a^b (v-u)^2\,dt, \end{aligned}$$

which is a contradiction. □

Another result in this direction uses a one-sided Lipschitz condition on the function f.

Theorem 4.4. *Let α_0, $\beta_0 \in \mathcal{C}([a,b])$ and E be defined from (4.3). Assume $g : E \to \mathbb{R}$ is an L^1-Carathéodory function such that $g(t,u,v)$ is non-increasing in u and non-decreasing in v and for some $L > \frac{\pi^2}{(b-a)^2}$, $g(t,u,v) - Lv$ is non-increasing in v. Assume α_0 and β_0 are coupled lower and upper quasi-solutions of (4.1) with*

$$f(t,u,v) = \tfrac{1}{2}[g(t,u,u) + g(t,u,v) - L(u-v)].$$

Then, both sequences $(\alpha_n)_n$ and $(\beta_n)_n$, defined for $n \geq 1$ by

$$\begin{aligned} \alpha_n'' &= f(t,\alpha_{n-1},\beta_{n-1}), \quad \alpha_n(a) = 0,\ \alpha_n(b) = 0, \\ \beta_n'' &= f(t,\beta_{n-1},\alpha_{n-1}), \quad \beta_n(a) = 0,\ \beta_n(b) = 0, \end{aligned}$$

converge to the same solution u of

$$u'' = g(t,u,u), \quad u(a) = u(b) = 0.$$

Proof. By Proposition 4.1, the limit functions

$$u = \lim_{n\to\infty} \alpha_n \quad \text{and} \quad v = \lim_{n\to\infty} \beta_n \geq u$$

exist and are such that

$$u'' = \tfrac{1}{2}[g(t,u,u) + g(t,u,v) - L(u-v)], \quad u(a) = 0,\ u(b) = 0,$$

$$v'' = \tfrac{1}{2}[g(t,v,v) + g(t,v,u) - L(v-u)], \quad v(a) = 0,\ v(b) = 0.$$

Hence, $w = v - u$ is a nonnegative solution of

$$w'' + Lw = h(t), \quad w(a) = w(b) = 0,$$

where $h(t) = \frac{1}{2}[g(t,v,v) - g(t,u,v) + g(t,v,u) - g(t,u,u)] \leq 0$. Such a nontrivial nonnegative solution does not exists if $L > \frac{\pi^2}{(b-a)^2}$ as otherwise we have the contradiction

$$\begin{aligned} 0 &< \left(L - \tfrac{\pi^2}{(b-a)^2}\right) \int_a^b w(s) \sin \tfrac{\pi}{b-a}(s-a)\, ds \\ &= \int_a^b (w''(s) + Lw(s)) \sin \tfrac{\pi}{b-a}(s-a)\, ds \\ &= \int_a^b h(s) \sin \tfrac{\pi}{b-a}(s-a)\, ds \leq 0. \end{aligned}$$

□

Chapter VI

Parametric Multiplicity Problems

1 Periodic Solutions of the Liénard Equation

Consider the periodic problem for a Liénard equation

$$\begin{aligned} u'' + g(u)u' + f(t,u) = s, \\ u(a) = u(b),\ u'(a) = u'(b), \end{aligned} \tag{1.1}$$

where $a < b$, $s \in \mathbb{R}$, $g : \mathbb{R} \to \mathbb{R}$ is continuous and $f : [a,b] \times \mathbb{R} \to \mathbb{R}$ is L^1-Carathéodory. This chapter is concerned with the Ambrosetti-Prodi problem. This reduces here to study the dependence of the number of solutions on the parameter s. We first state conditions such that problem (1.1) has at least one solution for s in an interval $]s_0, s_1]$ and no solution for $s < s_0$.

Theorem 1.1. *Let $g : \mathbb{R} \to \mathbb{R}$ be continuous, $f : [a,b] \times \mathbb{R} \to \mathbb{R}$ be L^1-Carathéodory and $s \in \mathbb{R}$. Assume there exist $s_1 \in \mathbb{R}$, $s_2 > s_1$ and $r > 0$ such that for all $u \le -r$ and a.e. $t \in [a,b]$*

$$f(t,0) \le s_1 < s_2 \le f(t,u).$$

Then there exists $s_0 \in [-\infty, s_1]$ such that
(i) for $s < s_0$, the problem (1.1) has no solution;
(ii) for $s \in]s_0, s_2] \cup \{s_1\}$, the problem (1.1) has at least one solution.

Proof. Notice first that for $s \in [s_1, s_2]$, $\alpha(t) = -r$ is a lower solution of (1.1) and $\beta(t) = 0$ an upper one. It follows then from Theorem I-6.9 that problem (1.1) with $s \in [s_1, s_2]$ has a solution $u_1 \in W^{2,1}(a,b)$.

Define next $s_0 \in [-\infty, s_1]$ to be the infimum of those s such that (1.1) has a solution. Hence, Claim (i) holds.

To prove Claim (ii), let us fix $s \in]s_0, s_2]$. There exists $s^* \in]s_0, s]$ such that (1.1) with $s = s^*$ has a solution u^*. Observe that $\beta(t) = u^*(t)$ is an upper solution of (1.1). Further $\alpha(t) = r^* \leq -r$ is a lower one and we can choose r^* small enough so that $\alpha(t) = r^* \leq \beta(t)$. The claim follows now from Theorem I-6.9. □

Remark It follows from the proof of the above theorem that if u_a is a solution corresponding to s_a and u_b is a solution corresponding to $s_b \geq s_a$, u_a is an upper solution of (1.1) with $s = s_b$ and we can choose $u_b \leq u_a$.

The following examples show that the maximal interval of existence of solutions can be the closed interval $[s_0, \infty[$ or the open one $]s_0, \infty[$. Dependence of solutions on s is represented in figure 1.

Example 1.1. Consider the problem

$$\begin{gathered} u'' + u^- = s, \\ u(a) = u(b), \ u'(a) = u'(b). \end{gathered}$$

It is easy to see from a phase plane analysis that this problem has a unique solution $u(t) = -s$ if $s > 0$, no solution if $s < 0$, and an infinite number of solutions, the nonnegative constants, for $s = 0$.

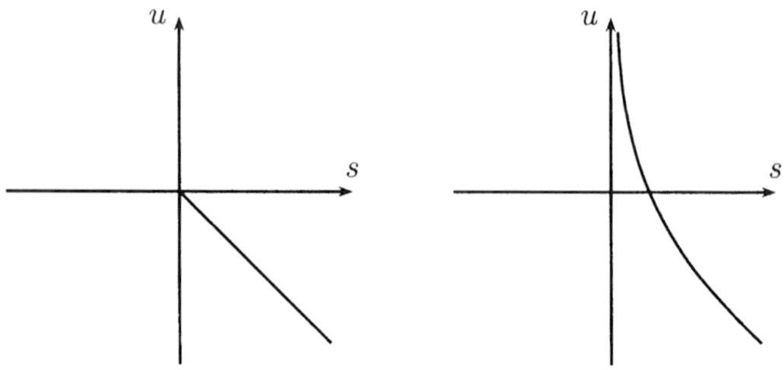

Fig. 1 : Examples 1.1 and 1.2

Example 1.2. The problem

$$\begin{gathered} u'' + \exp(-u) = s, \\ u(a) = u(b), \ u'(a) = u'(b), \end{gathered}$$

has a unique constant solution if $s > 0$ and no solution if $s \leq 0$. Uniqueness of the solution follows easily from the decreasingness of the nonlinearity $\exp(-u)$.

The next example shows that the maximal interval of existence of solutions can be bounded.

Example 1.3. The problem

$$u'' - \arctan u = s,$$
$$u(a) = u(b), \ \ u'(a) = u'(b),$$

has a unique constant solution for any $s \in]-\pi/2, \pi/2[$. Further, any solution is such that

$$s = -\frac{1}{b-a}\int_a^b \arctan u(t)\, dt,$$

i.e. there is no solution if $s \notin]-\pi/2, \pi/2[$.

Notice that the uniqueness of solutions in the above examples shows that we have to put additional assumptions to obtain multiplicity results, which is the case in the following theorem.

Theorem 1.2. *Let $s \in \mathbb{R}$, $g : \mathbb{R} \to \mathbb{R}$ be continuous and $f : [a, b] \times \mathbb{R} \to \mathbb{R}$ be an L^1-Carathéodory function such that*
(A) *for all $t_0 \in [a, b]$, $u_0 \in \mathbb{R}$ and $\epsilon > 0$, there exists $\delta > 0$ such that*

$$|t - t_0| < \delta, \, |u - u_0| < \delta \ \ \Rightarrow \ \ |f(t, u) - f(t, u_0)| < \epsilon.$$

Assume there exist $s_1 \in \mathbb{R}$, $s_2 > s_1$ and $r > 0$ such that for all $|u| \geq r$ and a.e. $t \in [a, b]$

$$f(t, 0) \leq s_1 < s_2 \leq f(t, u).$$

Then there exists $s_0 \in]-\infty, s_1]$ such that
(i) for $s < s_0$, the problem (1.1) *has no solution;*
(ii) for $s = s_0$, the problem (1.1) *has at least one solution;*
(iii) for $s \in]s_0, s_2]$, the problem (1.1) *has at least two ordered solutions.*

Remark As we already noticed after Proposition III-1.5, assumption (A) does not imply that f is continuous.

Proof. Let s_0 be defined as in Theorem 1.1.

Claim 1 – $s_0 \in \mathbb{R}$. Let $h_r \in L^1(a, b)$ be such that for any $|u| \leq r$, $|f(t, u)| \leq h_r(t)$. It follows then that for any $u \in \mathbb{R}$ and a.e. $t \in [a, b]$, $f(t, u) \geq \min\{f(t, 0), -h_r(t)\} = -h_r(t)$. Hence, direct integration of (1.1) proves that

$$s = \frac{1}{b-a}\int_a^b f(t, u(t))\, dt \geq -\frac{1}{b-a}\int_a^b h_r(t)\, dt =: s^*,$$

i.e. there is no solution for $s < s^*$. The claim follows.

Claim 2 – *For* $s \in]s_0, s_2]$, *the problem* (1.1) *has at least two solutions.* First observe that a solution u^* of (1.1) with $s = s^*$ is a strict $W^{2,1}$-upper solution of (1.1) with $s > s^*$, which follows from Proposition III-1.6. Moreover, for every A large enough, the constants $\alpha = A$ and $\overline{\alpha} = -A$ are lower solutions. Increasing A if necessary, we can assume $\overline{\alpha} \le u^*$ and $\alpha \not\le u^*$. The claim follows now from Theorem III-3.6.

Claim 3 – *For* $s = s_0$, *the problem* (1.1) *has a solution.* By definition of s_0, we can find a sequence $(s_n)_n$ and corresponding solutions u_n of (1.1) with $s = s_n$, such that $s_n \to s_0$. Notice first that for any n and some t_n, $|u_n(t_n)| < r$. Next, arguing as in Theorem III-3.2, we prove that the $\|u_n\|_\infty$ are *a priori* bounded. This result implies, as in Proposition I-4.3, an *a priori* bound on $\|u_n\|_{C^1}$ and the claim follows from Arzelà-Ascoli Theorem. □

This theorem has the obvious following corollary.

Corollary 1.3. *Let* $s \in \mathbb{R}$, $g : \mathbb{R} \to \mathbb{R}$ *be continuous and* $f : [a,b] \times \mathbb{R} \to \mathbb{R}$ *be an* L^1*-Carathéodory function such that for all* $t_0 \in [a,b]$, $u_0 \in \mathbb{R}$ *and* $\epsilon > 0$, *there exists* $\delta > 0$ *that verifies*

$$|t - t_0| < \delta,\ |u - u_0| < \delta \ \Rightarrow\ |f(t,u) - f(t,u_0)| < \epsilon.$$

Assume that for some $s_1 \in \mathbb{R}$ *and a.e.* $t \in [a,b]$, $f(t,0) \le s_1$, *and that*

$$\lim_{|u| \to \infty} f(t,u) = +\infty$$

uniformly in $t \in [a,b]$.

Then there exists $s_0 \in \mathbb{R}$ *such that*

(i) for $s < s_0$, *the problem* (1.1) *has no solution;*
(ii) for $s = s_0$, *the problem* (1.1) *has at least one solution;*
(iii) for $s > s_0$, *the problem* (1.1) *has at least two ordered solutions.*

The following examples show that the count of solution can be exact, see Example 1.4, or that the number of solutions can be larger as in Example 1.5.

Example 1.4. Consider the problem

$$\begin{gathered} u'' + u' + u^2 = s, \\ u(a) = u(b),\ u'(a) = u'(b). \end{gathered}$$

Multiplying the equation by u' and integrating gives $\|u'\|^2_{L^2} = 0$. Hence this problem has only the constant solutions $u = \pm\sqrt{s}$.

Example 1.5. In a similar way, we prove that the solutions of problem

$$u'' + u' + (u^2 - 1)^2 = s,$$
$$u(a) = u(b), \;\; u'(a) = u'(b),$$

are the constants $u = \pm\sqrt{1 \pm s^{1/2}}$.

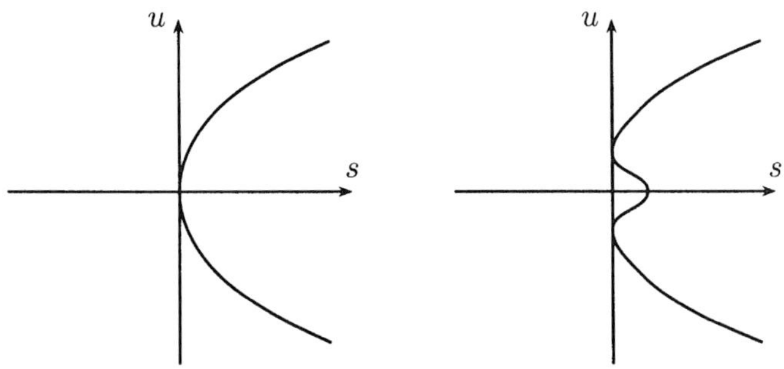

Fig. 2 : Examples 1.4 and 1.5

Exact count of solutions can be given for the frictionless problem

$$\begin{aligned} u'' + f(t,u) &= s, \\ u(a) = u(b), \;\; u'(a) &= u'(b), \end{aligned} \tag{1.2}$$

in case the nonlinearity is strictly convex.

Theorem 1.4. *Let $s \in \mathbb{R}$ and $f : [a,b] \times \mathbb{R} \to \mathbb{R}$ be a continuous, strictly convex function such that*

$$\lim_{|u| \to \infty} f(t,u) = +\infty$$

uniformly in $t \in [a,b]$. Assume moreover that for all $t \in [a,b]$ and $u \neq v$

$$\frac{f(t,u) - f(t,v)}{u - v} < \left(\frac{2\pi}{b-a}\right)^2.$$

Then there exists $s_0 \in \mathbb{R}$ such that
(i) for $s < s_0$, the problem (1.2) has no solution;
(ii) for $s = s_0$, the problem (1.2) has exactly one solution;
(iii) for $s > s_0$, the problem (1.2) has exactly two solutions which are ordered.

Proof. From Corollary 1.3, we only have to prove that for $s > s_0$ the problem (1.2) has at most two solutions and for $s = s_0$ at most one.

Claim 1 – Solutions of (1.2) *are strictly ordered.* Let u and v be two solutions which are not ordered. Extending these solutions by periodicity, we can find $t_1 \in [a,b]$ and $t_2 > t_1$, with $t_2 - t_1 \le (b-a)/2$, such that $w = u - v$ is one sign on $]t_1, t_2[$, say $w > 0$, and $w(t_1) = w(t_2) = 0$. Notice that

$$w'' + k(t)w = 0, \quad \text{with } k(t) = \frac{f(t,u(t)) - f(t,v(t))}{u(t) - v(t)}.$$

A Sturm-Liouville argument leads then to the contradiction

$$\begin{aligned} 0 &= \left[w'(t)\sin(\pi \tfrac{t-t_1}{t_2-t_1}) - \tfrac{\pi}{t_2-t_1} w(t)\cos(\pi \tfrac{t-t_1}{t_2-t_1}) \right] \Big|_{t_1}^{t_2} \\ &= \int_{t_1}^{t_2} \left[\left(\tfrac{\pi}{t_2-t_1} \right)^2 - k(t) \right] w(t) \sin(\pi \tfrac{t-t_1}{t_2-t_1})\, dt > 0. \end{aligned}$$

Notice at last that the convexity assumption implies f is locally lipschitzian. Hence, we deduce from uniqueness of solutions of the Cauchy problem, that solutions of (1.2) which are ordered are strictly ordered.

Claim 2 – Problem (1.2) *has at most two solutions.* Let $u_3 < u_2 < u_1$ be three solutions of (1.2). The functions $v_1 = u_1 - u_3$ and $v_2 = u_2 - u_3$ are such that

$$v_1'' + k_1(t)v_1 = 0 \text{ and } v_2'' + k_2(t)v_2 = 0,$$

where

$$k_1(t) := \frac{f(t,u_1(t)) - f(t,u_3(t))}{u_1(t) - u_3(t)} > k_2(t) := \frac{f(t,u_2(t)) - f(t,u_3(t))}{u_2(t) - u_3(t)}.$$

A Sturm-Liouville argument gives then the contradiction

$$0 = v_1(t)v_2'(t) - v_1'(t)v_2(t)\Big|_a^b = \int_a^b (k_1(t) - k_2(t))v_1(t)v_2(t)\, dt > 0.$$

Claim 3 – If problem (1.2) *with* $s = s^*$ *has two solutions, then* $s^* \ne s_0$. Let $u_1 < u_2$ be two solutions corresponding to s^*. Choose an $\tilde{s} < s^*$ near enough s^* so that

$$\frac{f(t,u_1) + f(t,u_2)}{2} \ge f(t,(u_1+u_2)/2) + s^* - \tilde{s}.$$

It follows then that $\beta = (u_1 + u_2)/2$ is a C^2-upper solution of (1.2) with $s = \tilde{s}$. Also, for r large enough, $\alpha = -r$ is a C^2-lower solution such that $\alpha \le \beta$. Hence, it follows from Theorem I-2.3 that problem (1.2), with $s = \tilde{s}$, has a solution and $s_0 \le \tilde{s} < s^*$. □

2 Periodic Solutions of the Rayleigh Equation

Parametric multiplicity can occur in other problems. In this section, we consider the periodic problem for a Rayleigh equation

$$\begin{aligned} u'' + g(u') + f(t,u,u') &= s, \\ u(a) = u(b), \;\; u'(a) &= u'(b). \end{aligned} \tag{2.1}$$

A first result concerns existence of one solution. In this section, for $u \in L^2(a,b)$, we write

$$u(t) = \tilde{u}(t) + \bar{u},$$

where $\bar{u} = \dfrac{1}{b-a}\displaystyle\int_a^b u(t)\,dt$ and $\displaystyle\int_a^b \tilde{u}(t)\,dt = 0$. Also, we shall use the following spaces :

$$\begin{aligned} \tilde{L}^p(a,b) &= \{u \in L^p(a,b) \mid \bar{u} = 0\}, \\ \tilde{C}^1_{per}([a,b]) &= \{u \in C^1([a,b]) \mid u(a) = u(b),\, u'(a) = u'(b),\, \bar{u} = 0\}. \end{aligned}$$

Theorem 2.1. *Let $s \in \mathbb{R}$, $g : \mathbb{R} \to \mathbb{R}$ be a continuous function and $f : [a,b] \times \mathbb{R}^2 \to \mathbb{R}$ be a Carathéodory function such that for some $h \in L^2(a,b)$, a.e. $t \in [a,b]$ and all $(u,v) \in \mathbb{R}^2$,*

$$|f(t,u,v)| \le h(t).$$

Then there exists a nonempty, bounded interval I such that
(i) if $s \notin I$, the problem (2.1) has no solution;
(ii) if $s \in I$, the problem (2.1) has at least one solution.

Proof. Let $\mathcal{M}$ be the set of s such that (2.1) has a solution.

Claim 1 : Let $R > \sqrt{b-a}\,||h||_{L^2}$. Then any solution u of (2.1) is such that

$$||u'||_\infty < R. \tag{2.2}$$

Multiplying (2.1) by u'' and integrating on $[a,b]$, we get

$$||u''||^2_{L^2} = -\int_a^b f(t,u,u')u''\,dt \le ||h||_{L^2}||u''||_{L^2}.$$

Hence,

$$||u''||_{L^2} \le ||h||_{L^2} \tag{2.3}$$

and if we choose t_0 such that $u'(t_0) = 0$, we have

$$|u'(t)| = |\int_{t_0}^t u''(s)\,ds| \le \sqrt{b-a}\,||u''||_{L^2} \le \sqrt{b-a}\,||h||_{L^2} < R.$$

A modified problem. Consider the function

$$\hat{g}(y) := g(\delta(y)),$$

with $\delta(y) = \max\{\min\{y, R\}, -R\}$. Repeating the proof of Claim 1, it is clear that any solution u of

$$\begin{aligned} u'' + \hat{g}(u') + f(t, u, u') &= s, \\ u(a) = u(b), \ u'(a) &= u'(b). \end{aligned} \tag{2.4}$$

satisfies (2.2). Hence u is a solution of (2.1) if and only if it is a solution of (2.4).

Claim 2 : $\mathcal{M}$ is nonempty. Consider the projector $P : L^2(a,b) \to \mathbb{R}$ defined by

$$Pu = \frac{1}{b-a} \int_a^b u(t)\, dt,$$

denote by M the compact inverse of the linear operator

$$L : C^1([a,b]) \to \tilde{L}^2(a,b), u \mapsto -u'',$$

with domain $Dom\, L = H^2(a,b) \cap \tilde{C}^1_{per}([a,b])$ and let

$$\hat{N} : C^1([a,b]) \to L^2(a,b)$$

be defined from $\hat{N}u = \hat{g}(u') + f(\cdot, u, u')$. The operator $\hat{T} = M(I-P)\hat{N}$ is completely continuous and bounded so that Schauder's Fixed Point Theorem provides a solution of the equation $u = \hat{T}u$. This last equation is equivalent to

$$\begin{aligned} -u'' = \hat{g}(u') + f(t, u, u') - P(\hat{g}(u') + f(\cdot, u, u')), \\ u(a) = u(b), \ u'(a) = u'(b), \ Pu = 0, \end{aligned}$$

which proves that $s := P(\hat{g}(u') + f(., u, u')) \in \mathcal{M}$.

Claim 3 : $\mathcal{M}$ is bounded. Direct integration of (2.4) shows that

$$|s| \leq ||\hat{g}||_\infty + \frac{||h||_{L^2}}{\sqrt{b-a}}.$$

Claim 4 : $\mathcal{M}$ is an interval. Consider $r_1, r_2 \in \mathcal{M}$ with $r_1 < r_2$ and let u_1, u_2 be the corresponding solutions of (2.1). For $s \in]r_1, r_2[$, the functions u_1 and u_2 are respectively $W^{2,1}$-upper and lower solutions of (2.4). Recall that $\hat{g}(v) + f(t, u, v)$ is bounded by a L^2-function. Hence, we can deduce from Theorems I-6.8 or III-3.1 (according as the lower and upper solutions are ordered or not) that problem (2.4), and therefore (2.1), has a solution. □

Example 2.1. Consider the problem

$$u'' + u'^3 - f(u) = s,$$
$$u(a) = u(b),\ u'(a) = u'(b),$$

where f is a continuous, bounded function. Multiplying the equation by u' and integrating, we prove that all solutions are constant. Particularizing $f(u)$, it is easy to find examples with different structures of the solution set.

Consider first $f(u) = \arctan u$. In this case there is a solution, which is unique, if and only if $s \in]-\pi/2, \pi/2[$.

If we choose $f(u) = ue^{-|u|}$, we have no solution if $s \notin [-e^{-1}, e^{-1}]$; one solution if $s = -e^{-1}$, 0 or e^{-1}; and two if $s \in]-e^{-1}, e^{-1}[\setminus\{0\}$.

Another example is $f(u) = \sin u$. Here there is no solution if $s \notin [-1, +1]$ and an infinite number of them for $s \in [-1, +1]$.

The next theorem gives conditions to have at least two solutions for some values of s. It describes a solution set such as in figure 3, where the vertical axis represents functions mod 2π.

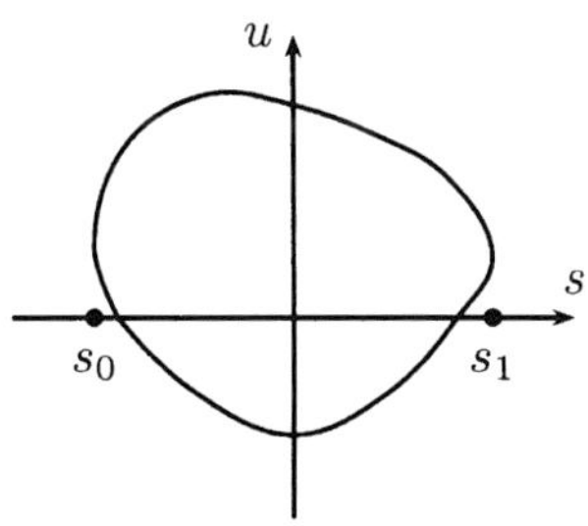

Fig. 3 : Solution set in Theorem 2.2

Theorem 2.2. *Let $s \in \mathbb{R}$, $g : \mathbb{R} \to \mathbb{R}$ be a continuous function and $f : [a,b] \times \mathbb{R}^2 \to \mathbb{R}$ be a Carathéodory function such that*
(a) for a.e. $t \in [a,b]$ and all $(u,v) \in \mathbb{R}^2$, $f(t,u,v) = f(t, u+2\pi, v)$;
(b) for all $t_0 \in [a,b]$, $(u_0, v_0) \in \mathbb{R}^2$ and $\epsilon > 0$, there exists $\delta > 0$ such that

$$|t - t_0| < \delta,\ |u - u_0| < \delta,\ |v - v_0| < \delta \Rightarrow |f(t,u,v) - f(t, u_0, v_0)| < \epsilon;$$

(c) for some $h \in L^2(a,b)$, a.e. $t \in [a,b]$ and all $(u,v) \in \mathbb{R}^2$,

$$|f(t,u,v)| \le h(t).$$

Then there exists a nonempty bounded interval $[s_0, s_1]$ such that
(i) if $s \notin [s_0, s_1]$, the problem (2.1) has no solution;
(ii) if $s \in \{s_0, s_1\}$, the problem (2.1) has at least one solution;
(iii) if $s \in]s_0, s_1[$, the problem (2.1) has at least two ordered solutions that do not differ by a multiple of 2π.

Proof. From Theorem 2.1, the set of s such that (2.1) has at least one solution is a nonempty, bounded interval I, i.e. $\text{cl}I = [s_0, s_1]$.

Step 1 - Claim : I is closed. Let $(s_n)_n$ be a sequence in I converging to a s and $(u_n)_n$ be the corresponding solutions of (2.1). From the periodicity, we can assume (adding a multiple of 2π to u_n if necessary) that $u_n(a) \in [0, 2\pi]$. It follows now from (2.2) that

$$|u_n(t)| = |u_n(a) + \int_a^t u_n'(r)\,dr| \le 2\pi + R(b-a)$$

which, using (2.3), gives that the sequence $(u_n)_n$ is bounded in $H^2(a,b)$. As $H^2(a,b)$ is compactly embedded in $C^1([a,b])$, a subsequence converges to some function $u \in C^1([a,b])$ and going to the limit in (2.1) (with $s = s_n$), it follows that u is a solution of (2.1) with the given s.

Step 2 - Existence of strict $W^{2,1}$-lower and upper solutions of (2.1) *for* $s \in]s_0, s_1[$. Consider the modified problem (2.4), where $R > 0$ is defined from Claim 1 of Theorem 2.1. Let u_{s_0} and u_{s_1} be the solutions of (2.1) with $s = s_0$ and $s = s_1$ respectively. As f is periodic, we can choose $k \in \mathbb{Z}$ such that $\alpha_1 := u_{s_1} < \beta_1 := u_{s_0} + 2k\pi$ and $\alpha_1(t_*) + 2\pi \ge \beta_1(t_*)$ for some $t_* \in [a,b]$. By Proposition III-1.6, the functions β_1 and $\beta_2 := \beta_1 + 2\pi$ are strict $W^{2,1}$-upper solutions of (2.4) for $s > s_0$. Similarly, we deduce from Proposition III-1.5 that α_1 and $\alpha_2 := \alpha_1 + 2\pi$ are strict $W^{2,1}$-lower solutions if $s < s_1$.

Step 3 - Claim : If $s \in]s_0, s_1[$, problem (2.1) *has at least two ordered solutions that do not differ by a multiple of* 2π. Recall that $\hat{g}(v) + f(t,u,v)$ is bounded by a L^2 function. It follows then from Theorem III-1.13 that the problems (2.4), and also (2.1), have at least two solutions u_1 and u_2 which are such that $\alpha_1 \le u_1 < \beta_1$, $u_1 \le u_2$, $u_2(t_1) > \beta_1(t_1)$ for some $t_1 \in [a,b]$ and $u_2(t_2) < \alpha_2(t_2)$ for some $t_2 \in [a,b]$. Hence $u_2 - u_1$ cannot be a multiple of 2π. $\square$

Example 2.2. In Theorem 2.2 the solution set can reduce to one point as in the problem

$$\begin{gathered} u'' + \sin(t+u) = s, \\ u(0) = u(2\pi),\ u'(0) = u'(2\pi). \end{gathered}$$

Let u be a solution, multiply the equation by $1 + u'$ and integrate, which gives $s = 0$ and $s_0 = s_1 = 0$.

A classical example of differential equation with periodic nonlinearity is the pendulum equation

$$\begin{gathered} u'' + g(u') + A\sin u = s + \tilde{p}(t), \\ u(a) = u(b),\ u'(a) = u'(b). \end{gathered} \tag{2.5}$$

The next result gives some estimates on the interval I of admissible s.

Proposition 2.3. *Let $g : \mathbb{R} \to \mathbb{R}$ be a continuous function, $A \in \mathbb{R}$ and $K \in]0, \pi[$. Then for all $\tilde{p} \in \tilde{L}^2(a,b)$ with $||\tilde{p}||_{L^2} \leq K \frac{6\sqrt{5}}{(b-a)\sqrt{b-a}}$, there exists $s_c \in \mathbb{R}$ (independent of A) with*

$$|s_c| \leq \max\{|g(v)| : |v| \leq \sqrt{\tfrac{b-a}{12}}\, ||\tilde{p}||_{L^2}\} \tag{2.6}$$

such that (2.5) *has at least one solution for any s with*

$$|s - s_c| \leq |A| \sin \tfrac{\pi - K}{2}. \tag{2.7}$$

Proof. Let $u^* \in H^2(a,b)$ be a solution of

$$\begin{aligned} u'' = \tilde{p}(t) - g(u') + Pg(u'), \\ u(a) = u(b),\ u'(a) = u'(b),\ Pu = 0, \end{aligned} \tag{2.8}$$

where $Pf = \frac{1}{b-a}\int_a^b f(t)\,dt$. Multiplying by $u^{*\prime\prime}$ and integrating we get $||u^{*\prime\prime}||_{L^2} \leq ||\tilde{p}||_{L^2}$. Further, we obtain from Propositions A-4.3 and A-4.1

$$||u^*||_\infty \leq \frac{(b-a)\sqrt{b-a}}{12\sqrt{5}} ||u^{*\prime\prime}||_{L^2} \leq \frac{(b-a)\sqrt{b-a}}{12\sqrt{5}} ||\tilde{p}||_{L^2} \leq \frac{K}{2}$$

and

$$||u^{*\prime}||_\infty \leq \sqrt{\tfrac{b-a}{12}}\, ||u^{*\prime\prime}||_{L^2} \leq \sqrt{\tfrac{b-a}{12}}\, ||\tilde{p}||_{L^2}.$$

Now we can modify (2.8) as in Claim 1 of Theorem 2.1 and obtain from Schauder's fixed point Theorem a solution u^* of (2.8). Let $s_c := Pg(u^{*\prime})$. Notice that (2.6) holds from the estimate on $||u^{*\prime}||_\infty$.

Assume $A \geq 0$ and observe that $\alpha(t) := u^*(t) + \frac{\pi}{2} \in [\frac{\pi}{2} - \frac{K}{2}, \frac{\pi}{2} + \frac{K}{2}]$ and $\beta(t) := u^*(t) + \frac{3\pi}{2} \in [\frac{3\pi}{2} - \frac{K}{2}, \frac{3\pi}{2} + \frac{K}{2}]$. We also fix s according to (2.7). It is then easy to check that α is a $W^{2,1}$-lower solution for (2.5) since

$$\begin{aligned} \alpha'' + g(\alpha') + A\sin\alpha - s - \tilde{p}(t) &= A\sin(u^* + \tfrac{\pi}{2}) - (s - s_c) \\ &\geq A(\sin(u^* + \tfrac{\pi}{2}) - \sin\tfrac{\pi - K}{2}) \geq 0. \end{aligned}$$

Similarly, we see that β is a $W^{2,1}$-upper solution so that the existence of a solution follows from Theorem I-6.8.

A similar argument holds if $A \leq 0$. □

If the friction force is linear, we can give an explicit estimate of the interval I of admissible s.

Corollary 2.4. *Let $c \in \mathbb{R}$, $A \in \mathbb{R}$ and $K \in]0, \pi[$. Then, for all $\tilde{p} \in \tilde{L}^2(a,b)$ with $||\tilde{p}||_2 \leq K \frac{6\sqrt{5}}{(b-a)\sqrt{b-a}}$ and for any s such that*

$$|s| \leq |A| \sin \tfrac{\pi - K}{2},$$

the problem

$$u'' + cu' + A\sin u = s + \tilde{p}(t),$$
$$u(a) = u(b),\ u'(a) = u'(b)$$

has at least one solution.

Proof. One has to see that in the proof of the above proposition $s_c = \frac{c}{b-a}\int_a^b u^{*\prime}(t)\,dt = 0$. □

Many results can be worked out for the pendulum equations. A variant of the previous result is the following.

Proposition 2.5. *Let $g : \mathbb{R} \to \mathbb{R}$ be a continuous function, $A \in \mathbb{R}$ and $\tilde{p} \in \tilde{L}^2(a,b)$. Let $u^* \in \tilde{C}^1_{per}([a,b])$ be a solution of*

$$u'' + g(u') = \tilde{p}(t),$$
$$u(a) = u(b),\ u'(a) = u'(b).$$

Assume $\Delta = \frac{1}{2}(\max_{t\in[a,b]} u^*(t) - \min_{t\in[a,b]} u^*(t)) \leq \frac{\pi}{2}$ and $|s| \leq |A|\cos\Delta$.

Then problem (2.5) *has at least one solution.*

Exercise 2.1. Prove the above proposition.

Example 2.3. As an example of application of the above proposition, consider the problem

$$u'' + \sin u = s + n\sin nt,$$
$$u(0) = u(2\pi),\ u'(0) = u'(2\pi),$$

which has a solution if $|s| \leq \cos(\frac{1}{n})$.

Our next result concerns the Ambrosetti-Prodi problem for the Rayleigh equation

$$\begin{aligned} u'' + g(u') + f(t,u,u') &= s + h(t,u,u'), \\ u(a) = u(b),\ u'(a) &= u'(b). \end{aligned} \tag{2.9}$$

Here the nonlinearity can be unbounded. More precisely, $h(t,u,v)$ is supposed to be uniformly bounded by a L^2-function and $f(t,u,v)$ satisfies a uniform limit as $u \to \infty$. Such a structure is modelled from a physical system with a restoring force $-f(t,u,v)$ and a forcing $h \in L^2(a,b)$.

Theorem 2.6. *Let $s \in \mathbb{R}$, $g : \mathbb{R} \to \mathbb{R}$ be continuous and $f, h : [a,b]\times\mathbb{R}^2 \to \mathbb{R}$ be Carathéodory functions. Assume that for all $t_0 \in [a,b]$, $(u_0,v_0) \in \mathbb{R}^2$ and $\epsilon > 0$, there exists a $\delta > 0$ such that*

$$|t - t_0| < \delta,\ |u - u_0| < \delta,\ |v - v_0| < \delta$$
$$\Rightarrow\quad |f(t,u,v) - f(t,u_0,v_0)| < \epsilon \text{ and } |h(t,u,v) - h(t,u_0,v_0)| < \epsilon.$$

Assume further

(a) there exist $d \geq c > 0$ *such that for all* $v \in \mathbb{R}$

$$c \leq \frac{g(v)}{v} \leq d;$$

(b) there exists a function $k_1 \in L^2(a,b)$ *such that for a.e.* $t \in [a,b]$ *and all* $(u,v) \in \mathbb{R}^2$

$$|h(t,u,v)| \leq k_1(t);$$

(c) for any $R \geq 0$, *there exists a function* $k_2 \in L^\infty(a,b)$ *such that for a.e.* $t \in [a,b]$ *and all* $(u,v) \in [-R,R] \times \mathbb{R}$,

$$|f(t,u,v)| \leq k_2(t);$$

(d) $\lim_{|u|\to\infty} f(t,u,v) = +\infty$ *uniformly in* t *and* v.

Then there exists $s_0 \in \mathbb{R}$ *such that*
(i) if $s < s_0$, *the problem* (2.9) *has no solution;*
(ii) if $s = s_0$, *the problem* (2.9) *has at least one solution;*
(iii) if $s > s_0$, *the problem* (2.9) *has at least two ordered solutions.*

Proof. Claim 1 : Given s^* *there exist* $R_0 > 0$ *and* $R_1 > 0$ *so that any solution* u *of* (2.9) *with* $s \leq s^*$ *is such that*

$$||u||_\infty < R_0, \quad ||u'||_\infty < R_1.$$

Let us pick $R > 0$ large enough so that for a.e. $t \in [a,b]$ and all $(u,v) \in \mathbb{R}^2$ with $|u| \geq R$, $f(t,u,v) \geq 0$ and choose k_2 according to assumption (c). Hence, we have $f(t,u,v) \geq -k_2(t)$ for a.e. t and all u, v. If we multiply the equation in (2.9) by $(u')^+$ and integrate, we obtain

$$\begin{aligned} c\int_a^b (u')^{+2}\,dt &\leq \int_a^b [u'' + g(u') + (f(t,u,u') + k_2(t))](u')^+\,dt \\ &= \int_a^b [s + h(t,u,u') + k_2(t)](u')^+\,dt \\ &\leq (|s^*|\sqrt{b-a} + ||k_1||_{L^2} + ||k_2||_{L^2})\,||(u')^+||_{L^2}. \end{aligned}$$

This gives a bound K_1 on $||(u')^+||_{L^2}$ and therefore $||(u')^+||_{L^1} \leq K_1\sqrt{b-a}$. Moreover, we have

$$0 = \int_a^b u'dt = ||(u')^+||_{L^1} - ||(u')^-||_{L^1},$$

and

$$||u'||_{L^1} = 2||(u')^+||_{L^1} \leq 2K_1\sqrt{b-a}.$$

Take now $K_2 := s^* + (||k_1||_{L^2} + 2dK_1)/\sqrt{b-a}$. If u is a solution of (2.9), direct integration gives

$$\int_a^b f(t,u,u')\,dt = s(b-a) + \int_a^b [h(t,u,u') - g(u')]\,dt$$
$$\leq s^*(b-a) + ||k_1||_{L^2}\sqrt{b-a} + d\int_a^b |u'|dt$$
$$\leq s^*(b-a) + ||k_1||_{L^2}\sqrt{b-a} + 2dK_1\sqrt{b-a} = K_2(b-a).$$

Next, by assumption (d), we can pick $r > 0$ such that for a.e. $t \in [a,b]$, all $v \in \mathbb{R}$ and every $|u| \geq r$, $f(t,u,v) > K_2$. Hence, given u solution of (2.9), we can choose t_0 such that $|u(t_0)| < r$ and we have

$$|u(t)| \leq |u(t_0)| + \int_a^b |u'(t)|\,dt < r + 2K_1\sqrt{b-a} =: R_0.$$

The bound on $\|u'\|_\infty$ follows now from Proposition I-4.1.

Claim 2 : Equation (2.9) *has a solution for some s large enough.* From Theorem 2.1, there exists some s_1 such that

$$\begin{aligned} u'' + g(u') + k_1(t) = s_1, \\ u(a) = u(b),\ u'(a) = u'(b) \end{aligned} \tag{2.10}$$

has a solution u, and therefore a family of solutions $u + C$, with $C \in \mathbb{R}$. Let u_0 be the element of this family with mean value zero. Multiplying (2.10) by u_0'' and integrating, we obtain $\|u_0''\|_{L^2} \leq \|k_1\|_{L^2}$ and from Propositions A-4.3 and A-4.1

$$\|u_0\|_\infty \leq C_0 := \tfrac{(b-a)\sqrt{b-a}}{12\sqrt{5}}||k_1||_{L^2},\ \ \|u_0'\|_\infty \leq C_1 := \sqrt{\tfrac{b-a}{12}}||k_1||_{L^2}.$$

Fix $s > s_1 + \sup\{f(t,u,v) \mid t \in [a,b], |u| \leq C_0, |v| \leq C_1\}$. Then, we claim that u_0 is an upper solution of (2.9). Indeed,

$$u_0'' + g(u_0') + f(t,u_0,u_0') - h(t,u_0,u_0')$$
$$\leq u_0'' + g(u_0') + f(t,u_0,u_0') + k_1(t) = s_1 + f(t,u_0,u_0') < s.$$

On the other hand, it is possible to get an ordered lower solution by considering the equation

$$\begin{aligned} u'' + g(u') - k_1(t) = s_2, \\ u(a) = u(b),\ u'(a) = u'(b). \end{aligned} \tag{2.11}$$

As above, it is clear there exists some s_2 such that equation (2.11) has a family of solutions $u + C$, with $C \in \mathbb{R}$. Let u_1 be an element of this family small enough so that for every $t \in [a, b]$,

$$u_1(t) < u_0(t) \quad \text{and} \quad s_2 + f(t, u_1(t), u_1'(t)) > s,$$

which is possible from (d). Then,

$$\begin{aligned} u_1'' + g(u_1') &+ f(t, u_1, u_1') - h(t, u_1, u_1') \\ &\geq u_1'' + g(u_1') + f(t, u_1, u_1') - k_1(t) = s_2 + f(t, u_1, u_1') > s. \end{aligned}$$

Claim 2 follows then from Theorem I-6.8, using the fact that u_1 and u_0 are ordered lower and upper solutions of (2.9).

Claim 3 : The set $\mathcal{M}$ of all the s such that (2.9) *has a solution is bounded below.* From Claim 2, $\mathcal{M}$ is not empty. Let $s_0 \in \mathcal{M}$ and u be a solution of (2.9) with $s \leq s_0$. From Claim 1, a direct integration of (2.9) leads to

$$s \geq -\frac{1}{b-a}\left(\int_a^b (k_2 + k_1)\, dt\right) - \sup\{|g(v)| \mid |v| < R_1\},$$

where k_2 is associated with R_0 in assumption (c).

Claim 4 : $\mathcal{M}$ is closed. Let $(s_n)_n$ be a sequence in $\mathcal{M}$ converging to s^* and $(u_n)_n$ be the corresponding solutions of (2.9). From Claim 1, the sequence $(u_n)_n$ is bounded in $C^1([a, b])$ and from Arzelà-Ascoli Theorem it converges to a solution of (2.9) with $s = s^*$.

Claim 5 : $\mathcal{M} = [s_0, \infty[$. Let $s_1 \in \mathcal{M}$, u_1 be the solution of (2.9) with $s = s_1$ and let $s > s_1$. The function u_1 is a $W^{2,1}$-upper solutions of (2.9) since

$$u_1'' + g(u_1') + f(t, u_1, u_1') - h(t, u_1, u_1') - s = s_1 - s < 0.$$

An ordered $W^{2,1}$-lower solution α can be obtained by the argument in Claim 2. It follows now from Theorem I-6.8 that $s \in \mathcal{M}$.

Claim 6 : If $s \in \text{int}\mathcal{M}$, equation (2.9) *has two ordered solutions.* Let $s_1 \in \mathcal{M}$ be such that $s_1 < s$ and u_1 be the corresponding solution. It follows from Proposition III-1.6 that u_1 is a strict $W^{2,1}$-upper solution. As in Claim 2 we can find a lower solution below u_1. The proof follows now from the argument in Claim 1 and Theorem III-1.14. □

3 A Two Parameters Dirichlet Problem

We can write the problem studied in the first section as

$$\begin{gathered} u'' + g(u)u' + f(t,u) = p(t) + s, \\ u(a) = u(b), \ u'(a) = u'(b), \end{gathered}$$

and think of it as studying the range of the operator $\mathcal{R} : L^1(a,b) \to L^1(a,b)$, defined by $\mathcal{R}u = u'' + g(u)u' + f(t,u)$ and with domain $\operatorname{Dom}\mathcal{R} = \{u \in W^{2,1}(a,b) \mid u(a) = u(b),\ u'(a) = u'(b)\}$. In that section, we gave conditions so that given $p \in L^1(a,b)$, the set of functions $h(t) = p(t) + s$, with $s \in \mathbb{R}$, intersects the range of $\mathcal{R}$ along a closed half-line and that on the corresponding open half-line there are at least two solutions. Given p and $q \in L^1(a,b)$, a similar problem could be worked out for the line $h(t) = p(t) + sq(t)$, where $s \in \mathbb{R}$. Consequently, we study the problem

$$\begin{gathered} u'' + g(u)u' + f(t,u) = p(t) + sq(t), \\ u(a) = u(b), \ u'(a) = u'(b). \end{gathered}$$

The choice $q(t) = 1$ is somewhat natural since this function is the first eigenfunction of the corresponding linear problem. However this is not fundamental. To compare two such approaches, we consider in this section a two parameters problem, i.e. we fix a plane in $L^1(a,b)$ and look for the intersection of the range of the operator with this plane. More precisely, we work the Dirichlet problem

$$\begin{gathered} u'' + u + f(t,u) = r + s\varphi(t), \\ u(0) = 0, \ u(\pi) = 0, \end{gathered} \tag{3.1}$$

where $\varphi(t) = \sqrt{\frac{2}{\pi}} \sin t$ and r, s are parameters. This can be seen as a generalized Ambrosetti-Prodi problem. Throughout the section, for $u \in L^1(0,\pi)$, we write

$$u(t) = \tilde{u}(t) + \bar{u}\varphi(t),$$

where $\bar{u} = \displaystyle\int_0^\pi u(t)\varphi(t)\,dt$ and $\displaystyle\int_0^\pi \tilde{u}(t)\varphi(t)\,dt = 0$.

Our first theorem proves the existence of a curve with equation $r = r_0(s)$ that bounds the range of the operator defined by the left-hand side of the equation. This result compares with Theorem 1.1 and is illustrated in figure 4.

Theorem 3.1. *Let f satisfy L^1-Carathéodory conditions together with*
(A-1) *there exist $\hat{s} \in \mathbb{R}$ and a function $h \in L^1(0,\pi)$ such that*

$$\bar{h} = \int_0^\pi h(t)\varphi(t)\,dt > \hat{s}$$

and

$$\liminf_{u \to -\infty} f(t,u) \geq h(t)$$

uniformly in t.

Then there exists a non-increasing Lipschitz function $r_0 : \mathbb{R} \to \mathbb{R} \cup \{-\infty\}$ *with* $2\sqrt{\frac{2}{\pi}}\, r_0(s) + s \leq \hat{s}$ *such that*
(i) if $r < r_0(s)$, *the problem* (3.1) *has no solution;*
(ii) if $r_0(s) < r \leq \frac{1}{2}\sqrt{\frac{\pi}{2}}\,(\hat{s} - s)$, *the problem* (3.1) *has at least one solution.*

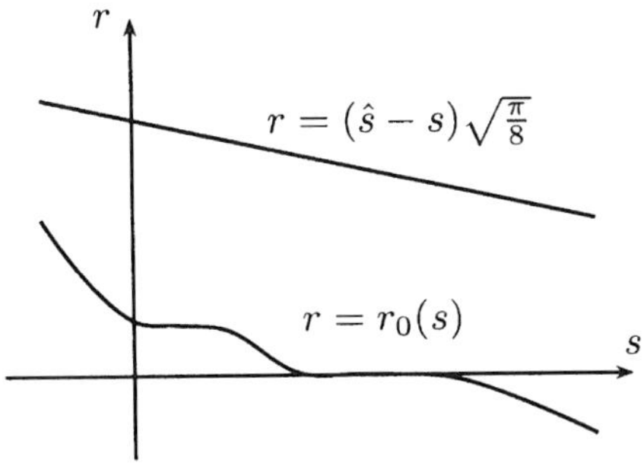

Fig. 4 : Illustration of Theorem 3.1

To prove this theorem we need the following lemma which provides lower solutions

Lemma 3.2. *Let* f *satisfy* L^1*-Carathéodory conditions and assume* (A-1) *holds. Then, for any* $z \in C_0^1([0,\pi])$ *and each* (r,s) *such that*

$$2\sqrt{\tfrac{2}{\pi}}r + s \leq \hat{s},$$

the problem (3.1) *has a* $W^{2,1}$*-lower solution* α *such that* $\alpha \leq z$.

Proof. Let us choose $\epsilon > 0$ so that

$$2\sqrt{\tfrac{2}{\pi}}\,\epsilon < \tfrac{1}{2}[\bar{h} - 2\sqrt{\tfrac{2}{\pi}}\,r - s].$$

From (A-1), we can pick $R \geq \|z\|_\infty$ such that for a.e. t and all $u \leq -R$, $f(t,u) \geq h(t) - \epsilon$. Also, from the Carathéodory conditions there exists $k \in L^1(0,\pi)$ such that for a.e. t and all $u \in]-\infty, R]$, $f(t,u) \geq k(t)$. Next, we choose $\delta > 0$ small enough so that

$$\int_{F_\delta} (h(t) - k(t))\varphi(t)\,dt < \frac{1}{2}[\bar{h} - 2\sqrt{\tfrac{2}{\pi}}\,r - s],$$

where $F_\delta := [0,\delta] \cup [\pi-\delta,\pi]$. The function

$$\begin{aligned} g(t) &:= k(t) - r, && \text{if } t \in F_\delta, \\ &:= h(t) - (r+\epsilon), && \text{if } t \in I_\delta := [\delta, \pi-\delta], \end{aligned}$$

is such that

$$\bar{g} = [\bar{h} - 2\sqrt{\tfrac{2}{\pi}}\, r] - \int_{F_\delta} (h(t) - k(t))\varphi(t)\, dt - \epsilon \int_{I_\delta} \varphi(t)\, dt > s.$$

Define now w to be the solution of

$$\begin{gathered} w'' + w + g(t) - \bar{g}\varphi(t) = 0, \\ w(0) = 0,\ w(\pi) = 0, \end{gathered}$$

and choose $A > 0$ large enough so that

$$\begin{aligned} \alpha(t) &:= -A\varphi(t) + w(t) \le z(t) && \text{on } F_\delta, \\ &:= -A\varphi(t) + w(t) \le -R && \text{on } I_\delta. \end{aligned}$$

Hence $\alpha \le z$ and since

$$\alpha'' + \alpha + f(t,\alpha) - r - s\varphi(t) \ge w'' + w + g(t) - s\varphi(t) = (\bar{g} - s)\varphi(t) \ge 0,$$

α is a $W^{2,1}$-lower solution. □

We can now proceed to the proof of Theorem 3.1.

Proof of Theorem 3.1 : Step 1 – Definition of r_0 satisfying (i) and (ii). Define $r_0(s) \in \mathbb{R} \cup \{-\infty\}$ by

$$r_0(s) := \min\{\tfrac{1}{2}\sqrt{\tfrac{\pi}{2}}(\hat{s} - s), \inf\{r \mid (3.1) \text{ has a solution for } (r,s)\}\}.$$

Notice that condition (i) holds by construction of r_0. Let now (r,s) be given such that

$$r_0(s) < r \le \tfrac{1}{2}\sqrt{\tfrac{\pi}{2}}\,(\hat{s} - s).$$

By definition, there exists $r_1 \in [r_0(s), r[$ and a solution u_1 of (3.1) for (r_1, s). The function u_1 is a $W^{2,1}$-upper solution of (3.1) for (r,s). From Lemma 3.2, we obtain a lower solution $\alpha \le u_1$ and the existence of a solution of (3.1) for the given (r,s) follows from Theorem II-2.4.

Step 2 – The function $r_0(s)$ is non-increasing and Lipschitz. Let s_2 be such that $r_0(s_2) < \frac{1}{2}\sqrt{\frac{\pi}{2}}\,(\hat{s} - s_2)$. For any $\eta > 0$, small enough, there exist

$$r_2 \in [r_0(s_2), r_0(s_2) + \eta] \subset [r_0(s_2), \tfrac{1}{2}\sqrt{\tfrac{\pi}{2}}\,(\hat{s} - s_2)[$$

and a solution u_2 of (3.1) for (r_2, s_2). Next, for any $s_1 < s_2$ close enough to s_2,

$$r_1 = r_2 + \sqrt{\tfrac{2}{\pi}}\,(s_2 - s_1)$$

is such that $2\sqrt{\frac{2}{\pi}}\,r_1 + s_1 \leq \hat{s}$. The function u_2 is a $W^{2,1}$-upper solution of (3.1) for (r_1, s_1) and from Lemma 3.2 there exists a $W^{2,1}$-lower solution $\alpha \leq u_2$. This result together with Theorem II-2.4 implies $r_0(s_1) \leq r_1$. It follows that

$$r_0(s_1) \leq r_1 \leq r_0(s_2) + \sqrt{\tfrac{2}{\pi}}\,(s_2 - s_1) + \eta$$

and as η is arbitrary,

$$r_0(s_1) \leq r_0(s_2) + \sqrt{\tfrac{2}{\pi}}\,(s_2 - s_1).$$

On the other hand, let s_1 be such that $r_0(s_1) < \frac{1}{2}\sqrt{\frac{\pi}{2}}\,(\hat{s} - s_1)$. Then, for any $\eta > 0$, we can find $r_1 \in]r_0(s_1), r_0(s_1) + \eta]$ so that there exists a solution u_1 of (3.1) for (r_1, s_1). If $s_2 > s_1$, this function is a $W^{2,1}$-upper solution of (3.1) for (r_1, s_2). Using the same arguments as above, it follows that

$$r_0(s_2) \leq r_0(s_1).$$

Hence, the claim follows. □

Example 3.1. Consider the linear problem

$$u'' + u = r + s\sqrt{\tfrac{2}{\pi}}\sin t$$
$$u(0) = 0, \;\; u(\pi) = 0,$$

which has a solution if and only if $2\sqrt{\frac{2}{\pi}}r + s = 0$. In this case $\hat{s}$ is any negative number and $r_0(s) = \frac{1}{2}\sqrt{\frac{\pi}{2}}(\hat{s} - s)$. Theorem 3.1 states that there is no solution if $r < r_0(s)$ and as $\hat{s}$ is any negative number, we have that there is no solution if $r < -\frac{1}{2}\sqrt{\frac{\pi}{2}}s$, which is the best one-sided condition we can give.

Example 3.2. The function $r_0(s)$ can take the value $-\infty$ as follows from the problem

$$u'' = r + s\sqrt{\tfrac{2}{\pi}}\sin t$$
$$u(0) = 0, \;\; u(\pi) = 0,$$

which has solutions for all values of r and s. Here $f(t, u) = -u$, $\hat{s}$ is any positive constant and $r_0(s) = -\infty$.

To rule out the case where $r_0(s) = -\infty$, we shall impose some lower bound on f.

Theorem 3.3. *Let f satisfy L^1-Carathéodory conditions,* (A-1) *and*
(A-2) *there exists $k \in L^1(0,\pi)$ such that for a.e. $t \in [0,\pi]$ and all $u \in \mathbb{R}$,*
$f(t,u) \geq k(t)$.

Then there exists a non-increasing Lipschitz function $r_0 : \mathbb{R} \to \mathbb{R}$ with $2\sqrt{\frac{2}{\pi}}\, r_0(s) + s \leq \hat{s}$ such that
(i) if $r < r_0(s)$, the problem (3.1) *has no solution;*
(ii) if $r_0(s) < r \leq \frac{1}{2}\sqrt{\frac{\pi}{2}}\,(\hat{s} - s)$, the problem (3.1) *has at least one solution.*

Proof. Let u be a solution of (3.1), multiply the equation by φ and integrate. If we fix s, this gives a lower bound on r

$$\int_0^\pi k(t)\varphi(t)\,dt \leq \int_0^\pi f(t,u(t))\varphi(t)\,dt = 2\sqrt{\tfrac{2}{\pi}}r + s.$$

The rest of the proof follows from Theorem 3.1. □

To make sure the region

$$r_0(s) < r \leq \tfrac{1}{2}\sqrt{\tfrac{\pi}{2}}\,(\hat{s} - s),$$

where there is at least one solution is nonempty, we reinforce the assumption on the nonlinearity for large negative values of u.

Theorem 3.4. *Let f satisfy L^1-Carathéodory conditions,* (A-2) *and*
(A-1*) $\lim_{u \to -\infty} f(t,u) = +\infty$ *uniformly in t.*

Then there exists a non-increasing Lipschitz function $r_0 : \mathbb{R} \to \mathbb{R}$ such that
(i) if $r < r_0(s)$, the problem (3.1) *has no solution;*
(ii) if $r_0(s) < r$, the problem (3.1) *has at least one solution.*

Proof. Claim 1 – For every $s \in \mathbb{R}$, problem (3.1) *has an upper solution if r is large enough.* Let $m \in L^1(0,\pi)$ be such that for a.e. $t \in [0,\pi]$ and all $u \in [-1,1]$, $|f(t,u)| \leq m(t)$. By Proposition A-4.4, there exists $K > 0$ such that for all $u \in W^{2,1}(0,\pi)$ with $u(0) = 0$, $u(\pi) = 0$ and $\int_0^\pi u'\varphi'\,dt = 0$, we have

$$\|u\|_\infty \leq K \int_0^\pi |u'' + u|\varphi\,dt.$$

Next, we choose $p \in C([0,\pi])$ such that $\|m - p\|_{L^1} < \frac{1}{2K}\sqrt{\frac{\pi}{2}}$ and define β as the solution of

$$\begin{gathered} u'' + u + m - p - (\bar{m} - \bar{p})\varphi = 0 \\ u(0) = 0,\ u(\pi) = 0,\ \bar{u} = 0. \end{gathered}$$

We have

$$\|\beta\|_\infty \le K \int_0^\pi |m - p - (\bar{m} - \bar{p})\varphi|\varphi \, dt \le 2K\sqrt{\frac{2}{\pi}}\|m - p\|_{L^1} \le 1$$

and therefore

$$\beta''(t)+\beta(t)+f(t,\beta(t)) \le \beta''(t)+\beta(t)+m(t) = p(t)+(\bar{m}-\bar{p})\varphi(t) \le r+s\varphi(t)$$

if r is large enough.

Claim 2 – For every $s \in \mathbb{R}$, problem (3.1) has a solution if r is large enough. Let $s \in \mathbb{R}$ and, according to Claim 1, choose $\bar{r}(s)$ large enough so that problem (3.1) for $(\bar{r}(s), s)$ has an upper solution β. Let $\hat{s} \ge 2\sqrt{\frac{2}{\pi}}\bar{r}(s)+ s$. According to (A-1*), the assumptions of Lemma 3.2 are satisfied and problem (3.1) for $(\bar{r}(s), s)$ has a lower solution $\alpha \le \beta$. We deduce then from Theorem II-2.4 the existence of a solution of (3.1) for $(\bar{r}(s), s)$.

Conclusion. From Theorem 3.3, the function $r_0(s)$ exists such that $r_0(s) \le \bar{r}(s)$ and
(i) if $r < r_0(s)$, the problem (3.1) has no solution;
(ii) if $r_0(s) < r \le \frac{1}{2}\sqrt{\frac{\pi}{2}}\,(\hat{s} - s)$, the problem (3.1) has at least one solution.
As $\hat{s}$ is arbitrarily large, the second conclusion holds for all $r > r_0(s)$. □

Example 3.3. Multiplicity results are not straightforward as follows from the problem

$$u'' + u^+ = r,$$
$$u(0) = 0, \;\; u(\pi) = 0,$$

which corresponds to $f(t, u) = u^-$. Here, there is no solution if $r < 0$, an infinite number if $r = 0$ and only one if $r > 0$.

To obtain multiplicity results, we will have to reinforce (A-1) assuming a control of the nonlinearity for large values of $|u|$. Also, we shall use degree theory, i.e. strict lower and upper solutions, which forces us to assume additional assumptions as in Section III-2. Here we consider condition (B), which is satisfied in particular if $f(t, u) = g(t, u) + h(t)$, where g is continuous and $h \in L^1(a, b)$.

Theorem 3.5. *Let f satisfy L^1-Carathéodory conditions together with*
(A-1**) $\lim_{|u|\to\infty} f(t, u) = +\infty$ *uniformly in t and*
(B) *given $t_0 \in [0, \pi]$, $u_0 \in \mathbb{R}$ and $\epsilon > 0$, there exists $\delta > 0$ such that for a.e. $t \in]t_0-\delta, t_0+\delta[\cap [0, \pi]$ and all $u_1 \in [u_0-\delta, u_0+\delta]$, $u_2 \in [u_1, u_1+\delta \sin t]$,*

$$f(t, u_2) - f(t, u_1) \le \epsilon.$$

Then, there exists a non-increasing Lipschitz function $r_0 : \mathbb{R} \to \mathbb{R}$ such that
(i) if $r < r_0(s)$, the problem (3.1) has no solution;
(ii) if $r = r_0(s)$, the problem (3.1) has at least one solution;
(iii) if $r_0(s) < r$, the problem (3.1) has at least two ordered solutions.

To prove this result, we use the following lemma which provides the necessary *a priori* bounds to apply Theorem III-2.12.

Lemma 3.6. *Suppose the assumptions of Theorem 3.5 are verified. Then, given s and r^*, there exists an $R > 0$ such that for all solutions u of*

$$\begin{gathered} u'' + u + f(t,u) \leq r^* + s\varphi(t), \\ u(0) = 0, \ u(\pi) = 0, \end{gathered} \tag{3.2}$$

we have

$$||u||_\infty < R.$$

Proof. Let u be a solution of (3.2). Write

$$u(t) = \bar{u}\varphi(t) + \tilde{u}(t),$$

where $\displaystyle\int_0^\pi \tilde{u}(t)\varphi(t)\,dt = \int_0^\pi \tilde{u}'(t)\varphi'(t)\,dt = 0$. By Proposition A-4.4 there exists $K > 0$ such that

$$||\tilde{u}||_\infty \leq K \int_0^\pi |u''(t) + u(t)|\varphi(t)\,dt.$$

Using (A-1**) and the Carathéodory conditions, we can find $k \in L^1(0,\pi)$ such that

$$\text{for a.e. } t \in [0,\pi] \text{ and all } u \in \mathbb{R}, \quad f(t,u) \geq k(t). \tag{3.3}$$

Hence, we compute

$$\begin{aligned} |u''(t) + u(t)| &\leq |u''(t) + u(t) + f(t,u(t)) - r^* - s\varphi(t)| \\ &\qquad + |f(t,u(t))| + |r^*| + |s|\varphi(t) \\ &\leq -(u''(t) + u(t) + f(t,u(t)) - r^* - s\varphi(t)) \\ &\qquad + f(t,u(t)) - k(t) + |k(t)| + |r^*| + |s|\varphi(t) \end{aligned}$$

and

$$\begin{aligned} &\int_0^\pi |u''(t) + u(t)|\varphi(t)\,dt \\ &\qquad \leq \int_0^\pi (r^* + s\varphi(t))\varphi(t)\,dt + 2\int_0^\pi |k(t)|\varphi(t)\,dt + 2\sqrt{\tfrac{2}{\pi}}|r^*| + |s|. \end{aligned}$$

Hence, we obtain

$$\begin{aligned}||\tilde{u}||_\infty &\leq K[2\sqrt{\tfrac{2}{\pi}}\,(r^* + |r^*|) + (s + |s|) + 2\int_0^\pi |k(t)|\varphi(t)\,dt] \\ &\leq 2K[2\sqrt{\tfrac{2}{\pi}}\,|r^*| + |s| + \int_0^\pi |k(t)|\varphi(t)\,dt].\end{aligned}$$

Assume now there exists a sequence $(u_k)_k = (\bar{u}_k\varphi + \tilde{u}_k)_k$ of solutions of (3.2) such that $||u_k||_\infty \to \infty$. From the first part of the proof, $||\tilde{u}_k||_\infty$ is bounded and $|\bar{u}_k| \to \infty$. Going to a subsequence, we have $u_k(t) \to \infty$ (or $u_k(t) \to -\infty$) for all $t \in]0,\pi[$. Hence, using (3.2), Fatou's Lemma and (A-1**), we obtain the contradiction

$$(2\sqrt{\tfrac{2}{\pi}}\,r^* + s) \geq \liminf_{k\to\infty} \int_0^\pi f(t, u_k(t))\varphi(t)\,dt = +\infty. \qquad \square$$

Proof of Theorem 3.5 : Let $r_0(s)$ be defined from Theorem 3.4 and (r,s) be such that $r_0(s) < r$. By definition, there exist $r_1 \in [r_0(s), r[$ and a solution u_1 of (3.1) for (r_1, s). By Proposition III-2.6 the function u_1 is a strict upper solution of (3.1) for (r,s). As (A-1) is satisfied for any $\hat{s}$, we can assume $2\sqrt{\frac{2}{\pi}}\,r + s \leq \hat{s}$ and using Lemma 3.2 we obtain a lower solution $\alpha \prec u_1$. From Lemma 3.6, there exists an $R > 0$, such that for all solutions u of

$$\begin{aligned}u'' + u + f(t,u) &\leq r + s\varphi(t),\\ u(0) = 0,\ u(\pi) &= 0,\end{aligned}$$

we have $||u||_\infty < R$. It follows then from Theorem III-2.12 that the problem (3.1) has at least two ordered solutions.

Let now (r,s) be such that $r = r_0(s)$. We can find $r_n > r_0(s)$ such that $\lim_{n\to\infty} r_n = r_0(s)$ and solutions u_n of (3.1) for (r_n, s). From Lemma 3.6, there exists an $R > 0$ such that $\|u_n\|_\infty \leq R$ and from Carathéodory conditions, we have $k \in L^1(0,\pi)$ so that for all $n \in \mathbb{N}$, $|u_n''(t)| \leq k(t)$. Using Arzelà-Ascoli Theorem, we find a subsequence $(u_{n_k})_k$ that converges in $C^1([0,\pi])$ to some function $u \in C^1([0,\pi])$. Going to the limit in (3.1), it is easy to see that u solves (3.1) for $(r,s) = (r_0(s), s)$. $\square$

The next theorem gives conditions for the function $r_0(s)$ to be decreasing.

Theorem 3.7. *Let f satisfy L^1-Carathéodory conditions together with assumptions* (A-1**) *and* (B). *Assume one of the functions $g(t,x) = f(t,x)$ or $g(t,x) = -f(t,x)$ is such that*

(A-3) *given $R > 0$, we have*

$$\begin{aligned}&\forall\epsilon > 0, \exists\delta > 0, (\forall u_0 : |u_0| \leq R\varphi(t)), (\forall u : |u| \leq R\varphi(t),\ \text{for a.e. } t \in [0,\pi],\\ &\qquad 0 \leq u_0 - u \leq \delta\varphi(t) \Rightarrow g(t,u) - g(t,u_0) \leq \epsilon\varphi(t).\end{aligned}$$

Then the function $r_0(s)$ is decreasing.

Proof. Let $s_0 \in \mathbb{R}$.
Step 1 – Claim: For all $s^* > s_0$, there exists $r^* < r_0(s_0)$ and a $W^{2,1}$-upper solution β of (3.1) with $r = r^*$ and $s = s^*$. Assume $g(t,x) = f(t,x)$ in (A-3). The proof is similar in case $g(t,x) = -f(t,x)$.

From Theorem 3.5, there is a solution u_0 of

$$u_0'' + u_0 + f(t,u_0) = r_0(s_0) + s_0\varphi(t),$$
$$u_0(0) = 0, \; u_0(\pi) = 0.$$

Let w be the solution of

$$w'' + w = -(1 - 2\sqrt{\tfrac{2}{\pi}}\varphi(t)),$$
$$w(0) = 0, \; w(\pi) = 0, \; \bar{w} = 0.$$

Choose then $\tilde{K} > 0$ and $R > 0$ such that for all $t \in [0,\pi]$,

$$|w(t)| \le \tilde{K}\varphi(t), \quad |u_0(t)| < \tfrac{R}{2}\varphi(t) \quad \text{and} \quad |w(t) - \tilde{K}\varphi(t)| < \tfrac{R}{2}\varphi(t).$$

Define $\epsilon > 0$ such that

$$s_0 + \epsilon < s^*$$

and pick $\delta > 0$ from assumption (A-3). Next, we choose $m \in]0,1]$ small enough so that

$$2m\tilde{K} \le \delta, \quad s_0 + 2\sqrt{\tfrac{2}{\pi}}\, m + \epsilon \le s^*.$$

and let $r^* := r_0(s_0) - m$. Define

$$\beta(t) := u_0(t) + m(w(t) - \tilde{K}\varphi(t))$$

and observe that

$$0 \le u_0(t) - \beta(t) \le 2m\tilde{K}\varphi(t) \le \delta\varphi(t).$$

Thus we compute

$$\begin{aligned}\beta''(t) + \beta(t) + f(t,\beta(t)) &= u_0''(t) + u_0(t) + m(w''(t) + w(t)) + f(t,\beta(t)) \\ &= r_0(s_0) + s_0\varphi(t) - m(1 - 2\sqrt{\tfrac{2}{\pi}}\,\varphi(t)) + f(t,\beta(t)) - f(t,u_0(t)) \\ &\le (r_0(s_0) - m) + (s_0 + 2\sqrt{\tfrac{2}{\pi}}\, m + \epsilon)\varphi(t) \\ &\le r^* + s^*\varphi(t).\end{aligned}$$

Hence β is a $W^{2,1}$-upper solution of (3.1) with $r = r^*$ and $s = s^*$.

Step 2 – As (A-1) is satisfied for any $\hat{s}$, we can assume $2\sqrt{\frac{2}{\pi}}\, r^* + s^* \le \hat{s}$ and it follows from Lemma 3.2 that there exists a $W^{2,1}$-lower solution $\alpha \le \beta$. We obtain then from Theorem II-2.4 the existence of a solution of (3.1) for the given (r^*, s^*). This result implies that $r_0(s^*) \le r^* < r_0(s_0)$. □

Remark The condition (A-3) is a one-sided continuity condition for the norm

$$\|u\| = \sup_{t\in[0,\pi]} \frac{|u(t)|}{\varphi(t)}.$$

Such a condition can be obtained from a continuity assumption together with a one-sided Lipschitz condition near zero.

Proposition 3.8. *Let f be a continuous function and*
(A-3*) *there exist $L > 0$ and $R > 0$ such that*

$$f(t,u) + Lu \text{ or } -f(t,u) + Lu$$

is non-decreasing for a.e. $t \in [0,\pi]$ and $u \in [-R,R]$.
Then assumption (A-3) *holds.*

Proof. Assume $f(t,u) + Lu$ is non-decreasing. Let $R_0 > 0$ be given and choose $\tau \in]0, \frac{\pi}{2}]$ so that $R_0\varphi(\tau) < R$. For $t \in F_\tau := [0,\tau] \cup [\pi - \tau, \pi]$ and all u_0, u in $[-R_0\varphi(t), R_0\varphi(t)]$, we have

$$0 \le u_0 - u \Rightarrow f(t,u) - f(t,u_0) \le L(u_0 - u).$$

If $t \in [\tau, \pi - \tau]$, we compute from the continuity of f

$$\forall \epsilon' > 0, \exists \delta_2 > 0, \quad |u_0 - u| < \delta_2 \Rightarrow f(t,u) - f(t,u_0) \le \epsilon' \le \frac{\epsilon'}{\varphi(\tau)}\varphi(t). \quad (3.4)$$

Now, given ϵ, we choose δ_1 and ϵ' small enough so that

$$L\delta_1 < \epsilon, \quad \frac{\epsilon'}{\varphi(\tau)} < \epsilon,$$

we choose δ_2 from (3.4) and we put $\delta := \min(\delta_1, \delta_2\sqrt{\frac{\pi}{2}})$. It is then easy to see that if $|u_0|, |u| < R_0\varphi(t)$ are such that $0 \le u_0 - u \le \delta\varphi(t)$, we have for $t \in F_\tau$,

$$f(t,u) - f(t,u_0) \le L\delta\varphi(t) < \epsilon\varphi(t),$$

and for $t \in [\tau, \pi - \tau]$, $0 < u_0 - u < \delta\varphi(t)$ implies

$$|u_0 - u| < \delta_2 \text{ and } f(t,u) - f(t,u_0) \le \tfrac{\epsilon'}{\varphi(\tau)}\varphi(t) < \epsilon\varphi(t). \qquad \square$$

The decreasing character of r_0 can also be deduced from a Hölder condition.

Problem 3.1. Let f be a continuous function that satisfies (A-1**) and

(A-4) either $g(t,u) = f(t,u)$ or $g(t,u) = -f(t,u)$ is such that for some $L \geq 0$, $\alpha \in]1/3, 1[$ and $r > 0$, for a.e. t and all $u, v \in [-r, r]$ with $0 \leq u - v \leq 1$, we have

$$g(t,u) - g(t,v) \leq L(u-v)^{\alpha}.$$

Prove then that the function $r_0(s)$ is decreasing.

Hint : See [88].

4 Strong Resonance Problems

In this section, we study strong resonance problems. These are problems which avoid resonance using a nonlinearity that can vanish at infinity.

As in the previous section, for $u \in L^1(0,\pi)$, we write

$$u(t) = \tilde{u}(t) + \bar{u}\varphi(t),$$

where $\varphi(t) = \sqrt{\frac{2}{\pi}} \sin t$, $\bar{u} = \int_0^\pi u(t)\varphi(t)\,dt$ and $\int_0^\pi \tilde{u}(t)\varphi(t)\,dt = 0$. Also, we shall use the following spaces

$$\begin{aligned} \tilde{L}^1(0,\pi) &= \{u \in L^1(0,\pi) \mid \bar{u} = 0\}, \\ \tilde{C}([0,\pi]) &= \{u \in C([0,\pi]) \mid \bar{u} = 0\}, \\ \tilde{C}_0^1([0,\pi]) &= \{u \in C_0^1([0,\pi]) \mid \bar{u} = 0\}. \end{aligned}$$

Our first result concerns the problem

$$\begin{aligned} u'' + u + f(t,u) &= s\varphi(t) + \tilde{p}(t), \\ u(0) = 0, \ u(\pi) &= 0, \end{aligned} \tag{4.1}$$

where f is an L^1-Carathéodory function.

Theorem 4.1. *Let $s \in \mathbb{R}$ and $\tilde{p} \in \tilde{L}^1(0,\pi)$. Assume that f is an L^1-Carathéodory function such that for some $k \in L^1(0,\pi)$, for a.e. $t \in [0,\pi]$ and all $u \in \mathbb{R}$,*

$$|f(t,u)| \leq k(t).$$

Then there exists a nonempty, bounded interval I such that

(i) if $s \notin I$, problem (4.1) *has no solution;*

(ii) if $s \in I$, problem (4.1) *has at least one solution.*

Proof. Let $\mathcal{M}$ be the set of those s such that (4.1) has at least one solution.

Step 1 – The set $\mathcal{M}$ is nonempty. Consider the projector $Q : L^1(0,\pi) \to \tilde{L}^1(0,\pi)$ defined by

$$Qu(t) = \tilde{u}(t) = u(t) - (\int_0^\pi u(r)\varphi(r)\,dr)\varphi(t)$$

and denote by M the inverse of the linear operator $L : \tilde{C}_0^1([0,\pi]) \to \tilde{L}^1(0,\pi)$, $u \mapsto u'' + u$, with domain $\operatorname{Dom} L = W^{2,1}(0,\pi) \cap \tilde{C}_0^1([0,\pi])$. The problem

$$\begin{aligned} &u'' + u = \tilde{p} - Qf(\cdot, u),\\ &u(0) = 0,\ u(\pi) = 0,\ \bar{u} = 0 \end{aligned}$$

can be written

$$u = \tilde{T}u := M(\tilde{p} - Qf(\cdot, u)).$$

The operator $\tilde{T} : \tilde{C}_0^1([0,\pi]) \to \tilde{C}_0^1([0,\pi])$ is completely continuous and bounded. Hence, using Schauder's Theorem, it has a fixed point $\tilde{u}$ which is a solution of (4.1) with

$$s = \int_0^\pi f(r, \tilde{u}(r))\varphi(r)\,dr.$$

Step 2 – The set $\mathcal{M}$ is bounded. Observe that if u is a solution of (4.1),

$$|s| = |\int_0^\pi f(t, u(t))\varphi(t)\,dt| \le \int_0^\pi k(t)\varphi(t)\,dt.$$

Step 3 – The set $\mathcal{M}$ is an interval. Let us define the real numbers $s_0 := \inf\{s \mid (4.1) \text{ has a solution}\}$ and $s_1 := \sup\{s \mid (4.1) \text{ has a solution}\}$, which are such that $\mathcal{M} \subset [s_0, s_1]$. Fix $s \in]s_0, s_1[$. There exists $a \in [s_0, s]$ and u_a solution of (4.1) with $s = a$. Also, there exists $b \in [s, s_1]$ and u_b solution of (4.1) with $s = b$. It is easy to see that u_a and u_b are upper and lower solutions of (4.1). The claim follows then from Theorems II-2.4 or III-3.8 according as $u_b \le u_a$ or $u_b \not\le u_a$. □

Remark 4.1. For any $a \in \mathbb{R}$, the problem

$$\begin{aligned} &u'' + u = \tilde{p} - Qf(\cdot, a\varphi + u),\\ &u(0) = 0,\ u(\pi) = 0,\ \bar{u} = 0, \end{aligned} \tag{4.2}$$

can be written

$$u = \tilde{T}_a u := M(\tilde{p} - Qf(\cdot, a\varphi + u)),$$

where M and Q are defined in Step 1 of the above proof. The operator $\tilde{T}_a : \tilde{C}_0^1([0,\pi]) \to \tilde{C}_0^1([0,\pi])$ is completely continuous and bounded thus

Schauder's Theorem applies. As further the bound is uniform in a, there is a $K > 0$ so that for any given $a \in \mathbb{R}$, the problem (4.2) has a solution $\tilde{u}_a \in \tilde{C}_0^1([0,\pi])$ and any such solution verifies $\|\tilde{u}_a\|_{C^1} < K$ and $|\tilde{u}_a(t)| \le K\varphi(t)$. Hence, for any value of $a \in \mathbb{R}$ the function $u = \tilde{u}_a + a\varphi$ solves (4.1) with

$$s = \int_0^\pi f(r, \tilde{u}_a(r) + a\varphi(r))\varphi(r)\,dr,$$

i.e. there is an unbounded branch of solutions.

Example 4.1. The above theorem does not rule out the linear resonance problem

$$\begin{gathered} u'' + u = s\varphi(t) + \tilde{p}(t), \\ u(0) = 0, \ u(\pi) = 0, \end{gathered}$$

which has solutions if and only if $s = 0$. Here, $I = \{0\}$.

Example 4.2. A typical example of application of Theorem 4.1 is

$$\begin{gathered} u'' + u + \arctan u = s\varphi(t) + \tilde{p}(t), \\ u(0) = 0, \ u(\pi) = 0. \end{gathered}$$

Using Remark 4.1, we know that solutions of this problem read $u = \tilde{u} + \bar{u}\varphi$ and correspond to

$$s = S(\tilde{u}, \bar{u}) = \int_0^\pi \arctan(\tilde{u}(t) + \bar{u}\varphi(t))\varphi(t)\,dt.$$

It is easy to see that the range of S is $]-\sqrt{2\pi}, \sqrt{2\pi}[$, which proves $I \subset]-\sqrt{2\pi}, \sqrt{2\pi}[$. On the other hand, we can find solutions $u_k = \tilde{u}_k + k\varphi$, $k \in \mathbb{Z}$, corresponding to $s_k = S(\tilde{u}_k, k)$. We can prove that the set $\{\|\tilde{u}_k\|_\infty \mid k \in \mathbb{Z}\}$ is bounded and, using Lebesgue's Dominated Convergence Theorem, that $\lim_{k\to\pm\infty} S(\tilde{u}_k, k) = \pm\sqrt{2\pi}$, which proves that solutions exist if and only if $s \in]-\sqrt{2\pi}, \sqrt{2\pi}[$.

Notice that solutions of this example are unique. If u and v are two such solutions, the function $w = u - v$ satisfies a linear problem

$$\begin{gathered} w'' + k(t)w = 0, \\ w(0) = 0, \ w(\pi) = 0, \end{gathered}$$

with $1 < k(t) \le 2$, which implies $w = 0$.

The above examples show that the interval of parameters s so that we have the existence of a solution can be reduced to one point or to an open interval. To rule out these possibilities, we impose some specific behaviour of f near infinity.

Theorem 4.2. *Let $s \in \mathbb{R}$ and $\tilde{p} \in \tilde{L}^1(0,\pi)$. Assume that f is an L^1-Carathéodory function such that*
(a) for some $k \in L^1(0,\pi)$, for a.e. $t \in [0,\pi]$ and all $u \in \mathbb{R}$,

$$|f(t,u)| \leq k(t).$$

(b) $\lim_{|u|\to\infty} f(t,u) = 0$ and $\liminf_{|u|\to\infty} uf(t,u) \geq m > 0$ uniformly in t.
Then there exist $s_0 < 0 < s_1$ such that
(i) if $s \notin [s_0, s_1]$, the problem (4.1) has no solution;
(ii) if $s \in [s_0, s_1]$, the problem (4.1) has at least one solution.

Proof. Let I be given from Theorem 4.1, $s_0 = \inf I$ and $s_1 = \sup I$.

Step 1 – $s_0 < 0 < s_1$. Let us first prove there exists $r > 0$ and $u = \tilde{u} + \bar{u}\varphi$ solution of (4.1) with $s = r$. Recall from Remark 4.1 that the solutions $\tilde{u}$ of (4.2) are such that $\|\tilde{u}\|_\infty \leq K$ and $|\tilde{u}(t)| \leq K\varphi(t)$ for some $K > 0$. Choose then $R > 0$ large enough so that for a.e. $t \in [0,\pi]$ and all $u \geq R$,

$$uf(t,u) \geq \tfrac{2m}{3} \quad \text{and} \quad |f(t,u)| \leq \tfrac{m}{3K}.$$

Next we pick a $\delta > 0$ small enough so that

$$-(R+K)\left[\int_0^\delta k(t)\,dt + \int_{\pi-\delta}^\pi k(t)\,dt\right] - \tfrac{m}{3}\pi + \tfrac{2m}{3}(\pi - 2\delta) > 0,$$

a $\bar{u}$ large enough such that

$$\bar{u}\,\varphi(\delta) \geq K + R,$$

and finally a solution $\tilde{u}$ of (4.2). Observe that these choices imply that $u(t) = \tilde{u}(t) + \bar{u}\varphi(t) \geq R\varphi(t)$ if $t \in [0,\pi]$ and $u(t) \geq R$ if $t \in [\delta, \pi-\delta]$. Let $A = \{t \in [0,\pi] \mid |u(t)| \leq R\} \subset [0,\delta] \cup [\pi-\delta, \pi]$ and $B = [0,\pi] \setminus A$. We compute then

$$\begin{aligned}\int_0^\pi f(t,u(t))u(t)\,dt &= \int_A f(t,u(t))u(t)\,dt + \int_B f(t,u(t))u(t)\,dt \\ &\geq -R\left[\int_0^\delta k(t)\,dt + \int_{\pi-\delta}^\pi k(t)\,dt\right] + \tfrac{2m}{3}(\pi - 2\delta), \\ \int_0^\pi f(t,u(t))\tilde{u}(t)\,dt &= \int_A f(t,u(t))\tilde{u}(t)\,dt + \int_B f(t,u(t))\tilde{u}(t)\,dt \\ &\leq K\left[\int_0^\delta k(t)\,dt + \int_{\pi-\delta}^\pi k(t)\,dt\right] + \tfrac{m}{3}\pi\end{aligned}$$

and

$$\bar{u}\int_0^\pi f(t,u(t))\varphi(t)\,dt = \int_0^\pi f(t,u(t))u(t)\,dt - \int_0^\pi f(t,u(t))\tilde{u}(t)\,dt$$
$$\geq -(R+K)\left[\int_0^\delta k(t)\,dt + \int_{\pi-\delta}^\pi k(t)\,dt\right] - \tfrac{m}{3}\pi + \tfrac{2m}{3}(\pi - 2\delta) > 0.$$

Hence, $r := \displaystyle\int_0^\pi f(t,u(t))\varphi(t)\,dt > 0$ and $u = \tilde{u}+\bar{u}\varphi$ solves (4.1) with $s = r$.

In a similar way, we prove there exists $r < 0$ and $u = \tilde{u} + \bar{u}\varphi$ solution of (4.1) with $s = r$.

Step 2 – $s_0 \in I$, $s_1 \in I$. Let us prove there exists a solution of (4.1) for $s = s_0$. Let $s_n \in]s_0, s_1[$ be such that $\lim_{n\to\infty} s_n = s_0$ and let u_n be the corresponding solution of (4.1). Multiplying (4.1) by φ and integrating we see that $s_n = \int_0^\pi f(t,u_n(t))\varphi(t)\,dt$, i.e. $u_n = \tilde{u}_n + \bar{u}_n\varphi$, where $\tilde{u}_n$ is solution of (4.2). Hence there exists $K > 0$ such that for every n, $\|\tilde{u}_n\|_\infty < K$. Let us prove there exists $M > 0$ such that for every n, $\|u_n\|_\infty < M$. Otherwise $|\bar{u}_n| \to \infty$ and going to a subsequence, we have $u_n(t) \to \infty$ (or $u_n(t) \to -\infty$) for all $t \in]0,\pi[$. Hence, as $\lim_{|u|\to\infty} f(t,u) = 0$, by Lebesgue's Dominated Convergence Theorem,

$$s_0 = \lim_{n\to\infty} s_n = \lim_{n\to\infty}\int_0^\pi f(t,u_n(t))\varphi(t)\,dt = 0,$$

which contradicts $s_0 < 0$. The claim follows then from Arzelà-Ascoli Theorem.

In a similar way, we prove $s_1 \in I$. □

Example 4.3. A typical example where Theorem 4.2 applies is

$$u'' + u + \frac{u}{1+u^2} = s\varphi(t) + \tilde{p}(t),$$
$$u(0) = 0,\ u(\pi) = 0.$$

To obtain a multiplicity result we use strict lower and upper solutions, which forces to assume additional assumptions as in Section III-2.

Theorem 4.3. *Let $s \in \mathbb{R}$ and $\tilde{p} \in \tilde{L}^1(0,\pi)$. Assume that f is an L^1-Carathéodory function such that for some $L \in L^1(a,b;\mathbb{R}^+)$ and for a.e. $t \in [0,\pi]$, $f(t,u) + L(t)u$ is non-decreasing in u and*
(a) for some $k \in L^1(0,\pi)$, for a.e. $t \in [0,\pi]$ and all $u \in \mathbb{R}$,

$$|f(t,u)| \leq k(t).$$

(b) $\lim_{|u|\to\infty} f(t,u) = 0$ *and* $\liminf_{|u|\to\infty} uf(t,u) \geq m > 0$ *uniformly in* t.

Then there exist $s_0 < 0 < s_1$ *such that*

(i) if $s \notin [s_0, s_1]$*, the problem* (4.1) *has no solution;*

(ii) if $s \in [s_0, s_1]$*, the problem* (4.1) *has at least one solution;*

(iii) if $s \in]s_0, s_1[\setminus\{0\}$*, the problem* (4.1) *has at least two ordered solutions.*

Proof. Define s_0 and s_1 from Theorem 4.2. Let $s^* \in]0, s_1[$. A similar argument holds if $s^* \in]s_0, 0[$.

Part 1 – Construction of a strict lower solution. The problem (4.1) with $s = a \in]s^*, s_1[$ has a solution u_a which is such that

$$u_a'' + u_a + f(t, u_a) - s^*\varphi(t) - \tilde{p}(t) = (a - s^*)\varphi(t) > 0,$$
$$u_a(0) = 0, \ u_a(\pi) = 0.$$

From Proposition III-2.7, u_a is a strict $W^{2,1}$-lower solution of (4.1) with $s = s^*$.

Part 2 – Construction of upper solutions. Let us choose $R > 0$ large enough so that for a.e. $t \in [0, \pi]$ and all $u \geq R$,

$$|f(t,u)| \leq \frac{s^*}{2\sqrt{2\pi}}.$$

Next, we pick $\delta > 0$ small enough so that

$$\sqrt{\tfrac{2}{\pi}}\left[\int_0^\delta k(t)\,dt + \int_{\pi-\delta}^{\pi} k(t)\,dt\right] < \tfrac{s^*}{2}.$$

Recall that there exists $K > 0$ so that for any $\bar{u}$, the solution $\tilde{u}$ of (4.2) satisfies $|\tilde{u}(t)| \leq K\varphi(t)$. Hence we can find $\bar{u}$ large enough so that $\bar{u}\varphi(t) + \tilde{u}(t) \geq R$ on $[\delta, \pi - \delta]$ and $\bar{u}\varphi + \tilde{u} \succ u_a$, where $\tilde{u}$ is the corresponding solution of (4.2). Then, we have

$$\begin{aligned} r &= \int_0^\pi f(t, u(t))\varphi(t)\,dt \\ &\leq \sqrt{\tfrac{2}{\pi}}\left[\int_0^\delta k(t)\,dt + \int_\delta^{\pi-\delta} \frac{s^*}{2\sqrt{2\pi}}\,dt + \int_{\pi-\delta}^{\pi} k(t)\,dt\right] < s^*. \end{aligned}$$

The function $u_1 = \tilde{u} + \bar{u}\varphi$ is a solution of (4.1) with $s = r$, i.e. it is a $W^{2,1}$-upper solution of (4.1) with $s = s^*$ such that $u_a \prec u_1$.

From the same argument but with $\bar{u}$ small enough, we obtain a $W^{2,1}$-upper solution u_2 such that $u_2 \prec u_a$.

Conclusion – The result follows from Theorem III-3.11. □

Problem 4.1. Prove the following result.

Let $s \in \mathbb{R}$ and $\tilde{p} \in \tilde{L}^1(0,\pi)$. Assume that f is an L^1-Carathéodory function such that for some $k \in L^\infty(0,\pi)$, for a.e. $t \in [0,\pi]$, and all $u \in \mathbb{R}$,

$$|f(t,u)| \leq k(t).$$

Assume further that for some $\mu \in [1,2[$

$$\lim_{|u|\to\infty} |u|^{\mu-1} f(t,u) = 0 \text{ and } \liminf_{|u|\to\infty} \operatorname{sgn}(u)|u|^{\mu} f(t,u) \geq m > 0$$

uniformly in t.

Then there exist $s_0 < 0 < s_1$ such that
(i) if $s \notin [s_0, s_1]$, the problem (4.1) has no solution;
(ii) if $s \in [s_0, s_1]$, the problem (4.1) has at least one solution.

If moreover, for some $L \in L^1(0,\pi;\mathbb{R}^+)$ and for a.e. $t \in [0,\pi]$, $f(t,u) + L(t)u$ is non-decreasing in u then,
(iii) if $s \in]s_0, s_1[\setminus\{0\}$, the problem (4.1) has at least two solutions.

Hint : Combine the ideas of Theorems 4.2, 4.3 and VII-5.3.

In the preceding theorems, resonance is avoided with a restoring force $f(t,u)$ depending on the position u. A similar result holds for a system with a nonlinearity that depends only on the derivative. More precisely, we consider the boundary value problem

$$\begin{aligned} u'' + u + g(u') &= s\varphi(t) + \tilde{p}(t), \\ u(0) = 0, \ u(\pi) &= 0, \end{aligned} \tag{4.3}$$

where g is a continuous function in $\mathbb{R}$ and $\tilde{p} \in \tilde{L}^1(0,\pi)$. The following theorem gives conditions so that the structure of the solution set is described by figure 5, where l is 0.

Theorem 4.4. *Let $g : \mathbb{R} \to \mathbb{R}$ be locally Lipschitz continuous and such that the limits $g(-\infty)$ and $g(+\infty)$ exist and are finite. Let $s \in \mathbb{R}$ and $\tilde{p} \in \tilde{L}^1(0,\pi)$.*

Then there exists real numbers $s_0 = s_0(\tilde{p}, g) \leq s_1 = s_1(\tilde{p}, g)$ such that

$$l := \sqrt{\tfrac{2}{\pi}}(g(-\infty) + g(+\infty)) \in [s_0, s_1] \tag{4.4}$$

and problem (4.3) has
(i) no solution if $s \notin [s_0, s_1]$;
(ii) at least one solution if $s \in]s_0, s_1[$, $s \in \{s_0, s_1\} \setminus \{l\}$ or $s = s_0 = s_1$;
(iii) at least two ordered solutions u_1 and u_2 if $s \in]s_0, s_1[\setminus\{l\}$.

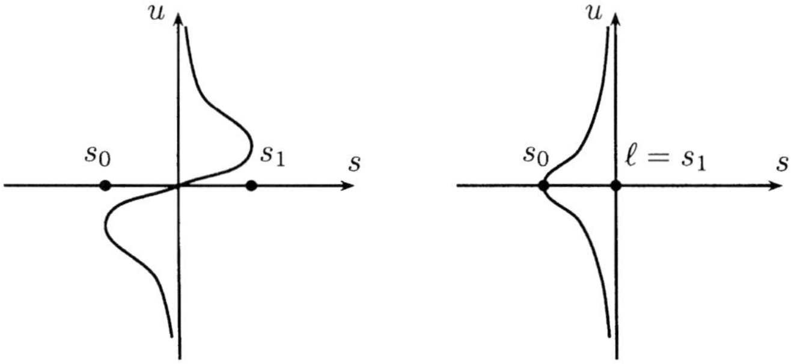

Fig. 5 : generic examples of Theorem 4.4

To prove this theorem, we shall use the following lemma which shows that v takes negative values arbitrarily near $t = 0$.

Lemma 4.5. *Let $m \in L^1(0,a)$ and $v \in W^{2,1}(0,a)$ satisfy*
(a) $v'' + m(t)v' + v < 0$ in $]0,a[$,
(b) $v(0) = 0$, $v'(0) = 0$.
Then there exists $a_1 \in]0,a[$ such that $v(a_1) < 0$.

Proof. Suppose that $v \geq 0$ in $]0,a[$. Let $M(t) = \int_0^t m(s)\,ds$. Assumption (a) can be written

$$(v'(t) \exp M(t))' + v(t) \exp M(t) < 0 \quad \text{in }]0,a[$$

and $v'(t) \exp M(t)$ is strictly decreasing in $[0,a]$. From (b) we infer that $v < 0$ in $]0,a[$, a contradiction. □

Proof of Theorem 4.4 : First we note that without loss of generality we can suppose $l = 0$, i.e. $g(-\infty) + g(+\infty) = 0$. In fact, replacing g, $\tilde{p}$ and s by $h(v) := g(v) - \frac{1}{2}\sqrt{\frac{\pi}{2}}\,l$, $\tilde{q}(t) = \tilde{p}(t) - \frac{1}{2}\sqrt{\frac{\pi}{2}}\,\tilde{l}(t)$ and $r = s - \frac{1}{2}\sqrt{\frac{\pi}{2}}\,\bar{l}$, where $l = \tilde{l}(t) + \bar{l}\varphi(t)$, we obtain an equivalent problem that satisfies the same assumptions as (4.3) with $l = 0$. Henceforth, from now on we assume this condition.

Claim 1 – There exists an s^* such that problem (4.3), *with $s = s^*$, has at least one solution.* We proceed as in Remark 4.1. The equivalent of (4.2) is

$$\begin{aligned} u'' + u &= \tilde{p} - Qg(a\varphi' + u'), \\ u(0) = 0,\ u(\pi) &= 0,\ \bar{u} = 0 \end{aligned} \tag{4.5}$$

which can be written as a fixed point problem

$$u = \tilde{T}_a u := M(\tilde{p} - Qg(a\varphi' + u')),$$

to which Schauder's Theorem applies. There exists a $K > 0$ so that given any $a \in \mathbb{R}$, the problem (4.5) has a solution $\tilde{u}_a$ with $\|\tilde{u}_a\|_{C^1} \leq K$ and the function $u = \tilde{u}_a + a\varphi$ solves (4.3), with $s = s^* := \int_0^\pi g(a\varphi'(r) + \tilde{u}_a'(r))\varphi(r)\,dr$.

Claim 2 – (i) holds. If u is a solution of (4.3),

$$|s| = |\int_0^\pi g(u'(t))\varphi(t)\,dt|$$

is bounded and we can define $s_0 := \inf\{s \mid (4.3) \text{ has a solution}\}$ and $s_1 := \sup\{s \mid (4.3) \text{ has a solution}\}$. Hence, assertion (i) is satisfied.

Claim 3 – $l = 0 \in [s_0, s_1]$. Using Lebesgue's Theorem and the uniform bound on $\|\tilde{u}_a\|_{C^1}$, we have

$$\lim_{a\to\infty} \int_0^\pi g(a\varphi'(r) + \tilde{u}_a'(r))\varphi(r)\,dr = 0.$$

Hence, choosing a sufficiently large, we obtain numbers

$$s := \int_0^\pi g(a\varphi'(r) + \tilde{u}_a'(r))\varphi(r)\,dr$$

arbitrarily small such that (4.3) has a solution, i.e. $s \in [s_0, s_1]$. Going to the limit we have $l = 0 \in [s_0, s_1]$.

Claim 4 – If $s \in]s_0, s_1[$, the problem (4.1) has at least one solution. Let $s \in]s_0, s_1[$. As in Step 3 of the proof of Theorem 4.1 it is easy to build lower and upper solutions of (4.3) so that the claim follows from Corollary II-2.2 or Theorem III-3.8 according as the lower and upper solutions are well-ordered or not.

Claim 5 – For any $\epsilon > 0$ there exists $K_0 > 0$ such that if u is a solution of (4.3) for some s with $|s| \geq \epsilon$, then $\|u\|_{C^1} \leq K_0$. Assume that there exists a sequence $(s_n)_n$ with $|s_n| \geq \epsilon$ and corresponding solutions u_n with $\|u_n\|_{C^1} \to \infty$. Write $u_n(t) = \bar{u}_n\varphi(t) + \tilde{u}_n(t)$. As in Claim 1 we see that $\|\tilde{u}_n\|_{C^1} \leq K$. Hence $|\bar{u}_n| \to \infty$. Also, we obtain from Lebesgue's Dominated Convergence Theorem

$$\lim_{n\to\infty} \left| \int_0^\pi g(\bar{u}_n\varphi'(\tau) + \tilde{u}_n'(\tau))\varphi(\tau)\,d\tau \right| = 0,$$

i.e. $\lim_{n\to\infty} s_n = 0$, which is a contradiction.

Claim 6 – If $s \in \{s_0, s_1\} \setminus \{l\}$, the problem (4.3) has at least one solution. This claim follows from Claim 5 and Arzelà-Ascoli Theorem.

Claim 7 – (iii) holds. Let $s \in]0, s_1[$. We build a strict lower solution α as in Part 1 of Theorem 4.3, using Proposition III-2.7 together with Remark III-2.1. Then we choose $\bar{u}$ large enough so that if $\tilde{u}$ is the corresponding solution of (4.5), we have for all $t \in [0, \pi]$

$$|\alpha(t) - \tilde{u}(t)| \leq \bar{u}\varphi(t)$$

and

$$\int_0^\pi g(\pm\bar{u}\varphi'(t) + \tilde{u}'(t))\varphi(t)\,dt < s.$$

In that way, $\pm\bar{u}\varphi + \tilde{u}$ are upper solutions of (4.3). We conclude applying Corollary II-2.2 and Theorem III-3.8. □

Example 4.4. A model example is

$$u'' + u + \arctan u' = s\varphi(t),$$
$$u(0) = 0, \ u(\pi) = 0,$$

which is quite different from Example 4.2 as we have here a set $]s_0, s_1[\backslash\{0\}$ so that the problem has two solutions. Here $s_0 < 0 < s_1$ as follows from Proposition 4.6 below.

Example 4.5. Notice also that Theorem 4.4 can be applied to the resonant problem

$$u'' + u = s\varphi(t),$$
$$u(0) = 0, \ u(\pi) = 0,$$

which is such that $s_0 = s_1 = l = 0$.

The proposition that follows gives conditions that rule out the situation of the last example.

Proposition 4.6. *Assume that $g : \mathbb{R} \to \mathbb{R}$ is of class C^1, $g(0) = 0$, the limits $g(-\infty)$ and $g(+\infty)$ exist and $g'(0) \neq 0$. Let $s \in \mathbb{R}$ and $\tilde{p} \in \tilde{C}([0,\pi])$. Then there exists $\epsilon > 0$ such that, if $\|\tilde{p}\|_\infty < \epsilon$, we have $s_0(\tilde{p}, g) < 0 < s_1(\tilde{p}, g)$.*

Proof. The mapping $\mathcal{F} : C^2([0,\pi]) \cap C_0^1([0,\pi]) \to C([0,\pi])$, $u \mapsto u'' + u + g(u')$ is differentiable at $u = 0$. Since the problem

$$v'' + v + cv' = 0,$$
$$v(0) = 0, \ v(\pi) = 0,$$

where c is a nonzero constant, has only the trivial solution, it follows that $\mathcal{F}'(0)$ is an isomorphism. Hence, by the Inverse Mapping Theorem, the range of $\mathcal{F}$ contains an open ball centered at the origin and we can conclude. □

We can also have that l coincides with one of the values s_0 or s_1 as in the following proposition.

Proposition 4.7. *Let $\tilde{p} \in \tilde{L}^1(0,\pi)$ and $g : \mathbb{R} \to \mathbb{R}$ be locally Lipschitz continuous such that the limits $g(-\infty)$ and $g(+\infty)$ exist and are finite. Assume that for all $v \in \mathbb{R}$*

$$g(v) < g(-\infty) = g(+\infty).$$

Then we have $s_0(\tilde{p}, g) < s_1(\tilde{p}, g) = l = 2\sqrt{\frac{2}{\pi}}\, g(+\infty)$.

Proof. Notice that any solution u of (4.3) is such that

$$s = \int_0^\pi g(u'(t))\varphi(t)\, dt < 2\sqrt{\frac{2}{\pi}}\, g(+\infty).$$

From Claim 1 of the proof of Theorem 4.4, we know there exists a solution u of (4.3) for some value of s and the claim follows. $\square$

Chapter VII

Resonance and Non-resonance

1 A Periodic Non-resonant Problem

Consider the problem

$$\begin{aligned} &u'' + f(t,u) = h(t), \\ &u(a) = u(b), \;\; u'(a) = u'(b). \end{aligned} \tag{1.1}$$

Our first result considers a nonlinearity with an asymptotic slope under the first eigenvalue.

Theorem 1.1. *Let $h \in L^1(a,b)$. Assume f satisfies L^1-Carathéodory conditions and*

$$\limsup_{|u|\to\infty} \frac{f(t,u)}{u} \le \gamma(t) \lneqq 0 \qquad \textit{uniformly in } t \in [a,b],$$

where $\gamma \in L^1(a,b)$.

Then the periodic problem (1.1) *has at least one solution.*

Proof. Let λ_1 be the first eigenvalue of the problem

$$\begin{aligned} &u'' + \gamma(t)\,u + \lambda\,u = 0, \\ &u(a) = u(b), \;\; u'(a) = u'(b). \end{aligned}$$

It is known (see Proposition A-3.9) that the corresponding eigenfunction φ_1 can be chosen positive.

Step 1 – $\lambda_1 > 0$. Observe that

$$\lambda_1 = \min_{u\in H\setminus\{0\}} \frac{\int_a^b (u'^2 - \gamma u^2)\,dt}{\int_a^b u^2\,dt} \ge 0,$$

where $H = \{u \in H^1(a,b) \mid u(a) = u(b)\}$ (see Theorem A-3.7). If $\lambda_1 = 0$, we compute

$$0 = \int_a^b (\varphi_1''(t) + \gamma(t)\varphi_1(t))\,dt = \int_a^b \gamma(t)\varphi_1(t)\,dt.$$

This equation contradicts the fact that $\gamma(t)\varphi_1(t) < 0$ on a set of positive measure.

Step 2 – Construction of an upper solution. Let ψ be the solution of the boundary value problem

$$\begin{gathered}\psi'' + (\gamma(t) + \tfrac{\lambda_1}{2})\,\psi = -1,\\ \psi(a) = \psi(b),\ \psi'(a) = \psi'(b).\end{gathered}$$

We deduce from the maximum principle, Theorem A-5.3, that $\psi(t) > 0$ on $[a,b]$.

By hypothesis, there exists $r \in L^1(a,b)$ such that for all $u \geq 0$ and a.e. $t \in [a,b]$,

$$f(t,u) \leq (\gamma(t) + \frac{\lambda_1}{2})\,u + r(t).$$

Let w be the solution of

$$\begin{gathered}w'' + (\gamma(t) + \tfrac{\lambda_1}{2})\,w + r(t) - h(t) = 0,\\ w(a) = w(b),\ w'(a) = w'(b).\end{gathered}$$

If k is large enough, the function $\beta(t) = w(t) + k\,\psi(t)$ is a positive upper solution of (1.1).

Step 3 – Conclusion. In the same way we construct a negative lower solution $\alpha \leq 0$ and the claim follows from Theorem I-3.1. □

Notice that we can always choose $\gamma \in L^\infty(a,b)$ since $\max\{\gamma, -1\}$ satisfies also the assumptions. Moreover, we can generalize this result using functions γ that can take positive values provided these are not too large. For such functions, we define $\gamma^+(t) = \max(\gamma(t), 0)$ and $\gamma^-(t) = \max(-\gamma(t), 0)$.

Theorem 1.2. *Let $h \in L^1(a,b)$. Assume f satisfies L^1-Carathéodory conditions and*

$$\limsup_{|u|\to\infty} \frac{f(t,u)}{u} \leq \gamma(t) \quad \textit{uniformly in } t \in [a,b],$$

where $\gamma \in L^1(a,b)$ is such that

$$\|\gamma^+\|_{L^1} < \tfrac{3}{b-a} \quad \textit{and} \quad \|\gamma^+\|_{L^1} - (1 - \sqrt{\tfrac{b-a}{3}}\,\|\gamma^+\|_{L^1}^{1/2})^2 \|\gamma^-\|_{L^1} < 0.$$

Then the periodic problem (1.1) has at least one solution.

Proof. Define λ_1 to be the first eigenvalue of the eigenvalue problem

$$\begin{gathered} u'' + \gamma(t)u + \lambda\, u = 0, \\ u(a) = u(b), \;\; u'(a) = u'(b), \end{gathered}$$

and φ_1 the corresponding eigenfunction which can be chosen positive (see Proposition A-3.9).

Step 1 – $\lambda_1 > 0$. Let us prove that for all $u \in H^1(a,b)$, $u \not\equiv 0$, $u(a) = u(b)$,

$$\int_a^b [u'^2(t) - \gamma(t)u^2(t)]dt > 0.$$

Otherwise, there exists $u \not\equiv 0$ such that

$$\int_a^b u'^2(t)\, dt \leq \int_a^b \gamma(t)u^2(t)\, dt \tag{1.2}$$

and

$$\|u'\|_{L^2}^2 \leq \|u\|_\infty^2 \|\gamma^+\|_{L^1}.$$

Let us write $u(t) = \bar{u} + \tilde{u}(t)$, where $\bar{u} = \frac{1}{b-a}\int_a^b u(t)\, dt$. We have

$$\|\tilde{u}\|_\infty \leq c\, \|u'\|_{L^2} \leq c\, \|u\|_\infty \|\gamma^+\|_{L^1}^{1/2}$$

with $c = \sqrt{\frac{b-a}{12}}$ (see Proposition A-4.1) and

$$\begin{aligned} |u(t)| &\geq |\bar{u}| - |\tilde{u}(t)| \geq \|u\|_\infty - \|\tilde{u}\|_\infty - |\tilde{u}(t)| \\ &\geq \|u\|_\infty - 2\|\tilde{u}\|_\infty \geq \|u\|_\infty (1 - 2\,c\, \|\gamma^+\|_{L^1}^{1/2}). \end{aligned}$$

It follows now that

$$\begin{aligned} \int_a^b \gamma(t)u^2(t)dt &= \int_a^b \gamma^+(t)u^2(t)dt - \int_a^b \gamma^-(t)u^2(t)dt \\ &\leq \|u\|_\infty^2 \left[\, \|\gamma^+\|_{L^1} - (1 - 2\,c\, \|\gamma^+\|_{L^1}^{1/2})^2 \|\gamma^-\|_{L^1}\right] < 0 \end{aligned}$$

and we have a contradiction with (1.2).

Step 2 – *Construction of an upper solution.* Notice first that there exists an $r > 0$ such that for all $u \geq r$ and a.e. $t \in [a,b]$,

$$f(t,u) \leq (\gamma(t) + \frac{\lambda_1}{2})\, u.$$

Define w to be the solution of

$$\begin{gathered} w'' + (\gamma(t) + \tfrac{\lambda_1}{2})w - h(t) = 0, \\ w(a) = w(b), \;\; w'(a) = w'(b). \end{gathered}$$

Then for $k > 0$ large enough, $\beta(t) = w(t) + k\varphi_1(t) \geq r$ is an upper solution.

Step 3 – Conclusion. In the same way, we construct a negative lower solution and we conclude from Theorem I-3.1. □

We can obtain similar results for a nonlinearity with asymptotic slope between the first and the second eigenvalue.

Theorem 1.3. *Let $h \in L^1(a,b)$. Assume f satisfies L^1-Carathéodory conditions and*

$$\gamma_1(t) \le \liminf_{|u|\to\infty} \frac{f(t,u)}{u} \le \limsup_{|u|\to\infty} \frac{f(t,u)}{u} \le \gamma_2(t) \quad \textit{uniformly in } t \in [a,b],$$

where γ_1 and $\gamma_2 \in L^1(a,b)$ are such that

$$\int_a^b \gamma_1(t)\,dt \ge 0, \quad \gamma_1 \not\equiv 0 \quad \textit{and} \quad \gamma_2(t) \lneqq (\tfrac{2\pi}{b-a})^2.$$

Then the periodic problem (1.1) has at least one solution.

Proof. Let λ_i be the i-th eigenvalue of

$$\begin{gathered} u'' + \gamma_1(t)u + \lambda u = 0, \\ u(a) = u(b),\ u'(a) = u'(b), \end{gathered}$$

and let $\varphi_1 > 0$ be an eigenfunction corresponding to λ_1 (see Proposition A-3.9).

Step 1 – $\lambda_1 < 0$. Recall that (see Theorem A-3.7)

$$\lambda_1 = \min_{u \in H\setminus\{0\}} \Phi(u),$$

where

$$\Phi(u) = \frac{\int_a^b (u'^2 - \gamma_1 u^2)\,dt}{\int_a^b u^2\,dt} \quad \text{and} \quad H = \{u \in H^1(a,b) \mid u(a) = u(b)\}.$$

It follows that $\lambda_1 \le \Phi(1) = -\frac{1}{b-a}\int_a^b \gamma_1(t)\,dt \le 0$. If $\lambda_1 = 0$, we have $\Phi(1) = 0$ which implies we can choose $\varphi_1(t) = 1$ (see Proposition A-3.11). Hence the equation satisfied by φ_1 implies $\gamma_1 = 0$, which contradicts the assumptions.

Step 2 – Construction of lower and upper solutions. Let $\lambda \in]\lambda_1, \min\{0, \lambda_2\}[$. It follows from the assumptions that for some $r > 0$ and all $u \ge r$, $f(t,u) \ge (\gamma_1 + \lambda)u$. Define next w to be the solution of

$$\begin{gathered} w'' + (\gamma_1(t) + \lambda)w - h(t) = 0, \\ w(a) = w(b),\ w'(a) = w'(b). \end{gathered}$$

Then for $k > 0$ large enough, $\alpha(t) = w(t) + k\varphi_1(t) \ge r$ is a lower solution.

In the same way, we construct an upper solution $\beta \le -r$.

Step 3 – Conclusion. Proposition III-3.4 proves that nontrivial solutions of the problem

$$\begin{gathered} u'' + p(t)u^+ - q(t)u^- = 0, \\ u(a) = u(b), \;\; u'(a) = u'(b), \end{gathered}$$

do not have zeros if $\gamma_1 \leq p \leq \gamma_2$ and $\gamma_1 \leq q \leq \gamma_2$. The claim follows now from Theorem III-3.3. □

2 Resonance Conditions for the Periodic Problem

Consider the periodic problem (1.1) and notice that by adding a constant to both the functions $f(t, u)$ and $h(t)$, we can suppose that

$$\int_a^b h(t)\, dt = 0.$$

A first result, which is of Landesman-Lazer type concerns a nonlinearity with asymptotic slope on the left of the first eigenvalue.

Theorem 2.1. *Let $h \in L^1(a, b)$ be such that $\int_a^b h(t)\, dt = 0$. Assume f satisfies L^1-Carathéodory conditions and*
(a) there exist $s_+ \in \mathbb{R}$ and $f_+ \in L^1(a, b)$ such that

$$f(t, u) \leq f_+(t) \quad \text{if } u \geq s_+$$

and

$$\int_a^b f_+(t)\, dt \leq 0;$$

(b) there exist $s_- \in \mathbb{R}$ and $f_- \in L^1(a, b)$ such that

$$f(t, u) \geq f_-(t) \quad \text{if } u \leq s_-$$

and

$$\int_a^b f_-(t)\, dt \geq 0.$$

Then the periodic problem (1.1) *has at least one solution.*

Proof. To build a lower solution α, we consider the solution w of

$$\begin{gathered} w'' + (\tilde{f}_- - h)(t) = 0, \\ w(a) = w(b), \;\; w'(a) = w'(b), \end{gathered}$$

with $\int_a^b w(t)\, dt = 0$. Then for k large enough, $\alpha(t) = -k + w(t)$ is a lower solution. In the same way we construct an upper solution $\beta > \alpha$ and the theorem follows from Theorem I-3.1. □

The next result uses the Landesman-Lazer conditions to consider a non-linearity with asymptotic slope between the two first eigenvalues.

Theorem 2.2. *Let $h \in L^1(a,b)$ be such that $\int_a^b h(t)\,dt = 0$. Assume f satisfies L^1-Carathéodory conditions and*
(a) there exist $s_+ \in \mathbb{R}$, $f_+ \in L^1(a,b)$ such that if $u \geq s_+$ we have

$$f(t,u) \geq f_+(t)$$

and

$$\int_a^b f_+(t)\,dt \geq 0;$$

(b) there exist $s_- \in \mathbb{R}$, $f_- \in L^1(a,b)$ such that if $u \leq s_-$ we have

$$f(t,u) \leq f_-(t)$$

and

$$\int_a^b f_-(t)\,dt \leq 0;$$

(c) $\limsup_{|u|\to\infty} \frac{f(t,u)}{u} \lneqq \left(\frac{2\pi}{b-a}\right)^2$ *uniformly in t.*

Then the periodic problem (1.1) has at least one solution.

Proof. The argument of the proof of Theorem 2.1 gives lower and upper solutions α and β such that $\alpha > \beta$. We conclude then as in Step 3 of Theorem 1.3. □

Another type of result uses a monotone nonlinearity.

Theorem 2.3. *Let $h \in L^1(a,b)$ be such that $\int_a^b h(t)\,dt = 0$ and f be a continuous, non-increasing function. Then the problem*

$$u'' + f(u) = h(t),$$
$$u(a) = u(b),\ u'(a) = u'(b)$$

has at least one solution if and only if $0 \in \operatorname{Im}(f)$.

Proof. The sufficient part of the claim can be deduced from Theorem 2.1. The necessary condition follows by direct integration of the equation. □

3 A Non-resonant Dirichlet Problem

Consider the Dirichlet problem

$$\begin{aligned} u'' + u + f(t,u) &= h(t), \\ u(0) = 0, \ u(\pi) &= 0. \end{aligned} \tag{3.1}$$

The first results consider a nonlinearity with asymptotic slope on the left of the first eigenvalue.

Theorem 3.1. *Let $h \in L^1(0,\pi)$. Assume f satisfies L^1-Carathéodory conditions and there exists $\gamma \in L^1(0,\pi)$ such that*

$$\limsup_{|u|\to\infty} \frac{f(t,u)}{u} \le \gamma(t) \lneqq 0 \text{ uniformly in } t.$$

Then the Dirichlet problem (3.1) *has at least one solution.*

Proof. Let λ_1 be the first eigenvalue of the problem

$$\begin{aligned} u'' + (1+\gamma(t))\, u + \lambda\, u &= 0, \\ u(0) = 0, \ u(\pi) &= 0. \end{aligned}$$

Arguing as in the proof of Theorem 1.1 and using Theorem A-3.2 we see that $\lambda_1 > 0$. Hence, we can find a function $\psi(t) > 0$ on $[0,\pi]$ such that

$$\psi'' + (1 + \gamma(t) + \frac{\lambda_1}{2})\, \psi = 0.$$

By hypothesis, there exists $a \in L^1(0,\pi)$ such that for all $u \ge 0$ and a.e. $t \in [0,\pi]$,

$$f(t,u) \le (\gamma(t) + \frac{\lambda_1}{2})\, u + a(t).$$

Let w be the solution of

$$\begin{aligned} w'' + (1 + \gamma(t) + \tfrac{\lambda_1}{2})\, w + a(t) - h(t) &= 0, \\ w(0) = 0, \ w(\pi) &= 0. \end{aligned}$$

If B is large enough, the function $\beta(t) = w(t) + B\,\psi(t)$ is a positive upper solution.

In the same way we construct a lower solution $\alpha \le 0$ and the theorem follows from Theorem II-2.4. □

If the variables u and t separate as in

$$\begin{aligned} u'' + u + f(u) &= h(t), \\ u(0) = 0, \ u(\pi) &= 0, \end{aligned} \tag{3.2}$$

we can use conditions on the potential $F(u) = \displaystyle\int_0^u f(s)\, ds$.

Theorem 3.2. *Let $h \in L^\infty(0,\pi)$. Assume $f \in C(\mathbb{R})$ is such that*

$$\liminf_{u\to+\infty} \frac{F(u)}{u^2} < 0 \text{ and } \liminf_{u\to-\infty} \frac{F(u)}{u^2} < 0,$$

where $F(u) = \int_0^u f(s)\,ds.$

Then the problem (3.2) *has at least one solution.*

Proof. Claim – There exists a nonnegative upper solution of (3.2). In case $u + f(u)$ is unbounded from below on $\mathbb{R}^+$ we choose $\beta \in \mathbb{R}^+$ such that $\beta + f(\beta) \le -\|h\|_\infty$. The constant function β verifies the claim.

Assume next $u + f(u)$ is bounded from below on $\mathbb{R}^+$ and consider the equation

$$u'' + u + f(u) + M = 0, \tag{3.3}$$

where $M \ge \|h\|_\infty$ and $u + f(u) + M \ge 1$ on $\mathbb{R}^+$. We deduce from the assumptions that there exist a sequence $(v_n)_n$ and an $\epsilon > 0$ such that $v_n \to +\infty$ and $F(v_n) \le -\epsilon v_n^2$. Define $V(u) = \frac{1}{2}u^2 + F(u) + Mu$ and notice that the sequence $(\frac{(1-\epsilon)}{2}v_n^2 - V(v_n))_n$ goes to infinity with n since

$$\tfrac{(1-\epsilon)}{2}v_n^2 - V(v_n) \ge \tfrac{\epsilon}{2}v_n^2 - Mv_n.$$

Therefore, we can find a $u_0 > 0$ such that $V'(u_0) = u_0 + f(u_0) + M \ge 1$ and for all $u \in [0, u_0]$

$$\tfrac{(1-\epsilon)}{2}u^2 - V(u) \le \tfrac{(1-\epsilon)}{2}u_0^2 - V(u_0).$$

Consider now the solution β of (3.3) such that $\beta(\pi/2) = u_0$, $\beta'(\pi/2) = 0$, whose energy is $E = V(u_0)$. This solution cannot take the value zero on the interval $]\frac{\pi}{2} - T, \frac{\pi}{2} + T[$, where

$$T = \tfrac{1}{\sqrt{2}} \int_0^{u_0} \frac{du}{\sqrt{V(u_0) - V(u)}} \ge \tfrac{1}{\sqrt{1-\epsilon}} \int_0^{u_0} \frac{du}{\sqrt{u_0^2 - u^2}} = \tfrac{\pi}{2\sqrt{1-\epsilon}} > \tfrac{\pi}{2}.$$

The claim follows.

Conclusion – In a similar way we build a negative lower solution and the theorem follows from Theorem II-2.4. □

Example 3.1. Consider the problem (3.2) where

$$f(u) = u\sin(\ln(u^2+1)) + \frac{u^3}{u^2+1}\cos(\ln(u^2+1)). \tag{3.4}$$

Theorem 3.1 does not apply since

$$\limsup_{|u|\to\infty} \frac{f(u)}{u} = \sqrt{2}.$$

However, we compute $F(u) = \frac{u^2}{2}\sin(\ln(u^2+1))$ and

$$\liminf_{u\to+\infty} \frac{F(u)}{u^2} = -\tfrac{1}{2} \quad \text{and} \quad \liminf_{u\to-\infty} \frac{F(u)}{u^2} = -\tfrac{1}{2},$$

so that existence follows from Theorem 3.2.

The next result considers a nonlinearity with asymptotic slope between the first and the second eigenvalue.

Theorem 3.3. *Let $h \in L^1(0,\pi)$. Assume f satisfies L^1-Carathéodory conditions and*

$$\gamma_1(t) \le \liminf_{|u|\to\infty} \frac{f(t,u)}{u} \le \limsup_{|u|\to\infty} \frac{f(t,u)}{u} \le \gamma_2(t) \text{ uniformly in } t \in [0,\pi],$$

where γ_1 and $\gamma_2 \in L^1(0,\pi)$ are such that

$$\int_0^\pi \gamma_1(t)\sin^2 t\,dt \ge 0,\ \gamma_1 \not\equiv 0 \quad \text{and} \quad \gamma_2(t) \lneqq 3.$$

Then the Dirichlet problem (3.1) *has at least one solution.*

Proof. Consider the eigenvalue problem

$$\begin{gathered} u'' + (1+\gamma_1(t))u + \lambda u = 0, \\ u(0) = 0,\ u(\pi) = 0. \end{gathered}$$

We show as in the proof of Theorem 1.3 and using Theorem A-3.2 that its first eigenvalue λ_1 is negative.

Next, we fix some $\lambda \in]\lambda_1, \min\{0,\lambda_2\}[$. By assumption, there exists $k \in L^1(0,\pi)$ so that for all $u \ge 0$ and a.e. $t \in [0,\pi]$ we have $f(t,u) \ge -k(t) + (\gamma_1(t)+\lambda)u$.

At last, we define w to be the solution of

$$\begin{gathered} w'' + (1+\gamma_1(t)+\lambda)w - k(t) - h(t) = 0, \\ w(a) = 0,\ w(b) = 0. \end{gathered}$$

As the first eigenfunction φ_1 can be chosen positive (see Proposition A-3.4), we can choose $A > 0$ large enough so that $\alpha(t) = w(t) + A\varphi_1(t) \ge 0$ is a lower solution.

In the same way, we construct an upper solution $\beta \le 0$ and the theorem follows from Theorem III-3.9 and the equivalent of Proposition III-3.4 for the Dirichlet problem. □

We can write an analog of Theorem 3.2 in case the nonlinearity has an asymptotic slope between the first and the second eigenvalue.

Theorem 3.4. *Let $h \in L^\infty(0,\pi)$. Assume $f \in C(\mathbb{R})$ is such that*

$$\limsup_{u\to+\infty} \frac{F(u)}{u^2} > 0 \text{ and } \limsup_{u\to-\infty} \frac{F(u)}{u^2} > 0,$$

where $F(u) = \int_0^u f(s)\,ds$ and

$$\limsup_{|u|\to\infty} \frac{f(u)}{u} < 3. \tag{3.5}$$

Then the problem (3.2) *has at least one solution.*

Proof. We deduce from the assumptions that there exist a sequence $(v_n)_n$ and an $\epsilon > 0$ such that $v_n \to +\infty$ and $F(v_n) \geq \epsilon v_n^2$. Define $V(u) = \frac{1}{2}u^2 + F(u) - \|h\|_\infty u$ and notice that the sequence $(\frac{(1+\epsilon)}{2}v_n^2 - V(v_n))_n$ goes to $-\infty$ with n since

$$\tfrac{(1+\epsilon)}{2}v_n^2 - V(v_n) \leq -\tfrac{\epsilon}{2}v_n^2 + \|h\|_\infty v_n.$$

Therefore, we can find $u_0 > 0$ which is such that for all $u \in [0, u_0]$

$$\tfrac{(1+\epsilon)}{2}u^2 - V(u) \geq \tfrac{(1+\epsilon)}{2}u_0^2 - V(u_0).$$

Observe that $V(u) < V(u_0)$ for all $u \in [0, u_0[$ and then $V'(u_0) = u_0 + f(u_0) - \|h\|_\infty \geq 0$. Consider now the solution w of

$$\begin{aligned} w'' + w + f(w) - \|h\|_\infty = 0,\\ w(0) = u_0,\ w'(0) = 0. \end{aligned}$$

There exists $T > 0$ such that $w(t) > 0$ on $[0, T[$, $w(T) = 0$ and $w'(t) < 0$ on $]0, T[$. We compute then

$$T = \tfrac{1}{\sqrt{2}} \int_0^{u_0} \frac{du}{\sqrt{V(u_0) - V(u)}} \leq \tfrac{1}{\sqrt{1+\epsilon}} \int_0^{u_0} \frac{du}{\sqrt{u_0^2 - u^2}} = \tfrac{\pi}{2\sqrt{1+\epsilon}} < \tfrac{\pi}{2}.$$

Hence, the function

$$\begin{aligned} \alpha(t) &= w(T-t), && \text{if } t \in [0, T[,\\ &= u_0, && \text{if } t \in [T, \pi - T[,\\ &= w(t - \pi + T), && \text{if } t \in [\pi - T, \pi] \end{aligned}$$

is a nonnegative lower solution of (3.2).

In the same way, we build a nonpositive upper solution β. If $u + f(u)$ is unbounded from below for $u \to +\infty$, the problem (3.2) has an upper solution $\bar{\beta} > \alpha$ and the result follows from Theorem II-2.4. We work in the same way the case $u + f(u)$ unbounded from above for $u \to -\infty$. Hence we can assume

$$\liminf_{|u|\to+\infty} \frac{u + f(u) - h(t)}{u} \geq 0$$

and we conclude by Theorem III-3.9 and the equivalent of Proposition III-3.4 for the Dirichlet problem. □

Example 3.2. Consider again the problem (3.2) with f defined in (3.4). We know that existence of a solution follows from Theorem 3.2. We can also apply Theorem 3.4 so that existence follows both from a non-resonance result on the left and on the right of the first eigenvalue.

Exercise 3.1. Prove, using the Three Solutions Theorem (Theorem III-2.11), that problem (3.2) with f defined in (3.4)and $h \in L^\infty(0, \pi)$ has at least three solutions.

We can generalize the above theorem, replacing (3.5) by a condition on the potential.

Exercise 3.2. Let $h \in L^\infty(0, \pi)$. Assume $f \in C(\mathbb{R})$ is such that

$$\limsup_{u\to+\infty} \frac{F(u)}{u^2} > 0 \quad \text{and} \quad \limsup_{u\to-\infty} \frac{F(u)}{u^2} > 0,$$

where $F(u) = \int_0^u f(s)\, ds$, and for some μ, ν such that $\frac{1}{\sqrt{\mu+1}} + \frac{1}{\sqrt{\nu+1}} = 1$,

$$\limsup_{u\to+\infty} \frac{f(u)}{u} \leq \mu \quad \text{and} \quad \limsup_{u\to-\infty} \frac{f(u)}{u} \leq \nu,$$
$$\liminf_{u\to+\infty} \frac{2F(u)}{u^2} < \mu \quad \text{and} \quad \liminf_{u\to-\infty} \frac{2F(u)}{u^2} < \nu.$$

Prove then that the problem (3.2) has at least one solution.

Hint : See Exercise III-3.3 or [79].

Our last result concerns the problem

$$\begin{gathered} u'' + (1 + \gamma(t))u + f(t, u) = h(t), \\ u(0) = 0, \ u(\pi) = 0, \end{gathered} \tag{3.6}$$

where the nonlinearity $f(t, u)$ is monotone.

Theorem 3.5. *Let $h \in L^1(0,\pi)$ and $\gamma \in L^1(0,\pi)$ be such that $\gamma(t) \lneqq 0$. Assume $f : [0,\pi] \times \mathbb{R} \to \mathbb{R}$ satisfies L^1-Carathéodory conditions, $f(t,0) \in L^\infty(0,\pi)$ and for each $t \in [0,\pi]$, $f(t,.)$ is a non-increasing function.*

Then the Dirichlet problem (3.6) has a unique solution.

Proof. Compute first

$$\limsup_{|u|\to\infty} \frac{f(t,u)+\gamma(t)u}{u} \le \lim_{|u|\to\infty} \frac{f(t,0)}{u} + \gamma(t) = \gamma(t),$$

uniformly in t. Hence by Theorem 3.1 we have the existence of a solution u. We only have to prove its uniqueness. Assume that u and v are two distinct solutions of (3.6) and let $w = u - v$. From the proof of Theorem 3.1, we know $\lambda_1 > 0$ and we compute

$$\begin{aligned} 0 = &-\int_0^\pi w'^2(s)\,ds + \int_0^\pi (1+\gamma(s))w^2(s)\,ds \\ &+ \int_0^\pi (f(s,u(s)) - f(s,v(s)))(u(s)-v(s))\,ds \\ \le &-\lambda_1 \int_0^\pi w^2(s)\,ds + \int_0^\pi (f(s,u(s)) - f(s,v(s)))(u(s)-v(s))\,ds < 0, \end{aligned}$$

which is a contradiction. □

4 Landesman-Lazer conditions for the Dirichlet problem

Consider the Dirichlet problem (3.1) and define $\varphi(t) = \sqrt{\frac{2}{\pi}}\sin t$. As in Section 2, we can assume

$$\int_0^\pi h(t)\varphi(t)\,dt = 0.$$

The first results consider a nonlinearity with asymptotic slope on the left of the first eigenvalue.

Proposition 4.1. *Let $h \in L^1(0,\pi)$ be such that $\int_0^\pi h(t)\sin t\,dt = 0$. Assume f is an L^1-Carathéodory function and*
(a) there exist $s_+ \in \mathbb{R}$ and $f_+ \in L^1(0,\pi)$ such that

$$f(t,u) \le f_+(t) \quad \text{if } u \ge s_+ \sin t$$

and

$$\int_0^\pi f_+(t)\sin t\,dt \le 0;$$

(b) there exist $s_- \in \mathbb{R}$ and $f_- \in L^1(0,\pi)$ such that

$$f(t,u) \geq f_-(t) \quad \textit{if } u \leq s_- \sin t$$

and

$$\int_0^\pi f_-(t) \sin t \, dt \geq 0.$$

Then the Dirichlet problem (3.1) has at least one solution.

Proof. Let us build a lower solution $\alpha(t) \leq s_- \sin t$. To this end, define first w to be the solution of

$$w'' + w + f_-(t) - h(t) - (\int_0^\pi f_-(s)\,\varphi(s)\,ds)\varphi(t) = 0,$$
$$w(0) = 0, \ w(\pi) = 0,$$

which satisfies $\int_0^\pi w(t)\,\varphi(t)\,dt = 0$. Next we choose $a < 0$ small enough so that $a\,\varphi(t) + w(t) \leq s_- \sin t$ and we verify that $\alpha(t) = a\,\varphi(t) + w(t)$ is a lower solution.

In the same way, we construct an upper solution $\beta(t) \geq s_+ \sin t$. As we have chosen $\alpha(t) \leq 0 \leq \beta(t)$, we have a solution of (3.1) by application of Theorem II-2.4. □

Remark 4.1. Notice that assumptions

$$u \leq s_- < 0 \Rightarrow f(t,u) \geq f_-(t) \quad \text{and} \quad u \geq s_+ > 0 \Rightarrow f(t,u) \leq f_+(t)$$

with

$$\int_0^\pi f_+(t) \sin t \, dt \leq 0, \int_0^\pi f_-(t) \sin t \, dt \geq 0,$$

as in Theorem 2.1, are not sufficient to prove the result in Proposition 4.1. Consider for example the problem (3.1) with $h = 0$, $f(t,u) = g(u) \sin t$, $g(u) = \cos u$ if $u \in [-\pi/2, \pi/2]$ and $g(u) = 0$ if $u \notin [-\pi/2, \pi/2]$. This problem has no solution since such a solution u verifies $\int_0^\pi g(u(t)) \sin^2 t \, dt = 0$, which is impossible.

The following is a result in this direction.

Theorem 4.2. *Let $h \in L^1(0,\pi)$ be such that $\int_0^\pi h(t) \sin t \, dt = 0$. Assume f is an L^1-Carathéodory function, and there exist f_+, $f_- \in L^1(0,\pi)$ such that* *(a)* $\limsup_{u\to+\infty} f(t,u) \le f_+(t)$ *uniformly in* t *and*

$$\int_0^\pi f_+(t)\,\sin t\,dt < 0;$$

(b) $\liminf_{u\to-\infty} f(t,u) \ge f_-(t)$ *uniformly in* t *and*

$$\int_0^\pi f_-(t)\,\sin t\,dt > 0.$$

Then the Dirichlet problem (3.1) *has at least one solution.*

Proof. We prove that assumption (a) implies the corresponding assumption in Proposition 4.1.

Observe first that (a) implies that there exist r and $\eta > 0$ such that for a.e. $t \in [0,\pi]$ and all $u \ge r$, we have

$$f(t,u) \le k_+(t) := f_+(t) + \eta$$

and

$$\int_0^\pi k_+(t)\,\sin t\,dt < 0.$$

Moreover, as f satisfies L^1-Carathéodory conditions, we have $a_r \in L^1(0,\pi)$ so that for a.e. $t \in [0,\pi]$ and all $u \ge 0$,

$$f(t,u) \le |k_+(t)| + a_r(t).$$

Let ϵ be such that

$$\int_0^\epsilon (|k_+(t)| + a_r(t))\,\sin t\,dt \\ + \int_\epsilon^{\pi-\epsilon} k_+(t)\,\sin t\,dt + \int_{\pi-\epsilon}^\pi (|k_+(t)| + a_r(t))\,\sin t\,dt \le 0$$

and s_+ such that $s_+ \sin t \ge r$ for all $t \in [\epsilon, \pi-\epsilon]$. Then the assumption (a) of Proposition 4.1 is fulfilled with

$$\begin{aligned} f_+(t) &= k_+(t), && \text{if } t \in [\epsilon, \pi-\epsilon], \\ &= |k_+(t)| + a_r(t), && \text{if } t \notin [\epsilon, \pi-\epsilon]. \end{aligned}$$

□

Example 4.1. A simple example of application of the above theorem is

$$\begin{gathered} u'' + u - \arctan u = k(t), \\ u(0) = 0,\ u(\pi) = 0, \end{gathered}$$

where $\overline{k} = \int_0^\pi k(t)\varphi(t)\,dt \in]-\sqrt{2\pi}, \sqrt{2\pi}[$. Here we take $h(t) = k(t) - \overline{k}\varphi(t)$ and $f(t,u) = -\arctan u - \overline{k}\varphi(t)$.

Remark 4.2. Theorem 4.2 is no more true if either $\int_0^\pi f_+(t)\,\sin t\,dt = 0$ or $\int_0^\pi f_-(t)\,\sin t\,dt = 0$. Consider for example

$$\begin{aligned} u'' + u + \exp(-u^2) = 0, \\ u(0) = 0,\ u(\pi) = 0. \end{aligned}$$

This problem has no solution (This fact can be proved multiplying the equation by $\sin t$ and integrating.) but

$$\liminf_{u\to-\infty} f(t,u) \geq f_-(t) = 0, \quad \limsup_{u\to+\infty} f(t,u) \leq f_+(t) = 0$$

and

$$\int_0^\pi f_+(t)\,\sin t\,dt = 0, \quad \int_0^\pi f_-(t)\,\sin t\,dt = 0.$$

Notice also that assumptions such as (a) and (b) in Proposition 4.1 or Theorem 4.2 cannot be necessary as they are essentially asymptotic. To build counter-examples, we can take any problem which has a solution and modify the equation for $|u| \geq s\sin t$, where s is large enough, so that the assumptions (a) and (b) do not hold.

Similar results can be obtained for systems with asymptotic slope between the first and the second eigenvalues. Here we use lower and upper solutions in the reversed order.

Proposition 4.3. *Let $h \in L^1(0,\pi)$ be such that $\int_0^\pi h(t)\sin t\,dt = 0$. Assume f is an L^1-Carathéodory function and*
(a) there exist $s_+ \in \mathbb{R}$, $f_+ \in L^1(0,\pi)$ such that

$$f(t,u) \geq f_+(t) \quad \text{if } u \geq s_+\ \sin t$$

and

$$\int_0^\pi f_+(t)\,\sin t\,dt \geq 0;$$

(b) there exist $s_- \in \mathbb{R}$, $f_- \in L^1(0,\pi)$ such that

$$f(t,u) \leq f_-(t) \quad \text{if } u \leq s_-\ \sin t$$

and

$$\int_0^\pi f_-(t)\,\sin t\,dt \leq 0;$$

(c) $\displaystyle\limsup_{|u|\to\infty} \frac{f(t,u)}{u} \lneqq 3$ *uniformly in t.*

Then the Dirichlet problem (3.1) *has at least one solution.*

Proof. Lower and upper solutions α and β are build as in the proof of Proposition 4.1 except that $\alpha = a\varphi + w \geq s_+\varphi$ and $\beta = b\varphi + w \leq s_-\varphi$. We conclude then using Theorem III-3.9. □

As in Theorem 4.2, we can impose conditions on $f(t,u)$ which are asymptotic as u goes to $\pm\infty$.

Theorem 4.4. *Let $h \in L^1(0,\pi)$ be such that $\int_0^\pi h(t)\sin t\,dt = 0$. Assume f is an L^1-Carathéodory function and there exist f_+, $f_- \in L^1(0,\pi)$ such that*

(a) $\liminf_{u\to+\infty} f(t,u) \geq f_+(t)$ *uniformly in t and*

$$\int_0^\pi f_+(t)\,\sin t\,dt > 0;$$

(b) $\limsup_{u\to-\infty} f(t,u) \leq f_-(t)$ *uniformly in t and*

$$\int_0^\pi f_-(t)\,\sin t\,dt < 0;$$

(c) $\limsup_{|u|\to\infty} \dfrac{f(t,u)}{u} \lneqq 3$ *uniformly in t.*

Then the Dirichlet problem (3.1) *has at least one solution.*

Proof. This proof repeats the argument in the proof of Theorem 4.2. □

5 Other Resonance Conditions

An alternative to the Landesman-Lazer conditions given in Theorems 4.2 and 4.4 is to control the resonance using repulsive forces in case the asymptotic slope is on the left of the first eigenvalue and attracting ones if it is on the right. Here, the assumption $\int_0^\pi h(t)\sin t\,dt = 0$ is a real restriction since we cannot add a constant to $f(t,u)$.

Theorem 5.1. *Let $h \in L^1(0,\pi)$ and f satisfy L^1-Carathéodory conditions. Assume*

(a) $\displaystyle\int_0^\pi h(t)\,\sin t\,dt = 0;$

(b) $u\,f(t,u) \leq 0$ *for a.e.* $t \in [0,\pi]$ *and all* $u \in \mathbb{R}$.

Then the Dirichlet problem (3.1) *has at least one solution.*

Proof. Let w be the solution of

$$w'' + w = h(t),$$
$$w(0) = 0, \ w(\pi) = 0,$$

and define $\alpha(t) = w(t) - a \sin t$, where a is large enough so that $\alpha \leq 0$. It is easy to see that α is a lower solution. Similarly, we define $\beta(t) = w(t) + b \sin t$ which is nonnegative if b is large enough and therefore is an upper solution. The proof follows then from Theorem II-2.4. □

Remark An alternative proof would be to deduce this theorem from Proposition 4.1. Notice however that we cannot apply Theorem 4.2. Consider for example the problem

$$u'' + u - u \exp(-u^2) = \sin 2t,$$
$$u(0) = 0, \ u(\pi) = 0.$$

Existence of at least one solution follows from Theorem 5.1. On the other hand, any function $f_+(t)$ such that

$$f(t,u) = -u \exp(-u^2) \leq f_+(t) \quad \text{if} \quad u \geq s_+ \sin t$$

has to be nonnegative and $\int_0^\pi f_+(t) \sin t \, dt \geq 0$. Hence, we cannot use Theorem 4.2 and Proposition 4.1 can only apply with $f_+(t) \equiv 0$.

If the asymptotic slope is on the right of the first eigenvalue we can prove a dual result.

Theorem 5.2. *Let $h \in L^1(0,\pi)$ and f satisfy L^1-Carathéodory conditions. Assume*

(a) $\displaystyle\int_0^\pi h(t) \sin t \, dt = 0;$

(b) $u\, f(t,u) \geq 0$ *for a.e.* $t \in [0,\pi]$ *and all* $u \in \mathbb{R}$;

(c) $\displaystyle\limsup_{|u|\to\infty} \frac{f(t,u)}{u} \lneqq 3$ *uniformly in* t.

Then the Dirichlet problem (3.1) *has at least one solution.*

Proof. Lower and upper solutions α and β are built as in the proof of Theorem 5.1 except that here $\alpha \geq 0$ and $\beta \leq 0$. We conclude then using Theorem III-3.9. □

Remark Here also, we can deduce this result from Proposition 4.3.

An alternative concerns strong resonance. Here $f(t,u)$ goes to zero as $u \to \pm\infty$ but we control the rate of decrease of $f(t,u)$.

Theorem 5.3. *Let $h \in L^1(0,\pi)$ and f be a Carathéodory function. Assume*

(a) $\displaystyle\int_0^\pi h(t)\sin t\,dt = 0$;

(b) for some $k \in L^\infty(0,\pi)$, for a.e. $t \in [0,\pi]$ and all $u \in \mathbb{R}$, $|f(t,u)| \le k(t)$;

(c) for some $\mu \in [1,2[$

$$\lim_{|u|\to\infty} |u|^{\mu-1} f(t,u) = 0 \quad \textit{and} \quad \liminf_{|u|\to\infty} \operatorname{sgn}(u)|u|^\mu f(t,u) \ge m > 0$$

uniformly in t.

Then the problem (3.1) *has at least one solution.*

Proof. The proof follows the line of Step 1 in the proof of Theorem VI-4.2. Let $\varphi(t) = \sqrt{\frac{2}{\pi}}\sin t$ and define K as in Remark VI-4.1. Hence, for every $a \in \mathbb{R}$, the solution $\tilde{u}$ of (VI-4.2) is such that $\|\tilde{u}\|_\infty < K$ and $|\tilde{u}(t)| \le K\varphi(t)$. It follows that there exists an $M > 0$ such that for every $a \ge K+1$ and the corresponding solution $\tilde{u}$ of (VI-4.2), we have

$$\left| \frac{(\tilde{u}(t) + a\varphi(t))^\mu - (a\varphi(t))^\mu}{(\tilde{u}(t) + a\varphi(t))^{\mu-1}} \right| \le M.$$

Choose then $R > 0$ large enough so that for a.e. $t \in [0,\pi]$ and all $u \ge R$,

$$u^{\mu-1} f(t,u) \le \tfrac{m}{2M} \quad \text{and} \quad u^\mu f(t,u) \ge \tfrac{3m}{4}.$$

Pick next $\delta > 0$ small enough so that

$$(R^{\mu-1}(M+R)\|k\|_\infty + \tfrac{3m}{4}) \int_0^\pi \frac{\chi(t)}{\varphi^{\mu-1}(t)}\,dt - \tfrac{m}{4}\int_0^\pi \frac{dt}{\varphi^{\mu-1}(t)} \le 0,$$

where χ is the characteristic function of $[0,\delta] \cup [\pi-\delta,\pi]$. Then we choose $a > 0$ large enough such that $a\sin\delta \ge K+R$ and define $\alpha(t) = \tilde{u}(t) + a\varphi(t)$, where $\tilde{u}$ is the corresponding solution of (VI-4.2). Notice that $\alpha(t) \ge R\varphi(t)$ if $t \in [0,\pi]$ and $\alpha(t) \ge R$ if $t \in [\delta, \pi-\delta]$. Let $A = \{t \in [0,\pi] \mid |\alpha(t)| \le R\} \subset [0,\delta] \cup [\pi-\delta,\pi]$ and $B = [0,\pi] \setminus A$. We compute then

$$\begin{aligned}
\int_0^\pi f(t,\alpha)a^\mu\varphi\,dt &= \int_A f(t,\alpha)\frac{\alpha^\mu}{\varphi^{\mu-1}}\,dt - \int_A f(t,\alpha)\frac{\alpha^\mu - (a\varphi)^\mu}{\varphi^{\mu-1}}\,dt \\
&\quad + \int_B f(t,\alpha)\frac{\alpha^\mu}{\varphi^{\mu-1}}\,dt - \int_B f(t,\alpha)\frac{\alpha^\mu - (a\varphi)^\mu}{\varphi^{\mu-1}}\,dt \\
&\ge -R^\mu \int_0^\pi \frac{\chi(t)k(t)}{\varphi^{\mu-1}(t)}\,dt - MR^{\mu-1}\int_0^\pi \frac{\chi(t)k(t)}{\varphi^{\mu-1}(t)}\,dt \\
&\quad + \tfrac{3m}{4}\int_0^\pi \frac{(1-\chi(t))}{\varphi^{\mu-1}(t)}\,dt - \tfrac{m}{2}\int_0^\pi \frac{dt}{\varphi^{\mu-1}(t)} \ge 0.
\end{aligned}$$

We deduce, as in Remark VI-4.1, that α solves

$$\begin{gathered} u'' + u + f(t,u) = s_0\varphi(t) + h(t), \\ u(0) = 0, \ u(\pi) = 0, \end{gathered}$$

with $s_0 = \int_0^\pi f(t,\alpha(t))\varphi(t)\,dt \geq 0$ and therefore α is a lower solution of (3.1). The construction of an upper solution is similar and we conclude from Theorem III-3.8. □

Remark The condition

$$\lim_{u\to\pm\infty} |u|^{\mu-1} f(t,u) = 0$$

alone does not ensure existence of a solution as follows from the example in Remark 4.2.

Example 5.1. An example of application of Theorem 5.3 is the following

$$\begin{gathered} u'' + u + \tfrac{u}{1+u^2} - e^{-|u|}\cos u = h(t), \\ u(0) = 0, \ u(\pi) = 0, \end{gathered}$$

where $\int_0^\pi h(t)\,\sin t\,dt = 0$. Here Landesman-Lazer's conditions (Proposition 4.1) do not hold since

$$\frac{u}{1+u^2} - e^{-|u|}\cos u \leq f_+(t), \qquad \text{for } u \geq s_+ \sin t$$

implies $f_+(t) > 0$ and $\int_0^\pi f_+(t)\sin t\,dt > 0$. Proposition 4.3 does not apply either since

$$\frac{u}{1+u^2} - e^{-|u|}\cos u \geq f_+(t), \qquad \text{for } u \geq s_+ \sin t$$

implies $f_+(t) \leq 0$, $f_+(t) \leq -1/2$ in a neighbourhood of $t = 0$, and therefore $\int_0^\pi f_+(t)\sin t\,dt < 0$. Also, Theorems 5.1 and 5.2 cannot be used since $uf(t,u)$ is not one sign.

6 Fredholm Alternative Results

The next theorem considers the inverse of the operator $Fu = u'' + u - qu^-$ defined on $W^{2,1}(0,\pi) \cap \mathcal{C}_0([0,\pi])$. More precisely, we consider the problem

$$\begin{aligned} u'' + u - qu^- &= h(t), \\ u(0) = 0, \ u(\pi) &= 0, \end{aligned} \tag{6.1}$$

where $q \neq 0$ and $h \in L^1(0,\pi)$. This corresponds to a Fredholm alternative for the problem

$$\begin{aligned} u'' + \mu u^+ - \nu u^- &= h(t), \\ u(0) = 0, \ u(\pi) &= 0, \end{aligned}$$

with $(\mu, \nu) = (1, 1+q)$ on the first curve of the Fučík spectrum.

Theorem 6.1. *Let $h \in L^1(0,\pi)$ and $q \neq 0$. Then the problem* (6.1) *has a solution if and only if*

$$q \int_0^\pi h(t) \sin t \, dt \leq 0. \tag{6.2}$$

Proof. Claim 1 – The condition (6.2) *is sufficient.* Let w be the solution of

$$\begin{aligned} w'' + w = h(t) - (\tfrac{2}{\pi} \int_0^\pi h(s) \sin s \, ds) \sin t, \\ w(0) = 0, \ w(\pi) = 0, \end{aligned}$$

such that $\int_0^\pi w(t) \sin t \, dt = 0$.

Assume $q > 0$ and define $\alpha = w + A \sin t$, with A large enough so that $\alpha \geq 0$. This function is a lower solution since

$$\alpha'' + \alpha = h(t) - (\tfrac{2}{\pi} \int_0^\pi h(s) \sin s \, ds) \sin t \geq h(t) = h(t) + q\alpha^-.$$

In a similar way, we build a nonpositive upper solution $\beta = w - B \sin t$, where B is large enough, and the claim follows from Theorem III-3.9.

In case $q < 0$, $\alpha = w - A \sin t$, with A large enough so that α is a nonpositive lower solution and $\beta = w + B \sin t$, with B large enough so that β is a nonnegative upper solution. The claim follows then from Theorem II-2.4.

Claim 2 – The condition (6.2) *is necessary.* Let u be a solution of (6.1). Multiplying this equation by $\sin t$ and integrating, we have

$$q \int_0^\pi h(t) \sin t \, dt = q \int_0^\pi (u'' + u - qu^-) \sin t \, dt = -q^2 \int_0^\pi u^- \sin t \, dt \leq 0. \quad \square$$

Some results can be obtained from a monotonicity assumption.

Theorem 6.2. *Let $h \in L^1(0,\pi)$ be such that $\int_0^\pi h(t) \sin t \, dt = 0$ and $f : [0,\pi] \times \mathbb{R} \to \mathbb{R}$ be an L^1-Carathéodory function. Assume that for each $t \in [0,\pi]$, $f(t,.)$ is a non-increasing function.*

Then the Dirichlet problem (3.1) *has at least one solution if and only if there exists* $\xi \in \mathbb{R}$ *such that*

$$\int_0^\pi f(t, \xi \sin t) \sin t \, dt = 0. \tag{6.3}$$

Proof. Claim 1 – The condition (6.3) *is sufficient.* Let w be a solution of

$$w'' + w = h - f(t, \xi \sin t),$$
$$w(0) = 0, \; w(\pi) = 0.$$

Choose then A and B large enough so that $\alpha := w - A \sin t \leq \xi \sin t \leq w + B \sin t =: \beta$. We can see that α is a lower solution and β an upper one so that the claim follows from Theorem II-2.4.

Claim 2 – The condition (6.3) *is necessary.* Let u be a solution of (3.1). Multiplying this equation by $\sin t$ and integrating, we obtain

$$\int_0^\pi f(t, u(t)) \sin t \, dt = 0.$$

On the other hand, there exist ξ_1 and ξ_2 such that for all $t \in [0, \pi]$,

$$\xi_1 \sin t \leq u(t) \leq \xi_2 \sin t.$$

Hence and as f is non-increasing we obtain

$$\int_0^\pi f(t, \xi_2 \sin t) \sin t \, dt \leq 0 \leq \int_0^\pi f(t, \xi_1 \sin t) \sin t \, dt$$

and conclude by an application of the Intermediate Value Theorem. □

Example 6.1. Notice that the function

$$f(t, u) = -\exp u + a$$

verifies the assumptions of Theorem 6.2 for any $a > 0$.

Remark 6.1. If f is decreasing, we can prove that this solution is unique. Let u and v be two distinct solutions of (3.1) and let $w = u - v$. We compute

$$\begin{aligned} 0 = -\int_0^\pi w'^2(s) \, ds + \int_0^\pi w^2(s) \, ds \\ + \int_0^\pi (f(s, u(s)) - f(s, v(s)))(u(s) - v(s)) \, ds \\ \leq \int_0^\pi (f(s, u(s)) - f(s, v(s)))(u(s) - v(s)) \, ds < 0, \end{aligned}$$

which is a contradiction.

If f is only non-increasing, uniqueness does not occur in general. For example, consider the case where $f(t,u) = 0$ if $(t,u) \in [0,\pi] \times [-1,1]$ and $h(t) \equiv 0$, then for all $a \in [-1,1]$, $u(t) = a \sin t$ is a solution. This example is generic as follows from the next result which completes the equivalent for Dirichlet problem of Theorem I-3.5.

Theorem 6.3. *Under the hypothesis of Theorem 6.2, the set $\mathcal{S}$ of solutions of (3.1) is convex.*

Moreover, there exists $\tilde{u} \in W^{2,1}(0,\pi)$ with $\int_0^\pi \tilde{u}(t) \sin t \, dt = 0$ and an interval $J \subset \mathbb{R}$ such that

$$\mathcal{S} = \{u = \tilde{u} + c \sin t : c \in J\}.$$

Proof. Let u and v be two solutions of (3.1). The function $w = u - v$ is such that

$$\begin{aligned} 0 &= -\int_0^\pi w'^2(t)\, dt + \int_0^\pi w^2(t)\, dt + \int_0^\pi (f(t,u(t)) - f(t,v(t)))w(t)\, dt \\ &\le -\int_0^\pi w'^2(t)\, dt + \int_0^\pi w^2(t)\, dt \le 0. \end{aligned}$$

Hence, $w(t) = a \sin t$ and we can find a set $J \subset \mathbb{R}$ such that

$$\mathcal{S} = \{u = \tilde{u} + c \sin t \mid c \in J\},$$

where $\tilde{u}(t) = u(t) - \frac{2}{\pi}(\int_0^\pi u(s) \sin s \, ds) \sin t$ and u is any solution of (3.1). We only have to prove that J is an interval. To this aim, let $c_1 < c_2$ be in J and $c \in]c_1, c_2[$. Observe that

$$\begin{aligned} h(t) &= \tilde{u}''(t) + \tilde{u}(t) + f(t, \tilde{u}(t) + c_2 \sin t) \\ &\le (\tilde{u}(t) + c \sin t)'' + (\tilde{u}(t) + c \sin t) + f(t, \tilde{u}(t) + c \sin t) \\ &\le \tilde{u}''(t) + \tilde{u}(t) + f(t, \tilde{u}(t) + c_1 \sin t) = h(t). \end{aligned}$$

This relation shows that $v(t) = \tilde{u}(t) + c \sin t$ is also a solution and completes the proof. □

Remark 6.2. Notice that without the monotonicity assumption on $f(t,.)$ the set of solutions might be non-convex. Consider for example the problem

$$\begin{aligned} u'' + u + (u^2 - \sin^2 t)^2 &= 0, \\ u(0) = 0, \ u(\pi) &= 0. \end{aligned}$$

It is easy to see that $u(t) = -\sin t$ and $u(t) = \sin t$ are solutions. Further, if u is another solution then

$$\int_0^\pi (u^2(t) - \sin^2 t)^2 \sin t \, dt = 0.$$

This equation implies that $u(t) = -\sin t$ and $u(t) = \sin t$ are the only two solutions.

Exercise 6.1. Prove the same kind of results for the Neumann boundary value problem.

7 Derivative Dependent Dirichlet Problem

Consider the problem

$$\begin{aligned} &u'' + f(t, u, u') = 0, \\ &u(0) = 0, \;\; u(\pi) = 0. \end{aligned} \tag{7.1}$$

First, we consider a non-resonance result which considers a one-sided linear control of the nonlinearity $f(t, u, v)$.

Theorem 7.1. *Let f satisfy L^1-Carathéodory conditions and assume that for any $r_0 > 0$, there exist p, $q \in [1, \infty]$ such that $\frac{1}{q} + \frac{1}{p} = 1$, $\varphi \in C(\mathbb{R}^+, \mathbb{R}_0^+)$ and $\psi \in L^p(0, \pi)$ with*

$$\int_0^\infty \frac{s^{1/q}}{\varphi(s)}\, ds = \infty,$$

and for a.e. $t \in [0, \pi]$, all $u \in [-r_0, r_0]$ and all $v \in \mathbb{R}$

$$|f(t, u, v)| \le \psi(t)\varphi(|v|).$$

Assume also there exist $b \in L^1(0, \pi; \mathbb{R}^+)$, δ, $\gamma \in \mathbb{R}^+$, and $r_1 \ge 0$ such that for a.e. $t \in [0, \pi]$ and all $(u, v) \in \mathbb{R}^2$, with $|u| \ge r_1$, we have

$$\operatorname{sgn}(u)\, f(t, u, v) \le b(t) + \gamma|u| + \delta|v|.$$

If further $\pi < \Gamma(\delta, \gamma)$, where

$$\begin{aligned} \Gamma(\delta,\gamma) &= \frac{4}{\sqrt{\delta^2-4\gamma}} \tanh^{-1}\left(\frac{\sqrt{\delta^2-4\gamma}}{\delta}\right), && \textit{if } \delta^2 - 4\gamma > 0, \\ &= \frac{4}{\sqrt{4\gamma-\delta^2}} \tan^{-1}\left(\frac{\sqrt{4\gamma-\delta^2}}{\delta}\right), && \textit{if } \delta^2 - 4\gamma < 0, \\ &= \frac{4}{\delta}, && \textit{if } \delta^2 - 4\gamma = 0, \end{aligned}$$

then the problem (7.1) has a solution.

Proof. Let y be a solution of

$$y'' + \delta|y'| + \gamma y = -b(t)$$

on $[0, \pi]$. Define v to be the solution of

$$\begin{gathered} v'' - \delta v' + \gamma v = 0, \\ v(\pi/2) = 1, \ v'(\pi/2) = 0, \end{gathered}$$

on $[\pi/2, \pi]$ and extend v by symmetry on $[0, \pi]$ so that $v(t) = v(\pi - t)$. Next, verify that for all values of δ and γ, the condition $\pi < \Gamma(\delta, \gamma)$ implies $v(t) > 0$ and $v'(t)(t - \pi/2) \le 0$, whence v satisfies the equation

$$v'' + \delta|v'| + \gamma v = 0.$$

It follows that if $k > 0$ is large enough $\beta = y + kv > r_1$ and we can compute

$$\begin{aligned} &\beta''(t) + f(t, \beta(t), \beta'(t)) \le \beta''(t) + b(t) + \delta|\beta'(t)| + \gamma\beta(t) \\ &\quad \le y''(t) + kv''(t) + b(t) + \delta(|y'(t)| + k|v'(t)|) + \gamma(y(t) + kv(t)) = 0. \end{aligned}$$

Hence, β is an upper solution. We build similarly a lower solution and conclude from Corollary II-2.2. □

Remark 7.1. If $\delta = 0$, the condition becomes $\gamma < 1$, which is best possible since 1 is the first eigenvalue of the corresponding linear problem

$$u'' + \lambda u = 0, \quad u(0) = 0, \ u(\pi) = 0.$$

Next, we consider a non-resonance result with a quadratic control of the nonlinearity $f(t, u, v)$.

Theorem 7.2. *Let f satisfy L^1-Carathéodory conditions and assume that for any $r_0 > 0$, there exist p, $q \in [1, \infty]$ such that $\frac{1}{q} + \frac{1}{p} = 1$, $\varphi \in C(\mathbb{R}^+, \mathbb{R}_0^+)$ and $\psi \in L^p(0, \pi)$ with*

$$\int_0^\infty \frac{s^{1/q}}{\varphi(s)}\, ds = \infty,$$

and for a.e. $t \in [0, \pi]$, all $u \in [-r_0, r_0]$ and all $v \in \mathbb{R}$

$$|f(t, u, v)| \le \psi(t)\varphi(|v|).$$

Assume also there exist some $r_1 > 0$, $b \in L^1(0, \pi)$, $0 < \gamma < 1$ and $h \in C(\mathbb{R})$ such that for a.e. $t \in [0, \pi]$, all $|u| \ge r_1$ and all $v \in \mathbb{R}$

$$\operatorname{sgn}(u) f(t, u, v) \le b(t) + \gamma|u| + h(u)v^2$$

and

$$\limsup_{|u| \to \infty} |u| h(u) < \frac{1 - \gamma}{\gamma}.$$

Then the problem (7.1) has a solution.

Proof. Construction of an upper solution. First, let us choose a $\gamma_a \in]\gamma, 1[$ and an $R > r_1$ such that

$$\forall u \geq R, \quad h(u) \leq \frac{1-\gamma_a}{\gamma_a}\frac{1}{u}.$$

Next, we choose $\gamma_b \in]\gamma, \gamma_a[$ and $\epsilon > 0$ such that

$$\frac{1-\gamma_a}{\gamma_a}(1+\epsilon) = \frac{1-\gamma_b}{\gamma_b}.$$

Define then w to be a solution of

$$w'' + \gamma w + b(t) = 0, \quad w \geq R,$$

and $Y(u) = u^{\gamma_b}$. This function solves

$$Y'' + \frac{1-\gamma_b}{\gamma_b}\frac{Y'^2}{Y} = 0.$$

At last, we choose k such that $\frac{\gamma}{\gamma_b} < k^2 < 1$, and define $v(t) = V_0 \cos k(t - \frac{\pi}{2})$. If V_0 is large enough, this function verifies

$$v'' + \frac{\gamma}{\gamma_b}v + Kw'^2(t)v^{1-\gamma_b} \leq 0, \quad v > 0,$$

where $K = \frac{1-\gamma_a}{\gamma_a\gamma_b}(1+\frac{1}{\epsilon})\frac{1}{R}$.

It is now easy to check that the function $\beta(t) = w(t) + Y(v(t)) \geq R$ is a positive upper solution since

$$\begin{aligned}\beta'' + f(t,\beta,\beta') &\leq w'' + Y''(v)v'^2 + Y'(v)v'' \\ &\quad + b + \gamma(w + Y(v)) + \tfrac{1-\gamma_a}{\gamma_a}[(1+\epsilon)\tfrac{Y'^2}{Y}v'^2 + (1+\tfrac{1}{\epsilon})\tfrac{w'^2}{R}] \leq 0.\end{aligned}$$

Conclusion. We obtain in a similar way a lower solution and the theorem follows from Corollary II-2.2. □

Remark 7.2. In case $h(u) \equiv 0$, we recover the non-resonance condition

$$\limsup_{|u|\to\infty} \frac{f(t,u)}{u} \leq \gamma < 1$$

uniformly in t.

Example 7.1. A non-strict inequality $\limsup_{|u|\to\infty} |u|h(u) \le \dfrac{1-\gamma}{\gamma}$ does not imply existence as follows from the problem

$$\begin{aligned} u'' + \gamma(1+|u|) + \tfrac{1-\gamma}{\gamma(1+|u|)}u'^2 = 0, \\ u(0) = 0,\ u(\pi) = 0, \end{aligned} \tag{7.2}$$

which has no solution. To verify this claim, let us introduce the solution $Y(u)$ of the Cauchy problem

$$Y' = \frac{Y+\gamma}{\gamma(1+u)}, \quad Y(0) = 0,$$

and verify that

$$Y'' = \frac{1-\gamma}{\gamma}\frac{Y'}{(1+u)}.$$

If u solves the problem (7.2), it is easy to see that $u \ge 0$ and the function $v(t) = Y(u(t))$ verifies

$$v'' + v + \gamma = 0, \quad v(0) = 0,\ v(\pi) = 0.$$

But as $\int_0^\pi \gamma \sin t\, dt \ne 0$, it is clear from Fredholm alternative that this last equation has no solution.

The same ideas can be used to study a resonance problem with a quadratic control. In case $h = 0$, it reduces to Landesman-Lazer conditions.

Theorem 7.3. *Let f be an L^1-Carathéodory function and assume that for any $r_0 > 0$, there exist p, $q \in [1,\infty]$ such that $\frac{1}{q} + \frac{1}{p} = 1$, $\varphi \in C(\mathbb{R}^+, \mathbb{R}_0^+)$ and $\psi \in L^p(0,\pi)$ with*

$$\int_0^\infty \frac{s^{1/q}}{\varphi(s)}\, ds = \infty,$$

and for a.e. $t \in [0,\pi]$, all $u \in [-r_0, r_0]$ and all $v \in \mathbb{R}$

$$|f(t,u,v)| \le \psi(t)\varphi(|v|).$$

Assume also that there exist an L^1-Carathéodory function $g(t,u)$ and a continuous function $h(u) \ge 0$ such that for a.e. $t \in [0,\pi]$ and all $(u,v) \in \mathbb{R}^2$,

$$\operatorname{sgn}(u) f(t,u,v) \le g(t,u) + h(u)v^2.$$

Define

$$H(y) = \Big|\int_0^y h(s)\, ds\Big|, \quad Q(y) = \int_0^y \exp H(r)\, dr$$

and

$$F(t,y) = \operatorname{sgn}(y)g(t,y)\exp H(y) - Q(y).$$

Assume at last there exists γ_+ *and* $\gamma_- \in L^1(0,\pi)$ *such that*

$$\gamma_+(t) \geq \limsup_{y\to+\infty} F(t,y), \quad \gamma_-(t) \leq \liminf_{y\to-\infty} F(t,y),$$

uniformly in t *and*

$$\int_0^\pi \gamma_+(t)\sin t\,dt < 0, \quad \int_0^\pi \gamma_-(t)\sin t\,dt > 0.$$

Then the problem (7.1) *has a solution.*

Proof. Notice first that the function $Q : \mathbb{R}^+ \to \mathbb{R}^+$ is an increasing homeomorphism whose inverse $Y(x) = Q^{-1}(x)$ is the solution of the problem

$$y'' + h(y)y'^2 = 0, \quad y(0) = 0, \; y'(0) = 1.$$

Define then $\beta(t) = Y(v(t))$, where the function $v(t) \geq 0$ will be chosen later. We compute

$$\begin{aligned}\beta'' + f(t,\beta,\beta') &\leq [Y''(v) + h(Y(v))Y'^2(v)]v'^2 + Y'(v)v'' + g(t,Y(v)) \\ &= Y'(v)[v'' + g(t,Y(v))\exp H(Y(v))],\end{aligned}$$

which proves that β is an upper solution if

$$v'' + g(t,Y(v))\exp H(Y(v)) = v'' + v + F(t,Y(v)) \leq 0.$$

Notice then that

$$\limsup_{v\to+\infty} F(t,Y(v)) = \limsup_{y\to+\infty} F(t,y) \leq \gamma_+(t).$$

Hence, following the proof of Theorem 4.2 and Proposition 4.1, we can find a function $v(t) \geq 0$ such that $v'' + v + F(t,Y(v)) \leq 0$.

A lower solution is obtained in a dual way and we conclude from Corollary II-2.2. □

Example 7.2. The simplest example to consider is a problem with a linear restoring force and a quadratic friction

$$\begin{aligned}u'' + b(t) + \gamma u + u'^2 = 0,\\ u(0) = 0, \; u(\pi) = 0.\end{aligned}$$

Here Theorem 7.2 does not apply since $h(u) = 1$ and

$$\limsup_{|u|\to+\infty} |u|h(u) = +\infty.$$

On the other hand, we can apply Theorem 7.3. We compute

$$F(t,y) = \operatorname{sgn}(y)[(|b(t)| + \gamma|y| - 1)e^{|y|} + 1].$$

If $\gamma < 0$, we have

$$\lim_{y\to\pm\infty} F(t,y) = \mp\infty.$$

If $\gamma = 0$ and $|b(t)| \lneqq 1$, we obtain

$$\limsup_{y\to+\infty} F(t,y) \le \gamma_+(t) \quad \text{with} \int_0^\pi \gamma_+(t)\sin t\,dt < 0$$

and

$$\liminf_{y\to-\infty} F(t,y) \ge \gamma_-(t) \quad \text{with} \int_0^\pi \gamma_-(t)\sin t\,dt > 0,$$

where $\gamma_+(t) = -\gamma_-(t) = (|b(t)| - 1)K + 1$ and K is large enough. In both cases, existence follows from Theorem 7.3.

Problem 7.1. Theorem 7.3 extends Theorem 4.2 to derivative dependent problems. This theorem is a resonance result at the left of the first eigenvalue. Consider a similar extension of Theorem 4.4 and obtain a resonance result at the right of the first eigenvalue for derivative dependent problems.

Hint : See [93].

In the following theorem, the existence result is obtained from the dependence in the derivative.

Theorem 7.4. *Let p, $q \in [1,\infty]$ be such that $\frac{1}{q} + \frac{1}{p} = 1$. Assume $f : [0,\pi]\times\mathbb{R}^2 \to \mathbb{R}$ satisfies L^p-Carathéodory conditions and that for any $r_0 > 0$, there exist $\varphi \in C(\mathbb{R}^+, \mathbb{R}_0^+)$ and $\psi \in L^p(0,\pi)$ with*

$$\int_0^\infty \frac{s^{1/q}}{\varphi(s)}\,ds = \infty,$$

and for a.e. $t \in [0,\pi]$, all $u \in [-r_0, r_0]$ and all $v \in \mathbb{R}$

$$|f(t,u,v)| \le \psi(t)\varphi(|v|).$$

Assume moreover there exist $r_1 \ge 0$, $h \in L^1([0,\pi],\mathbb{R}^+)$ and $g \in C(\mathbb{R},\mathbb{R}_0^+)$ such that

$$\int_0^\infty \frac{ds}{g(s)} > \|h\|_{L^1}$$

and, for a.e. $t \in [0, \pi]$ *and all* $(u, v) \in \mathbb{R}^2$, *with* $|u| \geq r_1$,

$$\operatorname{sgn}(u) f(t, u, v) \leq h(t) g(|v|).$$

Then the problem (7.1) *has at least one solution.*

Proof. Let z be the solution of

$$z' + h(t)g(z) = 0, \quad z(\pi) = 0.$$

It is easy to see that $\beta(t) = r_1 + \int_0^t z(s)\, ds \geq r_1$ is an upper solution. In a similar way, we compute a lower solution and existence of a solution follows from Corollary II-2.2. □

Example 7.3. Consider the boundary value problem

$$u'' = e^u(u'^2 + 1),$$
$$u(0) = 0, \; u(\pi) = 0.$$

It is easy to observe that the Nagumo condition in Theorem 7.4 is satisfied. The other conditions of this theorem follow with $g(v) = v^2 + 1$ and $h(t) = e^{-r_1}$ for r_1 large enough.

Chapter VIII

Positive Solutions

In this chapter, we consider positive solutions of the Dirichlet problem

$$\begin{aligned} u'' + f(t,u) &= 0, \\ u(a) = 0,\ u(b) &= 0, \end{aligned} \tag{0.1}$$

where $f : [a,b] \times \mathbb{R} \to \mathbb{R}$ is an L^1-Carathéodory function. Most of these results extend to other boundary value problems such as the separated boundary value problem or the periodic problem. Such extensions are left as exercises.

Let us recall that given $u, v \in C([a,b])$, we write $u \succ v$ or $v \prec u$ if

$$\begin{array}{ll} u(t) > v(t) & \text{on }]a,b[, \\ D_+u(a) > D^+v(a) & \text{if } u(a) = v(a), \\ D^-u(b) < D_-v(b) & \text{if } u(b) = v(b). \end{array}$$

Here, positive solutions $u : [a,b] \to \mathbb{R}$ are such that $u \succ 0$.

1 Existence of One Solution

A necessary condition for a positive solution of (0.1) to exist is that

$$\inf_{(t,u)\in[a,b]\times\mathbb{R}^+} \frac{f(t,u) - q(t)u}{u} \le \lambda_1(q) \le \sup_{(t,u)\in[a,b]\times\mathbb{R}^+} \frac{f(t,u) - q(t)u}{u}$$

where $q \in L^1(a,b)$ and $\lambda_1(q)$ is the first eigenvalue of the spectral problem

$$\begin{aligned} u'' + q(t)u + \lambda u &= 0, \\ u(a) = 0,\ u(b) &= 0. \end{aligned} \tag{1.1}$$

This condition can be obtained multiplying the equation (0.1) by $\varphi_1(t;q)$, the eigenfunction corresponding to $\lambda_1(q)$, and integrating. It motivates several results which consider assumptions that force

$$\frac{f(t,u)-q(t)u}{u} - \lambda_1(q)$$

to change sign when u goes from 0 to $+\infty$.

1.1 The Sublinear Case

Our first concern is a *sublinear case*, where the "slope" of $f - qu$ is greater than a first eigenvalue near zero and smaller near infinity.

Theorem 1.1. *Define $\varphi_1(\cdot;q)$ to be a positive eigenfunction corresponding to $\lambda_1(q)$, the first eigenvalue of the spectral problem* (1.1). *Assume $f : [a,b]\times\mathbb{R}\to\mathbb{R}$ is an L^1-Carathéodory function that satisfies*

(H_0^+) *there exist $\delta > 0$ and $q_0 \in L^1(a,b)$ such that $\lambda_1(q_0) \le 0$, and for all $u \in]0,\delta]$ and a.e. $t \in [a,b]$,*

$$f(t,u) - q_0(t)u \ge 0;$$

(H_∞^-) *there exist $\rho > 0$, f_∞ and $q_\infty \in L^1(a,b)$ such that $\lambda_1(q_\infty) \ge 0$,*

$$\int_a^b f_\infty(t)\varphi_1(t;q_\infty)\,dt < 0,$$

and for all $u \ge \rho$ and a.e. $t \in [a,b]$,

$$f(t,u) - q_\infty(t)u \le f_\infty(t).$$

Then the boundary value problem (0.1) *has at least one solution $u \succ 0$.*

Proof. Claim 1 – There exists $\alpha \in W^{2,1}(a,b)$, a $W^{2,1}$-lower solution of (0.1) *such that $\alpha \succ 0$, $\alpha(a) = 0$, $\alpha(b) = 0$.* Define

$$\alpha(t) = \delta\,\frac{\varphi_1(t;q_0)}{\|\varphi_1(\cdot;q_0)\|_\infty}.$$

The function $\alpha \in W^{2,1}(a,b)$ is such that $\alpha(t) \succ 0$, $\alpha(a) = 0$, $\alpha(b) = 0$ and

$$\alpha'' + f(t,\alpha) \ge \alpha'' + q_0(t)\alpha = -\lambda_1(q_0)\alpha \ge 0.$$

Claim 2 – The problem (0.1) *has a $W^{2,1}$-upper solution $\beta \in W^{2,1}(a,b)$ such that for all $t \in]a,b[$, $\beta(t) > \alpha(t)$.* Let $k \in L^1(a,b)$ be such that for a.e.

$t \in [a,b]$ and all $u \in [0,\rho]$, $|f(t,u)|+|q_\infty(t)|\rho+|f_\infty(t)| \le k(t)$. Choose next $\epsilon > 0$ small enough so that

$$\overline{(f_\infty + k\chi)} := \int_a^b (f_\infty(s) + k(s)\chi(s))\varphi_1(s;q_\infty)\,ds \le 0,$$

where χ is the characteristic function of $[a,a+\epsilon]\cup[b-\epsilon,b]$ and $\varphi_1(\cdot;q_\infty) \succ 0$ is chosen so that $\|\varphi_1(\cdot;q_\infty)\|_{L^2} = 1$. Define then w to be a solution of

$$\begin{gathered} w'' + (q_\infty(t) + \lambda_1(q_\infty))w + (f_\infty + k\chi)(t) - \overline{(f_\infty + k\chi)}\varphi_1(t;q_\infty) = 0, \\ w(a) = 0, \quad w(b) = 0, \end{gathered}$$

and pick $B > 0$ large enough so that $w + B\varphi_1(\cdot;q_\infty) > \alpha$ on $]a,b[$ and $w + B\varphi_1(\cdot;q_\infty) \ge \rho$ on $[a+\epsilon, b-\epsilon]$. It is easy to check now that $\beta = w + B\varphi_1(\cdot;q_\infty)$ is an upper solution since

$$\begin{aligned} \beta''(t) + f(t,\beta(t)) &\le \beta''(t) + q_\infty(t)\beta(t) + f_\infty(t) + k(t)\chi(t) \\ &\le -\lambda_1(q_\infty)\beta(t) + \overline{(f_\infty + k\chi)}\varphi_1(t;q_\infty) \le 0. \end{aligned}$$

Conclusion – The proof follows now from Theorem II-2.4. □

Corollary 1.2. *Assume $f : [a,b]\times\mathbb{R} \to \mathbb{R}$ is an L^1-Carathéodory function such that for some γ_0 and $\gamma_\infty \in L^1(a,b)$,*

$$\liminf_{u\to 0^+} \frac{f(t,u)}{u} \ge \gamma_0(t) \gneq \left(\frac{\pi}{b-a}\right)^2$$

and

$$\limsup_{u\to +\infty} \frac{f(t,u)}{u} \le \gamma_\infty(t) \lneq \left(\frac{\pi}{b-a}\right)^2$$

hold uniformly in t.

Then the problem (0.1) *has at least one solution $u \succ 0$.*

Proof. The assumptions imply $\lambda_1(\gamma_0) < 0$ and $\lambda_1(\gamma_\infty) > 0$. The proof follows then from Theorem 1.1 with $q_0 = \gamma_0 - \epsilon$, $q_\infty = \gamma_\infty + \epsilon$, $f_\infty = -1$ and $\epsilon > 0$ small enough. □

As an exercice the reader can prove the following result which extends Theorem 1.1 to separated boundary value problems. Here we define $\lambda_1(q)$ to be the first eigenvalue of the spectral problem

$$\begin{gathered} u'' + q(t)u + \lambda u = 0, \\ a_1 u(a) - a_2 u'(a) = 0, \\ b_1 u(b) + b_2 u'(b) = 0. \end{gathered} \tag{1.2}$$

Theorem 1.3. *Let* $a_1, b_1 \in \mathbb{R}$, $a_2, b_2 \in \mathbb{R}^+$ *with* $a_1^2+a_2^2 > 0$ *and* $b_1^2+b_2^2 > 0$. *Assume* $f : [a,b] \times \mathbb{R} \to \mathbb{R}$ *is an* L^1*-Carathéodory function that satisfies* (H_0^+) *and* (H_∞^-), *where* $\varphi_1(\cdot; q)$ *is a positive eigenfunction corresponding to* $\lambda_1(q)$, *the first eigenvalue of the spectral problem* (1.2).

Then the boundary value problem

$$\begin{aligned} u'' + f(t,u) &= 0, \\ a_1 u(a) - a_2 u'(a) &= 0, \\ b_1 u(b) + b_2 u'(b) &= 0, \end{aligned} \tag{1.3}$$

has at least one solution $u \succ 0$.

An alternative uses the comparison spectral problem

$$\begin{aligned} u'' + \mu r(t) u &= 0, \\ u(a) = 0, \ u(b) &= 0. \end{aligned} \tag{1.4}$$

Let us write $\mu_1(r)$ and $\psi_1(t;r)$, the corresponding eigenvalues and eigenfunctions. The following result parallels Theorem 1.1.

Theorem 1.4. *Define* $\psi_1(\cdot; r)$ *to be a positive eigenfunction corresponding to* $\mu_1(r)$, *the first eigenvalue of the spectral problem* (1.4), *and assume* $f : [a,b] \times \mathbb{R} \to \mathbb{R}$ *is an* L^1*-Carathéodory function that satisfies*

$(\tilde{H}_0^+)$ *there exist* $\delta > 0$ *and* $r_0 \in L^1(a,b)$, $r_0(t) > 0$ *a.e. on* $[a,b]$, *such that* $\mu_1(r_0) \le 1$ *and for all* $u \in]0,\delta]$ *and a.e.* $t \in [a,b]$,

$$f(t,u) - r_0(t) u \ge 0;$$

$(\tilde{H}_\infty^-)$ *there exist* $\rho > 0$, f_∞ *and* $r_\infty \in L^1(a,b)$, $r_\infty(t) > 0$ *a.e. on* $[a,b]$, *such that* $\mu_1(r_\infty) \ge 1$,

$$\int_a^b f_\infty(t) \psi_1(t; r_\infty)\, dt < 0,$$

and for all $u \ge \rho$ *and a.e.* $t \in [a,b]$,

$$f(t,u) - r_\infty(t) u \le f_\infty(t).$$

Then the boundary value problem (0.1) *has at least one solution* $u \succ 0$.

Proof. Verify that

$$\alpha(t) = \delta \, \frac{\psi_1(t; r_0)}{\|\psi_1(\cdot; r_0)\|_\infty}$$

is a lower solution and build an upper solution as in Claim 2 of the proof of Theorem 1.1. □

Remark 1.1. We must notice that Theorem 1.1 is equivalent to Theorem 1.4 if $q_0(t) > 0$ and $q_\infty(t) > 0$ a.e. on $[a,b]$. The first result is however somewhat more general since it allows cases where q_0 or q_∞ change sign.

Recall first that (see Theorem A-3.2), for $q \in L^1(a,b)$, $q(t) > 0$ a.e. on $[a,b]$, we can write

$$\mu_1(q) = \min_{u \in H_0^1 \setminus \{0\}} \frac{\int_a^b u'^2 \, dt}{\int_a^b q u^2 \, dt} \quad \text{and} \quad \lambda_1(q) = \min_{u \in H_0^1 \setminus \{0\}} \frac{\int_a^b (u'^2 - q u^2) \, dt}{\int_a^b u^2 \, dt}.$$

Hence it is easy to see that

$$\mu_1(q) \le 1 \Leftrightarrow \lambda_1(q) \le 0 \quad \text{and} \quad \mu_1(q) \ge 1 \Leftrightarrow \lambda_1(q) \ge 0.$$

This result proves in particular the equivalence of (H_0^+) and $(\tilde{H}_0^+)$, in case $q_0(t) > 0$ a.e. on $[a,b]$.

Next, if f satisfies (H_∞^-) with $q_\infty(t) > 0$ a.e. on $[a,b]$, we have for all $u \ge \rho$ and a.e. $t \in [a,b]$,

$$f(t,u) - r_\infty(t) u \le f_\infty(t),$$

with $r_\infty(t) = q_\infty(t) + \lambda_1(q_\infty)$. Observe that $\mu_1(r_\infty) = 1$, $\psi_1(t, r_\infty) = \varphi_1(t, r_\infty) = \varphi_1(t, q_\infty)$ and $(\tilde{H}_\infty^-)$ is satisfied as

$$\int_a^b f_\infty(t) \psi_1(t, r_\infty) \, dt = \int_a^b f_\infty(t) \varphi_1(t, q_\infty) \, dt < 0.$$

We prove in the same way that $(\tilde{H}_\infty^-)$ implies (H_∞^-).

1.2 The Superlinear Case

The *superlinear case* is a "reversed" case, where the nonlinearity is over the first eigenvalue near $+\infty$ and under it near zero. The next theorem presents an existence result under such assumptions that refer to the spectral problem (1.1).

Theorem 1.5. *Let $f : [a,b] \times \mathbb{R} \to \mathbb{R}$ be an L^1-Carathéodory function such that $f(t,0) = 0$ on $[a,b]$, define $\varphi_1(\cdot; q)$ to be a positive eigenfunction corresponding to $\lambda_1(q)$, the first eigenvalue of the spectral problem* (1.1) *and assume*

(H_0^-) *there exist $\delta > 0$ and $q_0 \in L^1(a,b)$, $q_0(t) \ge 0$ a.e. on $[a,b]$, such that $\lambda_1(q_0) \ge 0$ and for all $u \in \,]0,\delta]$ and a.e. $t \in [a,b]$,*

$$f(t,u) - q_0(t)\, u \le 0;$$

(H_∞^+) *there exist* $\rho > 0$, q_∞ *and* $f_\infty \in L^1(a,b)$, *such that* $\lambda_1(q_\infty) \le 0$,

$$\int_a^b f_\infty(t)\varphi_1(t; q_\infty)\, dt > 0$$

and for all $u \ge \rho$ *and a.e.* $t \in [a,b]$,

$$f(t,u) - q_\infty(t)\, u \ge f_\infty(t).$$

Then the boundary value problem (0.1) *has a nontrivial solution* $u \ge 0$.
If, moreover, there exists $h \in L^1(a,b)$ *such that for all* $u \in [0,\delta]$ *and a.e.* $t \in [a,b]$,

$$|f(t,u)| \le h(t)u, \tag{1.5}$$

then $u \succ 0$.

Notice that assumption (H_0^-) implies that (1.5) follows from the condition $f(t,u) \ge -h(t)u$.

Proof. If for some $B > 0$, $B\varphi_1(t;q_0)$ is a solution of (0.1), the theorem is proved. Let us assume then such a solution does not exist.

Claim 1 – There exists $\beta \in W^{2,1}(a,b)$*, a strict* $W^{2,1}$*-upper solution of* (0.1) *such that* $\beta \succ 0$, $\beta(a) = 0$ *and* $\beta(b) = 0$. The function

$$\beta(t) = \delta \frac{\varphi_1(t;q_0)}{\|\varphi_1(\cdot;q_0)\|_\infty}.$$

is such that $\beta \in W^{2,1}(a,b)$, $\beta \succ 0$, $\beta(a) = 0$ and $\beta(b) = 0$. Further, for a.e. $t \in [a,b]$ and all $u \in [0, \beta(t)]$

$$\beta''(t) + f(t,u) \le \beta''(t) + q_0(t)u = -\lambda_1(q_0)\beta(t) + q_0(t)(u - \beta(t)) \le 0.$$

The claim follows then from Proposition III-2.4.

Claim 2 – The problem (0.1) *has a* $W^{2,1}$*-lower solution* $\alpha \in W^{2,1}(a,b)$ *such that for all* $t \in]a,b[$, $\alpha(t) > \beta(t)$. The proof follows the argument of Claim 2 in the proof of Theorem 1.1.

Conclusion – Notice that $\bar\alpha = 0$ is a lower solution. From Claim 1 we find a strict upper solution β and from Claim 2 a lower solution α such that $\bar\alpha < \beta < \alpha$ on $]a,b[$. It follows now from Theorem III-3.10 that there exists a solution $u \ge 0$ such that $u \not\le \beta$. Hence u is a nontrivial solution.

If moreover (1.5) is satisfied, then $\bar\alpha = 0$ is a solution and we deduce from the uniqueness of the solution of the Cauchy problem with zero initial conditions that $u \succ 0$. □

Remark Notice that assumptions (H_0^-) and (H_∞^+) alone do not imply the existence of a solution. Consider for example the problem

$$u'' + 4u - 2 - \sin 2t = 0,$$
$$u(0) = 0, \ u(\pi) = 0.$$

From Fredholm alternative, one proves this problem has no solution although $f(t,u) = 4u - 2 - \sin 2t$ satisfies (H_0^-) and (H_∞^+) with $q_0 = q_\infty = 1$ and $f_\infty = 1$.

As in Corollary 1.2, we can write the following consequence of Theorem 1.5.

Corollary 1.6. *Let $f : [a,b] \times \mathbb{R} \to \mathbb{R}$ be an L^1-Carathéodory function such that $f(t,0) = 0$ on $[a,b]$. Assume that for some γ_0 and $\gamma_\infty \in L^1(a,b)$,*

$$\limsup_{u \to 0+} \frac{f(t,u)}{u} \leq \gamma_0(t) \lneqq \left(\frac{\pi}{b-a}\right)^2$$

and

$$\liminf_{u \to +\infty} \frac{f(t,u)}{u} \geq \gamma_\infty(t) \gneqq \left(\frac{\pi}{b-a}\right)^2$$

hold uniformly in t.

Then the problem (0.1) has at least one nontrivial solution $u \geq 0$.

If moreover, there exists $h \in L^1(a,b)$ such that for some $\delta > 0$, all $u \in [0,\delta]$ and a.e. $t \in [a,b]$,

$$|f(t,u)| \leq h(t)u,$$

then $u \succ 0$.

Proof. We proceed as in the proof of Corollary 1.2 noticing that we can assume $\inf_{[a,b]} \gamma_0 \geq \epsilon$ so that $q_0 = \gamma_0 - \epsilon \geq 0$. □

Theorem 1.5 can be extended to the separated boundary value problem.

Theorem 1.7. *Let a_1, $b_1 \in \mathbb{R}$, a_2, $b_2 \in \mathbb{R}^+$ with $a_1^2 + a_2^2 > 0$ and $b_1^2 + b_2^2 > 0$. Assume $f : [a,b] \times \mathbb{R} \to \mathbb{R}$ is an L^1-Carathéodory function that satisfies $f(t,0) = 0$ on $[a,b]$, (H_0^-) and (H_∞^+), where $\varphi_1(\cdot;q)$ is a positive eigenfunction corresponding to $\lambda_1(q)$, the first eigenvalue of the spectral problem (1.2).*

Then the boundary value problem (1.3) has at least one nontrivial solution $u \geq 0$.

If, moreover, there exists $h \in L^1(a,b)$ such that for all $u \in [0,\delta]$ and a.e. $t \in [a,b]$,

$$|f(t,u)| \leq h(t)\, u,$$

then $u \succ 0$.

As for Theorem 1.5 we can use assumptions that refer to the spectral problem (1.4). Such assumptions give a result which is equivalent to Theorem 1.5 if $q_\infty \geq 0$ as follows from Remark 1.1.

Theorem 1.8. *Let $f : [a,b] \times \mathbb{R} \to \mathbb{R}$ be an L^1-Carathéodory function such that $f(t,0) = 0$ on $[a,b]$, define $\psi_1(\cdot; r)$ to be a positive eigenfunction corresponding to $\mu_1(r)$, the first eigenvalue of the spectral problem (1.4), and assume*

($\tilde{H}_0^-$) *there exist $\delta > 0$ and $r_0 \in L^1(a,b)$, $r_0(t) > 0$ a.e. on $[a,b]$, such that $\mu_1(r_0) \geq 1$, and for all $u \in]0,\delta]$ and a.e. $t \in [a,b]$,*

$$f(t,u) - r_0(t)u \leq 0;$$

($\tilde{H}_\infty^+$) *there exist $\rho > 0$, f_∞ and $r_\infty \in L^1(a,b)$, $r_\infty(t) > 0$ a.e. on $[a,b]$, such that $\mu_1(r_\infty) \leq 1$,*

$$\int_a^b f_\infty(t)\psi_1(t; r_\infty)\, dt > 0,$$

and for all $u \geq \rho$ and a.e. $t \in [a,b]$,

$$f(t,u) - r_\infty(t)u \geq f_\infty(t).$$

Then the boundary value problem (0.1) has at least one nontrivial solution $u \geq 0$.

If, moreover, there exists $h \in L^1(a,b)$ such that for all $u \in [0,\delta]$ and a.e. $t \in [a,b]$,

$$|f(t,u)| \leq h(t)u,$$

then $u \succ 0$.

Exercise 1.1. Prove the above theorem.

In the above results, we can replace the condition on $\frac{f(t,u)}{u}$ at infinity by a condition on $\frac{2}{u^2}\displaystyle\int_0^u f(t,s)\, ds$.

Theorem 1.9. *Let $f : [a,b] \times \mathbb{R} \to \mathbb{R}$ be an L^1-Carathéodory function that satisfies for some $a_+ \in L^1(a,b)$, $\liminf_{u\to+\infty} \frac{f(t,u)}{u} \geq a_+(t)$ uniformly in $t \in [a,b]$. Assume $f(t,0) = 0$ on $[a,b]$, (H_0^-) and*

($\hat{H}_\infty^+$) *there exist $\rho > 0$ and $\gamma_\infty \in L^1(a,b)$ such that for a.e. $t \in [a,b]$ and all $u > \rho$*

$$\frac{2F(t,u)}{u^2} \geq \gamma_\infty(t) \gneqq \left(\frac{\pi}{b-a}\right)^2,$$

where $F(t,u) = \displaystyle\int_0^u f(t,s)\, ds.$

Then the boundary value problem (0.1) *has at least one nontrivial solution* $u \geq 0$.

If, moreover, there exists $h \in L^1(a,b)$ *such that for all* $u \in [0,\delta]$ *and a.e.* $t \in [a,b]$,

$$|f(t,u)| \leq h(t)\,u,$$

then $u \succ 0$.

Proof. Construction of a lower solution. Let $\varphi_1(\cdot,\gamma_\infty)$ be a positive eigenfunction corresponding to $\lambda_1(\gamma_\infty)$, the first eigenvalue of (1.1) with $q = \gamma_\infty$. We deduce then from the assumptions that $\lambda_1(\gamma_\infty) < 0$. Next, we can find $k \in L^1(a,b)$ so that for a.e. $t \in [a,b]$ and all $u \geq 0$

$$F(t,u) \geq \gamma_\infty(t)\tfrac{u^2}{2} - k(t).$$

Hence for any $u \in H_0^1(a,b)$, $u \geq 0$, we have

$$\Phi(u) = \int_a^b [\tfrac{u'^2(t)}{2} - F(t,u(t))]\,dt \leq \int_a^b [\tfrac{u'^2(t)}{2} - \gamma_\infty(t)\tfrac{u^2(t)}{2} + k(t)]\,dt$$

and we can find K large enough so that

$$\Phi(K\varphi_1(t;\gamma_\infty)) \leq \int_a^b [\tfrac{1}{2}\lambda_1(\gamma_\infty)K^2\varphi_1^2(t;\gamma_\infty) + k(t)]\,dt < \Phi(0).$$

Let $r > K\|\varphi_1(\cdot;\gamma_\infty)\|_\infty$ and define

$$\begin{aligned} \tilde{f}(t,u) &= 0, && \text{if } u < 0, \\ &= f(t,u), && \text{if } 0 \leq u < r, \\ &= \min\{f(t,r), f(t,u)\}, && \text{if } u \geq r, \end{aligned}$$

$$\tilde{F}(t,u) = \int_0^u \tilde{f}(t,s)\,ds$$

and

$$\tilde{\Phi}(u) = \int_a^b (\tfrac{u'^2}{2} - \tilde{F}(t,u))\,dt.$$

It is easy to see that

$$\begin{aligned} -\tilde{F}(t,u) &= 0, && \text{if } u < 0, \\ &\geq -\ell(t)u, && \text{if } u \geq r, \end{aligned}$$

for some $\ell \in L^1(a,b)$. This implies $\tilde{\Phi}(u)$ is coercive on $H_0^1(a,b)$. Hence, it has a minimum α which is such that

$$\begin{gathered} \alpha'' + \tilde{f}(t,\alpha) = 0, \\ \alpha(a) = 0, \ \alpha(b) = 0. \end{gathered}$$

As $\min \tilde{\Phi}(u) \le \tilde{\Phi}(K\varphi_1(\cdot;\gamma_\infty)) = \Phi(K\varphi_1(\cdot;\gamma_\infty)) < \Phi(0) = \tilde{\Phi}(0)$, it is clear that $\alpha \neq 0$. Also, $\alpha \ge 0$ as $\tilde{f}(t,u) = 0$ for $u \le 0$. At last, we compute $0 = \alpha'' + \tilde{f}(t,\alpha) \le \alpha'' + f(t,\alpha)$, which proves that α is a lower solution of (0.1).

Conclusion – Notice that $\alpha_1(t) = 0$ is a lower solution and $\alpha_1(t) = 0 \le \alpha(t)$. Next, proceeding as in Claim 1 of Theorem 1.5 we can find a strict $W^{2,1}$-upper solution of (0.1), $\beta(t) = \delta \frac{\varphi_1(t;q_0)}{\|\varphi_1(\cdot;q_0)\|_\infty}$, such that $\alpha_1 = 0 \le \beta$ and $\alpha \not\le \beta$. The proof follows now from Theorem III-3.10. □

1.3 The Semipositone Problem

In this section, we consider positive solutions of the parametric problem

$$\begin{aligned} u'' + sf(u) = 0, \\ u(a) = 0, \ u(b) = 0, \end{aligned} \tag{1.6}$$

where $f(0)$ can be negative. This problem is called the *semipositone problem.*

Theorem 1.10. *Let $s \in](\frac{\pi}{b-a})^2, (\frac{2\pi}{b-a})^2[$. Assume $f : \mathbb{R} \to \mathbb{R}$ is a continuous function such that*

$$\limsup_{u \to +\infty} \frac{f(u)}{u} \le \frac{1}{4} \tag{1.7}$$

and

$$f(u) \ge u - m \quad on \quad [0, \sigma m],$$

with $m > 0$ and $\sigma = 1 + \dfrac{1}{|\cos(\sqrt{s}(\frac{b-a}{2}))|}$.

Then the problem (1.6) *has a solution $u_s \succ 0$.*

Proof. Claim 1 – *There exists a C^2-lower solution $\alpha \succ 0$ of* (1.6). The function

$$\alpha(t) = m\left(1 - \frac{\cos(\sqrt{s}(t - \frac{a+b}{2}))}{\cos(\sqrt{s}(\frac{b-a}{2}))}\right)$$

is such that $\alpha \succ 0$, $\alpha(a) = \alpha(b) = 0$, $\alpha(t) \le \sigma m$ and

$$\alpha''(t) + sf(\alpha(t)) \ge \alpha''(t) + s(\alpha(t) - m) = 0.$$

Claim 2 – *The problem* (1.6) *has a C^2-upper solution $\beta \ge \alpha$.* We deduce from (1.7) that assumption (H_∞^-) is satisfied with $q_\infty = \frac{s}{4} + \epsilon$, $f_\infty = -1$ and $\epsilon > 0$ small enough. The upper solution β is obtained then as in Claim 2 of the proof of Theorem 1.1.

Conclusion – The proof follows from Theorem II-1.5. □

Exercise 1.2. Extend Theorem 1.10 to deal with a function f such that

$$f(u) \geq pu - m \quad \text{on} \quad [0, \sigma m],$$

for some $p > 0$.

Remark With the assumptions of Theorem 1.10 and if $f(0) < 0$, the problem (1.6) has no positive solution for $s > 0$ small. More precisely, if $f(u) \leq ku$ on $\mathbb{R}^+$ and $u \succ 0$ is a solution of (1.6), we compute

$$\begin{aligned} 0 &= \int_a^b (u'' + sf(u)) \sin\left(\tfrac{\pi}{b-a}\, t\right) dt \\ &\leq \int_a^b (u'' + sku) \sin\left(\tfrac{\pi}{b-a}\, t\right) dt = \left(sk - \left(\tfrac{\pi}{b-a}\right)^2\right) \int_a^b u \sin\left(\tfrac{\pi}{b-a}\, t\right) dt, \end{aligned}$$

which is a contradiction if $s > 0$ is small enough.

For large values of s, we can write the following result which gives large solutions.

Theorem 1.11. *Let $f : \mathbb{R} \to \mathbb{R}$ be a continuous function that satisfies*

$$\limsup_{u \to +\infty} \frac{f(u)}{u} \leq 0$$

and for some $r > 0$ and $m > 0$

$$f(u) \geq m \qquad on \quad [r, +\infty[.$$

Then, for s large enough, (1.6) has a solution $u_s \succ 0$ such that

$$\lim_{s \to \infty} \|u_s\|_\infty = +\infty.$$

Proof. Claim 1 – For each s large enough the problem (1.6) has a C^2-lower solution $\alpha_s \succ 0$ such that $\|\alpha_s\|_\infty \to \infty$ as $s \to \infty$. Let $k \geq 0$ be such that for any $u \in [0, r]$, $f(u) \geq -k$. Define

$$\tilde{a} = a + \sqrt{\tfrac{2r}{sk}} \quad \text{and} \quad \tilde{b} = b - \sqrt{\tfrac{2r}{sk}}.$$

The claim is then verified with

$$\begin{aligned} \alpha_s(t) &= \tfrac{sk}{2}(t-a)^2, && \text{if } a \leq t < \tilde{a}, \\ &= r + 2\tfrac{\sqrt{2rsk}}{b-a}(t - \tilde{a})(\tilde{b} - t), && \text{if } \tilde{a} \leq t < \tilde{b}, \\ &= \tfrac{sk}{2}(b-t)^2, && \text{if } \tilde{b} \leq t \leq b. \end{aligned}$$

Claim 2 – For s as in Claim 1, the problem (1.6) has a C^2-upper solution $\beta_s \geq \alpha_s$. We proceed as in Claim 2 of the proof of Theorem 1.10.

Conclusion – The proof follows from Theorem II-1.5. □

1.4 A Problem with Indefinite Weight

Consider now the problem

$$\begin{aligned} u'' + sq(t)g(u) &= 0, \\ u(a) = 0, \ u(b) &= 0, \end{aligned} \tag{1.8}$$

where g is nonnegative and q can change sign.

Theorem 1.12. *Assume $g : \mathbb{R}^+ \to \mathbb{R}^+$ is a continuous function such that $g(0) > 0$ and $q \in L^1(a, b)$ satisfies*

(i) $\int_a^b G(t,r)q(r)\,dr > 0$ *for* $t \in]a, b[$,
where $G(t,r) = \frac{(r-a)(b-t)}{b-a}$ *if* $r \in [a, t[$ *and* $G(t,r) = \frac{(t-a)(b-r)}{b-a}$ *if* $r \in [t, b]$*;*

(ii) $\int_a^b (b-r)q(r)\,dr > 0, \quad \int_a^b (r-a)q(r)\,dr > 0.$

Then for $s > 0$ small enough, (1.8) has a solution $u_s \succ 0$ such that

$$\lim_{s\to 0} \|u_s\|_\infty = 0.$$

Proof. Step 1 – Construction of a $W^{2,1}$-lower solution of (1.8) for small values of s. Let

$$v_0(t) = \int_a^b G(t,r)|q(r)|\,dr \quad \text{and} \quad u_0(t) = \int_a^b G(t,r)q(r)\,dr.$$

We have

$$-u_0'(b) = \int_a^b \frac{r-a}{b-a}\,q(r)\,dr > 0$$

and

$$u_0'(a) = \int_a^b \frac{b-r}{b-a}\,q(r)\,dr > 0.$$

Hence, we can fix $\epsilon > 0$ small enough so that

$$\alpha(t) = s(g(0)u_0(t) - \epsilon v_0(t)) \succ 0.$$

Next, we choose $s > 0$ small enough, i.e. α small enough, so that for $t \in]a, b[$

$$\alpha'' + sq(t)g(\alpha) = s[q(t)(g(\alpha) - g(0)) + |q(t)|\epsilon] > 0.$$

It follows that α is a $W^{2,1}$-lower solution of (1.8).

Step 2 – Construction of a $W^{2,1}$-upper solution of (1.8). The function $\beta(t) = s^r v_0(t) > 0$, with $r \in]0,1[$, is an upper solution for small values of s since in this case

$$\beta'' + sq(t)g(\beta) \le s^r(-|q(t)| + s^{1-r}q(t)g(\beta)) < 0.$$

Conclusion – Notice at last that, for small values of s,

$$\alpha(t) \le sg(0)u_0(t) \le sg(0)v_0(t) \le s^r v_0(t) = \beta(t).$$

The theorem follows then from Theorem II-2.4. □

Remark The hypothesis of Theorem 1.12 are almost optimal. If there exists a family of positive solutions $u_s(t)$ of (1.8) that tends to zero with s, we have

$$\int_a^b G(t,r)q(r)g(u_s(r))\,dr = \frac{u_s(t)}{s} \ge 0$$

and going to the limit as s goes to zero, we obtain

$$\int_a^b G(t,r)q(r)g(0)\,dr \ge 0.$$

It follows that $u_0 \ge 0$ which implies

$$-u_0'(b) = \int_a^b \frac{r-a}{b-a}\,q(r)\,dr \ge 0$$

and

$$u_0'(a) = \int_a^b \frac{b-r}{b-a}\,q(r)\,dr \ge 0.$$

Hence, the conditions (i) and (ii) with non-strict inequalities are necessary.

Another case of interest concerns radial solutions of the Laplacian. Here we have to consider a mixed problem

$$\begin{aligned} (t^n u')' + st^n q(t)g(u) = 0, \\ u'(0) = 0,\ u(1) = 0. \end{aligned} \tag{1.9}$$

Exercise 1.3. Let $g : \mathbb{R}^+ \to \mathbb{R}^+$ be a continuous function such that $g(0) > 0$, $n \in \mathbb{N}$ and assume $q \in L^1(0,1)$ satisfies

(i) $\displaystyle\int_t^1 \frac{1}{r^n}\Big(\int_0^r \sigma^n q(\sigma)\,d\sigma\Big)\,dr > 0$ for $t \in [0,1[$;

(ii) $\displaystyle\int_0^1 r^n q(r)\,dr > 0.$

Then prove that for $s > 0$ small enough, (1.9) has a solution $u_s \succ 0$ such that

$$\lim_{s\to 0} \|u_s\|_\infty = 0.$$

2 Some Multiplicity Results

2.1 The Sub-superlinear Case

Multiplicity results can be obtained if the nonlinearity crosses the first eigenvalue twice. The *sub-superlinear case* concerns a nonlinearity with a "slope" which is greater than the first eigenvalue both at 0 and at $+\infty$. Such an assumption alone does not imply the existence of solutions as can be seen from the example

$$\begin{gathered} u'' + 4u + 2 + \sin 2t = 0, \\ u(0) = 0, \ u(\pi) = 0. \end{gathered}$$

We have to add assumptions that force a "double crossing" of the first eigenvalue.

Theorem 2.1. *Let $f : [a,b] \times \mathbb{R} \to \mathbb{R}$ satisfy L^1-Carathéodory conditions, (H_0^+) and (H_∞^+), where $\varphi_1(\cdot; q)$ is a positive eigenfunction corresponding to $\lambda_1(q)$, the first eigenvalue of the spectral problem* (1.1). *Assume there exists a strict $W^{2,1}$-upper solution $\beta \succ 0$ of* (0.1).

Then the problem (0.1) *has at least two solutions $u_1 \succ 0$ and $u_2 \gneqq u_1$.*

Proof. Claim 1 – There exists α, a $W^{2,1}$-lower solution of (0.1) *such that $0 \prec \alpha \le \beta$.* Define

$$\alpha(t) = A \frac{\varphi_1(t; q_0)}{\|\varphi_1(\cdot; q_0)\|_\infty},$$

with $0 < A < \delta$ small enough so that $\alpha \le \beta$. We know that $\varphi_1(\cdot; q_0) \succ 0$, i.e. $\alpha \succ 0$, and

$$\alpha'' + f(t, \alpha) \ge \alpha'' + q_0(t)\alpha = -\lambda_1(q_0)\alpha \ge 0.$$

Claim 2 – There exists $k \ge \beta$ such that for all solution $u \ge \alpha$ of

$$\begin{gathered} u'' + f(t,u) \le 0, \\ u(a) = 0, \ u(b) = 0, \end{gathered} \tag{2.1}$$

we have $u(t) < k$ on $[a,b]$. Suppose that, for any n, there exists a solution $u_n \ge \alpha$ of (2.1) such that $\max_{[a,b]} u_n(t) \ge n$. Let us write $u_n(t) =$

$\tilde{u}_n(t)+\bar{u}_n\varphi_1(t;q_\infty)$, where $\int_a^b(\tilde{u}_n'(t)\varphi_1'(t;q_\infty)-q_\infty(t)\tilde{u}_n(t)\varphi_1(t;q_\infty))\,dt = 0$. From Proposition A-4.5, we can write, for some $K \geq 0$,

$$\begin{aligned}\|\tilde{u}_n\|_\infty &\leq K\int_a^b |\tilde{u}_n'' + q_\infty\tilde{u}_n + \lambda_1(q_\infty)\tilde{u}_n|\varphi_1(\cdot;q_\infty)\,dt\\ &\leq K\int_a^b |u_n'' + q_\infty u_n + \lambda_1(q_\infty)u_n|\varphi_1(\cdot;q_\infty)\,dt\\ &\leq K[2\int_a^b (u_n'' + q_\infty u_n + \lambda_1(q_\infty)u_n)^+\varphi_1(\cdot;q_\infty)\,dt\\ &\qquad\qquad - \int_a^b (u_n'' + q_\infty u_n + \lambda_1(q_\infty)u_n)\varphi_1(\cdot;q_\infty)\,dt]\\ &\leq 2K\int_a^b (u_n'' + q_\infty u_n + \lambda_1(q_\infty)u_n)^+\varphi_1(\cdot;q_\infty)\,dt.\end{aligned}$$

As

$$u_n'' + q_\infty(t)u_n + f_\infty(t) \leq -f(t,u_n) + q_\infty(t)u_n + f_\infty(t)$$

we obtain

$$\begin{aligned}u_n'' + q_\infty(t)u_n + f_\infty(t) &\leq 0, && \text{if } u_n(t) \geq \rho,\\ &\leq h_\rho(t) + |q_\infty(t)|\rho + |f_\infty(t)|, && \text{if } u_n(t) \in [\alpha(t),\rho[,\end{aligned}$$

and

$$\|\tilde{u}_n\|_\infty \leq 2K\int_a^b [h_\rho(t) + |q_\infty(t)|\rho + |f_\infty(t)|]\varphi_1(t;q_\infty)\,dt.$$

It follows that $\bar{u}_n \to +\infty$ and $u_n(t) \to +\infty$ for all $t \in]a,b[$. Since

$$\begin{aligned}&f(t,u_n(t)) - q_\infty(t)u_n(t) - \lambda_1(q_\infty)u_n(t)\\ &\qquad\geq -h_\rho(t) - |q_\infty(t)|\rho - |\lambda_1(q_\infty)|\rho, && \text{if } u_n(t) \in [\alpha(t),\rho],\\ &\qquad\geq f_\infty(t) - \lambda_1(q_\infty)u_n(t) \geq f_\infty(t), && \text{if } u_n(t) \geq \rho,\end{aligned}$$

we can use Fatou's Lemma and obtain the contradiction

$$\begin{aligned}0 &= \lim_{n\to\infty}\int_a^b -[u_n''(t) + q_\infty(t)u_n(t) + \lambda_1(q_\infty)u_n(t)]\varphi_1(t;q_\infty)\,dt\\ &\geq \lim_{n\to\infty}\int_a^b [f(t,u_n(t)) - q_\infty(t)u_n(t) - \lambda_1(q_\infty)u_n(t)]\varphi_1(t;q_\infty)\,dt\\ &\geq \int_a^b \liminf_{n\to\infty}[f(t,u_n(t)) - q_\infty(t)u_n(t) - \lambda_1(q_\infty)u_n(t)]\varphi_1(t;q_\infty)\,dt\\ &\geq \int_a^b f_\infty(t)\varphi_1(t;q_\infty)\,dt > 0.\end{aligned}$$

Conclusion – From Theorem III-2.12, problem (0.1) has at least two solutions $u_1 \gneqq u_2 \geq \alpha \succ 0$. □

For systems controlled by nonnegative coefficients q_0 and q_∞, we can reformulate Theorem 2.1 using the spectral problem (1.4).

Theorem 2.2. *Let $f : [a,b] \times \mathbb{R} \to \mathbb{R}$ satisfy L^1-Carathéodory conditions, $(\tilde{H}_0^+)$ and $(\tilde{H}_\infty^+)$, where $\psi_1(\cdot;q)$ is a positive eigenfunction corresponding to $\mu_1(q)$, the first eigenvalue of the spectral problem (1.4). Assume there exists a strict $W^{2,1}$-upper solution $\beta \succ 0$ of (0.1).*

Then the problem (0.1) has at least two solutions $u_1 \succ 0$ and $u_2 \gneqq u_1$.

The next result gives conditions under which we have the required strict upper solution.

Proposition 2.3. *Let $f : [a,b] \times \mathbb{R} \to \mathbb{R}$ satisfy L^1-Carathéodory conditions. Let $r \in L^1(a,b)$, $r(t) > 0$ a.e. on $[a,b]$, and assume that $\mu_1(r)$, the first eigenvalue of (1.4) satisfies $\mu_1(r) \geq 1$. Suppose that there exist $M > m > 0$ such that for a.e. $t \in]a,b[$ and all $u \in]m,M[$,*

$$f(t,u) \leq r(t)(u-m).$$

Then the problem (0.1) has a strict $W^{2,1}$-upper solution β such that $\beta(t) \in [m,M]$ for every $t \in [a,b]$.

Proof. Let $\psi_1(\cdot;r)$ be a positive eigenfunction of (1.4) corresponding to $\mu_1(r)$ such that $\|\psi_1(\cdot;r)\|_\infty \leq M-m$. Then $\beta = \psi_1(\cdot;r) + m$ is a strict upper solution. Indeed, for a.e. $t \in [a,b]$ and all $u \in [m, \beta(t)]$

$$\beta'' + f(t,u) \leq \psi_1''(\cdot;r) + r(t)(u-m) \leq -r(t)[\mu_1(r)\psi_1(\cdot;r) - (u-m)] \leq 0$$

and $\beta(a) = \beta(b) = m > 0$. The proof follows now from Proposition III-2.4. □

The next result gives a simple alternative condition that implies the existence of a strict upper solution.

Proposition 2.4. *Let $f : [a,b] \times \mathbb{R} \to \mathbb{R}$ satisfy L^1-Carathéodory conditions and assume there exist $R > 0$, $D > 0$ such that for a.e. t and all $s \in [0,R]$, we have*

$$f(t,s) \leq D$$

with

$$(b-a)^2 \leq \frac{8R}{D}.$$

Then

$$\beta(t) = \frac{4R}{(b-a)^2}(t-a)(b-t),$$

is either a strict $W^{2,1}$-upper solution of (0.1) or a solution.

Proof. The proof follows immediately from Proposition III-2.4. □

As above, the reader can prove as an exercice the following extension to the separate boundary value problem.

Theorem 2.5. *Let a_1, $b_1 \in \mathbb{R}$, a_2, $b_2 \in \mathbb{R}^+$ with $a_1^2+a_2^2 > 0$ and $b_1^2+b_2^2 > 0$. Assume $f : [a,b] \times \mathbb{R} \to \mathbb{R}$ satisfies L^1-Carathéodory conditions, (H_0^+) and (H_∞^+), where $\varphi_1(\cdot;q)$ is a positive eigenfunction corresponding to $\lambda_1(q)$, the first eigenvalue of the spectral problem (1.2). Assume further there exists a strict $W^{2,1}$-upper solution $\beta \succ 0$ of (1.3).*

Then the problem (1.3) has at least two solutions $u_1 \succ 0$ and $u_2 \gneqq u_1$.

2.2 The Super-sublinear Case

We can work out a situation symmetric to the preceeding one, the *super-sublinear case*, where the "slope" of f is smaller than the first eigenvalue both at 0 and at $+\infty$. In this case also we have to impose assumptions that force a real double crossing of the eigenvalue if we want a multiplicity result. This remark is clear from the example

$$u'' + \tfrac{1}{4}u = 0, \quad u(0) = 0, \ u(\pi) = 0.$$

Using the idea of Section 2.1, we shall assume existence of a strict lower solution. Such an assumption will not be sufficient to ensure multiple solutions as follows from the following example

$$u'' - 2|u-2| = -2, \quad u(0) = 0, \ u(T) = 0.$$

It can be seen from a phase plane analysis that there exists $T_0 < \sqrt{2}\pi$ so that for $T \in]T_0, \sqrt{2}\pi[$ this problem has two positive solutions and for $T > \sqrt{2}\pi$ only one. Nevertheless, there exists a strict lower solution. For example, if $\sqrt{2}\pi < T < 2\sqrt{2}\pi$, we can use $\alpha(t) = \max\{1 - \cos\sqrt{2}t, 1 - \cos\sqrt{2}(T-t)\}$.

To obtain multiple positive solutions, we will assume further $f(t,0) = 0$. Notice that (H_0^-) already implies $f(t,0) \le 0$.

Theorem 2.6. *Let $f : [a,b] \times \mathbb{R} \to \mathbb{R}$ be an L^1-Carathéodory function such that $f(t,0) = 0$. Assume (H_0^-), (H_∞^-) hold with $\lambda_1(q)$ and $\varphi_1(\cdot;q)$ the first eigenvalue and the corresponding positive eigenfunction of the spectral problem (1.1). Assume further there exists a strict $W^{2,1}$-lower solution $\alpha \gneqq 0$ of (0.1).*

Then the problem (0.1) has two solutions u_1 and u_2 such that

$$0 \lneqq u_1 \lneqq u_2 \quad \textit{and} \quad u_2 \succ 0.$$

Proof. Step 1 – Lower solutions. Notice that $\alpha_1 = 0$ and $\alpha_2 = \alpha$ are $W^{2,1}$-lower solutions, α_2 being strict, and $\alpha_1 \leq \alpha_2$.

Step 2 – Upper solutions. As in Claim 1 of the proof of Theorem 1.5, we deduce from (H_0^-) the existence of a strict $W^{2,1}$-upper solution of (0.1) $\beta_1 \in W^{2,1}(a,b)$ such that $\beta_1 \geq \alpha_1 = 0$ and $\beta_1 \ngeq \alpha_2$.

Also, as in Claim 2 of the proof of Theorem 1.1, we deduce from (H_∞^-) the existence of a $W^{2,1}$-upper solution $\beta_2 \in W^{2,1}(a,b)$ such that $\beta_2 \geq \alpha_2$ and $\beta_2 \geq \beta_1$.

Conclusion – From Theorem III-2.11, problem (0.1) has three solutions u_0, u_1 and u_2 such that

$$0 \leq u_0 \lneqq u_1 \lneqq u_2.$$

Notice that u_0 can be the zero solution so that we only proved the existence of two nontrivial solutions. □

Corollary 2.7. *Let $f : [a,b] \times \mathbb{R} \to \mathbb{R}$ be an L^1-Carathéodory function such that*

$$\lim_{u \to 0^+} \frac{f(t,u)}{u} = 0 \quad \text{and} \quad \lim_{u \to +\infty} \frac{f(t,u)}{u} = 0 \tag{2.2}$$

uniformly in t. Assume moreover there exist $M^* > M > m > 0$ such that for a.e. $t \in [a,b]$ and all $u \in [0,m]$

$$f(t,u) \geq 0$$

and for a.e. $t \in [a,b]$ and all $u \in [m, M^*]$

$$f(t,u) > \tfrac{2}{(b-a)^2} \tfrac{(2M-m)^2}{M-m}.$$

Then the problem (0.1) has at least two solutions $u_1 \succ 0$ and $u_2 \gneqq u_1$.

Proof. The assumptions $f(t,0) = 0$, (H_0^-) and (H_∞^-) follow from (2.2). Next, let $\tau = \frac{b-a}{2} \frac{m}{2M-m}$, $\tilde{a} = a + \tau$ and $\tilde{b} = b - \tau$. We deduce then from Proposition III-2.3 that

$$\begin{aligned} \alpha(t) &= \tfrac{2(2M-m)}{b-a}(t-a), && \text{if } a \leq t < \tilde{a}, \\ &= \tfrac{2(2M-m)}{b-a}\big[\tau + \tfrac{(t-\tilde{a})(\tilde{b}-t)}{\tilde{b}-\tilde{a}}\big], && \text{if } \tilde{a} \leq t < \tilde{b}, \\ &= \tfrac{2(2M-m)}{b-a}(b-t), && \text{if } \tilde{b} \leq t \leq b, \end{aligned}$$

is a strict $W^{2,1}$-lower solution. The existence of two solutions $0 \lneqq u_1 \lneqq u_2$ follows now from Theorem 2.6. Clearly, $u = 0$ is a solution. However, (2.2) implies uniqueness of solutions of the Cauchy problem with initial conditions $u(t_0) = u'(t_0) = 0$ so that $u_1(t) \succ 0$. □

As an exercise, we can prove the following extension of Theorem 2.6.

Theorem 2.8. *Let a_1, $b_1 \in \mathbb{R}$, a_2, $b_2 \in \mathbb{R}^+$ with $a_1^2 + a_2^2 > 0$ and $b_1^2 + b_2^2 > 0$. Assume $f : [a,b] \times \mathbb{R} \to \mathbb{R}$ is an L^1-Carathéodory function such that $f(t,0) = 0$ and (H_0^-), (H_∞^-) hold with $\lambda_1(q)$ and $\varphi_1(\cdot;q)$ the first eigenvalue and the corresponding positive eigenfunction of the spectral problem* (1.2). *Assume further there exists a strict $W^{2,1}$-lower solution $\alpha \gneqq 0$ of* (1.3).

Then the problem (1.3) *has two solutions u_1 and u_2 such that*

$$0 \lneqq u_1 \lneqq u_2 \quad \text{and} \quad u_2 \succ 0.$$

2.3 Applications of Variational Equations

Using the variational equation of (0.1) along a solution u we can obtain other types of multiplicity results.

Theorem 2.9. *Let $f : [a,b] \times \mathbb{R} \to \mathbb{R}$ be a continuous function and assume there exist α and $\beta \geq \alpha$ which are C^2-lower and upper solutions of* (0.1) *but not solutions of* (0.1).

Assume also there exists a solution u_0 of (0.1), *with $\alpha \leq u_0 \leq \beta$, so that $\frac{\partial f}{\partial u}(t,u)$ exists and is continuous in a neighbourhood of $\{(t, u_0(t)) \mid t \in [a,b]\}$. Assume at last that the solution of*

$$y'' + \frac{\partial f}{\partial u}(t, u_0(t))\, y = 0, \quad y(a) = 0,\ y'(a) = 1, \tag{2.3}$$

has a zero on $]a,b[$.

Then the boundary value problem (0.1) *has, aside from u_0, at least two additional solutions u_1 and u_2 with $\alpha \prec u_1 \prec u_0 \prec u_2 \prec \beta$.*

Remark Define M to be such that $\frac{\partial f}{\partial u}(t, u_0(t)) + M > 0$. To require that the solution of (2.3) has a zero on $]a,b[$ is equivalent to require that the first eigenvalue λ_1 of

$$\begin{aligned} -u'' + Mu &= \lambda\left(\frac{\partial f}{\partial u}(t, u_0(t)) + M\right)u, \\ u(a) &= 0,\ u(b) = 0, \end{aligned}$$

satisfies $\lambda_1 < 1$.

Let φ_1 be the eigenfunction corresponding to λ_1 and y be the solution of (2.3). For any $t \in]a,b]$, we have

$$\begin{aligned} y'(t)\varphi_1(t) - y(t)\varphi_1'(t) &= \int_a^t (y''\varphi_1 - \varphi_1'' y)(s)\, ds \\ &= (\lambda_1 - 1)\int_a^t \left(\frac{\partial f}{\partial u}(s, u_0(s)) + M\right) y(s)\varphi_1(s)\, ds. \end{aligned} \tag{2.4}$$

If y is positive on $]a,b[$, evaluating (2.4) for $t=b$ we obtain $\lambda_1 \geq 1$ and if y has a first zero $t_1 \in]a,b[$, an evaluation of (2.4) for $t=t_1$ leads to $\lambda_1 < 1$. The result follows.

Proof of Theorem 2.9 : As α and β are not solutions, arguing as in Proposition III-2.7, we can prove that $\alpha \prec u_0 \prec \beta$.

Next, using the above remark, we can find $\epsilon > 0$ and $\delta > 0$ such that $\frac{\partial f}{\partial u}(t, u_0(t)) + M - \epsilon > 0$, the first eigenvalue $\lambda_1(\epsilon)$ of

$$\begin{aligned} -y'' + My = \lambda_1(\epsilon)(\tfrac{\partial f}{\partial u}(t, u_0(t)) + M - \epsilon)y, \\ y(a) = 0, \ y(b) = 0, \end{aligned} \tag{2.5}$$

satisfies $\lambda_1(\epsilon) < 1$ and for all v with $|v| < \delta$ and all $t \in [a,b]$, we have

$$\frac{\partial f}{\partial u}(t, u_0(t) + v) \geq \frac{\partial f}{\partial u}(t, u_0(t)) - \epsilon.$$

Denote by $\psi_1 \succ 0$ the eigenfunction of (2.5) corresponding to $\lambda_1(\epsilon)$ with $\|\psi_1\|_\infty = 1$. It is then easy to see, using Propositions III-2.3 and III-2.4, that $\alpha_1 = u_0 + \frac{\delta}{2}\psi_1$ and $\beta_1 = u_0 - \frac{\delta}{2}\psi_1$ are strict lower and upper solutions of (0.1). Decreasing δ if necessary, we have $\alpha \prec \beta_1 \prec \alpha_1 \prec \beta$ and we conclude by Theorem III-2.11. □

When we have a lower and an upper solution $\alpha \leq \beta$, another way to "cross" the first eigenvalue is to ask that the minimal solution $u_{\min}$ and the maximal solution $u_{\max}$ satisfy $u_{\min} \prec u_{\max}$. We deduce from Theorem VII-3.5 that problem (0.1) has a unique solution if the slope of f is always strictly smaller than λ_1. On the other hand, if the slope of f is always strictly greater than λ_1, we have $\alpha = \beta$ and hence $u_{\min} = \alpha = \beta = u_{\max}$.

Theorem 2.10. *Let $f : [a,b] \times \mathbb{R} \to \mathbb{R}$ be continuous and α and β be lower and upper solutions of (0.1) such that $\alpha \leq \beta$. Assume the minimal and maximal solutions $u_{\min}$ and $u_{\max}$ are such that $\alpha \prec u_{\min} \prec u_{\max} \prec \beta$ and let $\frac{\partial f}{\partial u}(t,u)$ be continuous in a neighbourhood of $\{(t, u_{\min}(t)) \mid t \in [a,b]\} \cup \{(t, u_{\max}(t)) \mid t \in [a,b]\}$. Then the boundary value problem (0.1) has at least three solutions $u_{\min} \prec u \prec u_{\max}$ provided the boundary value problem*

$$\begin{aligned} y'' + \tfrac{\partial f}{\partial u}(t, v(t))y = 0, \\ y(a) = 0, \ y(b) = 0, \end{aligned}$$

has only the trivial solution for $v = u_{\min}$ and $u_{\max}$.

Proof. Let M be such that $\frac{\partial f}{\partial u}(t, u_{\min}(t)) + M > 0$, $\frac{\partial f}{\partial u}(t, u_{\max}(t)) + M > 0$. By assumption, the eigenvalue problem

$$\begin{aligned} -y'' + My = \lambda(\tfrac{\partial f}{\partial u}(t, u_{\min}(t)) + M)y, \\ y(a) = 0, \ y(b) = 0, \end{aligned}$$

is such that $\lambda_1(u_{\min}) \neq 1$.

If $\lambda_1(u_{\min}) < 1$, we prove arguing as in Theorem 2.9 that the problem (0.1) has a solution $\bar{u}$ such that

$$\alpha \leq \bar{u} \prec u_{\min}.$$

This result contradicts the definition of $u_{\min}$ and hence $\lambda_1(u_{\min}) > 1$.

In the same way we prove that $\lambda_1(u_{\max}) > 1$.

As $\lambda_1(u_{\min}) > 1$ and $\lambda_1(u_{\max}) > 1$, we construct as in Theorem 2.9, α_1 and β_1 strict lower and upper solutions of (0.1) such that

$$\alpha \prec u_{\min} \prec \beta_1 \prec \alpha_1 \prec u_{\max} \prec \beta.$$

Existence of a third solution u follows now from Theorem III-2.11. Further $u_{\min} \prec u$ since otherwise, there exists $t_0 \in]a, b[$ so that $u(t_0) = u_{\min}(t_0)$ and $u'(t_0) = u'_{\min}(t_0)$. From uniqueness of solutions of the Cauchy problem, we come to the contradiction $u = u_{\min}$. Similarly, we have $u \prec u_{\max}$. □

Remark This result is also true for two ordered solutions $u_1 \prec u_2$ (not necessarily the maximum and minimum ones) but then we loose the localization $u_1 \prec u \prec u_2$.

3 Parametric Problems

In this section, we study the parametric problem

$$\begin{aligned} u'' + sf(t, u) &= 0, \\ u(0) = 0, \ u(1) &= 0, \end{aligned} \tag{3.1}$$

in case f and s are nonnegative. With such an assumption, nontrivial solutions are positive, i.e. $u \succ 0$. We noticed in Section II-4 that for such a Dirichlet problem it is natural to allow f to be singular, i.e. to satisfy $\mathcal{A}$-Carathéodory conditions. This section is written with this setting. Our first result establishes existence of at least one solution for small values of the parameter s.

Theorem 3.1. *Let $f : [0,1] \times \mathbb{R} \to \mathbb{R}^+$ satisfy $\mathcal{A}$-Carathéodory conditions and assume $f(t,0) \not\equiv 0$.*

Then there exists $s_1 \in]0, \infty[\cup \{\infty\}$ such that
(a) for any $s \in]0, s_1[$, (3.1) has at least one positive solution;
(b) for any $s > s_1$, (3.1) has no solution.

Proof. Claim 1 : for any $s > 0$ small enough, there is a positive solution u_s of (3.1). Let $h \in \mathcal{A}$ be such that

$$\text{for a.e.}\, t \in [0,1],\ \forall u \in [0,1], \quad |f(t,u)| \le h(t).$$

Define β_1 to be the solution of

$$\begin{gathered} \beta_1'' + h(t) = 0, \\ \beta_1(0) = 0,\ \beta_1(1) = 0. \end{gathered}$$

From Proposition II-4.1, we know that

$$\beta_1(t) \ge 0 \quad \text{and} \quad ||\beta_1||_\infty \le ||h||_{\mathcal{A}}.$$

Next, we choose $\hat{s} > 0$ small enough so that $\hat{s}||h||_{\mathcal{A}} \le 1$ and define for $s \in]0, \hat{s}]$, $\beta := s\beta_1$. We compute then

$$\beta'' + sf(t,\beta) = s(f(t,\beta) - h(t)) \le 0.$$

Hence, $\alpha = 0$ and $\beta \ge \alpha$ are $W^{2,1}$-lower and upper solutions for (3.1) and the claim follows from Theorem II-4.2.

Claim 2 : Define $s_1 = \sup\{s \mid (3.1)$ *has a solution*$\} \in]0,\infty[\cup \{\infty\}$. *Then for any $\bar{s} \in]0, s_1[$ problem* (3.1) *has a positive solution.* From Claim 1, it follows that $s_1 > 0$. Let us fix $\bar{s} \in]0, s_1[$ and $s^* \in [\bar{s}, s_1]$ such that (3.1), with $s = s^*$, has a solution u^*, which has to be positive. Notice that $\alpha = 0$ and $\beta = u^*$ are $W^{2,1}$-lower and upper solutions for (3.1), with $s = \bar{s}$, and the claim follows from Theorem II-4.2. □

Remarks (a) If in the proof of Claim 1 we let s go to zero, we have that $||\beta||_\infty$ and hence $||u_s||_\infty$ goes to zero. This fact shows that there is, in the space (u, s), a "branch" of solutions that emanates from the origin.

Further, if we assume $f \in C^1([0,1] \times \mathbb{R})$, we can apply the implicit function theorem to the equation

$$\Phi(s,u) = u'' + sf(t,u) = 0$$

and prove that this "branch" is actually a curve parametrized by s.

(b) If $s_1 < s_2$, it follows from the proof of Claim 2 that there exist corresponding solutions u_1 and u_2 which are ordered : $u_1 \le u_2$. It can also be proved that for any $r \in]0, s_1[$ there exists a minimal positive solution u_r which is an increasing function of r, i.e. $r_1 \le r_2$ implies that for all $t \in [0,1]$, $u_{r_1}(t) \le u_{r_2}(t)$.

Example 3.1. If we consider the example

$$u'' + \frac{s}{t} = 0,$$
$$u(0) = 0, \; u(1) = 0,$$

whose solution is $u(t) = st\ln(1/t)$, it is clear that solutions do not have, in general, bounded derivatives.

Example 3.2. Notice that the condition $f(t,0) \not\equiv 0$ is essential. The theorem does not hold for the problem

$$u'' + s|u| = 0,$$
$$u(0) = 0, \; u(1) = 0.$$

Example 3.1 has a positive solution for any $s > 0$. Hence without additional assumptions we might have $s_1 = +\infty$. The next result gives conditions so that $s_1 \in \mathbb{R}^+$.

Theorem 3.2. *Let $f : [0,1] \times \mathbb{R} \to \mathbb{R}^+$ satisfy $\mathcal{A}$-Carathéodory conditions. Assume $f(t,0) \not\equiv 0$ and there exists $k \in L^\infty(0,1)$ such that $k \geq 0$, $k(t) \not\equiv 0$ and*

$$\text{for a.e. } t \in [0,1],\ \forall u \geq 0, \quad f(t,u) \geq k(t)u.$$

Then there exists $s_1 \in \mathbb{R}_0^+$ such that
(a) for any $s \in]0, s_1[$, (3.1) has at least one positive solution;
(b) for any $s > s_1$, (3.1) has no solution.

Proof. From Theorem 3.1, it is enough to prove that for $s > 0$ large enough, there is no solution of (3.1). Consider the eigenvalue problem

$$u'' + \lambda k(t)u = 0,$$
$$u(0) = 0, \; u(1) = 0,$$

and let $\lambda_1 > 0$ and φ_1 be its first eigenvalue and eigenfunction. Assume there exists a solution u of (3.1) with $s > \lambda_1$. We compute

$$0 = \int_0^1 \frac{d}{dt}(u'(t)\varphi_1(t) - u(t)\varphi_1'(t))\,dt \leq \int_0^1 (\lambda_1 - s)k(t)u(t)\varphi_1(t)\,dt < 0,$$

which is impossible. □

Remarks (a) Notice there is no point to assume that $k \in \mathcal{A}$, since we can replace $k(t)$ by $\min\{k(t), 1\}$, which is in $L^\infty(0,1)$.

(b) It follows from the proof that $s_1 \leq \lambda_1$. Hence, in case k is constant, $\lambda_1 = \pi^2/k$ and there is no solution if $s \in]\pi^2/k, \infty[$.

Example 3.3. The type of problem we consider can have unique solutions as it is clear from the example

$$\begin{gathered} u'' + s(u+1)_+ = 0, \\ u(0) = 0, \ u(1) = 0, \end{gathered}$$

whose solution

$$u(t) = \frac{\cos(\sqrt{s}(t-1/2))}{\cos(\sqrt{s}/2)} - 1,$$

is unique and defined for $s \in]0, \pi^2[$.

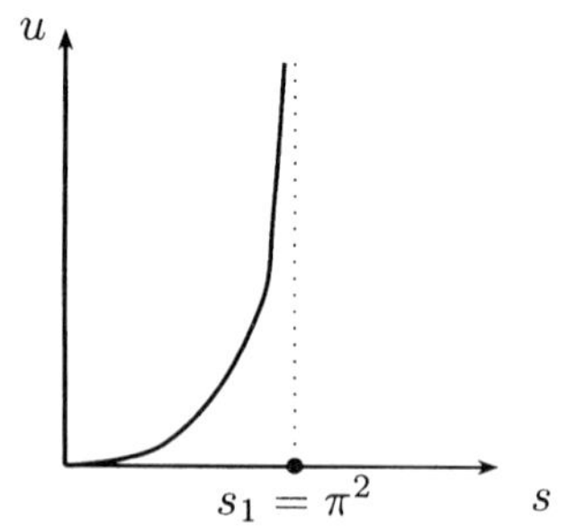

Fig. 1 : Example 3.3

To better understand the role of the lower bound on $f(t,u)$, we can write the following modification of Theorem 3.2.

Proposition 3.3. *Let $f : [0,1] \times \mathbb{R} \to \mathbb{R}^+$ satisfy $\mathcal{A}$-Carathéodory conditions. Assume $f(t,0) \not\equiv 0$ and there exists a continuous function $g : \mathbb{R} \to \mathbb{R}^+$ such that*

(i) $\displaystyle\sup_{u_0>0} \int_0^{u_0} \frac{du}{\sqrt{G(u_0) - G(u)}} < +\infty$, *where* $\displaystyle G(u) := \int_0^u g(v)\,dv$;

(ii) $\displaystyle\liminf_{u \to 0+} \frac{g(u)}{u} > 0$;

(iii) for a.e. $t \in [0,1]$, $\forall u \geq 0$, $\quad f(t,u) \geq g(u)$.

Then there exists $s_1 \in \mathbb{R}_0^+$ such that

(a) for any $s \in]0, s_1[$, (3.1) has at least one positive solution;

(b) for any $s > s_1$, (3.1) has no solution.

Proof. Assumption (ii) is such that for a small enough and s large enough, $\alpha(t) = a \sin \pi t$ is a $W^{2,1}$-lower solution of

$$\begin{gathered} u'' + sg(u) = 0, \\ u(0) = 0, \ u(1) = 0. \end{gathered} \tag{3.2}$$

Any solution u of (3.1) is a $W^{2,1}$-upper solution for (3.2). Hence if $s_1 = \infty$, we deduce from Theorem II-4.2 that the problem (3.2) has solutions for any large value of s. But this is impossible since assumption (i) and a standard time-map argument imply that for s large enough, (3.2) has no solution. □

We can obtain more precise results for linearly bounded nonlinearities.

Proposition 3.4. *Let $f : [0,1] \times \mathbb{R} \to \mathbb{R}^+$ satisfy L^∞-Carathéodory conditions. Assume $f(t,0) \not\equiv 0$ and for some $a > 0$, $b > 0$ and all $u \geq 0$*

$$f(t,u) \leq a + bu.$$

Then, for $s \in \,]0, \frac{\pi^2}{b}[$, (3.1) has a positive solution.

Proof. Take

$$\beta(t) = B(\cos\sqrt{sb}(t - \tfrac{1}{2}) - \cos\tfrac{\sqrt{sb}}{2}) \geq 0$$

and compute for B large enough

$$\begin{aligned}\beta''(t) &+ sf(t,\beta(t)) \\ &\leq -sbB\cos\sqrt{sb}(t-\tfrac{1}{2}) + s[a + bB(\cos\sqrt{sb}(t-\tfrac{1}{2}) - \cos\tfrac{\sqrt{sb}}{2})] \\ &= s[a - bB\cos\tfrac{\sqrt{sb}}{2}] < 0.\end{aligned}$$

The proof follows now from Theorem II-2.4 with $\alpha(t) = 0$. □

The situation is somewhat different if f is asymptotically linear near the origin as follows from the following result.

Proposition 3.5. *Let $f : [0,1] \times \mathbb{R} \to \mathbb{R}^+$ satisfy $\mathcal{A}$-Carathéodory conditions. Assume that for some $c \in [0,\infty[$, $d \in [0,\infty[\cup\{+\infty\}$, for a.e. $t \in [0,1]$ and all $u > 0$*

$$c \leq \frac{f(t,u)}{u} \leq d.$$

Then, for $s \in \,]0, \frac{\pi^2}{d}[\,\cup\,]\frac{\pi^2}{c}, \infty[$, (3.1) has no positive solution.

Proof. Assume u is a positive solution of (3.1) and let $v(t) = \sin\pi t$. If $sc > \pi^2$, we have the contradiction

$$0 = \int_0^1 \frac{d}{dt}(u'(t)v(t) - u(t)v'(t))\,dt = \int_0^1 \Big(\pi^2 - s\frac{f(t,u(t))}{u(t)}\Big)u(t)v(t)\,dt < 0.$$

Similarly, if $sd < \pi^2$, we obtain

$$0 = \int_0^1 \frac{d}{dt}(u'(t)v(t) - u(t)v'(t))\,dt = \int_0^1 \Big(\pi^2 - s\frac{f(t,u(t))}{u(t)}\Big)u(t)v(t)\,dt > 0.$$

□

Notice that the nonexistence of positive solutions for $s \in]\frac{\pi^2}{c}, \infty[$ is essentially the remark (b) after Theorem 3.2.

Example 3.4. Consider first the problem

$$\begin{aligned} u'' + s(u + h(t)e^{-u})_+ &= 0, \\ u(0) = 0, \ u(1) &= 0, \end{aligned}$$

where $h(t) \in]0,1]$ and $u_+ = \max(u,0)$. We have here, with the notations of Propositions 3.4 and 3.5, $a = 1$, $b = 1$, $c = 1$, $d = \infty$ so that these propositions imply there exists a solution if $s \in]0, \pi^2[$ and no solution if $s \in]\pi^2, \infty[$, i.e. $s_1 = \pi^2$.

Example 3.5. Consider next the problem

$$\begin{aligned} u'' + s\max(1, u-1) &= 0, \\ u(0) = 0, \ u(1) &= 0. \end{aligned}$$

Here $a = 1$, $b = 1$, $c = 1/2$, $d = \infty$ so that we can deduce from the same Propositions 3.4 and 3.5 there is a solution if $s \in]0, \pi^2[$ and there is no solution if $s \in]2\pi^2, \infty[$, i.e. $s_1 \in [\pi^2, 2\pi^2]$.

Example 3.6. Another example is

$$\begin{aligned} u'' + s(u-1)_+ &= 0, \\ u(0) = 0, \ u(1) &= 0, \end{aligned}$$

Here $c = 0$, $d = 1$ and there is no solution if $s \in]0, \pi^2[$.

In the next theorem, we obtain more solutions. To this end, some additional assumptions on the nonlinearity f are necessary.

Theorem 3.6. *Let $f : [0,1] \times \mathbb{R} \to \mathbb{R}^+$ satisfy $\mathcal{A}$-Carathéodory conditions and assume that given $r > 0$ and $\eta > 1$, there exists some $\epsilon > 0$ such that*

$$\begin{aligned} &\forall u_1, u \in [0,r] \text{ and for a.e. } t \in [0,1], \\ &\qquad u_1 \le u \le u_1 + \epsilon \Rightarrow f(t,u) \le \eta f(t,u_1). \end{aligned} \tag{3.3}$$

Suppose $f(t,0) \not\equiv 0$ and further that for some $k \in L^\infty(0,1)$ such that $k \ge 0$, $k(t) \not\equiv 0$, and $a > 0$, we have

$$\text{for a.e. } t \in [0,1], \forall u \ge 0, \quad f(t,u) \ge k(t)(1 + u^a)u.$$

Then there exists $s_1 \in \mathbb{R}_0^+$ such that

(a) for any $s \in]0, s_1[$, (3.1) has at least two positive solutions which are ordered;

(b) for $s = s_1$, (3.1) has at least one positive solution;

(c) for any $s > s_1$, (3.1) has no solution.

Proof. Define s_1 from Theorem 3.2, fix $\bar{s} \in]0, s_1[$ and choose $r_1 \in]\bar{s}, s_1[$. Let u_1 be a solution of (3.1) with $s = r_1$ and $\epsilon \in]0, 1]$ be small enough so that for all $u, v \in [0, ||u_1||_\infty + 1]$ and for a.e. $t \in [0, 1]$

$$v \le u \le v + \epsilon \Rightarrow f(t, u) \le \frac{r_1}{\bar{s}} f(t, v).$$

Arguing as in Proposition III-2.3, we prove that $\beta = u_1 + \epsilon$ is a strict $W^{2,1}$-upper solution of (3.1) with $s = \bar{s}$.

Claim 1 : There exists $R_0 > 0$ such that for all positive solutions u of

$$\begin{aligned} u'' + \bar{s}k(t)u^{1+a} &= 0, \\ u(0) = 0, \ u(1) &= 0, \end{aligned} \tag{3.4}$$

we have $||u||_\infty \le R_0$. Assume by contradiction there exists a sequence $(u_n)_n$ of solutions of (3.4) such that $A_n := \|u_n\|_\infty \ge n$. Notice that $u_n(t) \ge A_n \varphi(t)$ with $\varphi(t) = \min\{t, 1 - t\}$. Hence, for n large enough, we have the contradiction

$$\begin{aligned} 0 = -\int_0^1 (u_n''(t) + \pi^2 u_n(t)) \sin \pi t \, dt \\ = \int_0^1 (\bar{s}k(t)u_n^{1+a}(t) - \pi^2 u_n(t)) \sin \pi t \, dt \\ \ge \left[\bar{s}(\int_0^1 k(t)\varphi^{1+a}(t) \sin \pi t \, dt) \, A_n^a - \pi^2\right] A_n > 0. \end{aligned}$$

Claim 2 : There exists $R > 0$ such that solutions u of

$$\begin{aligned} u'' + \bar{s}f(t, u) &\le 0, \\ u(0) = 0, \ u(1) &= 0, \end{aligned} \tag{3.5}$$

satisfy $||u||_\infty \le R$. Assume the claim is wrong. Hence, there exist solutions u_n of (3.5), such that $||u_n||_\infty \to +\infty$ as $n \to +\infty$.

Let $\mu_1 > 0$ and φ_1 be the first eigenvalue and eigenfunction of the problem

$$\varphi_1'' + \mu_1 k(t) \varphi_1 = 0, \quad \varphi_1(1/4) = \varphi_1(3/4) = 0.$$

Extend φ_1 on $[0, 1]$ so that $\varphi_1(t) = 0$ on $[0, 1/4] \cup [3/4, 1]$ and define $\alpha_0 \in W^{2,\infty}(0, 1)$ to be such that its graph is the concave envelope of the graph of φ_1 on $[0, 1]$. Hence there exist $1/4 < t_0 < t_1 < 3/4$ such that α_0 is linear on $[0, t_0]$ and on $[t_1, 1]$, and $\alpha_0(t) = \varphi_1(t)$ on $[t_0, t_1]$. Consider now $\alpha = A\alpha_0$. It is easy to see that if A is large enough

$$\alpha''(t) + \bar{s}k(t)\alpha^{1+a}(t) = Ak(t)\varphi_1(t)(\bar{s}A^a \varphi_1^a(t) - \mu_1) > 0, \quad \text{on } [t_0, t_1].$$

Hence, α is a $W^{2,1}$-lower solution for (3.4). Choose now u_n such that $u_n \geq \alpha$ and notice that u_n is a $W^{2,1}$-upper solution for (3.4). If $\|\alpha\|_\infty > R_0$, we get, from Theorem II-2.4, a solution of (3.4) with maximum larger than R_0, which contradicts Claim 2.

Claim 3 : Problem (3.1), *with* $s = \bar{s}$, *has at least two solutions.* This result follows from Theorem III-2.13 using $\beta = u_1 + \epsilon$ as strict upper solution and $\alpha = 0$ as a lower one.

Claim 4 : for $s = s_1$, (3.1) *has at least one solution.* Consider a sequence $(s_n)_n \subset [0, s_1]$ such that $\lim_{n\to\infty} s_n = s_1$ and corresponding solutions u_n of (3.1) with $s = s_n$. From Claim 2 and the $\mathcal{A}$-Carathéodory conditions on $f(t,u)$, there exists $h \in \mathcal{A}$ such that

$$\begin{aligned}|u_n'(t)| &= s_n\Big[-\int_0^t rf(r,u_n(r))\,dr + \int_t^1 (1-r)f(r,u_n(r))\,dr\Big]\\ &\leq s_n\Big[\int_0^t rh(r)\,dr + \int_t^1 (1-r)h(r)\,dr\Big] \in L^1(0,1).\end{aligned}$$

Therefore, from Arzelà-Ascoli Theorem, there exists a subsequence $(u_{n_i})_i$ which converges in $\mathcal{C}([0,1])$ to some function v_0. From the closedness of the derivative, $v_0 \in W^{2,1}_{loc}(0,1)$ and satisfies (3.1), with $s = s_1$. □

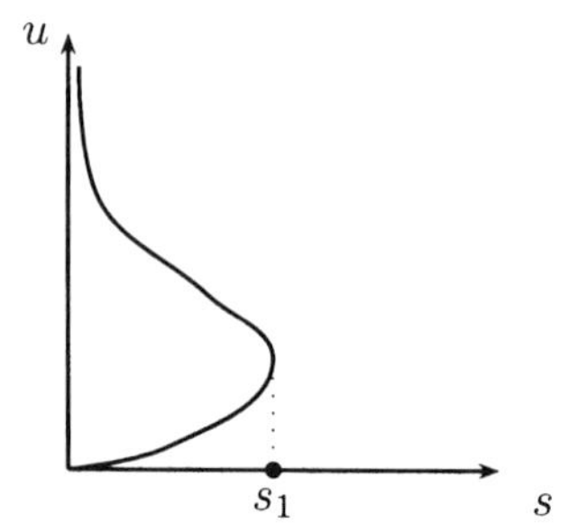

Fig. 2 : Theorem 3.6

Remarks (a) Notice that condition (3.3) is satisfied if $f(t,u)$ is continuous and $f(t,u) > 0$ on $[0,1] \times [0,+\infty[$. However condition (3.3) is somewhat restrictive if $f(t,u)$ takes zero value as follows from the example $f(t,u) = \sin^2 2\pi t + u(1+u^2)$ which does not satisfy (3.3).

(b) If $f \in \mathcal{C}^1([0,1]\times\mathbb{R})$, we can arrange the two solutions u_s and v_s such that $\|u_s\|_\infty \to 0$ and $\|v_s\|_\infty \to \infty$ as $s \to 0$. Indeed, if some subsequence of $(\|v_s\|_\infty)_s$ is bounded as $s \to 0$, then for some sub-subsequence $v_s \to v$ in

$\mathcal{C}([0,1])$ and v satisfies $v''(t) = 0$, $v(0) = v(1) = 0$. Hence $v = 0$, but in a neighbourhood of the origin, we deduce from the implicit function Theorem that there exists a single branch of solutions which has to be u_s.

(c) The Theorem 3.6 establishes a structure of the solution set that can be illustrated with figure 2.

We must notice that this set can be more complicate as follows from the study of the model example

$$\begin{gathered} u'' + sh(t)e^u = 0, \\ u(0) = 0, \; u(1) = 0, \end{gathered}$$

(see [125]).

(d) An exact count of solutions can be given in case the nonlinearity is strictly convex. Such a proposition with the relevant additional assumptions is in [125]. Notice that a related result was already obtained for another problem in Theorem VI-1.4.

Another multiplicity result can be obtained from the type of assumptions used in Section 2.2. Here, by opposition with the first part of this section, the multiplicity takes place for large values of s and the solutions are bounded with $s \to \infty$.

Theorem 3.7. *Let $f : [0,1] \times \mathbb{R} \to \mathbb{R}$ be an L^1-Carathéodory function such that for some $r > 0$ and $k \in L^1(0,1;\mathbb{R}^+)$,*

$$0 \le u_1 \le u_2 \le r \text{ implies } f(t,u_2) - f(t,u_1) \ge -k(t)(u_2 - u_1). \tag{3.6}$$

Assume moreover

(a) $f(t,0) = 0$ for a.e. $t \in [0,1]$;

(b) $\limsup_{u \to 0+} \frac{f(t,u)}{u} \le 0$ holds uniformly in t;

(c) $f(t,r) \le 0$ for a.e. $t \in [0,1]$;

(d) there exists an open interval $I_0 \subset [0,1]$, $\eta > 0$ and $u_0 \in [0,r]$ such that for all $t \in I_0$

$$\int_0^{u_0} f(t,s)\,ds \ge \eta.$$

Then, for $s > 0$ large enough, there exist at least two solutions $u_1 \succ 0$ and $u_2 \gneqq u_1$ of (3.1).

Proof. Step 1 – Lower and upper solutions of (3.1). Notice that $\alpha_1 = 0$ and $\beta_2 = r$ are respectively $W^{2,1}$-lower and upper solutions of (3.1).

Step 2 – Construction of a second lower solution $\alpha_2 \succ \alpha_1 = 0$. Let $\epsilon > 0$ and $\gamma(u) = \max\{0, \min\{u, r\}\}$. Consider then the modified problem

$$\begin{gathered} u'' + sf(t, \gamma(u)) - \epsilon u = 0, \\ u(0) = 0, \ u(1) = 0, \end{gathered}$$

together with the associated functional $\bar{\Phi} : H_0^1(0,1) \to \mathbb{R}$ defined by

$$\bar{\Phi}(u) = \int_0^1 [\tfrac{u'^2(t)}{2} - s\bar{F}(t, u(t)) + \epsilon \tfrac{u^2(t)}{2}]\, dt,$$

where $\bar{F}(t,u) = \int_0^u f(t, \gamma(s))\, ds$. As there exists a function $h \in L^1(0,1)$ so that $|f(t, \gamma(u))| \leq h(t)$, the functional $\bar{\Phi}(u)$ is coercive. Hence, it has a minimum α_2 such that

$$\alpha_2''(t) + sf(t, \gamma(\alpha_2(t))) - \epsilon \alpha_2(t) = 0, \quad \alpha_2(0) = 0, \ \alpha_2(1) = 0.$$

It is now standard to see that $0 \leq \alpha_2(t) \leq r$ so that

$$\alpha_2''(t) + sf(t, \alpha_2(t)) = \epsilon \alpha_2(t) \geq 0, \quad \alpha_2(0) = 0, \ \alpha_2(1) = 0.$$

Next, using assumption (d), we can find a function $\varphi \in H_0^1(0,1)$ with support in I_0 so that

$$\int_0^1 \bar{F}(t, \varphi(t))\, dt > 0.$$

It follows that for s large enough $\bar{\Phi}(\varphi) < 0$. Hence, for those values of s, $\bar{\Phi}(\alpha_2) < \bar{\Phi}(0) = 0$, i.e. $\alpha_2 \neq \alpha_1 = 0$. By Proposition III-2.7, α_2 is a strict lower solution of (3.1) and we deduce from (3.6) that $\alpha_2 \succ 0$.

Step 3 – Construction of a strict upper solution β_1 *of* (3.1) *so that* $\alpha_1 = 0 \leq \beta_1 \leq \beta_2$ *and* $\alpha_2 \not\leq \beta_1$. Assumption (H_0^-) in Theorem 1.5 holds with $q_0 = \epsilon > 0$ small enough. We deduce then as in Claim 1 of this theorem the existence of a strict $W^{2,1}$-upper solution $\beta_1 \geq 0$ which we can choose small enough so that $\alpha_2 \not\leq \beta_1$.

Conclusion – The proof follows now from Theorem III-2.11. □

Remark Such a result can be obtained using variational arguments as in Example IV-1.1.

Chapter IX

Problems with Singular Forces

1 A Periodic Problem

In this chapter, we are mainly interested in problems which are singular as u goes to zero and, as we are dealing with scalar equations, this assumption implies solutions are one-signed, say positive. To begin with we shall consider the periodic problem

$$\begin{aligned} u'' + g(u)u' + f(t,u) &= h(t), \\ u(a) = u(b), \ \ u'(a) &= u'(b). \end{aligned} \tag{1.1}$$

The first result we examine deals with a nonlinearity $f(t,u)$ which represents an attractive force. Here a singularity such as $f(t,u) = \frac{1}{u}$ can take place and it helps in finding lower solutions. However such a singularity is not necessary, we only need the nonlinearity to be large enough near the origin (see condition (b) in the following theorem).

Theorem 1.1. *Let $g \in C(\mathbb{R}^+)$, $h \in L^1(a,b)$ and let $f : [a,b] \times \mathbb{R}_0^+ \to \mathbb{R}$ satisfy Carathéodory conditions and*

(a) for any $0 < r < s$, there exists a function $k \in L^1(a,b)$ such that for a.e. $t \in [a,b]$ and all $u \in [r,s]$, we have

$$|f(t,u)| \leq k(t).$$

Assume moreover that

(b) for some $\alpha > 0$ and a.e. $t \in [a,b]$, we have

$$f(t,\alpha) - h(t) \geq 0;$$

(c) there exist $R > \alpha$ and $f_0 \in L^1(a,b)$ such that for a.e. $t \in [a,b]$ and all $u \geq R$,

$$f(t,u) \leq f_0(t)$$

and

$$\int_a^b f_0(t)\,dt \le \int_a^b h(t)\,dt.$$

Then the problem (1.1) *has at least one positive solution.*

Proof. Condition (b) implies $\alpha(t) = \alpha$ is a lower solution of (1.1).

To construct an upper solution, we introduce the function

$$\phi(t) = f_0(t) - h(t) - \tfrac{1}{b-a}\int_a^b (f_0(t) - h(t))\,dt \in L^1(a,b)$$

and prove that for all $c \in \mathbb{R}$, the problem

$$\begin{gathered} u''(t) + g(|u(t)+c|)u'(t) + \phi(t) = 0, \\ u(a) = u(b),\ u'(a) = u'(b), \end{gathered} \tag{1.2}$$

has a solution $\beta_0 \in W^{2,1}(a,b)$ such that $\bar\beta_0 := \frac{1}{b-a}\int_a^b \beta_0(t)\,dt = 0$. To this aim consider the homotopy

$$\begin{gathered} u''(t) + \lambda[g(|u(t)+c|)u'(t) + \phi(t)] = 0, \\ u(a) = u(b),\ u'(a) = u'(b). \end{gathered} \tag{1.3}$$

Solutions of this problem with mean value zero are fixed points of the operator

$$T_\lambda : \tilde{C}^1([a,b]) \to \tilde{C}^1([a,b]), u \mapsto T_\lambda u,$$

where $\tilde{C}^1([a,b]) = \{u \in C^1([a,b]) \mid \int_a^b u(s)\,ds = 0\}$,

$$T_\lambda u = \lambda \int_a^b G(t,s)[g(|u(s)+c|)u'(s) + \phi(s)]\,ds$$

and $G(t,s)$ is the corresponding Green function.

Let us prove that the fixed points of T_λ are *a priori* bounded in $C([a,b])$ and then, arguing as in Proposition I-4.3, in $C^1([a,b])$. Multiply (1.3) by u and integrate, we obtain by Proposition A-4.1

$$\|u'\|_{L^2}^2 = \lambda \int_a^b \phi(t)u(t)\,dt \le \|\phi\|_{L^1}\|u\|_\infty \le \sqrt{\tfrac{b-a}{12}}\|\phi\|_{L^1}\|u'\|_{L^2}.$$

This inequality implies that $\|u'\|_{L^2} \le \sqrt{\frac{b-a}{12}}\|\phi\|_{L^1}$ and $\|u\|_\infty \le \frac{b-a}{12}\|\phi\|_{L^1}$.

Hence, by the invariance of the degree along the homotopy, the problem (1.2) has a solution β_0 such that $\bar\beta_0 = 0$ and $\|\beta_0\|_\infty \le \frac{b-a}{12}\|\phi\|_{L^1}$.

Now it is easy to observe that if we choose $c = \frac{b-a}{12}\|\phi\|_{L^1} + R$ and β_0 the corresponding solution of (1.2),

$$\beta(t) = \beta_0(t) + c \geq R$$

is the desired upper solution. We conclude then by application of Theorem I-6.9. □

Examples Consider the problem

$$\begin{aligned} u'' + g(u)u' + \tfrac{k}{\sqrt{u}} &= h, \\ u(a) = u(b),\ u'(a) &= u'(b), \end{aligned}$$

where $g : \mathbb{R}^+ \to \mathbb{R}$ is a positive continuous function, k and h are positive constants. This problem has a positive solution and it can be proved from Theorem 1.1.

Let us change now the restoring force into a repulsive one and study the problem

$$\begin{aligned} u'' + g(u)u' - \tfrac{k}{\sqrt{u}} &= -h, \\ u(a) = u(b),\ u'(a) &= u'(b). \end{aligned} \tag{1.4}$$

The above theorem does not apply though there exists a positive solution $u = (k/h)^2$. From a phase plane analysis, it is clear that this feature is due to the fact that this solution is not a saddle point so that we cannot use the method of lower and upper solutions as in the above theorem.

As we already pointed out, Theorem 1.1 does not really use the fact that $f(t, u)$ is singular for $u = 0$. This observation is clear from the further modification of the example

$$\begin{aligned} u'' + g(u)u' + k\tfrac{1}{1+u^2} &= h, \\ u(a) = u(b),\ u'(a) &= u'(b). \end{aligned}$$

If $k > h$, this problem has a positive solution and it can be found from the above theorem.

Exercise 1.1. Consider the problem

$$\begin{aligned} u'' + f(u, u') &= s + \tilde{p}(t), \\ u(a) = u(b),\ u'(a) &= u'(b). \end{aligned}$$

Assume $f :]0, +\infty[\times \mathbb{R} \to \mathbb{R}$ is a continuous function, Lipschitz in the second variable and such that

$$\lim_{u \to 0^+} f(u, 0) = +\infty.$$

Assume furthermore that there exist an $r > 0$, $R \geq r$ and a continuous function $g : \mathbb{R} \to \mathbb{R}$ such that for any $u \in [r, R]$ and $v \in \mathbb{R}$,

$$f(u, v) \leq g(v).$$

At last let $\tilde{p} \in L^2(a, b)$ be bounded from above and satisfy

$$\int_a^b \tilde{p}(s)\, ds = 0 \quad \text{and} \quad \|\tilde{p}\|_{L^2} \leq \frac{6\sqrt{5}(R - r)}{(b - a)^{3/2}}.$$

Prove then that there exists a solution for any s large enough.

The next result parallels Theorem 1.1 but for a repulsive force. A model example is

$$\begin{array}{c} u'' - \frac{1}{u^3} + \gamma(t)u = h(t), \\ u(a) = u(b), \ u'(a) = u'(b). \end{array} \tag{1.5}$$

Theorem 1.2. *Let $g \in C(\mathbb{R}^+)$, $h \in L^1(a, b)$ and assume $f : [a, b] \times \mathbb{R}_0^+ \to \mathbb{R}$ satisfies Carathéodory conditions together with*
(a) for any $0 < r < s$, there exists a function $k \in L^1(a, b)$ such that for a.e. $t \in [a, b]$ and all $u \in [r, s]$, we have

$$|f(t, u)| \leq k(t).$$

Assume moreover that
(b) for some $\beta > 0$ and a.e. $t \in [a, b]$, we have

$$f(t, \beta) - h(t) \leq 0;$$

(c) (strong force) *there exist $\rho \in]0, \beta[$, $\ell \in L^1(a, b)$ and $\hat{f} \in C(]0, \rho])$ with*

$$\int_0^\rho \hat{f}^-(u)\, du = +\infty \quad \textit{and} \quad \hat{f}^-(u) = \max\{-\hat{f}(u), 0\},$$

such that for all $u \in]0, \rho]$ and a.e. $t \in [a, b]$

$$f(t, u) \leq \hat{f}(u) \quad \textit{and} \quad f(t, u) \leq \ell(t);$$

(d) there exist $R > \beta$ and $f_0 \in L^1(a, b)$ such that, for a.e. $t \in [a, b]$ and all $u \geq R$,

$$f(t, u) \geq f_0(t)$$

and

$$\int_a^b f_0(t)\, dt \geq \int_a^b h(t)\, dt;$$

(e) there exists a function $\gamma \in L^1(a,b)$ *such that* $\gamma(t) \le (\frac{\pi}{b-a})^2$ *a.e. on* $[a,b]$ *with strict inequality on a subset of positive measure and*

$$\limsup_{u\to+\infty} \frac{f(t,u)}{u} \le \gamma(t),$$

uniformly in $t \in [a,b]$.

Then the problem (1.1) *has at least one positive solution.*

Remarks As will be clear from the proof, assumptions (b) and (d) provide upper and lower solutions.

The strong force condition (c) gives a lower *a priori* bound on solutions. It is important to notice that here we do not impose that $f(t,u)$ goes to $-\infty$ as u goes to zero, but that the bound $\hat{f}$ has a negative part whose primitive is unbounded as u goes to zero. Notice that some control of the singularity such as the strong force assumption (c) is necessary. This fact is clear from [199] where it is proved that the problem

$$\begin{gathered} u'' - \tfrac{1}{\sqrt{u}} = h(t), \\ u(a) = u(b),\ u'(a) = u'(b), \end{gathered}$$

has no solution for some negative $h \in \mathcal{C}([a,b])$.

At last, assumption (e) forces the nonlinearity to be, for large values of u, "under" the asymptote of the first nontrivial Fučík curve for the periodic problem, which is also the first eigenvalue of the Dirichlet problem

$$u'' + \lambda u = 0, \quad u(a) = 0,\ u(b) = 0.$$

Such a condition is somewhat natural if we realize that large solutions of the periodic problem (1.5) look like solutions of a Dirichlet problem. Also, such a condition cannot be avoided. This fact is clear from [42] where it is proved that for some $h \in \mathcal{C}([0,2\pi])$ the problem

$$\begin{gathered} u'' - \tfrac{1}{u^3} + \tfrac{1}{4}u = h(t), \\ u(0) = u(2\pi),\ u'(0) = u'(2\pi), \end{gathered}$$

has no solution.

Example 1.1. It is easy to see from Theorem 1.2 that the model example (1.5) has a solution if $\gamma \in L^1(a,b)$ is such that

$$0 \lneqq \gamma(t) \lneqq (\tfrac{\pi}{b-a})^2,$$

and $h \in L^1(a,b)$ is lower bounded.

Proof of Theorem 1.2 : The modified problem – Let $\delta \in]0, \rho]$, define the truncated function

$$\begin{aligned} f_\delta(t,u) &= f(t,u), \text{ if } u \geq \delta, \\ &= f(t,\delta), \text{ if } u < \delta, \end{aligned}$$

and consider the modified problem

$$\begin{aligned} u'' + g(|u|)u' + f_\delta(t,u) = h(t), \\ u(a) = u(b), \; u'(a) = u'(b). \end{aligned} \tag{1.6}$$

Claim 1 – there exists a solution u of (1.6) in

$$\mathcal{S} = \{u \in C([a,b]) \mid \exists t_1, t_2 \in [a,b], \; u(t_1) \geq \beta(t_1), \; u(t_2) \leq \alpha(t_2)\}.$$

We first deduce from (b) that the constant function $\beta(t) = \beta \geq \delta$ is an upper solution for (1.6).

Next, we obtain a lower solution from condition (d) and the argument used in the proof of Theorem 1.1 to define the upper solution.

Notice at last that for some $k \in L^1(a,b)$, all $u \in \mathbb{R}$ and a.e. $t \in [a,b]$, we have

$$f_\delta(t,u) \geq \min\{\min_{\delta \leq u \leq R} f(t,u), f_0(t)\} \geq k(t).$$

We deduce now the claim from Theorem III-3.2.

Claim 2 – There exists an $L^+ > 0$ so that any solution $u \in \mathcal{S}$ of (1.6) with $0 < \delta \leq \rho$ satisfies $\max_t u(t) \leq L^+$. Let $u \in \mathcal{S}$ be a solution of (1.6) such that $\max_t u(t) > \bar{R} = \max\{R, \|\alpha\|_\infty\}$. From the definition of $\mathcal{S}$, there exists $t_2 \in [a,b]$ so that $u(t_2) \leq \alpha(t_2) \leq \bar{R}$. Hence, extending u by periodicity if necessary, we can find $a' < b'$ so that $b' - a' < b - a$, $u(a') = u(b') = \bar{R}$, $u(t) > \bar{R}$ on $]a', b'[$ and $\max_{a' \leq t \leq b'} u(t) = \max_{a \leq t \leq b} u(t)$.

Let now $\varepsilon > 0$ be given by Proposition A-4.6 and choose $M > \bar{R}$ such that for all $u \geq M$ and a.e. $t \in [a,b]$

$$\frac{f_\delta(t,u)}{u} = \frac{f(t,u)}{u} \leq \gamma(t) + \frac{\varepsilon}{2}.$$

Define then

$$\begin{aligned} p(t,u) &= \frac{f(t,u)}{u}, \quad \text{for } u \geq M, \\ &= \frac{f(t,M)}{M}, \quad \text{for } u < M, \end{aligned}$$

and

$$q_\delta(t,u) = f_\delta(t,u) - p(t,u)u.$$

These functions are such that for some $k \in L^1(a,b)$

$$p(t,u) \leq \gamma(t) + \frac{\varepsilon}{2},$$

$$q_\delta(t,u) \leq \max\{\ell(t), \max_{\rho \leq u \leq M} |f(t,u)|\} + |f(t,M)| \leq k(t).$$

The function $v = u - \bar{R}$ is nonnegative on $[a', b']$ and solves the problem

$$v'' + g(v + \bar{R})v' + p(t, u(t))(v + \bar{R}) + q_\delta(t, u(t)) = h(t),$$
$$v(a') = 0, \; v(b') = 0.$$

Hence, we compute

$$\begin{aligned}\int_{a'}^{b'} v'^2(t)\,dt - \int_{a'}^{b'} (\gamma(t) + \tfrac{\varepsilon}{2})(v(t) + \bar{R})v(t)\,dt \\ \leq -\int_{a'}^{b'} v(t)(v''(t) + g(v(t) + \bar{R})v'(t) + p(t, u(t))(v(t) + \bar{R}))\,dt \\ = \int_{a'}^{b'} (q_\delta(t, u(t)) - h(t))v(t)\,dt\end{aligned}$$

and using Proposition A-4.6 we obtain

$$\begin{aligned}\tfrac{\varepsilon}{2}\|v\|^2_{H^1(a',b')} &\leq \varepsilon\|v\|^2_{H^1(a',b')} - \tfrac{\varepsilon}{2}\|v\|^2_{L^2(a',b')} \\ &\leq \int_{a'}^{b'} (v'^2(t) - (\gamma(t) + \tfrac{\varepsilon}{2})v^2(t))\,dt \\ &\leq \int_{a'}^{b'} (q_\delta(t, u(t)) - h(t) + (\gamma(t) + \tfrac{\varepsilon}{2})\bar{R})v(t)\,dt, \\ &\leq \int_{a'}^{b'} (k(t) - h(t) + (\gamma(t) + \tfrac{\varepsilon}{2})\bar{R})v(t)\,dt,\end{aligned}$$

i.e.

$$\|v\|^2_{H^1(a',b')} \leq K\|v\|_\infty,$$

for some $K > 0$ independent of δ. It follows that for all $t \in [a', b']$

$$v(t) = \int_{a'}^{t} v'(s)\,ds \leq K^{1/2}\|v\|_\infty^{1/2}(b - a)^{1/2}$$

so that

$$\|v\|_\infty \leq K(b - a)$$

and

$$\max_t u(t) \leq R + K(b - a) =: L^+.$$

Claim 3 – There exists an L^- *so that any solution* $u \in \mathcal{S}$ *of* (1.6) *with* $0 < \delta \leq \rho$ *satisfies* $\min_t u(t) \geq L^-$. Let L^+ be given from Claim 2. We deduce from (a) and (c) that there exists $k \in L^1(a, b)$ so that for all $u \in]0, L^+[$ and a.e. $t \in [a, b]$

$$f(t, u) \leq k(t).$$

Using Claim 2, it follows that we can write for a.e. $t \in [a', b']$

$$f_\delta(t, u(t)) \leq k(t)$$

if $u \in \mathcal{S}$ is a solution of (1.6) with $0 < \delta \leq \rho$. Next, integrating (1.6) or multiplying this equation by u and integrating, we obtain

$$\int_a^b (f_\delta(t, u(t)) - h(t))\, dt = 0$$

and

$$\int_a^b u'^2(t)\, dt = \int_a^b (f_\delta(t, u(t)) - h(t))u(t)\, dt.$$

It follows then that

$$\begin{aligned} \|u'\|_{L^2}^2 &= \int_a^b (f_\delta(t, u(t)) - h(t))(\|u\|_\infty + u(t))\, dt \\ &\leq 2\int_a^b |k(t) - h(t)|\, \|u\|_\infty\, dt \leq 2(\|h\|_{L^1} + \|k\|_{L^1})\|u\|_\infty, \end{aligned}$$

i.e.

$$\|u'\|_{L^2} \leq (2(\|h\|_{L^1} + \|k\|_{L^1}))^{1/2}\|u\|_\infty^{1/2}.$$

Define now $t_1 \in [a, b]$ to be such that $u(t_1) \geq \beta(t_1) \geq -\|\beta\|_\infty$. Extending u by periodicity, we can write for $t \in [t_1, t_1 + b - a]$

$$u(t) = u(t_1) + \int_{t_1}^t u'(s)\, ds \geq -\|\beta\|_\infty - (2(b-a)(\|h\|_{L^1} + \|k\|_{L^1}))^{1/2}\|u\|_\infty^{1/2}.$$

It is now easy to obtain L^- so that $u(t) \geq L^-$ on $[a, b]$.

Claim 4 – There exists an $N > 0$ so that any solution $u \in \mathcal{S}$ of (1.6) *with $0 < \delta \leq \rho$ satisfies $\|u'\|_\infty \leq N$.* As in Claim 3, we can find $k \in L^1(a, b)$ so that for all solution $u \in \mathcal{S}$ of (1.6) with $0 < \delta \leq \rho$, and a.e. $t \in [a, b]$,

$$f_\delta(t, u(t)) \leq k(t).$$

Let then $\bar{t} \in [a, b]$ be such that $u'(\bar{t}) = 0$. Hence, for $t \in [\bar{t}, \bar{t} + b - a]$ we can write

$$\begin{aligned} u'(t) &= \int_{\bar{t}}^t (h(s) - f_\delta(s, u(s)) - g(|u(s)|)u'(s))\, ds \\ &\geq -(\|h\|_{L^1} + \|k\|_{L^1} + 2 \max_{L^- \leq u \leq L^+} |G(u)|), \end{aligned}$$

where $G(u) = \int_0^u g(|s|)\, ds$. Similarly, we have for $t \in [\bar{t} - b + a, \bar{t}]$

$$u'(t) \leq \|h\|_{L^1} + \|k\|_{L^1} + 2 \max_{L^- \leq u \leq L^+} |G(u)| =: N.$$

Claim 5 – There exists a $\xi \in]0, \rho]$ *so that any solution* $u \in \mathcal{S}$ *of* (1.6) *with* $0 < \delta \leq \rho$, *is such that* $u(t) \geq \xi$ *on* $[a, b]$. Define $\xi > 0$ to be so that

$$\int_\xi^\rho \hat{f}^-(u)\, du > (\|h\|_{L^1} + \|\ell\|_{L^1})N + (\tfrac{1}{2} + (b-a) \max_{L^- \leq u \leq L^+} |g(|u|)|)N^2.$$

Let $u \in \mathcal{S}$ be a solution of (1.6). Define the set $A = \{t \in [a, b] \mid u'(t) \geq 0\}$ and suppose by contradiction that there exist t_1, $t_2 \in A$ so that $u(t_1) = \xi$, $u(t_2) = \rho$ and $u(t) \leq \rho$ on $[t_1, t_2]$. Multiplying (1.6) by u' and integrating on $B = [t_1, t_2] \cap A$, we get

$$\frac{u'^2(t_2)}{2} - \frac{u'^2(t_1)}{2} + \int_B g(|u(s)|)u'^2(s)\, ds + \int_B f_\delta(s, u(s))u'(s)\, ds = \int_B h(s)u'(s)\, ds.$$

From the *a priori* bounds on $\|u\|_\infty$ and $\|u'\|_\infty$, we obtain then

$$\int_B g(|u(s)|)u'^2(s)\, ds \leq (b-a)N^2 \max_{L^- \leq u \leq L^+} |g(|u|)|$$

and it follows that

$$-\int_B f_\delta(s, u(s))u'(s)\, ds \leq N\|h\|_{L^1} + \tfrac{N^2}{2} + (b-a)N^2 \max_{L^- \leq u \leq L^+} |g(|u|)|.$$

On the other hand, let $B^- = \{t \in B \mid \hat{f}(u(t)) \leq 0\}$ and $B^+ = B \setminus B^-$. We compute then

$$\int_{B^+} f_\delta(s, u(s))u'(s)\, ds \leq N\|\ell\|_{L^1}$$

and

$$-\int_{B^-} f_\delta(s, u(s))u'(s)\, ds \geq -\int_{B^-} \hat{f}(u(s))u'(s)\, ds = \int_B \hat{f}^-(u(s))u'(s)\, ds \geq \int_{t_1}^{t_2} \hat{f}^-(u(s))u'(s)\, ds = \int_\xi^\rho \hat{f}^-(u)\, du.$$

We deduce then the contradiction

$$\int_\xi^\rho \hat{f}^-(u)\, du \leq N\|\ell\|_{L^1} + N\|h\|_{L^1} + \tfrac{N^2}{2} + (b-a)N^2 \max_{L^- \leq u \leq L^+} |g(|u|)|.$$

Conclusion – From Claim 1, we know that problem (1.6) with $\delta = \xi$ has a solution. According to Claim 5 this solution solves (1.1). □

The next result deals with

$$\begin{aligned} u'' + f(t,u) &= h(t), \\ u(a) = u(b), \ u'(a) &= u'(b). \end{aligned} \tag{1.7}$$

Here the nonlinearity allows resonance with the first eigenvalue of the Dirichlet problem as u goes to infinity. The non-resonance condition (c) implies a lower bound on solutions so that the strong force condition is no more necessary.

Theorem 1.3. *Let $h \in L^1(a,b)$ and assume $f : [a,b] \times \mathbb{R}_0^+ \to \mathbb{R}$ satisfies Carathéodory conditions together with*

(a) for any $0 < r < s$, there exists a function $k \in L^1(a,b)$ such that for a.e. $t \in [a,b]$ and all $u \in [r,s]$, we have

$$|f(t,u)| \leq k(t).$$

Assume moreover that

(b) there exist an $R > 0$ and $f_0 \in L^1(a,b)$ such that for a.e. $t \in [a,b]$ and all $u \geq R$,

$$f(t,u) \geq f_0(t)$$

and

$$\int_a^b f_0(t)\,dt \geq \int_a^b h(t)\,dt;$$

(c) there exists $\delta > 0$ such that for all $u > 0$ and a.e. $t \in [a,b]$

$$h(t) - f(t,u) + (\tfrac{\pi}{b-a})^2 u \geq \delta.$$

Then the problem (1.7) has at least one positive solution.

Proof. The modified problem – Let $\beta_0 \in]0, (\frac{b-a}{\pi})^2\delta[$ be fixed. For $r > \beta_0$, we define the truncated function

$$\begin{aligned} f_r(t,u) &= f(t,\beta_0) + (\tfrac{\pi}{b-a})^2(u-\beta_0), && \text{if } u \leq \beta_0, \\ &= f(t,u), && \text{if } \beta_0 < u \leq r, \\ &= f(t,r), && \text{if } r < u, \end{aligned}$$

and consider the modified problem

$$\begin{aligned} u'' + f_r(t,u) &= h(t), \\ u(a) = u(b), \ u'(a) &= u'(b). \end{aligned} \tag{1.8}$$

Notice that $\beta(t) = \beta_0$ is an upper solution of (1.8). Next, we deduce a lower solution as in the proof of Theorem 1.2. At last, we notice that for some $k \in L^1(a,b)$, all $u \in]0,+\infty[$ and a.e. $t \in [a,b]$,

$$f_r(t,u) \leq \sup_{\beta_0 \leq u \leq r} f(t,u) \leq k(t).$$

We deduce now from Theorem III-3.1 the existence of a solution $u \in \mathcal{S}$ of (1.8).

Claim – There exists an $M > 0$ so that for any r large enough, solutions $u \in \mathcal{S}$ of (1.8) *are such that $\beta_0 \leq u \leq M$.* Notice first that if u is a solution of (1.8)

$$\begin{gathered}(u-\beta_0)'' + (\tfrac{\pi}{b-a})^2(u-\beta_0) > 0,\\ u(a) = u(b),\ u'(a) = u'(b).\end{gathered}$$

It follows now from the anti-maximum principle (Corollary A-6.3) that $u \geq \beta_0$.

To obtain an upper bound on the solutions, let $k \in L^1(a,b)$ be so that for all $r > R$, $u \geq \beta_0$ and a.e. $t \in [a,b]$

$$h(t) - f_r(t,u) \leq h(t) - \min\{\min_{\beta_0 \leq u \leq R} f(t,u), f_0(t)\} \leq k(t).$$

It follows as in Theorem 1.2 that

$$\|u'\|_\infty \leq \|k\|_{L^1}.$$

Next, as $u \in \mathcal{S}$, there exists some $\bar{t} \in [a,b]$ so that $u(\bar{t}) \leq \|\alpha\|_\infty$. Hence, we can write

$$u(t) = u(\bar{t}) + \int_{\bar{t}}^t u'(s)\,ds \leq \|\alpha\|_\infty + \|k\|_{L^1}(b-a) =: M.$$

Conclusion – It follows now from the previous steps that problem (1.8), with $r \geq M$ large enough, has a solution u with $\beta_0 \leq u \leq M$. This solution solves (1.7). □

Example 1.2. Let $h \in L^1(a,b)$ be bounded below by a positive constant and $\nu > 0$. It is then easy to see from Theorem 1.3 that the problem

$$\begin{gathered}u'' - \tfrac{1}{u^\nu} + (\tfrac{\pi}{b-a})^2 u = h(t),\\ u(a) = u(b),\ u'(a) = u'(b),\end{gathered}$$

has at least one positive solution.

Notice also that for $k > 0$ and $h(t) > 0$ small enough, the existence of the solution of

$$\begin{gathered} u'' - \tfrac{k}{\sqrt{u}} = -h(t), \\ u(a) = u(b), \ u'(a) = u'(b), \end{gathered}$$

can be deduced from Theorem 1.3.

The following result is an alternative to Theorem 1.3 which present an explicit non-resonance condition on h.

Theorem 1.4. *Let $h \in L^1(a,b)$ and assume $f : [a,b] \times \mathbb{R}_0^+ \to \mathbb{R}$ satisfies Carathéodory conditions together with*

(a) for any $0 < r < s$, there exists a function $k \in L^1(a,b)$ such that for a.e. $t \in [a,b]$ and all $u \in [r,s]$, we have

$$|f(t,u)| \leq k(t).$$

Assume moreover that

(b) for all $u > 0$ and a.e. $t \in [a,b]$,

$$f(t,u) \leq (\tfrac{\pi}{b-a})^2 u;$$

(c) there exist $R > 0$ and $f_0 \in L^1(a,b)$ such that for a.e. $t \in [a,b]$ and all $u \geq R$,

$$f(t,u) \geq f_0(t)$$

and

$$\int_a^b f_0(t)\,dt \geq \int_a^b h(t)\,dt;$$

(d) there exists $\delta > 0$ so that for any $t \in [a,b]$

$$\int_t^{t+b-a} h(s) \sin \pi \tfrac{s-t}{b-a}\,ds \geq \delta.$$

Then the problem (1.7) has at least one positive solution.

Proof. The modified problem – For $r > \beta_0 = \frac{b-a}{2\pi}\delta$, we consider the modified problem (1.8). The function $\beta(t) = w(t) - B$ is an upper solution of (1.8) if $B > 0$ is large enough and w solves the problem

$$\begin{gathered} w'' = h(t) - f(t,\beta_0) - \tfrac{1}{b-a}\int_a^b (h(s) - f(s,\beta_0))\,ds, \\ w(a) = w(b), \ w'(a) = w'(b). \end{gathered}$$

As in Theorem 1.2, we deduce from (c) the existence of a lower solution of the form $\alpha(t) = v(t) + A$. At last, we see as in Theorem 1.3 that $f_r(t,u)$ is upper bounded by a L^1-function. Existence of a solution in $\mathcal{S}$ of the modified problem (1.8) follows then from Theorem III-3.1.

Claim – Solutions u of the modified problem (1.8) *are such that* $u \geq \frac{b-a}{2\pi}\delta$.
Let t_0 be such that $u(t_0) = \min_{t\in[a,b]} u(t)$. Multiplying (1.8) by $\sin \pi \frac{t-t_0}{b-a}$ and integrating on $[t_0, t_0 + b - a]$, we obtain

$$\begin{aligned}\delta \leq \int_{t_0}^{t_0+b-a} h(t) \sin \pi \tfrac{t-t_0}{b-a}\, dt &= \int_{t_0}^{t_0+b-a} [u''(t) + f_r(t, u(t))] \sin \pi \tfrac{t-t_0}{b-a}\, dt \\ &\leq \int_{t_0}^{t_0+b-a} [u''(t) + (\tfrac{\pi}{b-a})^2 u(t)] \sin \pi \tfrac{t-t_0}{b-a}\, dt = \tfrac{2\pi}{b-a} u(t_0).\end{aligned}$$

Conclusion – An upper bound on the solutions of (1.8) in $\mathcal{S}$ is obtained as in the proof of Theorem 1.3. Hence, for $r > 0$ large enough, the solutions of the modified problem solve (1.7). □

Example 1.3. Consider once again Example 1.2, with

$$h(t) = -1, \text{ if } t \in [a, a+\varepsilon], \quad h(t) = 1, \text{ if } t \in [a+\varepsilon, b].$$

We cannot use Theorem 1.3 to prove existence of positive solutions but Theorem 1.4 applies if $\varepsilon > 0$ is small enough.

We can also replace the assumption (e) in Theorem 1.2 by a condition on the friction term $g(u)$. Consider the problem

$$\begin{aligned} &u'' + g(u)u' + f(u) = h(t), \\ &u(a) = u(b), \; u'(a) = u'(b). \end{aligned} \tag{1.9}$$

Theorem 1.5. *Let* $g \in C(\mathbb{R}^+)$, $h \in L^2(a, b)$ *and* $f \in C(\mathbb{R}_0^+)$. *Assume moreover that*
(a) for some $\beta > 0$ *and a.e.* $t \in [a, b]$, *we have*

$$f(\beta) - h(t) \leq 0;$$

(b) (strong force) *there exist a* $\rho \in]0, \beta[$ *and* $\ell \in \mathbb{R}$ *such that*

$$\int_0^\rho f^-(u)\, du = +\infty,$$

where $f^-(u) = \max\{-f(u), 0\}$ *and for all* $u \in]0, \rho]$

$$f(u) \leq \ell;$$

(c) there exists an $R > \beta$ *such that if* $u \geq R$,

$$f(u) \geq \frac{1}{b-a} \int_a^b h(t)\, dt;$$

(d) there exists an $A > 0$ *such that* $|g(u)| \geq A$ *for every* $u > 0$.

Then the problem (1.9) *has at least one positive solution.*

Proof. For $\delta \in]0, \rho]$, we define

$$\begin{aligned} f_\delta(u) &= f(u), \ \text{if } u \geq \delta, \\ &= f(\delta), \ \text{if } u < \delta, \end{aligned}$$

and consider the modified problem

$$\begin{aligned} u'' + g(|u|)u' + f_\delta(u) = h(t), \\ u(a) = u(b), \ u'(a) = u'(b). \end{aligned} \tag{1.10}$$

As in the proof of Theorem 1.2, we have the existence of a lower solution α and an upper solution $\beta \leq \alpha$, and we deduce from Theorem III-3.2 the existence of a solution of (1.10) in $\mathcal{S}$.

Claim – There exists an $L > 0$ *so that any solution* $u \in \mathcal{S}$ *of* (1.10) *with* $0 < \delta \leq \rho$ *is such that* $\max_t u(t) \leq L^+$. Assume $g(u) \geq A$. The proof is similar if $g(u) \leq -A$. Multiplying (1.10) by u' and integrating, we obtain

$$A\|u'\|_{L^2}^2 \leq \int_a^b g(|u|)u'^2 \, dt = \int_a^b hu' \, dt \leq \|h\|_{L^2}\|u'\|_{L^2},$$

i.e.

$$\|u'\|_{L^2} \leq \frac{1}{A}\|h\|_{L^2}.$$

As $u \in \mathcal{S}$, there exists a t_2 so that $u(t_2) \leq \alpha(t_2)$. This fact implies that for any $t \in [a, b]$

$$u(t) = u(t_2) + \int_{t_2}^t u'(s) \, ds \leq \|\alpha\|_\infty + \tfrac{\sqrt{b-a}}{A}\|h\|_{L^2} =: L^+.$$

Conclusion – We conclude the proof as in Theorem 1.2. □

2 A Sublinear Dirichlet Problem

In Section II-4, we noticed that a natural frame to study the Dirichlet problem

$$\begin{aligned} u'' + f(t, u) = 0, \\ u(a) = 0, \ u(b) = 0, \end{aligned} \tag{2.1}$$

considers nonlinearities which are $\mathcal{A}$-Carathéodory and defines solutions to be in $W^{2,\mathcal{A}}(a, b)$. Here, we are interested in problem (2.1) with a singularity at $u = 0$. In this case, f cannot satisfy $\mathcal{A}$-Carathéodory conditions. However, solutions can still be in $W^{2,\mathcal{A}}(0, \pi)$ as is clear from the example

$$\begin{aligned} u'' + \tfrac{1}{(\pi - t)t} \tfrac{1}{u} = 0, \\ u(0) = 0, \ u(\pi) = 0. \end{aligned} \tag{2.2}$$

Its solution reads

$$u(t) = \tfrac{2}{\pi}\sqrt{t(\pi - t)}$$

and is in $W^{2,\mathcal{A}}(0,\pi)$. Notice that $f(t,u) = \frac{1}{(\pi-t)t}\frac{1}{u}$ is only $\mathcal{A}$-Carathéodory on sets of the form $[0,\pi]\times[\epsilon,+\infty[$, with $\epsilon > 0$ and not on $[0,\pi]\times\,]0,+\infty[$. As the boundary conditions impose the solution to go into the singularity $u = 0$, there is no hope to obtain a positive solution from a simple application of the theory in Section II-4.

The next theorem considers a sublinear problem where the nonlinearity is over the first eigenvalue near the origin (see (a)) and under it near infinity (see (b)). Our first result is written in the continuous case so that solutions are C^2.

Theorem 2.1. *Let $f :]a,b[\times\mathbb{R}_0^+ \to \mathbb{R}$ be a continuous function and assume that for every compact set $L \subset \mathbb{R}_0^+$, there is an $h_L \in \mathcal{A}\cap C(]a,b[)$ such that for all $t \in]a,b[$ and $u \in L$,*

$$|f(t,u)| \le h_L(t).$$

Assume further
(a) there exists a $k > \pi/(b-a)$ and for any compact set $K \subset]a,b[$, there is $\epsilon > 0$ such that for all $t \in K$, $u \in]0,\epsilon]$,

$$f(t,u) \ge k^2 u;$$

(b) for some $M > 0$ and $\gamma \in]0,\pi/(b-a)[$, there is $h \in \mathcal{A}\cap C(]a,b[)$ such that for all $t \in]a,b[$ and $u \in [M,\infty[$,

$$f(t,u) \le \gamma^2 u + h(t).$$

Then the problem (2.1) *has at least one solution*

$$u \in C([a,b],\mathbb{R}^+)\cap C^2(]a,b[,\mathbb{R}_0^+).$$

Remark Assumption (a) is equivalent to assume there exist $k > \frac{\pi}{b-a}$ and a function $a_1 \in C_0^2([a,b],\mathbb{R}^+)$ such that
(i) $t \in]a,b[$ implies $a_1(t) > 0$;
(ii) $f(t,u) \ge k^2 u$, for all $t \in]a,b[$ and $0 < u \le a_1(t)$;
(iii) $a_1''(t) > 0$, for all $t \in [a,(2a+b)/3]\cup[(a+2b)/3,b]$.

Proof. Step 1 – Construction of lower solutions. Consider k_2 such that $\frac{\pi}{b-a} < k_2 < \min\{k,\frac{3\pi}{b-a}\}$ and the function

$$\alpha_2(t) = A_2\cos k_2\left(t - \frac{a+b}{2}\right),$$

where A_2 is chosen small enough so that

$$f(t,u) \geq k^2 u, \text{ for all } t \in]\tfrac{a+b}{2} - \tfrac{\pi}{2k_2}, \tfrac{a+b}{2} + \tfrac{\pi}{2k_2}[\text{ and } u \in]0, \alpha_2(t)].$$

Next, we choose a_1 from the remark and let

$$\alpha_1(t) = A_1 a_1(t),$$

where $A_1 \in]0,1]$ is small enough so that for some points $t_1 \in]a, \frac{2a+b}{3}[$, $t_2 \in]\frac{a+2b}{3}, b[$, we have
(a-1) $\alpha_1(t) \geq \alpha_2(t)$, for all $t \in [a, t_1] \cup [t_2, b]$;
(b-1) $\alpha_2(t) \geq \alpha_1(t)$, for all $t \in [t_1, t_2]$.

Notice that for any $f_* :]a,b[\times\mathbb{R}_0^+ \to \mathbb{R}$ such that

$$f_*(t,u) \geq f(t,u), \text{ for all } (t,u) \in]a,b[\times\mathbb{R}_0^+,$$

we have
(a-2) $\alpha_1''(t) + f_*(t, \alpha_1(t)) \geq \alpha_1''(t) + k^2\alpha_1(t) > 0$, for all $t \in [a, t_1] \cup [t_2, b]$;
(b-2) $\alpha_2''(t) + f_*(t, \alpha_2(t)) \geq -k_2^2\alpha_2(t) + k^2\alpha_2(t) > 0$, for all $t \in [t_1, t_2]$.

Step 2 – Approximation problems. We define for each $n \in \mathbb{N}$, $n \geq 1$,

$$\eta_n(t) = \max\{a + \tfrac{1}{2^{n+1}}, \min\{t, b - \tfrac{1}{2^{n+1}}\}\}, \quad t \in]a,b[$$

and set

$$\tilde{f}_n(t,u) = \max\{f(\eta_n(t), u), f(t,u)\}.$$

We have that for each index n, $\tilde{f}_n :]a,b[\times\mathbb{R}_0^+ \to \mathbb{R}$ is continuous and

$$\tilde{f}_n(t,u) = f(t,u), \text{ for all } (t,u) \in K_n \times \mathbb{R}_0^+,$$

where

$$K_n = [a + \tfrac{1}{2^{n+1}}, b - \tfrac{1}{2^{n+1}}].$$

Hence, the sequence of functions $\{\tilde{f}_n\}$ converges to f uniformly on any set $K \times \mathbb{R}_0^+$, where K is an arbitrary compact subset of $]a,b[$.

Next we define

$$f_n(t,u) = \min\{\tilde{f}_1(t,u), \cdots, \tilde{f}_n(t,u)\}.$$

Each of the functions f_i is a continuous function defined on $]a,b[\times\mathbb{R}_0^+$; moreover

$$f_1(t,u) \geq f_2(t,u) \geq \cdots \geq f_n(t,u) \geq f_{n+1}(t,u) \geq \cdots \geq f(t,u)$$

and the sequence $(f_n)_n$ converges to f uniformly on the compact subsets of $]a,b[\times\mathbb{R}_0^+$ since

$$f_n(t,u) = f(t,u), \text{ for all } t \in K_n \text{ and } u \in \mathbb{R}_0^+.$$

Define now a decreasing sequence $(\epsilon_n) \subset \mathbb{R}_0^+$ such that

$$\lim_{n\to\infty} \epsilon_n = 0,$$
$$f(t,u) \geq k^2 u, \text{ for all } t \in K_n \text{ and } u \in]0,\epsilon_n],$$

and consider the sequence of approximation problems

$$\begin{gathered} u'' + f_n(t,u) = 0, \\ u(a) = \epsilon_n, \ u(b) = \epsilon_n. \end{gathered} \tag{2.3}$$

Step 3 – ϵ_n is a lower solution of (2.3). It is clear that for any $c \in]0,\epsilon_n]$

$$\tilde{f}_n(t,c) \geq f(\eta_n(t),c) \geq k^2 c > 0.$$

As the sequence (ϵ_n) is decreasing, we also have

$$f_n(t,\epsilon_n) = \min_{1\leq i\leq n} \tilde{f}_i(t,\epsilon_n) \geq k^2\epsilon_n > 0,$$

which proves the claim.

Step 4 – Existence of a solution u_1 of (2.3), *with* $n = 1$, *such that*

$$\max\{\alpha_1(t),\alpha_2(t),\epsilon_1\} \leq u_1(t).$$

Notice first that $\max\{\alpha_1(t),\alpha_2(t),\epsilon_1\}$ is a C^2-lower solution of (2.3) with $n = 1$. Next, we can find from assumption (b), $M \geq \max\{\alpha_1(t),\alpha_2(t),\epsilon_1\}$ and $h \in \mathcal{A} \cap C(]a,b[)$, $h \geq 0$, such that for all $t \in]a,b[$ and $u \in [M,\infty[$

$$f(t,u) \leq \gamma^2 u + h(t).$$

Hence, we can write

$$f_1(t,u) = \max\{f(\eta_1(t),u), f(t,u)\} \leq \gamma^2 u + h(t) + h(\eta_1(t)).$$

Choose $\beta \in C([a,b]) \cap C^2(]a,b[)$ such that

$$\begin{gathered} \beta'' + \gamma^2\beta + h(t) + h(\eta_1(t)) = 0, \\ \beta(a) = M, \ \beta(b) = M, \end{gathered}$$

i.e.

$$\beta(t) = M + \int_a^b G(t,s)[h(s) + h(\eta_1(s)) + \gamma^2 M]\, ds \geq M,$$

where

$$G(t,s) := \frac{\sin\gamma(s-a)\sin\gamma(b-t)}{\gamma\sin\gamma(b-a)}, \quad \text{if } s<t,$$
$$:= \frac{\sin\gamma(t-a)\sin\gamma(b-s)}{\gamma\sin\gamma(b-a)}, \quad \text{if } s\geq t.$$

Notice that β is well-defined and bounded since $\tilde{h} = h\circ\eta_1 + h \in \mathcal{A}$ and

$$\int_a^t \sin\gamma(s-a)\sin\gamma(b-t)\,\tilde{h}(s)\,ds$$
$$\leq \int_a^{(a+b)/2} \sin\gamma(s-a)\,\tilde{h}(s)\,ds + \int_{(a+b)/2}^b \sin\gamma(b-s)\,\tilde{h}(s)\,ds < \infty,$$
$$\int_t^b \sin\gamma(t-a)\sin\gamma(b-s)\,\tilde{h}(s)\,ds$$
$$\leq \int_a^{(a+b)/2} \sin\gamma(s-a)\,\tilde{h}(s)\,ds + \int_{(a+b)/2}^b \sin\gamma(b-s)\,\tilde{h}(s)ds < \infty.$$

It is easy to see now that

$$\beta'' + f_1(t,\beta) \leq \beta'' + \gamma^2\beta + h(t) + h(\eta_1(t)) = 0.$$

Consider now the change of variables $x = u - \epsilon_1$ and apply Theorem II-4.2 to obtain a solution u_1 of (2.3), with $n=1$, that verifies

$$\max\{\alpha_1(t), \alpha_2(t), \epsilon_1\} \leq u_1(t) \leq \beta(t).$$

Step 5 – The problem (2.3) *has at least one solution u_n such that*

$$\max(\alpha_1(t), \alpha_2(t), \epsilon_n) \leq u_n(t) \leq u_{n-1}(t).$$

Let us notice that u_{n-1} is an upper solution of (2.3) since

$$0 = u''_{n-1}(t) + f_{n-1}(t, u_{n-1}(t)) \geq u''_{n-1}(t) + f_n(t, u_{n-1}(t))$$

and

$$u_{n-1}(a) = u_{n-1}(b) = \epsilon_{n-1} \geq \epsilon_n.$$

The claim follows by Theorem II-4.2.

Step 6 – Existence of a solution of (2.1). Consider now the pointwise limit

$$\tilde{u}(t) = \lim_{n\to\infty} u_n(t).$$

It is clear that for any $n \geq 1$,

$$\max\{\alpha_1(t), \alpha_2(t)\} \leq \tilde{u}(t) \leq u_n(t), \quad \forall t \in]a,b[.$$

Let now $K \subset]a, b[$ be a compact interval. There is an index $n^* = n^*(K)$ such that $K \subset K_n$ for all $n \geq n^*$ and therefore for these $n \geq n^*$,

$$0 = u_n''(t) + f_n(t, u_n(t)) = u_n''(t) + f(t, u_n(t)), \quad \forall t \in K.$$

Hence the function u_n is a solution of the equation in (2.1) for all $t \in K$ and $n \geq n^*$. Moreover

$$\sup\{|f(t,u)| \mid t \in K, \max\{\alpha_1(t), \alpha_2(t)\} \leq u \leq u_{n^*}(t)\} < +\infty.$$

Then by Arzelá-Ascoli Theorem it is standard to conclude that $\tilde{u}$ is a solution of (2.1) on the interval K. Since K was arbitrary, we find that $\tilde{u} \in \mathcal{C}^2(]a, b[, \mathbb{R}_0^+)$ and for all $t \in]a, b[$,

$$\tilde{u}''(t) + f(t, \tilde{u}(t)) = 0.$$

Since

$$\tilde{u}(a) = \tilde{u}(b) = \lim_{n \to +\infty} \epsilon_n = 0,$$

it remains only to check the continuity of $\tilde{u}$ at $t = a$ and $t = b$.

Let $\epsilon > 0$ be given. Take n_ϵ such that $u_{n_\epsilon}(0) < \epsilon$. By the continuity of $u_{n_\epsilon}(t)$ at $t = a$, we can find a constant $\delta = \delta_\epsilon > 0$ such that for any $t \in]a, a + \delta[$

$$0 < u_{n_\epsilon}(t) < \epsilon.$$

Hence, we obtain for such t

$$0 \leq \tilde{u}(t) < \epsilon.$$

The same argument works in proving the continuity of $\tilde{u}(t)$ at $t = b$. □

Example 2.1. Consider the generalized Emden-Fowler equation with Dirichlet boundary conditions

$$\begin{aligned} u'' + f(t)\tfrac{1}{u^\sigma} &= h(t), \\ u(a) = 0, \ u(b) &= 0, \end{aligned}$$

where $\sigma > 0$ and $f, h \in \mathcal{C}(]a, b[)$. If we assume $f > 0$ on $]a, b[$ and

$$\int_a^b (t-a)(b-t)(f(t) + |h(t)|)\, dt < +\infty,$$

existence of a solution follows from Theorem 2.1. Notice that in this example h can change sign, f and h can be singular or zero at $t = a$ and $t = b$.

The next result improves condition (b) in Theorem 2.1. It replaces the eigenvalue condition to build the upper solution by condition (2.4) which is reminiscent of the variational approach and allows a time-mapping argument.

Theorem 2.2. *Let $f :]a,b[\times\mathbb{R}_0^+ \to \mathbb{R}$ be a continuous function. Assume that for every compact set $L \subset \mathbb{R}_0^+$, there is $h_L \in \mathcal{A} \cap C(]a,b[)$ such that for all $t \in]a,b[$ and $u \in L$,*

$$|f(t,u)| \le h_L(t).$$

and condition (a) of Theorem 2.1 is satisfied.

Assume moreover there exist $h \in \mathcal{A} \cap C(]a,b[)$ and a continuous, non-decreasing function $\varphi : [0,+\infty[\to \mathbb{R}_0^+$ such that

$$\liminf_{u\to+\infty} \frac{2\Phi(u)}{u^2} < \left(\tfrac{\pi}{b-a}\right)^2, \tag{2.4}$$

where $\Phi(u) = \int_0^u \varphi(s)\,ds$, and for some $M > 0$, all $t \in]a,b[$ and all $u \ge M$,

$$f(t,u) \le \varphi(u) + h(t). \tag{2.5}$$

Then the problem (2.1) has at least one solution

$$u \in C([a,b],\mathbb{R}^+) \cap C^2(]a,b[,\mathbb{R}_0^+).$$

Proof. The argument is the same as in Theorem 2.1 except for Step 4.

Step 4 – Existence of a solution u_1 of (2.3), with $n = 1$, such that

$$\max\{\alpha_1(t), \alpha_2(t), \epsilon_1\} \le u_1(t).$$

We already know that $\max\{\alpha_1(t), \alpha_2(t), \epsilon_1\}$ can be used as a lower solution. Hence, we only have to find an upper solution β, which we will write

$$\beta(t) = w(t) + \beta_0(t).$$

We can assume $M \ge \max\{\alpha_1(t), \alpha_2(t), \epsilon_1\}$. Hence, using (2.5), we can write for $t \in]a,b[$ and $u \ge M$

$$f_1(t,u) = \max\{f(\eta_1(t),u), f(t,u)\} \le \varphi(u) + h(t) + h(\eta_1(t)).$$

The function $w(t)$. Define $w \in C([a,b],\mathbb{R}^+) \cap C^2(]a,b[,\mathbb{R}_0^+)$ to be such that

$$\begin{gathered} w''(t) + h(t) + h(\eta_1(t)) = 0, \\ w(a) = 0, \ w(b) = 0. \end{gathered}$$

We have for all $t \in [a,b]$

$$w(t) = \int_a^b G(t,s)[h(s) + h(\eta_1(s))]\,ds \geq 0,$$

where $G(t,s)$ is the corresponding Green function and h can be assumed to be nonnegative.

An auxiliary problem. Consider the function

$$\begin{aligned} \psi(u) &= \varphi(M) + \epsilon_0 u^+, && \text{for all } u < M - \|w\|_\infty, \\ &= \varphi(u + \|w\|_\infty) + \epsilon_0 u^+, && \text{for all } u \geq M - \|w\|_\infty, \end{aligned}$$

with $u^+ = \max\{0, u\}$, and choose $\epsilon_0 > 0$ small enough so that

$$\liminf_{u \to +\infty} \frac{2\Psi(u)}{u^2} < \left(\tfrac{\pi}{b-a}\right)^2, \tag{2.6}$$

where $\Psi(u) = \int_0^u \psi(s)\,ds$. The function $\psi : \mathbb{R} \to \mathbb{R}_0^+$ is continuous and verify

$$\lim_{u \to +\infty} \psi(u) = +\infty. \tag{2.7}$$

We consider now the auxiliary problem

$$\begin{gathered} u'' + \psi(u) = 0, \\ u(\tfrac{a+b}{2}) = L, \; u'(\tfrac{a+b}{2}) = 0, \end{gathered} \tag{2.8}$$

where $L \in \mathbb{R}^+$.

The function β_0. From the condition (2.6), we can find a sequence $(u_i)_i$ going to infinity and such that for some $\delta \in]0, \frac{\pi}{b-a}[$,

$$\lim_{i \to \infty} (2\Psi(u_i) - \delta^2 u_i^2) = -\infty.$$

Hence, we can find a sequence $(L_i)_i$ going to infinity and such that

$$2(\Psi(L_i) - \Psi(u)) < \delta^2(L_i^2 - u^2), \quad \forall u \in [0, L_i[.$$

Next, we deduce from the conservation of energy that any solution u of (2.8) is such that

$$\frac{1}{\sqrt{2}} \int_{u(t)}^L \frac{du}{\sqrt{\Psi(L) - \Psi(u)}} = |t - \tfrac{a+b}{2}|.$$

Let us choose then L_i large enough so that

$$\frac{1}{\sqrt{2}} \int_M^{L_i} \frac{du}{\sqrt{\Psi(L_i) - \Psi(u)}} \geq \frac{1}{\delta} \int_{M/L_i}^1 \frac{ds}{\sqrt{1 - s^2}} > \tfrac{b-a}{2}.$$

We choose β_0 to be the solution of (2.8) for $L = L_i$. It follows then that

$$\frac{1}{\sqrt{2}}\int_{\beta_0(t)}^{L_i}\frac{dy}{\sqrt{\Psi(L_i)-\Psi(y)}} = |t-\tfrac{b+a}{2}| \leq \tfrac{b-a}{2} < \frac{1}{\sqrt{2}}\int_{M}^{L_i}\frac{du}{\sqrt{\Psi(L_i)-\Psi(u)}}.$$

Hence, β_0 is defined on $[a,b]$ and $\beta_0 \geq M$.

The upper solution. Now, we have $\beta(t) = \beta_0(t) + w(t) \geq M$ and

$$\begin{aligned}\beta''(t) + f_1(t,\beta(t)) &= f_1(t,\beta(t)) - \psi(\beta_0(t)) - h(t) - h(\eta_1(t))\\ &\leq f_1(t,\beta(t)) - \varphi(\beta_0(t) + \|w\|_\infty) - h(t) - h(\eta_1(t)) \leq 0,\end{aligned}$$

which proves the claim. □

Example 2.2. Consider the boundary value problem

$$\begin{aligned}u'' + \lambda\, u(1+\cos u) + \tfrac{1}{t(\pi-t)} &= 0,\\ u(0) = 0,\ u(\pi) &= 0,\end{aligned}$$

where $\lambda \geq 0$. We can prove existence of a solution from Theorem 2.2 if $\lambda \in]1/2, 1[$, but no existence result follows from Theorem 2.1.

3 A Derivative Dependent Dirichlet Problem

Next, we consider a problem with derivative dependent nonlinearity

$$\begin{aligned}u'' + f(t,u,u') &= 0,\\ u(a) = 0,\ u(b) &= 0.\end{aligned} \tag{3.1}$$

In the previous section, we imposed the slope $\frac{f(t,u)}{u}$ to be larger than the first eigenvalue λ_1 as u goes to zero and smaller than λ_1 as u goes to infinity. To extend these sublinear assumptions to (3.1) we consider the following auxiliary piecewise linear problem

$$\begin{aligned}u'' + B|u'| + Cu &= 0,\\ u(0) = 1,\ u'(0) &= 0,\end{aligned}$$

with B, $C > 0$. Its solution is positive on $]-\frac{\Gamma(B,C)}{2}, \frac{\Gamma(B,C)}{2}[$, where

$$\begin{aligned}\Gamma(B,C) &= \tfrac{4}{\sqrt{B^2-4C}}\tanh^{-1}\left(\tfrac{\sqrt{B^2-4C}}{B}\right), && \text{if } B^2-4C>0,\\ &= \tfrac{4}{\sqrt{4C-B^2}}\tan^{-1}\left(\tfrac{\sqrt{4C-B^2}}{B}\right), && \text{if } B^2-4C<0,\\ &= \tfrac{4}{B}, && \text{if } B^2-4C=0.\end{aligned}$$

We can now write the following theorem where we consider Carathéodory conditions in order to diversify the framework.

Theorem 3.1. *Let* $f :]a,b[\times \mathbb{R}_0^+ \times \mathbb{R} \to \mathbb{R}$ *satisfy Carathéodory conditions and assume that for each compact set* $L \subset \mathbb{R}_0^+ \times \mathbb{R}$ *there exists* $h_L \in L^1_{loc}(a,b)$ *such that for a.e.* $t \in]a,b[$ *and all* $(u,v) \in L$,

$$|f(t,u,v)| \leq h_L(t).$$

Assume further

(a) there exists B_1, $C_1 \geq 0$ *with* $b - a > \Gamma(B_1, C_1)$ *and for any compact set* $K \subset]a,b[$, *there is* $\epsilon > 0$ *such that for a.e.* $t \in K$, *all* $u \in]0,\epsilon]$ *and* $v \in \mathbb{R}$,

$$f(t,u,v) \geq B_1|v| + C_1 u;$$

(b) there exists B_2, $C_2 \geq 0$ *with* $b - a < \Gamma(B_2, C_2)$, $M > 0$ *and* $h \in \mathcal{A}$ *such that for a.e.* $t \in [a,b]$ *and all* $(u,v) \in [M,\infty[\times \mathbb{R}$,

$$f(t,u,v) \leq B_2|v| + C_2 u + h(t);$$

(c) for any compact set $L \subset \mathbb{R}_0^+$, *there exist a non-decreasing function* $\varphi \in C(\mathbb{R}^+, \mathbb{R}_0^+)$ *and* $\psi \in L^1_{loc}(a,b)$ *such that for a.e.* $t \in [a,b]$, *all* $u \in L$ *and* $v \in \mathbb{R}$ *we have*

$$|f(t,u,v)| \leq \psi(t)\varphi(|v|)$$

and for some $C > 2$,

$$\int_0^\infty \frac{ds}{\varphi(s)} = \infty,$$

$$\Phi^{-1}(C|\int_{\frac{a+b}{2}}^t \psi(s)\,ds|) \in L^1(a,b),$$

where $\Phi(u) = \displaystyle\int_0^u \frac{ds}{\varphi(s)}$.

Then the problem (3.1) *has at least one solution*

$$u \in C([a,b]) \cap C^1(]a,b[, \mathbb{R}_0^+) \cap W^{2,1}_{loc}(a,b).$$

Remark 3.1. Assumption (a) is equivalent to assume there exist B_1, $C_1 \geq 0$, with $b - a > \Gamma(B_1, C_1)$, and a function $a_1 \in C_0^2([a,b], \mathbb{R}^+)$ such that

(i) $t \in]a,b[$ implies $a_1(t) > 0$;

(ii) $f(t,u,v) \geq B_1|v| + C_1 u$, for all $t \in]a,b[$, $0 < u \leq a_1(t)$ and $v \in \mathbb{R}$;

(iii) $a_1''(t) > 0$, for all $t \in [a, (2a+b)/3] \cup [(a+2b)/3, b]$.

Proof of Theorem 3.1 : This proof paraphrases the proof of Theorem 2.1. Therefore, we will just outline the argument and refer to the corresponding steps of this proof.

Step 1 – Construction of lower solutions. Decreasing B_1 and C_1 if necessary, we can assume that $\Gamma(B_1, C_1) > (b-a)/3$. We define then $\alpha_2(t) = A_2 a_2(t)$, where a_2 is the solution of

$$u'' + B_1|u'| + C_1 u = 0,$$
$$u((b+a)/2) = 1, \ u'((b+a)/2) = 0,$$

and A_2 is chosen small enough so that

$$f(t, u, v) \geq B_1|v| + C_1 u$$

for a.e. $t \in [\frac{b+a}{2} - \frac{\Gamma(B_1,C_1)}{2}, \frac{b+a}{2} + \frac{\Gamma(B_1,C_1)}{2}[$, all $0 < u \leq \alpha_2(t)$ and all $v \in \mathbb{R}$. Next, we choose a_1 from Remark 3.1 and let

$$\alpha_1(t) = A_1 a_1(t),$$

where $A_1 \in]0, 1]$ is small enough as in the corresponding Step 1 in Theorem 2.1.

Step 2 – Approximation problems. We define the sequence of approximation problems

$$u'' + f_n(t, u, u') = 0,$$
$$u(a) = \epsilon_n, \ u(b) = \epsilon_n \tag{3.2}$$

as in the proof of Theorem 2.1. The functions f_i are defined on $]a, b[\times \mathbb{R}_0^+ \times \mathbb{R}$, satisfy

$$f_1(t, u, v) \geq f_2(t, u, v) \geq \cdots \geq f_n(t, u, v) \geq f_{n+1}(t, u, v) \geq \cdots \geq f(t, u, v)$$

and the sequence $(f_n)_n$ converges to f uniformly on compact subsets of $]a, b[\ \times \mathbb{R}_0^+ \times \mathbb{R}$. The sequence $(\epsilon_n)_n \subset \mathbb{R}_0^+$ is such that

$$\lim_{n\to\infty} \epsilon_n = 0,$$

$$f(t, u, v) \geq B_1|v| + C_1 u, \ \text{ for all } t \in [a + \tfrac{1}{2^{n+1}}, b - \tfrac{1}{2^{n+1}}], \ u \in]0, \epsilon_n], \ v \in \mathbb{R}.$$

Step 3 – $\alpha_3(t) := \epsilon_n$ is a lower solution of (3.2).

Step 4 – Existence of a solution u_1 of (3.2), *with $n = 1$, such that*

$$\max(\alpha_1(t), \alpha_2(t), \epsilon_1) \leq u_1(t).$$

Notice first that $\max(\alpha_1(t), \alpha_2(t), \epsilon_1)$ is a $W^{2,1}$-lower solution of (3.2) with $n = 1$.

Next, we prove that for some $h \in \mathcal{A}$

$$f_1(t, u, v) = \max\{f(\eta_1(t), u, v), f(t, u, v)\} \leq B_2|v| + C_2 u + h(t),$$

and define β such that

$$\begin{gathered}\beta'' + B_2|\beta'| + C_2\beta + h(t) = 0,\\ \beta(\tfrac{a+b}{2}) = R, \quad \beta'(\tfrac{a+b}{2}) = 0,\end{gathered}$$

with $R > 0$ large enough to enforce $\beta \geq M$. It turns out that β is an upper solution of (3.2), with $n = 1$.

At last, we verify that

$$|f_1(t,u,v)| \leq \varphi(|v|)(\psi(t) + \psi(\eta_1(t)))$$

and that for ϵ such that $\frac{2}{1-\epsilon} \leq C$, we can write

$$\begin{aligned}
&\Phi^{-1}(2|\int_{\frac{a+b}{2}}^t (\psi(s) + \psi(\eta_1(s)))\,ds|)\\
&\quad \leq \Phi^{-1}(2|\int_{\frac{a+b}{2}}^t \psi(s)\,ds| + 2|\int_{\frac{a+b}{2}}^t \psi(\eta_1(s))\,ds|)\\
&\quad \leq (1-\epsilon)\Phi^{-1}(\tfrac{2}{1-\epsilon}|\int_{\frac{a+b}{2}}^t \psi(s)\,ds|) + \epsilon\Phi^{-1}(\tfrac{2}{\epsilon}|\int_{\frac{a+b}{2}}^t \psi(\eta_1(s))\,ds|)\\
&\quad \leq (1-\epsilon)\Phi^{-1}(C|\int_{\frac{a+b}{2}}^t \psi(s)\,ds|) + \epsilon\Phi^{-1}(\tfrac{2}{\epsilon}|\int_{\frac{a+b}{2}}^t \psi(\eta_1(s))\,ds|).
\end{aligned}$$

Hence, $\Phi^{-1}(2|\int_{\frac{a+b}{2}}^t (\psi(s) + \psi(\eta_1(s)))\,ds|) \in L^1(a,b)$.

The claim follows now from Theorem II-4.6.

Step 5 – The problem (3.2) *has at least one solution* u_n *such that*

$$\max(\alpha_1(t), \alpha_2(t), \epsilon_n) \leq u_n(t) \leq u_{n-1}(t).$$

Step 6 – Existence of a solution of (3.1).

The proof of these steps repeats the corresponding argument used in the proof of Theorem 2.1 and uses Proposition II-4.5. □

4 A Multiplicity Result

In this section, we consider a multiplicity result for the problem

$$\begin{gathered}u'' + f(t,u) = 0,\\ u(a) = 0, \ u(b) = 0.\end{gathered} \tag{4.1}$$

The idea is, as in Section VIII-2.1, to impose conditions so that the nonlinearity crosses twice the first eigenvalue. A first crossing is ensured by the existence of lower and upper solutions. The second one follows from conditions at infinity. Our first result assumes such sub-superlinear conditions and allows some resonance in the second crossing.

Theorem 4.1. *Let α and $\beta \in C([a,b])$ be respectively a $W^{2,1}$-lower solution and a strict $W^{2,1}$-upper solution of (4.1) with $0 \le \alpha(t) \le \beta(t)$ for $t \in]a,b[$, $\beta(a) > 0$ and $\beta(b) > 0$.*

Let $E = \{(t,u) \in [a,b] \times \mathbb{R} \mid \alpha(t) \le u\}$ and $f : E \to \mathbb{R}$ be an $\mathcal{A}$-Carathéodory function.

Assume further there exist $\rho > 0$ and c, $d \in \mathcal{A}$, $d > 0$ on $[a,b]$ such that
(a) for a.e. $t \in [a,b]$ and all $u \ge \rho$,

$$f(t,u) \ge d(t)\,u - c(t);$$

(b) the first eigenvalue μ_1 of

$$\begin{aligned} u'' + \mu\, d(t)u = 0,\\ u(a) = 0,\ u(b) = 0, \end{aligned} \tag{4.2}$$

is such that $\mu_1 \le 1$;

(c) $\displaystyle\int_a^b \liminf_{u\to\infty}[f(t,u) - \mu_1 d(t)\,u]\psi_1(t)\,dt > 0$,
where ψ_1 is a positive eigenfunction of (4.2) associated with μ_1.

Then the problem (4.1) has at least two solutions u_1 and $u_2 \in W^{2,\mathcal{A}}(a,b)$ such that for all $t \in [a,b]$

$$\alpha(t) \le u_1(t) < \beta(t),\ u_1(t) \le u_2(t).$$

Proof. *Claim – There exists $R > 0$ such that every solution $u \ge \alpha$ of*

$$\begin{aligned} u'' + f(t,u) \le 0,\\ u(a) = 0,\ u(b) = 0, \end{aligned} \tag{4.3}$$

satisfies

$$\max_{t\in[a,b]} u(t) < R.$$

Suppose that for any n, u_n is a solution of

$$\begin{aligned} u_n'' + f(t,u_n) \le 0,\\ u_n(a) = 0,\ u_n(b) = 0, \end{aligned}$$

with $u_n \ge \alpha$ and $\max_{t\in[a,b]} u_n(t) \ge n$. The function u_n can be written as

$$u_n(t) = \bar{u}_n(t) + \tilde{u}_n(t),$$

where $\bar{u}_n(t) = a_n\psi_1(t)$, $a_n \in \mathbb{R}$, and $\int_a^b \tilde{u}_n'(t)\psi_1'(t)\,dt = 0$.

From Proposition A-4.4, we know

$$\begin{aligned}\|\tilde{u}_n\|_\infty &\le K\int_a^b |\tilde{u}_n'' + \mu_1 d\tilde{u}_n|\psi_1\,dt = K\int_a^b |u_n'' + \mu_1 du_n|\psi_1\,dt \\ &\le K[2\int_a^b (u_n'' + \mu_1 du_n)^+\psi_1\,dt - \int_a^b (u_n'' + \mu_1 du_n)\psi_1\,dt].\end{aligned}$$

Using Lemma A-3.15, we have

$$\int_a^b (u_n'' + \mu_1 du_n)\psi_1\,dt = \int_a^b (\psi_1'' + \mu_1 d\psi_1)u_n\,dt = 0.$$

As f is an $\mathcal{A}$-Carathéodory function, there exists an $h \in \mathcal{A}$ so that for a.e. $t \in [a,b]$ and all $u \in [\alpha(t), \rho]$, $|f(t,u)| \le h(t)$. Hence,

$$\begin{aligned}u_n'' + \mu_1 d(t)u_n &\le -f(t,u_n) + \mu_1 d(t)u_n \\ &\le \mu_1 d(t)(\rho - u_n) + h(t) + |c(t)| + \mu_1 d(t)u_n \\ &\le h(t) + |c(t)| + \mu_1 d(t)\rho,\end{aligned}$$

and we obtain, using Proposition A-3.14,

$$\begin{aligned}\|\tilde{u}_n\|_\infty &\le 2K\int_a^b (h(t) + |c(t)| + \mu_1 d(t)\rho)\psi_1(t)\,dt \\ &\le 2K\,K_1\int_a^b (h(t) + |c(t)| + \mu_1 d(t)\rho)(t-a)(b-t)\,dt.\end{aligned}$$

As $\|u_n\|_\infty \to \infty$, we have $a_n \to \infty$ and $u_n(t) \to \infty$ for all $t \in]a,b[$. Notice next that for a.e. $t \in [a,b]$ and all $u \in \mathbb{R}_0^+$,

$$(f(t,u) - \mu_1 d(t)u)\psi_1(t) \ge (-h(t) - |c(t)| - \mu_1 d(t)\rho)\psi_1(t) \in L^1.$$

Using Fatou's lemma, we obtain the contradiction

$$\begin{aligned}0 &= \lim_{n\to\infty}\int_a^b -(u_n'' + \mu_1 du_n)\psi_1\,dt \\ &\ge \lim_{n\to\infty}\int_a^b (f(t,u_n) - \mu_1 du_n)\psi_1\,dt \\ &\ge \int_a^b \liminf_{n\to\infty}(f(t,u_n) - \mu_1 du_n)\psi_1\,dt > 0,\end{aligned}$$

where we have used (c).

Conclusion – The proof follows now from Theorem III-2.13. □

In very much the same way we can prove the following which is a corollary of the preceding theorem. Here we do not allow resonance at infinity.

Theorem 4.2. *Let α and $\beta \in C([a,b])$ be respectively a $W^{2,1}$-lower solution and a strict $W^{2,1}$-upper solution of (4.1) with $0 \le \alpha(t) \le \beta(t)$ for $t \in]a,b[$, $\beta(a) > 0$ and $\beta(b) > 0$.*

Let $E = \{(t,u) \in [a,b] \times \mathbb{R} \mid \alpha(t) \le u\}$ and $f : E \to \mathbb{R}$ be an $\mathcal{A}$-Carathéodory function.

Assume further there exist a $\rho > 0$ and c, $d \in \mathcal{A}$, $d > 0$ on $[a,b]$ such that
(a) for a.e. $t \in [a,b]$ and all $u \ge \rho$,

$$f(t,u) \ge d(t)\,u - c(t);$$

(b) the first eigenvalue μ_1 of (4.2) is such that $\mu_1 < 1$.

Then the problem (4.1) has at least two solutions u_1 and $u_2 \in W^{2,\mathcal{A}}(a,b)$ such that for all $t \in [a,b]$

$$\alpha(t) \le u_1(t) < \beta(t), \;\; u_1(t) \le u_2(t).$$

We can obtain lower and upper solutions from the following propositions.

Proposition 4.3. *Let $d \in \mathcal{A}$, $d(t) > 0$ a.e. on $[a,b]$ and assume that the first eigenvalue of (4.2) satisfies $\mu_1 \ge 1$. Suppose that there exist some $M > m > 0$ such that for a.e. $t \in [a,b]$ and all $u \in]m,M[$,*

$$f(t,u) \le d(t)(u-m).$$

Then the problem (4.1) has a strict $W^{2,1}$-upper solution β such that $\beta(t) \in [m,M]$ for every $t \in [a,b]$.

Proof. Let β_1 be a positive eigenfunction of (4.2) that corresponds to μ_1 and such that $\max_t \beta_1(t) < M - m$. Then $\beta(t) = \beta_1(t) + m$ is a strict $W^{2,1}$-upper solution. In fact, for $t_0 \in]a,b[$, let $\delta_0 = \min\{t_0 - a, b - t_0\}$, $I_0 =]t_0 - \frac{\delta_0}{2}, t_0 + \frac{\delta_0}{2}[$ and $\epsilon_0 = \min_{t \in \bar{I}_0} \beta_1(t)$. For a.e. $t \in I_0$ and any $u \in [\beta(t) - \epsilon_0, \beta(t)]$, we have

$$\beta'' + f(t,u) \le \beta_1'' + d(t)(u-m) \le 0$$

and

$$\beta(a) = \beta(b) = m > 0. \qquad \square$$

Proposition 4.4. *Let $d \in \mathcal{A}$, $d(t) > 0$ a.e. on $[a,b]$ and assume that the first eigenvalue of (4.2) satisfies $\mu_1 \leq 1$. Suppose that there exists $l > 0$ such that for a.e. $t \in [a,b]$ and all $u \in]0,l[$,*

$$f(t,u) \geq d(t)\,u.$$

Then the problem (4.1) has a $W^{2,1}$-lower solution α such that $\alpha(t) \in [0,l]$ for every $t \in [a,b]$.

Proof. Let α be a positive eigenfunction of (4.2) that corresponds to μ_1 and such that $\max_t \alpha(t) < l$. Then $\alpha(t)$ is a $W^{2,1}$-lower solution. □

Remark 4.1. If $l < m$, Propositions 4.3 and 4.4 give the existence of ordered lower and strict upper solutions of problem (4.1).

Remark 4.2. Notice that the lower solution α given by Proposition 4.4 is concave. Hence, there exists a $k > 0$ such that

$$\alpha(t) \geq k\,(t-a)(b-t).$$

This result can be used to verify the regularity assumption in Theorem 4.1. For example, if we consider $f :]a,b[\, \times \mathbb{R}_0^+ \to \mathbb{R}$ defined by

$$f(t,u) = \frac{1}{(t-a)(b-t)\,u^\sigma}$$

with $0 \leq \sigma < 1$, then, for any $t \in]a,b[$ and any $u \in [\alpha(t), R]$, we have

$$|f(t,u)| \leq \frac{1}{k^\sigma (t-a)^{1+\sigma}(b-t)^{1+\sigma}} \in \mathcal{A}.$$

Example 4.1. Consider the problem

$$\begin{gathered} u'' + \tfrac{3}{t(\pi-t)}\left(\tfrac{1}{u^2} - \tfrac{2}{u} + u\right) = 0, \\ u(0) = 0,\ u(\pi) = 0. \end{gathered}$$

The main feature of this example is that each of the terms $\frac{1}{u^2}$, $-\frac{2}{u}$, u provides one of the basic assumptions in Theorem 4.1. The function α can be build from the singularity $\frac{1}{u^2}$ near the origin. The upper solution follows from the term $-\frac{2}{u}$ which dominates on a bounded interval away from the origin. The term u implies the appropriate behaviour at infinity, which is clear if we check the following.

(a) The function

$$\alpha(t) = \alpha_0\, t^{1/3}(\pi - t)^{1/3}$$

is a lower solution if $\alpha_0 > 0$ is small enough.
(b) The constant function

$$\beta(t) = 0,9$$

is a strict upper solution.
(c) Given an $R > 0$, if $u \in [\alpha(t), R]$ we have

$$\begin{aligned} |f(t,u)| &= \left| \tfrac{3}{t(\pi-t)} \left(\tfrac{1}{u^2} - \tfrac{2}{u} + u \right) \right| \\ &\le \tfrac{3}{t(\pi-t)} \left(\tfrac{1}{\alpha_0^2 t^{2/3} (\pi-t)^{2/3}} + \tfrac{2}{\alpha_0 t^{1/3} (\pi-t)^{1/3}} + \alpha_0 R \right) \\ &=: h_R(t) \end{aligned}$$

and $h_R \in \mathcal{A}$.
(d) Take $d(t) = \frac{2}{t(\pi-t)}$. Easy computations show that the first eigenvalue of

$$\begin{aligned} u'' + \mu d(t) u &= 0, \\ u(0) = 0, \ u(\pi) &= 0 \end{aligned}$$

is $\mu_1 = 1$ with eigenfunction $\psi_1(t) = t(\pi - t)$. Then the assumptions of Theorem 4.1 are satisfied since

$$\liminf_{u \to +\infty} [f(t,u) - \mu_1 d(t) u] = \liminf_{u \to +\infty} \frac{1}{t(\pi - t)} [3(\frac{1}{u^2} - \frac{2}{u} + u) - 2u] = +\infty.$$

Chapter X

Singular Perturbations

1 Boundary Layers at One End Point

This section deals with problems such as

$$\begin{gathered} \varepsilon\, u'' + u' - 1 = 0, \\ u(0) = 0, \;\; u(1) = 0, \end{gathered}$$

where $\varepsilon > 0$ is a small parameter. This problem has the solution

$$u_\varepsilon(t) = t - \frac{1 - e^{-t/\varepsilon}}{1 - e^{-1/\varepsilon}}$$

which tends to $u_0(t) = t - 1$ as $\varepsilon \to 0$, uniformly on compact subsets of $]0, 1]$. The convergence is however nonuniform at the left endpoint $t = 0$. In such a case, the solution is said to have a *boundary layer* at the end point $t = 0$. A symmetric situation holds for the problem

$$\begin{gathered} \varepsilon\, u'' - u' + 1 = 0, \\ u(0) = 0, \;\; u(1) = 0. \end{gathered}$$

Here, the convergence is uniform on compact subsets of $[0, 1[$ and the boundary layer takes place at the right endpoint $t = 1$. At last, let us recall that a richer example, which exhibits several types of boundary layers,

$$\begin{gathered} \varepsilon\, u'' + uu' - u = 0, \\ u(0) = A, \;\; u(1) = B, \end{gathered}$$

has been worked out in Section II-1.3.

The following theorems consider the boundary value problem

$$\begin{aligned} \varepsilon u'' + f(t,u,u') &= 0, \\ u(a) = 0, \ u(b) &= 0. \end{aligned} \tag{1.1}$$

They describe cases where a boundary layer takes place at the left end point $t = a$. Symmetric results with a boundary layer at the right endpoint $t = b$ can easily be deduced.

Theorem 1.1. *Let $u_0 \in \mathcal{C}^2([a,b])$ and let*

$$F = \{(t,u) \mid a \le t \le b, \ r_1(t) \le u - u_0(t) \le r_2(t)\},$$

where r_1 and r_2 are two continuous functions on $[a,b]$ such that for all $t \in [a,b]$

$$r_1(t) < 0 < r_2(t)$$

and

$$u_0(a) + r_1(a) < 0 < u_0(a) + r_2(a).$$

Consider a function $f \in \mathcal{C}^1(F \times \mathbb{R})$ such that

$$\forall (t,u,v) \in F \times \mathbb{R}, \qquad |f(t,u,v)| \le \varphi(|v|), \tag{1.2}$$

where $\varphi \in \mathcal{C}(\mathbb{R}^+)$ satisfies

$$\int_0^\infty \frac{s\,ds}{\varphi(s)} = \infty. \tag{1.3}$$

Assume
(a) u_0 is a solution of the reduced problem

$$f(t,u_0(t),u_0'(t)) = 0, \quad u_0(b) = 0;$$

(b) there exists a $\mu > 0$ such that for all $(t,u,v) \in F \times \mathbb{R}$,

$$\frac{\partial f}{\partial v}(t,u,v) \ge \mu > 0.$$

Then for ε small enough, there exists a solution u_ε of (1.1) such that

$$u_\varepsilon(t) = u_0(t) + O(\varepsilon + \exp(-\tfrac{\mu(t-a)}{\varepsilon})).$$

Condition (b) can be interpreted as a stability assumption on the boundary layer equation

$$\frac{dv}{ds} + f(t,u,v) = 0,$$

where $(t,u) \in F$ are fixed parameters. This assumption is essential to have the boundary layer at the initial point $t = a$. If we reverse the inequalities, i.e. $\frac{\partial f}{\partial v}(t,u,v) \le -\mu < 0$, the boundary layer takes place at the terminal point $t = b$.

Next, we show that an asymptotic result on the derivative can be obtained provided we reinforce the assumption on $\frac{\partial f}{\partial v}(t,u,v)$.

Theorem 1.2. *Under the assumptions of Theorem 1.1 and if* $\frac{\partial f}{\partial v}(t,u,v)$ *is bounded on* $F \times \mathbb{R}$*, we have*

$$u'_\varepsilon(t) = u'_0(t) + O(\varepsilon + \tfrac{1}{\varepsilon}\exp(-\tfrac{\mu(t-a)}{\varepsilon})).$$

Proof of Theorem 1.1: Part A – The reduced problem.

Claim 1 – There exists a function $h \in \mathcal{C}^1(F)$ *such that*

(i) *for any* $(t,u,v) \in F \times \mathbb{R}$, $f(t,u,v) = 0$ *if and only if* $v = h(t,u)$*;*

(ii) $\frac{\partial h}{\partial u}(t,u) = -\frac{\partial f}{\partial u}(t,u,h(t,u))\left[\frac{\partial f}{\partial v}(t,u,h(t,u))\right]^{-1}$.

For any $(t,u) \in F$, assumption (b) implies $f(t,u,.)$ is a strictly increasing function such that

$$\lim_{v \to +\infty} f(t,u,v) = +\infty \text{ and } \lim_{v \to -\infty} f(t,u,v) = -\infty.$$

It follows that we can define $h(t,u)$ to be the unique v such that $f(t,u,v) = 0$. From implicit derivation, $h \in \mathcal{C}^1(F)$ and (ii) is satisfied.

Claim 2 – For some $Q > 0$, $\varepsilon_0 > 0$ *small enough and any* ε *with* $|\varepsilon| \le \varepsilon_0$*, the solution* $v_0(t,\varepsilon)$ *of the Cauchy problem*

$$u' = h(t,u), \quad u(b) = \varepsilon \tag{1.4}$$

is defined on $[a,b]$ *and we have*

$$|v_0(t,\varepsilon) - u_0(t)| < |\varepsilon| \exp \tfrac{Q}{\mu}(b-t) = O(\varepsilon),$$
$$|v'_0(t,\varepsilon) - u'_0(t)| < \tfrac{Q|\varepsilon|}{\mu} \exp \tfrac{Q}{\mu}(b-t) = O(\varepsilon).$$

Notice first that as $u_0 \in \mathcal{C}^1([a,b])$ and $f \in \mathcal{C}^1(F \times \mathbb{R})$, there exist $R > 0$ and $Q > 0$ such that for all $(t,u) \in F$ and all v with $|v - u'_0(t)| \le R$,

$$|\tfrac{\partial f}{\partial u}(t,u,v)| < Q.$$

Hence, in a neighbourhood of $u_0(t)$, $h(t,u)$ satisfies a Lipschitz condition with respect to u with constant Q/μ. The proof follows then from an application of Gronwall Lemma.

Part B – Construction of $\mathcal{C}^2$*-lower and upper solutions of* (1.1) *in case* $u_0(a) \ge 0$.

Choose K_1 and K_2 such that

$$-r_1(a) > K_1 > u_0(a) \quad \text{and} \quad K_2 > \tfrac{1}{Q}\, \|u_0''\|_\infty$$

and define λ_1 and λ_2 to be the greater and the smaller roots of $\varepsilon\lambda^2 - \mu\lambda + Q = 0$. We have

$$\lambda_1 = \frac{\mu}{\varepsilon} - \lambda_2, \quad \lambda_2 = \frac{2Q}{\mu + \sqrt{\mu^2 - 4\varepsilon Q}} = \frac{Q}{\mu} + O(\varepsilon).$$

Notice that for $\varepsilon > 0$ small enough the functions

$$\alpha(t) = v_0(t, -\varepsilon) - K_1 \exp(-\lambda_1(t-a)) - \varepsilon K_2[\exp(\lambda_2(b-t)) - 1],$$
$$\beta(t) = v_0(t, \varepsilon) + \varepsilon K_2[\exp(\lambda_2(b-t)) - 1],$$

have their graph in F.

Claim 1 – $\beta > \alpha$. It is clear that

$$\beta(t) - \alpha(t) > v_0(t, \varepsilon) - v_0(t, -\varepsilon).$$

Also, as $v_0(b, \varepsilon) > v_0(b, -\varepsilon)$ we deduce from the uniqueness of the solutions of the Cauchy problem (1.4) that for all $t \in [a, b]$, $v_0(t, \varepsilon) > v_0(t, -\varepsilon)$, i.e. $\beta > \alpha$.

Claim 2 – *For* $\varepsilon > 0$ *small enough,*

$$\varepsilon\, \alpha''(t) + f(t, \alpha(t), \alpha'(t)) \geq 0 \quad \textit{and} \quad \varepsilon\, \beta''(t) + f(t, \beta(t), \beta'(t)) \leq 0.$$

Let us write $v_0(t)$ for $v_0(t, -\varepsilon)$. Observe then that for ε small enough,

$$\{(t, v_0(t) + \xi(\alpha(t) - v_0(t))) \mid t \in [a, b],\ \xi \in [0, 1]\} \subset F$$

and

$$f(t, \alpha, \alpha') = f(t, v_0, v_0') + \tfrac{\partial f}{\partial u}(t, v_0 + \xi_1(\alpha - v_0), v_0')(\alpha - v_0)$$
$$+ \tfrac{\partial f}{\partial v}(t, \alpha, v_0' + \xi_2(\alpha' - v_0'))(\alpha' - v_0'),$$

for some $0 < \xi_i = \xi_i(t) < 1$, $i = 1, 2$. Hence, we have

$$\begin{aligned}\varepsilon\, \alpha'' + f(t, \alpha, \alpha') = {} & \varepsilon\, v_0'' - \varepsilon K_1 \lambda_1^2 \exp(-\lambda_1(t-a)) - \varepsilon^2 K_2 \lambda_2^2 \exp(\lambda_2(b-t)) \\ & + f(t, v_0, v_0') \\ & - \tfrac{\partial f}{\partial u}(t, v_0 + \xi_1(\alpha - v_0), v_0')[K_1 \exp(-\lambda_1(t-a)) + \varepsilon K_2(\exp(\lambda_2(b-t)) - 1)] \\ & + \tfrac{\partial f}{\partial v}(t, \alpha, v_0' + \xi_2(\alpha' - v_0'))[K_1 \lambda_1 \exp(-\lambda_1(t-a)) + \varepsilon K_2 \lambda_2 \exp(\lambda_2(b-t))]\end{aligned}$$

and we obtain

$$\begin{aligned}\varepsilon\,\alpha''(t) + f(t,\alpha(t),\alpha'(t)) \geq \varepsilon\, v_0''(t) + f(t, v_0(t), v_0'(t)) + \varepsilon K_2 Q \\ +K_1 \exp(-\lambda_1(t-a))[-\varepsilon\lambda_1^2 + \lambda_1\mu - Q] \\ +\varepsilon K_2 \exp(\lambda_2(b-t))[-\varepsilon\lambda_2^2 + \lambda_2\mu - Q].\end{aligned}$$

By definition of λ_i and as $f(t, v_0(t), v_0'(t)) = 0$, we can write

$$\varepsilon\,\alpha''(t) + f(t,\alpha(t),\alpha'(t)) \geq \varepsilon[v_0''(t) + K_2 Q] = \varepsilon[u_0''(t) + K_2 Q + o(1)] \geq 0.$$

In the same way we prove

$$\varepsilon\,\beta''(t) + f(t,\beta(t),\beta'(t)) \leq 0.$$

Claim 3 – For ε small enough, $\alpha(a) \leq 0 \leq \beta(a)$, $\alpha(b) \leq 0 \leq \beta(b)$. We compute

$$\begin{aligned}\alpha(a) &= v_0(a,-\varepsilon) - K_1 - \varepsilon K_2[e^{\lambda_2(b-a)} - 1] \\ &\leq u_0(a) - K_1 \leq 0, \\ \alpha(b) &= -\varepsilon - K_1 e^{-\lambda_1(b-a)} \leq 0,\end{aligned}$$

and obtain similarly

$$\beta(a) \geq 0 \quad \text{and} \quad \beta(b) \geq 0.$$

Part C – The case $u_0(a) < 0$. The same arguments hold with

$$\alpha(t) = v_0(t,-\varepsilon) - \varepsilon K_2[\exp(\lambda_2(b-t)) - 1],$$
$$\beta(t) = v_0(t,\varepsilon) + K_1 \exp(-\lambda_1(t-a)) + \varepsilon K_2[\exp(\lambda_2(b-t)) - 1],$$

and appropriate K_1 and K_2.

Part D – Conclusion. The proof follows now from Theorem II-1.3. □

Proof of Theorem 1.2 : Claim 1 – $|u_\varepsilon'(a) - u_0'(a)| = O(\frac{1}{\varepsilon})$. Let u_ε be the solution of (1.1) given by Theorem 1.1. It follows from this theorem that $z_\varepsilon(t) = u_\varepsilon(t) - u_0(t) = O(1)$ and we verify that

$$\varepsilon\, z_\varepsilon'' = -\varepsilon u_0'' - \frac{\partial f}{\partial u}(t, u_0 + \xi_1 z_\varepsilon, u_0') z_\varepsilon - \frac{\partial f}{\partial v}(t, u_\varepsilon, u_0' + \xi_2 z_\varepsilon') z_\varepsilon',$$

where $\xi_1 = \xi_1(t)$, $\xi_2 = \xi_2(t) \in [0,1]$. Hence, we can find $M > 0$ large enough so that

$$\begin{aligned}\varepsilon\,|z_\varepsilon''| &\leq \sup_F \left|\tfrac{\partial f}{\partial u}(t,u,u_0'(t))\right| |z_\varepsilon| + \sup_{F\times\mathbb{R}} \tfrac{\partial f}{\partial v}(t,u,v)|z_\varepsilon'| + \varepsilon\|u_0''\|_\infty \\ &< M(|z_\varepsilon'| + 1). \qquad (1.5)\end{aligned}$$

Let $N_0 = 2r/(b-a)$, where $\|z_\varepsilon\|_\infty \le r$. If we define N_ε to be such that

$$\int_{N_0}^{N_\varepsilon} \frac{s}{1+s}\, ds = \frac{2rM}{\varepsilon},$$

we deduce from (1.5) that $|z'_\varepsilon(t)| \le N_\varepsilon$. As the function

$$\psi(r) = \int_0^r \frac{s}{1+s}\, ds$$

is convex, we compute

$$\frac{2rM}{\varepsilon} = \psi(N_\varepsilon) - \psi(N_0) \ge \psi'(N_0)(N_\varepsilon - N_0) \ge \frac{N_0}{1+N_0}(N_\varepsilon - N_0)$$

and it follows that

$$|z'_\varepsilon(t)| \le N_\varepsilon \le N_0 + \tfrac{M}{\varepsilon}(1+N_0)(b-a) = O(\tfrac{1}{\varepsilon}).$$

Conclusion – We know that

$$|z_\varepsilon(t)| \le S(t,\varepsilon) := L(\varepsilon + e^{-\frac{\mu(t-a)}{\varepsilon}})$$

for some $L > 0$ and we have also that

$$\varepsilon z''_\varepsilon = -\frac{\partial f}{\partial u}(t, u_0 + \xi_1 z_\varepsilon, u'_0) z_\varepsilon - \frac{\partial f}{\partial v}(t, u_\varepsilon, u'_0 + \xi_2 z'_\varepsilon) z'_\varepsilon - \varepsilon u''_0.$$

This result implies

$$\begin{aligned} \varepsilon z''_\varepsilon &\le QS(t,\varepsilon) - \mu z'_\varepsilon + \varepsilon K, \qquad &&\text{if } z'_\varepsilon \ge 0,\\ \varepsilon z''_\varepsilon &\ge -QS(t,\varepsilon) - \mu z'_\varepsilon - \varepsilon K, \qquad &&\text{if } z'_\varepsilon < 0, \end{aligned}$$

where Q is defined as in the proof of Theorem 1.1 and $K > 0$. It follows that

$$\varepsilon \frac{d}{dt}|z'_\varepsilon| \le QS(t,\varepsilon) - \mu|z'_\varepsilon| + \varepsilon K,$$

which implies using Claim 1

$$\begin{aligned} |z'_\varepsilon(t)| &\le |z'_\varepsilon(a)| \exp -\tfrac{\mu(t-a)}{\varepsilon} + \tfrac{Q}{\varepsilon}\int_a^t S(s,\varepsilon) \exp -\tfrac{\mu(t-s)}{\varepsilon}\, ds + O(\varepsilon),\\ &\le O(\varepsilon + \tfrac{1}{\varepsilon} \exp -\tfrac{\mu(t-a)}{\varepsilon}). \end{aligned}$$

□

The boundary layers can appear in higher order terms only. This is the case in the Robin problem

$$\begin{aligned} &\varepsilon u'' + f(t,u,u') = 0,\\ &u(a) = 0,\ u'(b) = 0. \end{aligned}$$

More generally, we consider the boundary value problem

$$\begin{aligned} \varepsilon u'' + f(t,u,u') &= 0, \\ a_1 u(a) - a_2 u'(a) &= A, \\ b_1 u(b) + b_2 u'(b) &= B. \end{aligned} \tag{1.6}$$

As above the assumption (b) will ensure that the boundary layer takes place at the initial point $t = a$. Further, we assume that at this point the boundary condition is not of Dirichlet type, i.e. $a_2 \neq 0$. This assumption implies that the boundary layer that appears in the function u_ε is not of first order, i.e.

$$u_\varepsilon(t) = u_0(t) + O(\varepsilon).$$

Theorem 1.3. *Let A, $B \in \mathbb{R}$, a_1, $b_1 \in \mathbb{R}$, $a_2 > 0$, $b_2 \geq 0$, $b_1^2 + b_2^2 > 0$. Let $u_0 \in C^2([a,b])$, F be a closed neighbourhood of the graph of u_0 and $f \in C^1(F \times \mathbb{R})$. Assume*
(a) u_0 is a solution of the reduced problem

$$f(t, u_0(t), u_0'(t)) = 0, \quad b_1 u_0(b) + b_2 u_0'(b) = B;$$

(b) there exists $\mu > 0$ such that for all $(t,u,v) \in F \times \mathbb{R}$,

$$\frac{\partial f}{\partial v}(t,u,v) \geq \mu > 0;$$

(c) $\Delta = b_1 \frac{\partial f}{\partial v}(b, u_0(b), u_0'(b)) - b_2 \frac{\partial f}{\partial u}(b, u_0(b), u_0'(b)) > 0$.

Then for ε small enough, there exists a solution u_ε of (1.6) *such that*

$$\begin{aligned} u_\varepsilon(t) &= u_0(t) + O(\varepsilon), \\ u_\varepsilon'(t) &= u_0'(t) + O(\varepsilon + \exp(-\tfrac{\mu(t-a)}{\varepsilon})). \end{aligned}$$

Proof. Part A – The reduced problem.
As in the proof of Theorem 1.1, we define the function $h \in C^1(F)$ such that $f(t,u,v) = 0$ if and only if $v = h(t,u)$.

Next for $\eta_0 > 0$ small enough and any η with $|\eta| \leq \eta_0$, we can prove that the solution $v_0(t,\eta)$ of the Cauchy problem

$$u' = h(t,u), \quad u(b) = u_0(b) + \eta \tag{1.7}$$

is defined on $[a,b]$ and verifies

$$\begin{aligned} |v_0(t,\eta) - u_0(t)| &< |\eta| \exp \tfrac{Q}{\mu}(b-t) = O(\eta), \\ |v_0'(t,\eta) - u_0'(t)| &< \tfrac{Q|\eta|}{\mu} \exp \tfrac{Q}{\mu}(b-t) = O(\eta), \end{aligned}$$

for some $Q \geq 0$.

At last, we compute

$$b_1 v_0(b,\eta) + b_2 v_0'(b,\eta) - B = b_1(u_0(b)+\eta) + b_2 h(b, u_0(b)+\eta) - B$$
$$= b_1\eta + b_2 \tfrac{\partial h}{\partial u}(b, u_0(b)+\xi\eta)\eta,$$

where $\xi \in [0,1]$. As $h \in \mathcal{C}^1(F)$, we have

$$b_1 v_0(b,\eta) + b_2 v_0'(b,\eta) - B = [b_1 + b_2 \tfrac{\partial h}{\partial u}(b, u_0(b))]\eta + o(\eta)$$
$$= \Delta[\tfrac{\partial f}{\partial v}(b, u_0(b), h(b, u_0(b)))]^{-1}\eta + o(\eta).$$

Part B – Construction of $\mathcal{C}^2$-lower and upper solutions of (1.6).

Choose K_1, K_2 and $K_3 > 0$ such that

$$K_1 > \tfrac{1}{a_2\mu}|a_1 u_0(a) - a_2 u_0'(a) - A|,$$
$$K_2 > \tfrac{1}{Q}\|u_0''\|_\infty, \quad K_3 > \tfrac{K_2 b_2 Q}{\Delta\mu}\tfrac{\partial f}{\partial v}(b, u_0(b), h(b, u_0(b))),$$

and let $\lambda_1 = \frac{\mu}{\varepsilon} + O(1)$ and $\lambda_2 = \frac{Q}{\mu} + O(\varepsilon)$ be as in the proof of Theorem 1.1. Define now

$$\alpha(t) = v_0(t, -\varepsilon K_3) - \varepsilon K_1 \exp(-\lambda_1(t-a)) - \varepsilon K_2[\exp(\lambda_2(b-t)) - 1],$$
$$\beta(t) = v_0(t, \varepsilon K_3) + \varepsilon K_1 \exp(-\lambda_1(t-a)) + \varepsilon K_2[\exp(\lambda_2(b-t)) - 1].$$

If ε is small enough, these functions have their graph in F.

First, we prove as previously that $\beta > \alpha$.

Next, repeating the argument in the proof of Theorem 1.1 we have for small values of $\varepsilon > 0$

$$\varepsilon\,\alpha''(t) + f(t, \alpha(t), \alpha'(t)) \geq \varepsilon[v_0''(t, -\varepsilon K_3) + K_2 Q] \geq 0$$

and

$$\varepsilon\,\beta''(t) + f(t, \beta(t), \beta'(t)) \leq 0.$$

At last we compute for $\varepsilon > 0$ small enough

$$a_1\alpha(a) - a_2\alpha'(a) - A = a_1 u_0(a) - a_2 u_0'(a) - A - K_1 a_2\mu + O(\varepsilon) \leq 0,$$
$$b_1\alpha(b) + b_2\alpha'(b) - B = b_1 v_0(b, -\varepsilon K_3) + b_2 v_0'(b, -\varepsilon K_3) - B$$
$$-\varepsilon K_1(b_1 - b_2\lambda_1)e^{-\lambda_1(b-a)} + \varepsilon K_2 b_2 \lambda_2$$
$$= -\varepsilon K_3 \Delta[\tfrac{\partial f}{\partial v}(b, u_0(b), u_0'(b))]^{-1} + \varepsilon K_2 b_2 \tfrac{Q}{\mu} + o(\varepsilon) \leq 0$$

and in a similar way we obtain

$$a_1\beta(a) - a_2\beta'(a) - A \geq 0,$$
$$b_1\beta(b) + b_2\beta'(b) - B \geq 0.$$

Condition (b) implies a one-sided Nagumo condition which together with the boundary conditions implies a bound on $u'(a)$, so that existence of a solution $u_\varepsilon \in [\alpha, \beta]$ follows from Theorem II-3.2.

Part C – Estimate on the derivative u'_ε.

First we deduce from the boundary conditions that

$$u'_\varepsilon(a) - u'_0(a) = O(1).$$

Second, we know that $|u_\varepsilon(a) - u_0(a)| \leq S(t, \varepsilon) = L\varepsilon$ for some $L > 0$, and we deduce as in the proof of Theorem 1.2

$$u'_\varepsilon(t) - u'_0(t) = O(\varepsilon + \exp -\tfrac{\mu}{\varepsilon}(t-a)). \qquad \square$$

Example 1.1. Consider the problem

$$\begin{gathered} \varepsilon u'' + \tfrac{1}{2} - u + \tfrac{1}{2}(t + \tfrac{1}{2})(u' + u'^3) = 0, \\ u'(0) = 0, \ u(1) = 2. \end{gathered}$$

The hypothesis of Theorem 1.3 are satisfied with $u_0(t) = t+1$, $\frac{\partial f}{\partial v}(t, u, v) \geq \frac{1}{4}$ and $\Delta = 3$.

2 Boundary Layers at Both End Points

Consider the problem

$$\begin{gathered} \varepsilon\, u'' + f(t, u) = 0, \\ u(a) = A, \ u(b) = B, \end{gathered} \tag{2.1}$$

where ε is a small positive parameter. The following theorem uses a stability assumption

$$\frac{\partial f}{\partial u}(t, u) \leq -m < 0$$

which produces boundary layers at both end points.

Theorem 2.1. *Let* $u_0 \in C^2([a, b])$ *and let*

$$E = \{(t, u) \mid a \leq t \leq b, \ r_1(t) \leq u - u_0(t) \leq r_2(t)\},$$

where r_1 *and* r_2 *are two continuous functions on* $[a, b]$ *such that for all* $t \in [a, b]$

$$r_1(t) < 0 < r_2(t)$$

and

$$\begin{gathered} u_0(a) + r_1(a) < A < u_0(a) + r_2(a), \\ u_0(b) + r_1(b) < B < u_0(b) + r_2(b). \end{gathered}$$

Consider a function $f \in C^1(E)$ and assume

(a) u_0 is a solution of the reduced problem

$$f(t, u_0(t)) = 0;$$

(b) there exists $m > 0$ such that for all $(t, u) \in E$,

$$\frac{\partial f}{\partial u}(t, u) \leq -m < 0.$$

Then for ε small enough, there exists a solution u_ε of (2.1) such that for any $t \in [a, b]$

$$\begin{aligned}|u_\varepsilon(t) - u_0(t)| \leq\, & |u_0(a) - A| \exp(-\sqrt{\tfrac{m}{\varepsilon}}(t-a)) \\ & +|u_0(b) - B| \exp(-\sqrt{\tfrac{m}{\varepsilon}}(b-t)) + \tfrac{\|u_0''\|_\infty}{m}\varepsilon.\end{aligned}$$

Proof. Consider for example the case where $u_0(a) < A$, $u_0(b) > B$, the other cases are treated in a similar way.

It is easy to see that

$$\alpha(t) = u_0(t) - (u_0(b) - B) \exp(-\sqrt{\tfrac{m}{\varepsilon}}(b-t)) - \tfrac{\|u_0''\|_\infty}{m}\varepsilon$$

is a C^2-lower solution of (2.1),

$$\beta(t) = u_0(t) + (A - u_0(a)) \exp(-\sqrt{\tfrac{m}{\varepsilon}}(t-a)) + \tfrac{\|u_0''\|_\infty \varepsilon}{m}$$

is a C^2-upper solution and $\alpha(t) \leq \beta(t)$. Whence, we can apply Theorem II-1.5. □

Remark Theorem 2.1 implies that u_ε goes to u_0 with $\varepsilon > 0$, uniformly on compact subsets of $]a, b[$.

Boundary layers at both end points can occur in systems with derivative dependence. Consider for example the problem

$$\begin{aligned}&\varepsilon u'' - tu'|u'| - u = 0 \\ &u(-1) = 1, \ u(1) = 2.\end{aligned}$$

Its solution satisfies

$$\lim_{\varepsilon \to 0+} u_\varepsilon(t) = 0$$

if $t \neq \pm 1$, with uniform convergence on compact subsets of $]-1, 1[$. The limiting function $u_0(t) = 0$ is the continuous solution of the reduced equation

$$tu'|u'| + u = 0.$$

Exercise 2.1. Prove the above statements using lower and upper solutions.

3 Angular Solutions

Here we consider a problem

$$\begin{aligned} &\varepsilon u'' + f(t, u, u') = 0, \\ &u(a) = A, \ u(b) = B, \end{aligned} \tag{3.1}$$

and work out conditions so that it has a solution that converges uniformly to a function $u_0 \in \mathcal{C}([a,b])$ which is only piecewise $\mathcal{C}^2$ but satisfies both boundary conditions. These are the so-called *angular solutions.* In this case a boundary layer in the derivative u'_ε takes place at points t_0 where the derivative u'_0 is discontinuous. Notice that boundary layers also appear in the function u_0 but only in terms of higher order with respect to ε. A first result of this type is the following.

Theorem 3.1. *Let $u_0 \in \mathcal{C}([a,b])$ be such that for some $t_0 \in]a,b[$,*

$$u_0 \in \mathcal{C}^2([a,t_0]), \ u_0 \in \mathcal{C}^2([t_0,b]) \quad \text{and} \quad D_l u_0(t_0) < D_r u_0(t_0).$$

Let also E be a neighbourhood of

$$\{(t,u,v) \mid t \in [a,b] \setminus \{t_0\}, \ u = u_0(t), \ v = u'_0(t) \\ \text{or} \ \ t = t_0, \ u = u_0(t_0), \ D_l u_0(t_0) \le v \le D_r u_0(t_0)\}.$$

Consider a function $f \in \mathcal{C}^1(E)$ and assume

(a) u_0 is a solution of the reduced problem

$$f(t, u_0(t), u'_0(t)) = 0, \quad \text{if } t \in [a,b] \setminus \{t_0\}, \\ u_0(a) = A, \quad u_0(b) = B;$$

(b) $\frac{\partial f}{\partial v}(t, u_0(t), D_l u_0(t)) < 0$ if $a \le t \le t_0$,
$\frac{\partial f}{\partial v}(t, u_0(t), D_r u_0(t)) > 0$ if $t_0 \le t \le b$;

(c) $f(t_0, u_0(t_0), v) < 0$ if $D_l u_0(t_0) < v < D_r u_0(t_0)$.

Then, for $\varepsilon > 0$ small enough, the problem (3.1) has at least one solution u_ε such that for any $\delta > 0$,

$$\begin{aligned} &u_\varepsilon(t) = u_0(t) + O(\varepsilon), \quad u'_\varepsilon(t) = u'_0(t) + O(\varepsilon), \ \text{if } t \in [a,b] \setminus [t_0 - \delta, t_0 + \delta], \\ &u_\varepsilon(t) = u_0(t) + O(\varepsilon^{1/2}), \qquad u'_\varepsilon(t) = O(1), \qquad \text{if } t \in [t_0 - \delta, t_0 + \delta]. \end{aligned}$$

Proof. Part 1 – A modified problem. Since $\frac{\partial f}{\partial v}(t, u_0(t), u'_0(t)) \ne 0$, we can extend the restriction of u_0 on $[a, t_0]$ into $u_a \in \mathcal{C}^1([a, t_0 + \delta_0])$ such that $f(t, u_a(t), u'_a(t)) = 0$ and $\delta_0 > 0$. In the same way, restricting δ_0 if necessary, we extend u_0 into $u_b \in \mathcal{C}^1([t_0 - \delta_0, b])$ such that

$$f(t, u_b(t), u'_b(t)) = 0 \quad \text{and} \quad 4\delta_0 < u'_b(t_0) - u'_a(t_0).$$

Next, we consider the sets

$$\begin{aligned}\mathcal{R}_a &= \{(t,u,v) \mid a \le t \le t_0 + \delta_0,\ |u - u_a(t)| \le \delta_0,\ |v - u_a'(t)| \le \delta_0\},\\ \mathcal{R}_b &= \{(t,u,v) \mid t_0 - \delta_0 \le t \le b,\ |u - u_b(t)| \le \delta_0,\ |v - u_b'(t)| \le \delta_0\},\\ \mathcal{R}_{t_0} &= \{(t,u,y) \mid t_0 - \delta_0 \le t \le t_0 + \delta_0,\\ &\qquad |u - u_0(t)| \le \delta_0,\ u_a'(t) + \delta_0 \le v \le u_b'(t) - \delta_0\}.\end{aligned}$$

If $\delta_0 > 0$ is small enough, it is clear from assumptions (b) and (c) that there exist constants $\mu > 0$ and $Q > 0$ such that

$$|\tfrac{\partial f}{\partial u}(t,u,v)| < Q, \quad \text{if } (t,u,v) \in \mathcal{R}_a \cup \mathcal{R}_b;$$
$$\tfrac{\partial f}{\partial v}(t,u,v) < -\mu, \quad \text{if } (t,u,v) \in \mathcal{R}_a; \quad \tfrac{\partial f}{\partial v}(t,u,v) > \mu, \quad \text{if } (t,u,v) \in \mathcal{R}_b;$$
$$f(t,u,v) < -\mu, \quad \text{if } (t,u,v) \in \mathcal{R}_{t_0}.$$

We transform now f outside of $\mathcal{R}_a \cup \mathcal{R}_b \cup \mathcal{R}_{t_0}$ so that the modified function $\bar{f}$ satisfies a Nagumo condition and consider the modified problem

$$\begin{aligned}&\varepsilon\, u'' + \bar{f}(t,u,u') = 0,\\ &u(a) = A,\ u(b) = B.\end{aligned} \tag{3.2}$$

Part 2 – Construction of C^2-lower and upper solutions of (3.2). Let λ_1 be the root of the equation $\varepsilon\lambda^2 - \mu\lambda + Q = 0$ such that

$$\lambda_1 = (\mu - \sqrt{\mu^2 - 4\varepsilon Q})/2\varepsilon = Q/\mu + O(\varepsilon) > 0.$$

and choose a $K > 0$ large enough so that

$$\begin{aligned}QK &> (\exp(\lambda_1(t_0 - a)) - 1) \max_{t\in[a,t_0]} |u_0''(t)|,\\ QK &> (\exp(\lambda_1(b - t_0)) - 1) \max_{t\in[t_0,b]} |u_0''(t)|.\end{aligned}$$

Let also

$$\lambda_2 = \frac{(u_b'(t_0) - u_a'(t_0) - 2\delta_0)}{\varepsilon^{1/2}}$$

which is positive for δ_0 small enough and such that for $\varepsilon > 0$ small enough

$$\varepsilon\lambda_2^2 - \mu\lambda_2 + Q < 0.$$

Define now

$$\begin{aligned}\alpha(t) &= u_0(t) - \varepsilon K \frac{e^{\lambda_1(t-a)} - 1}{e^{\lambda_1(t_0-a)} - 1}, && \text{if } a \le t \le t_0,\\ &= u_0(t) - \varepsilon K \frac{e^{\lambda_1(b-t)} - 1}{e^{\lambda_1(b-t_0)} - 1}, && \text{if } t_0 \le t \le b\end{aligned}$$

and

$$\beta(t) = u_0(t) + \varepsilon K \frac{e^{\lambda_1(t-a)} - 1}{e^{\lambda_1(t_0-a)} - 1} + \varepsilon^{1/2} e^{\lambda_2(t-t_0)}, \qquad \text{if } a \leq t \leq t_0,$$

$$= u_0(t) + \varepsilon K \frac{e^{\lambda_1(b-t)} - 1}{e^{\lambda_1(b-t_0)} - 1} + \varepsilon^{1/2} e^{\lambda_2(t_0-t)}, \qquad \text{if } t_0 \leq t \leq b.$$

We can verify then that, for $\varepsilon > 0$ small enough, these functions are respectively $\mathcal{C}^2$-lower and upper solutions of (3.2), which is a tedious but straightforward calculation. For example, we compute for $t \leq t_0$,

$$\begin{aligned}
&\varepsilon\alpha'' + \bar{f}(t, \alpha, \alpha') \\
&\quad = \varepsilon\alpha'' + \bar{f}(t, u_0, u_0') + \tfrac{\partial \bar{f}}{\partial u}(t, x, y)(\alpha - u_0) + \tfrac{\partial \bar{f}}{\partial v}(t, x, y)(\alpha' - u_0') \\
&\quad \geq \varepsilon\alpha'' + Q(\alpha - u_0) - \mu(\alpha' - u_0') \\
&\quad \geq \varepsilon[-\max_{t\in[a,t_0]} |u_0''(t)| + Q\frac{K}{e^{\lambda_1(t_0-a)} - 1}] > 0,
\end{aligned}$$

where (x, y) is on the segment joining (α, α') and (u_0, u_0'). Similarly, we have to check

$$\varepsilon\beta'' + \bar{f}(t, \beta, \beta') \geq 0$$

separately for $(t, \beta(t), \beta'(t))$ in each of the regions $\mathcal{R}_a$, $\mathcal{R}_b$ and $\mathcal{R}_{t_0}$. We also have to notice that $D^-\beta(t_0) > D_+\beta(t_0)$.

At last, we deduce from Theorem II-1.3 that the modified problem of (3.2) has a solution u_ε with

$$\alpha \leq u_\varepsilon \leq \beta.$$

Part 3 – u_ε solves (3.1). First, we deduce from the definition of α and β that there exists $k > 0$ such that

$$|u_\varepsilon(t) - u_0(t)| = O(\varepsilon + \varepsilon^{1/2} \exp(-\tfrac{k}{\varepsilon^{1/2}}|t - t_0|)). \tag{3.3}$$

To obtain a bound on u_ε', consider the function $z_\varepsilon = u_\varepsilon - u_0$ on $[a, t_0]$. It satisfies an equation of the form

$$\varepsilon z_\varepsilon'' + p_\varepsilon(t) z_\varepsilon' = q_\varepsilon(t),$$

where

$$p_\varepsilon(t) = \tfrac{\partial \bar{f}}{\partial v}(t, u_\varepsilon(t), u_0'(t) + \xi_1(t)(u_\varepsilon'(t) - u_0'(t))),$$

$$q_\varepsilon(t) = -\varepsilon u_0''(t) - \tfrac{\partial \bar{f}}{\partial u}(t, u_0(t) + \xi_2(t)(u_\varepsilon(t) - u_0(t)), u_0'(t)) z_\varepsilon(t),$$

and $\xi_1(t)$, $\xi_2(t) \in [0, 1]$. If

$$(t, u_\varepsilon(t), u_0'(t)) \quad \text{and} \quad (t, u_\varepsilon(t), u_\varepsilon'(t)) \in \mathcal{R}_a$$

we have

$$p_\varepsilon(t) \le -\mu < 0 \quad \text{and} \quad |q_\varepsilon(t)| = O(\varepsilon + \varepsilon^{1/2} \exp(-\tfrac{k}{\varepsilon^{1/2}}|t - t_0|)).$$

We deduce then from (3.3), there exist $k_1 > 0$ and $t_1 \in [t_0 - 2k_1\varepsilon^{1/4}, t_0[$ such that

$$\|z_\varepsilon\|_\infty \le k_1\varepsilon^{1/2} \quad \text{and} \quad |z'_\varepsilon(t_1)| \le \varepsilon^{1/4}.$$

This result implies that, for all $t \in [a, t_1]$ and ε small enough, $|z'_\varepsilon(t)| < \delta_0$. Assume by contradiction there exists $t'_1 \in [a, t_1[$ so that

$$\forall t \in [t'_1, t_1],\ |z'_\varepsilon(t)| < \delta_0 \quad \text{and} \quad |z'_\varepsilon(t'_1)| = \delta_0.$$

Notice that on $[t'_1, t_1]$, $q_\varepsilon(t) = O(\varepsilon)$, so that for ε small enough, we compute

$$|u'_\varepsilon(t) - u'_0(t)| \le \varepsilon^{1/4} \exp(-\mu\tfrac{(t_1-t)}{\varepsilon}) + O(\varepsilon) < \delta_0 \tag{3.4}$$

and come to a contradiction for $t = t'_1$. It follows that (3.4) holds on $[a, t_1]$. In the same way, we have $t_2 \in [t_0, t_0 + 2k_1\varepsilon^{1/4}]$ such that for all $t \in [t_2, b]$,

$$|u'_\varepsilon(t) - u'_0(t)| \le \varepsilon^{1/4} \exp(-\mu\tfrac{(t-t_2)}{\varepsilon}) + O(\varepsilon) < \delta_0. \tag{3.5}$$

For ε small enough, $[t_1, t_2] \subset [t_0 - \delta_0, t_0 + \delta_0]$ and outside $[t_1, t_2]$ the announced bounds on u'_ε follow from (3.4) and (3.5). Let us now consider the interval $[t_1, t_2]$. On this interval

$$|u_\varepsilon(t) - u_0(t)| \le \delta_0$$

and

$$u'_\varepsilon(t_1) \le u'_a(t_1) + \delta_0 \le u'_b(t_1) - \delta_0.$$

Suppose that there exists $\bar{t} \in [t_1, t_2]$ such that

$$u'_\varepsilon(\bar{t}) \ge u'_b(\bar{t}) + \delta_0. \tag{3.6}$$

Then we have $\bar{t}_2 \in [t_1, \bar{t}]$ such that

$$|u'_\varepsilon(\bar{t}_2) - u'_b(\bar{t}_2)| \le \tfrac{1}{2}\delta_0.$$

Hence, using the above argument

$$|u'_\varepsilon(t) - u'_b(t)| \le \tfrac{1}{2}\delta_0 \exp(-\mu\tfrac{(t-\bar{t}_2)}{\varepsilon}) + O(\varepsilon^{1/2}) < \delta_0$$

holds for any $t \ge \bar{t}_2$ which contradicts (3.6). It follows that for all $t \ge t_1$

$$u'_\varepsilon(t) < u'_b(t) + \delta_0.$$

Similarly we prove that for all $t \leq t_2$

$$u'_\varepsilon(t) > u'_a(t) - \delta_0.$$

Hence, we have $(t, u_\varepsilon(t), u'_\varepsilon(t)) \in \mathcal{R}_a \cup \mathcal{R}_b \cup \mathcal{R}_{t_0}$ for all $t \in [a, b]$, which implies that u_ε is a solution of (3.1) and that the required asymptotic estimates hold. □

Example 3.1. Theorem 3.1 applies to the boundary value problem

$$\begin{gathered} \varepsilon\, u'' + u'^2 - 1 = 0, \\ u(-1) = 1, \;\; u(1) = 1. \end{gathered}$$

For $\varepsilon > 0$ small enough, this problem has a solution u_ε such that for any $\delta > 0$,

$$\begin{aligned} &u_\varepsilon(t) = |t| + O(\varepsilon), \quad u'_\varepsilon(t) = \operatorname{sgn}(t) + O(\varepsilon), \text{ if } t \in [-1,1] \setminus [t_0 - \delta, t_0 + \delta], \\ &u_\varepsilon(t) = |t| + O(\varepsilon^{1/2}), \qquad u'_\varepsilon(t) = O(1), \qquad \text{if } t \in [t_0 - \delta, t_0 + \delta]. \end{aligned}$$

An alternative result using a Nagumo condition can be worked out.

Theorem 3.2. *Let $u_0 \in \mathcal{C}([a, b])$ be such that for some $t_0 \in]a, b[$,*

$$u_0 \in \mathcal{C}^2([a, t_0]), \; u_0 \in \mathcal{C}^2([t_0, b]) \quad \textit{and} \quad D_l u_0(t_0) < D_r u_0(t_0).$$

Define

$$E = \{(t, u, v) \in [a, b] \times \mathbb{R}^2 \mid |u - u_0(t)| \leq \delta_0\}$$

for some $\delta_0 > 0$ and let F be a neighbourhood of

$$\begin{aligned} \{(t, u, v) \mid t \in [a, b] \setminus \{t_0\},\; & u = u_0(t),\; v = u'_0(t) \\ \textit{or}\;\; t = t_0,\; & u = u_0(t_0),\; D_l u_0(t_0) \leq v \leq D_r u_0(t_0)\}. \end{aligned}$$

Consider a function $f \in \mathcal{C}^1(E)$ that satisfies the Nagumo condition (1.2) *together with* (1.3). *Assume*

(a) u_0 is a solution of the reduced problem

$$\begin{gathered} f(t, u_0(t), u'_0(t)) = 0, \quad \textit{if } t \in [a, b] \setminus \{t_0\}, \\ u_0(a) = A, \quad u_0(b) = B; \end{gathered}$$

(b) for all $(t, u, v) \in F$,
$\frac{\partial f}{\partial v}(t, u, v) \leq 0$ *if* $a \leq t < t_0$,
$\frac{\partial f}{\partial v}(t, u, v) \geq 0$ *if* $t_0 < t \leq b$;

(c) for all $(t, u, v) \in F$, $\frac{\partial f}{\partial u}(t, u, v) \leq -m < 0$.

Then, for $\varepsilon > 0$ *small enough, the problem* (3.1) *has at least one solution* u_ε *such that for any* $\delta > 0$,

$$\begin{aligned} u_\varepsilon(t) &= u_0(t) + O(\varepsilon), & \textit{if } t \in [a,b] \setminus [t_0-\delta, t_0+\delta], \\ u_\varepsilon(t) &= u_0(t) + O(\varepsilon^{1/2}), & \textit{if } t \in [t_0-\delta, t_0+\delta]. \end{aligned}$$

Sketch of the proof : Let

$$M > m^{-1} \sup\{|u_0''(t)| \mid t \neq t_0\} \quad \text{and} \quad N = \frac{D_r u_0(t_0) - D_l u_0(t_0)}{m^{1/2}}.$$

Then the functions

$$\begin{aligned} \alpha(t) &= u_0(t) - \varepsilon M, \\ \beta(t) &= u_0(t) + \varepsilon M + \varepsilon^{1/2} N \exp(-\sqrt{\tfrac{m}{\varepsilon}}|t-t_0|) \end{aligned}$$

are respectively lower and upper solutions of (3.1). □

Example 3.2. Consider the problem

$$\begin{gathered} \varepsilon u'' + \tfrac{1}{2} - u + \tfrac{1}{2}(t - \tfrac{1}{2})(u' + u'^2) = 0, \\ u(0) = \tfrac{1}{2}, \ u(1) = 1. \end{gathered}$$

The reduced problem has the angular solution

$$\begin{aligned} u_0(t) &= \tfrac{1}{2}, & \text{if } t \le \tfrac{1}{2}, \\ &= t, & \text{if } t > \tfrac{1}{2}. \end{aligned}$$

We can apply Theorem 3.2 if we choose

$$F = \{(t,u,v) \mid 0 \le t \le 1, 0 \le u \le \tfrac{3}{2}, -\tfrac{1}{2} \le v \le \tfrac{3}{2}\}.$$

Further examples with interior layers or several angles can be found in Section II-1.3.

Exercise 3.1. Adapt Theorem 3.2 to deal with the case

$$D_l u_0(t_0) > D_r u_0(t_0).$$

Hint : See [169].

4 Algebraic Boundary Layers

Another type of singular perturbation takes place in problem

$$\begin{gathered}(t-a+\varepsilon)^2 u'' + f(t,u) = 0,\\ u(a) = A,\ u(b) = B,\end{gathered} \tag{4.1}$$

where ε is a small positive parameter. The main feature in this case is that the reduced problem

$$\begin{gathered}(t-a)^2 u'' + f(t,u) = 0,\\ u(b) = B,\end{gathered} \tag{4.2}$$

is a singular second order equation rather than an equation of lower order. Here the boundary layer turns out to be an $0((\frac{\varepsilon}{t-a+\varepsilon})^k)$. Such a boundary layer is called *algebraic.*

Theorem 4.1. *Let $u_0 \in C^2([a,b])$ and let*

$$E = \{(t,u) \mid a \le t \le b,\ r_1(t) \le u - u_0(t) \le r_2(t)\},$$

where r_1 and r_2 are two continuous functions on $[a,b]$ such that for all $t \in [a,b]$

$$r_1(t) < 0 < r_2(t)$$

and

$$u_0(a) + r_1(a) < A < u_0(a) + r_2(a).$$

Consider a function $f \in C^1(E)$ and assume

(a) u_0 is a solution of the reduced problem (4.2);
(b) there exists an $m > 0$ such that for all $(t,u) \in E$,

$$\frac{\partial f}{\partial u}(t,u) \le -m < 0.$$

Then for ε small enough, there exists a solution u_ε of (4.1) such that for any $t \in [a,b]$

$$|u_\varepsilon(t) - u_0(t)| \le \tfrac{3}{m}\|u_0''\|_\infty \varepsilon + |u_0(0) - A|(\tfrac{\varepsilon}{t-a+\varepsilon})^k,$$

where $k > 0$ satisfies $k(k+1) \le m$.

Proof. Consider the case where $u_0(a) \ge A$, the other one is similar.

Let $k > 0$ satisfies $k(k+1) \le m$ and

$$\alpha(t) = u_0(t) - \tfrac{(2(b-a)+1)\|u_0''\|_\infty}{m}\varepsilon - (\tfrac{\varepsilon}{t-a+\varepsilon})^k(u_0(a) - A).$$

Observe that there exists some $\theta(t) \in]0,1[$ such that

$$\begin{aligned}
&(t-a+\varepsilon)^2\alpha''(t) + f(t,\alpha(t)) \\
&\quad = (t-a+\varepsilon)^2\, u_0''(t) + f(t,u_0(t)) \\
&\qquad -[k(k+1) + \tfrac{\partial f}{\partial u}(t,u_0(t)+\theta(t)(\alpha(t)-u_0(t)))](\tfrac{\varepsilon}{t-a+\varepsilon})^k(u_0(a)-A) \\
&\qquad -\tfrac{\partial f}{\partial u}(t,u_0(t)+\theta(t)(\alpha(t)-u_0(t)))\tfrac{(2(b-a)+1)\|u_0''\|_\infty}{m}\,\varepsilon \\
&\quad \geq \varepsilon[(2(t-a)+\varepsilon)u_0''(t) + (2(b-a)+1)\|u_0''\|_\infty] \geq 0.
\end{aligned}$$

This show that α is a C^2-lower solution. In the same way, let

$$\beta(t) = u_0(t) + \tfrac{(2(b-a)+1)\|u_0''\|_\infty}{m}\,\varepsilon$$

and observe that, for some $\theta(t) \in]0,1[$,

$$\begin{aligned}
&(t-a+\varepsilon)^2\,\beta''(t) + f(t,\beta(t)) \\
&\quad = (t-a+\varepsilon)^2\, u_0''(t) + f(t,u_0(t)) \\
&\qquad +\tfrac{\partial f}{\partial u}(t,u_0(t)+\theta(t)(\beta(t)-u_0(t)))\tfrac{(2(b-a)+1)\|u_0''\|_\infty}{m}\,\varepsilon \\
&\quad \leq \varepsilon[(2(t-a)+\varepsilon)u_0''(t) - (2(b-a)+1)\|u_0''\|_\infty] \leq 0.
\end{aligned}$$

Hence, β is a C^2-upper solution. As further $\alpha < \beta$ the proof follows from Theorem II-1.5. □

Example 4.1. Consider the problem

$$\begin{gathered}
(t+\varepsilon)^2\, u'' - u(1+e^u) = 0, \\
u(0) = 1,\ u(1) = 0.
\end{gathered}$$

The reduced problem has the solution

$$u_0(t) = 0$$

and

$$\frac{\partial f}{\partial u}(t,u) = -[1+(1+u)e^u] \leq -[1-e^{-2}] < 0.$$

Hence Theorem 4.1 applies and there is a solution u_ε such that

$$|u_\varepsilon(t)| \leq (\tfrac{\varepsilon}{t+\varepsilon})^k,$$

with $k = \frac{1}{2}(\sqrt{5-4e^{-2}} - 1)$.

Algebraic boundary layers can appear in other problems. For example, the following theorem deals with angular solutions. Consider the problem

$$\begin{aligned} (|t|+\varepsilon)^2\, u'' + f(t,u) &= 0, \\ u(a) = A,\ u(b) &= B. \end{aligned} \tag{4.3}$$

Theorem 4.2. *Let $a < 0 < b$, $u_0 \in \mathcal{C}([a,b])$ and let*

$$E = \{(t,u) \mid a \le t \le b,\ r_1(t) \le u - u_0(t) \le r_2(t)\},$$

where r_1 and r_2 are two continuous functions on $[a,b]$ such that for all $t \in [a,b]$

$$r_1(t) < 0 < r_2(t).$$

Consider a function $f \in \mathcal{C}^1(E)$ and assume

(a) the function u_0 is a solution of the reduced problem

$$\begin{aligned} t^2\, u'' + f(t,u) = 0, \quad t \neq 0, \\ u(a) = A,\ u(b) = B, \end{aligned}$$

such that $u_0 \in \mathcal{C}([a,b])$, $u_0 \in \mathcal{C}^2([a,0])$, $u_0 \in \mathcal{C}^2([0,b])$ and $D_l u_0(0) \le D_r u_0(0)$;

(b) there exists an $m > 0$ such that for all $(t,u) \in E$,

$$\frac{\partial f}{\partial u}(t,u) \le -m < 0.$$

Then there exists a $K > 0$ so that for ε small enough, there exists a solution u_ε of (4.3) such that for any $t \in [a,b]$

$$|u_\varepsilon(t) - u_0(t)| \le K\varepsilon.$$

Proof. Define $k > 0$ to be such that $k(k+1) = m$. The proof uses then the functions

$$\alpha(t) = u_0(t) - N\varepsilon$$

and

$$\beta(t) = u_0(t) + N\varepsilon + L\tfrac{\varepsilon^{k+1}}{(|t|+\varepsilon)^k}.$$

It is easy to see that if $L > 0$ and $N > 0$ are large enough these functions are lower and upper solutions. Hence Theorem II-1.5 applies. □

Example 4.2. Consider the problem

$$\begin{aligned} (|t|+\varepsilon)^2\, u'' + 2(|t| - u) &= 0, \\ u(-1) = 1,\ u(1) &= 1. \end{aligned}$$

The reduced problem has the solution

$$u_0(t) = |t|$$

and

$$\frac{\partial f}{\partial u}(t, u) = -2 < 0.$$

Hence Theorem 4.2 applies.

Chapter XI

Bibliographical Notes

1 Notes on Chapters 1 and 2

In the two first chapters, we present several notions of lower and upper solutions. Definitions I-1.1, or the equivalent for other boundary value problems, are the simplest ones to use in practical problems. These can be traced back as early as 1931 in G. Scorza Dragoni [281] and are often used in a first presentation of the method such as in P.B. Bailey, L.F. Shampine, and P.E. Waltman [26], R.E. Gaines and J. Mawhin [123], S. Fučík [121], L.C. Piccinini, G. Stampacchia and G. Vidossich [255], J. Mawhin [213]. Definitions I-2.1, I-5.1 and II-1.1 of $\mathcal{C}^2$-lower and upper solutions are inspired by M. Nagumo [228] and K. Akô [6]. A first order condition, similar to $D_-\alpha(t_0) < D^+\alpha(t_0)$, appears in H. Knobloch [191]. The $W^{2,1}$-lower and upper solutions introduced in Definitions I-3.1, I-6.1 and II-2.1 can be found in C. De Coster [85] (see also [87, 88, 89, 91, 92]). In Section II-5, we present straightforward modifications of Definitions I-1.1 in order to deal with the other boundary value problems we consider in this section. A variety of different, but strongly related, notions of lower and upper solutions can be found in the literature, see for example H. Epheser [108], L.K. Jackson [173], I.T. Kiguradze [185] (see also [187]), V.V. Gudkov and A.Ja. Lepin [139], P. Habets and M. Laloy [144], P. Hess [159] (see also G. Stampacchia[292]), A. Adje [2], C. Fabry and P. Habets [114], P. Habets and L. Sanchez [149].

For the Dirichlet problem, the result corresponding to Theorem I-1.1 is due to G. Scorza-Dragoni [280, 281]. For the periodic case, this theorem follows from the work of H.W. Knobloch [191]. Theorem I-5.3 is due to J. Mawhin [213]. The ideas of the proof of Theorem I-6.1 can be found in J. Deuel and P. Hess [105], P. Hess [160], T. Kura [197] and M.X. Wang,

J.J. Nieto and A. Cabada [305] while Theorems I-6.6 and II-2.1 are due to C. De Coster [86] for the Dirichlet problem and C. De Coster and P. Habets [92] for the periodic problem using ideas of H. Epheser [108] and I.T. Kiguradze and B.L. Shekhter [187]. Other interesting but more complicated proofs which require α, $\beta \in W^{1,\infty}(a,b)$ can be found in H. Epheser [108] and I.T. Kiguradze [185]. Theorem I-6.8 extends [152] while Theorem I-6.9 is due to [41].

Theorem I-3.2, I-6.10, II-2.5 can be found in C. De Coster and P. Omari [94] for the parabolic problem and in C. De Coster and P. Habets [92] for the ODE case using ideas of G.M. Troianiello [301].

Existence of extremal solutions as in Theorems I-2.4, I-5.6, II-1.6 were studied by G. Peano [244] and O. Perron [246] for first order ODE, by K. Akô [5] for elliptic PDE, and W. Mlak [221] for parabolic problems. The proof used here is inspired by the recent paper of J.A. Cid [68]. For the classical proof, defining maximal solutions as maximum of lower solutions, and in case solutions are $\mathcal{C}^2$, we can mention K. Akô [5] or K. Schmitt [275].

For what concerns the structure of the solution set in case of monotonicity, i.e. Theorem I-2.5, I-5.7, we refer to T. Satō [267] and I. Hirai and K. Akô [163].

The Nagumo condition given in Proposition I-4.4, is due to M. Nagumo [224] in case

$$\int_0^\infty \frac{s\,ds}{\varphi(s)} = +\infty$$

and the condition

$$\int_r^\infty \frac{s\,ds}{\varphi(s)} > \max_t \beta(t) - \min_t \alpha(t),$$

where $r = \max\{\frac{\bar\beta(b)-\bar\alpha(a)}{b-a}, \frac{\bar\beta(a)-\bar\alpha(b)}{b-a}\}$ can be found in K. Akô [7]. Propositions I-4.5, I-4.6 concerning one-sided Nagumo conditions are due to I.T. Kiguradze [183] and K. Schrader [276]. Propositions I-4.7, I-4.8 and I-4.9 on the generalization to the Carathéodory case appear in I.T. Kiguradze [183]. Variants of these results can be found in [108, 184, 185]. The first counter-example showing that the Nagumo condition cannot be deleted is due to M. Nagumo [228] for the Dirichlet case. Example I-4.1 for the periodic case is due to P. Habets and R. Pouso [148].

The study of Dirichlet problems with singular nonlinearities presented in Section II-4 was initiated by A. Rosenblatt in 1933 [262]. Theorems II-4.2, II-4.3, and II-4.4 continue the work of G. Prodi [256] who used, in 1953, lower and upper solutions to deal with a singular parabolic problem. These theorems come in the continuous case and with an alternative definition of lower and upper solutions from P. Habets and F. Zanolin [153, 154]. We refer to C. De Coster, M. R. Grossinho, and P. Habets [87] for the Carathéodory case. Theorems II-4.6 and II-4.7 are due to C. De Coster [86].

Proposition I-3.3 extends a result of L. Sanchez [265]. Proposition II-1.7 improves a result of A. Granas, R.B. Guenther, and J. W. Lee [135] while Proposition II-2.9 comes from P. Habets and P. Omari [145].

In Section II-5 which concerns other boundary value problems, Theorem II-5.2 is due to F.I. Njoku, P. Omari, and F. Zanolin [230]; Theorem II-5.3 is related to L.J. Grimm and K. Schmitt [136, 137]; Theorem II-5.4 comes from V.C. Hutson [171]. The periodic problem has been studied in K. Schmitt [273]. Theorem II-5.5 refers to D. Dunninger [107]. Theorems II-5.6 and II-5.7 are due to K. Schmitt [270] and Theorem II-5.8 can be found in C. De Coster and P. Habets [90].

Section II-6 deals with extensions to PDE. Theorem II-6.3 and Theorem II-6.5 are particular cases of Theorem II-6.6 that comes from [93]. The proof of Theorem II-6.5 given here is slightly different from the one of [93]. Theorem II-6.7 is closely related, with another proof, to the result of [159].

2 Notes on Chapter 3

The idea to associate a degree to a pair of strict lower and upper solutions goes back to H. Amann [14, 15] (see also [67]) in 1972. The first results of this type had supposed the nonlinearity to be continuous. In 1994, C. De Coster [84] (see also [87, 145, 88, 31]), studied a Dirichlet problem with a nonlinearity satisfying Carathéodory conditions but independent of u'. A study of the derivative dependent case for a Rayleigh equation has appeared in P. Habets and P. Torres [152], for the Liénard equation in D. Bonheure and C. De Coster [41] and for the general case in C. De Coster and P. Habets [91, 92].

Section III-1 concerns the periodic problem. Propositions III-1.3, III-1.4 are taken from C. De Coster and P. Habets [92]. Some of the ideas of these results can already be found in [84]. Propositions III-1.5, III-1.6 and III-1.7 appear in C. De Coster and P. Habets [92]. Theorem III-1.11 extends P. Habets and P. Torres [152] and Theorem III-1.12 is due to D. Bonheure and C. De Coster [41].

The Dirichlet problem is studied in Section III-2. Propositions III-2.3, III-2.4 III-2.5 and III-2.6 come from C. De Coster and P. Habets [88]. Propositions III-2.7 is taken from M. Cherpion, C. De Coster, and P. Habets [64].

Results similar to Theorems III-1.8 and III-2.8 can be found in different forms and for several other problems in [84, 88, 31, 91, 92].

The abstract idea of the Three Solutions Theorem – Theorems III-1.13 and III-2.11 – goes back to Y.S. Kolesov [194] in 1970 and was extended by H. Amann [13] in 1971. A degree theoretical proof appears in H. Amann [15] (see also [17]). The first proof of this result in the derivative dependent case had been made by H. Amann [16] via parabolic problems and by H. Amann and M. Crandall [20] using a degree argument. Extension of the order relations between the two pairs of lower and upper solutions had also been given by Bongsoo Ko [192]. The results presented here come from C. De Coster and P. Habets [92] in the derivative independent case (see also [31]).

The second type of multiplicity result – Theorems III-1.14, III-1.15 and III-2.12 – goes back to K.J. Brown and H. Budin [47] and a more precise study of it can be found in C. De Coster [84, 88, 31]. Theorem III-2.10 comes from C. De Coster, M. R. Grossinho, and P. Habets [87] with another definition of strict lower and upper solutions. The ideas of Theorem III-2.13 can be found in this last paper.

The observation that the nonlinearity has to be "on the left of the first eigenvalue" in case the lower solution is smaller than the upper one comes from J.L. Kazdan and F.W. Warner [177]. Theorem III-3.1 is related to I. Rachůnková and M. Tvrdý [260]. Theorems III-3.2 and III-3.6 are due to D. Bonheure and C. De Coster [41]. Theorem III-3.3 comes from C. De Coster and M. Tarallo [95] and is related to C. De Coster and M. Henrard [93] where an elliptic version of Theorems III-3.9 and III-3.10 are developed. Exercise III-3.2 can be found in C. De Coster and M. Tarallo [95]. We refer also to C. De Coster and P. Omari [94] for other related results. Theorem III-3.7 is due to A. Cabada and L. Sanchez [53].

3 Notes on Chapter 4

Existence of a minimum of the related functional had first been proved independently by K.C. Chang [58, 59] and D.G. de Figueiredo and S. Solimini [102]. Theorem IV-1.3 is due to P. Omari and F. Zanolin [239] and Problem IV-1.3 to P. Habets and P. Omari [146].

Theorems IV-2.1 and IV-2.2 which relates minimum in the C_0^1-topology and in the H_0^1-one is based on H. Brezis and L. Nirenberg [44]. Theorem IV-2.4 is related to K.J. Brown and H. Budin [48] (see also [161, 98, 202]).

The idea to use invariant sets of the minus gradient flow in C^1 as in Section IV-3 seems to go back to K.C. Chang [58, 59]. Developments of this idea can be found in [28, 60, 82, 164, 200, 203, 204, 92]. A different point of view, defining the flow in H^1, had been used by M. Conti, L. Merizzi, and S. Terracini [74]. The result presented here are extensions of C. De Coster and P. Habets [92]. Theorem IV-3.5 presents an alternative to the result of M. Conti, L. Merizzi, and S. Terracini [74]. Theorems IV-3.7, IV-3.8 and IV-3.9 are classical results and can be found for example (with another proof) in H. Höfer [164].

E. Serra and M. Tarallo [287, 288] obtained lower and upper solutions from the real function

$$\varphi(\xi) = \min_{\bar{u}=\xi} J(u),$$

where J is the functional corresponding to the boundary value problem at hand. This method is developed in Section IV-4 whose content can be found in C. De Coster and M. Tarallo [95] except for Theorems IV-4.10 and IV-4.11 that come from E. Serra and M. Tarallo [288]. Theorems IV-4.8 and IV-4.9 answer and extend a conjecture of J. Mawhin [212]. The results of this section use weak lower and upper solutions. This notion can already be found in P. Hess [159] and J. Deuel and P. Hess [105] (see also G. Stampacchia [292]).

4 Notes on Chapter 5

Section V-1 concerns the abstract formulation of monotone schemes. We present in that section variations of the abstract results in L. Kantorovich [175] and E. Zeidler [309].

Theorems V-2.2, V-2.6, V-2.9 and Corollary V-2.3 come from M. Cherpion, C. De Coster, and P. Habets [65] following an idea of G.V. Gendzhoyan [127] while Theorems V-2.4 and V-2.7 are due to P. Omari [234].

Monotone iterative methods for problems with lower and upper solutions in the reversed order, i.e. $\alpha \geq \beta$ are considered in Section V-3. Theorem V-3.1, which concerns periodic solutions, comes from P. Omari and M. Trombetta [236]. Theorem V-3.2 refers to the same problem with derivative dependence of the nonlinearity and comes from M. Cherpion, C. De Coster, and P. Habets [65]. Theorem V-3.3 is an adaptation for the periodic problem of a result of A. Cabada, P. Habets, and S. Lois [51] for the Neumann problem. It can be found in M. Cherpion [63]. The corresponding results for the Neumann problem i.e. Theorems V-3.4, V-3.5 and V-3.6 are respectively due to A. Cabada and L. Sanchez [53], M. Cherpion, C. De Coster, and P. Habets [65] and A. Cabada, P. Habets, and S. Lois [51].

The results of Section V-4 come from M. Cherpion, C. De Coster, and P. Habets [64]. Proposition V-4.1 is inspired by the first mémoire of E. Picard [249]; Theorem V-4.2 is inspired by D. Guo [140]; Theorem V-4.3 improves V. Šeda [285]; and Theorem V-4.4 improves M. Khavanin and V. Lakshmikantham [181].

5 Notes on Chapter 6

As already observed, except for Theorem VI-1.4, Section VI-1 is an extension to the Carathéodory case of the work of C. Fabry, J. Mawhin, and M.N. Nkashama [115]. In this paper, the authors also give some results for the more general case

$$\begin{gathered} u'' + f(t, u, u') = s, \\ u(a) = u(b), \; u'(a) = u'(b). \end{gathered}$$

Theorem VI-1.4 extends to the periodic problem a result of H. Berestycki [32].

Except for Proposition VI-2.5, Section VI-2 comes from P. Habets and P. Torres [152]. Proposition VI-2.5 extends a result that can be found in J. Mawhin [213].

Section VI-3 is due to C. De Coster and P. Habets [88]. A different point of view was used by G. Harris in [155], who considers a problem with a parameter in the boundary conditions.

Theorem VI-4.2 and VI-4.3 are due to C. De Coster and M. Henrard [93] for an elliptic PDE. They improved a result of A. Ambrosetti and G. Mancini [23] although some previous steps are due to J.L. Kazdan and F.W. Warner [177]. Using lower and upper solutions, these authors consider Theorem VI-4.2 in a case where the nonlinearity is such that $\lim_{|u|\to\infty} f(t, u) = 0$.

Theorem VI-4.4 is due to P. Habets and L. Sanchez [150] with another proof. A proof via lower and upper solutions can be found in C. De Coster and M. Henrard [93]. Proposition VI-4.6 comes from P. Habets and L. Sanchez [150] and Proposition VI-4.7 can be found in A. Cañada and P. Drábek [54]. Other results concerning systems with a nonlinearity that depends only on the derivative appear in A. Cañada and P. Drábek [54] and J. Mawhin [216] where other boundary conditions are also considered.

6 Notes on Chapter 7

Some first important results concerning non-resonance problems are due to C.L. Dolph [106] and those on resonance problems to E.M. Landesman and A.C. Lazer [198] who introduced the so-called Landesman-Lazer condition.

J.L. Kazdan and F.W. Warner [177] in 1975 seem to be the first ones to use lower and upper solutions to consider such problems at the left of the first eigenvalue. Extensions to nonlinearities between the two first eigenvalues had been considered mainly by J.-P. Gossez and P. Omari [132] and P. Habets and P. Omari [145] with the help of non-ordered lower and upper solutions.

Theorem VII-1.1 improves a result that can be found in J. Mawhin [210] which adapts to the periodic case a result of J.L. Kazdan and F.W. Warner [177] while Theorem VII-1.2 is due to J.P. Gossez [129] with another proof. A proof using lower and upper solutions can be found in C. De Coster [85]. Theorem VII-1.3 improves J. Mawhin and J.R. Ward [218].

Theorem VII-2.1 is an adaptation to the periodic problem of J.L. Kazdan and F.W. Warner [177] that can be found in C. De Coster [85]. Theorem VII-2.3 is due to J.P. Gossez [129].

Theorem VII-3.1 can be found in C. De Coster [85] and extends J.L. Kazdan and F.W. Warner [177]. Theorem VII-3.2 is due to M.L. Fernandes, P. Omari, and F. Zanolin [116] with another proof and to A. Fonda, J.P. Gossez, and F. Zanolin [119] for a proof based on lower and upper solutions. Theorem VII-3.3 is an adaptation of J. Mawhin and J.R. Ward [218]. Theorem VII-3.4 is a variation of M. Cuesta, J.P. Gossez, and P. Omari [79] and Theorem VII-3.5 can be found in C. De Coster [85] and adapts J. Mawhin [210].

Proposition VII-4.1 and Theorem VII-4.2 come from J.L. Kazdan and F.W. Warner [177] and Theorem VII-4.4 from J.P. Gossez and P. Omari [132]. All these results were written for elliptic PDE.

Theorem VII-5.2 is due to D.G. de Figueiredo and W.N. Ni [101] in the bounded case and had been extended to a class of unbounded nonlinearities by R. Iannacci and M.N. Nkashama [172]. A proof of this result by lower and upper solution appears in J.P. Gossez and P. Omari [132]. Theorem VII-5.3 comes from C. De Coster and M. Henrard [93] for an elliptic PDE. It improves a result of A. Ambrosetti and G. Mancini [23].

Theorem VII-6.1 extends a result of L. Aguinaldo and K. Schmitt [3]. The use of lower and upper solutions simplifies the proof quite a lot. Theorem VII-6.2 is due to J.L. Kazdan and F.W. Warner [177] although some ideas already appears in K. Klingelhöfer [189]. Theorem VII-6.3 can be found in C. De Coster [85] and is an adaptation to the Dirichlet problem of a result of N. Nieto [229] and V. Šeda [286].

Theorem VII-7.1 generalizes H. Epheser [108] and R.E. Gaines [122] (see also [123]). Theorem VII-7.2 is due to J.B.M. Xavier [308] for elliptic PDE. Theorem VII-7.3 extends in case of ODE's a result of J.K. Kazdan and R.J. Kramer [176]. The corresponding result between the two first eigenvalues can be found in C. De Coster and M. Henrard [93]. Theorem VII-7.4 extends R.E. Gaines [122] and can be found in C. De Coster [85]. Such a result already appears in M. Nagumo [226] (see also K. Akô [7]).

7 Notes on Chapter 8

The results concerning the sublinear, superlinear, sub-super and super-sublinear cases presented in Subsections VIII-1.1, VIII-1.2, VIII-2.1, VIII-2.2 are closely related to K. Ben-Naoum and C. De Coster [31] (see also C. De Coster [84] for the sub-superlinear case) except Theorem VIII-1.9 which is related to C. De Coster and M. Henrard [93].

Subsection VIII-1.3 is concerned with the semipositone problem and present results related to A. Castro, J.B. Garner, and R. Shivaji [57].

The results of Subsection VIII-1.3 on problems with indefinite weight are due to M. Gaudenzi, P. Habets, and F. Zanolin [126].

About the use of variational equations, Theorem VIII-2.9 is due to K. Schmitt [274] with another proof and Theorem VIII-2.10 is due to H. Amann [14, 15]. See also [275].

Except for Theorem VIII-3.7, Section VIII-3 is due to M. Gaudenzi and P. Habets [125]. Theorem VIII-3.7 extend in the particular case of ODE a result of P. Rabinowitz [258] with an other proof.

8 Notes on Chapter 9

Theorem IX-1.1 concerns attractive forces for the periodic problem. It improves P. Habets and L. Sanchez [149]. The repulsive case was much more studied, see for example [149, 104, 117, 237, 310, 261, 41, 208]. Theorem IX-1.2 is due to D. Bonheure and C. De Coster [41], Theorem IX-1.3 can be found in I. Rachůnková, M. Tvrdý, and I. Vrkoč [261], Theorem IX-1.4 comes from D. Bonheure and C. De Coster [41] and Theorem IX-1.5 is related to P. Habets and L. Sanchez [149] although some ideas appears already in M. Nagumo [227].

Concerning the Dirichlet problem, we present in Section IX-2 the results of P. Habets and F. Zanolin [154]. The result of Section IX-3 is due to C. De Coster [86]. Section IX-4 is devoted to multiplicity results that can be found in C. De Coster, M.R. Grossinho, and P. Habets [87].

9 Notes on Chapter 10

The results of Section X-1 and X-2 come from N.I. Briš [46]. We refer to P. Habets and M. Laloy [144] for a detailed proof. See also Y.P. Boglaev [40] and F.A. Howes [167] for more results in this direction. Theorem X-3.1 is due to S. Haber and N. Levinson [143]. Theorems X-3.1 and X-3.2 are related to F.A. Howes [169] who uses the lower and upper solutions method. For Theorem X-4.1, we refer to F.A. Howes [168] or C. De Coster [85].

10 Notes on the Appendix

Except for Theorem A-1.7, the results on degree theory presented here are classical and can be found for example in [263, 205, 103].

The variational methods are another of the basic tools in nonlinear analysis. We refer to [259, 214, 219, 99, 294, 60] for a general overview of the subject as well as the particular references cited in the Appendix.

The spectral theory can be found in several textbook. Here for the abstract formulation, we use the presentation of [307] (see also [83, vol 5]). The proof of the positivity of the first eigenfunction presented here can be found in [96]. An alternative consists to apply the Krein-Rutman theorem. The regularity results for the eigenfunctions of the Dirichlet problem with singularities as presented in Section A-3.4 comes from C. De Coster, M.R. Grossinho, and P. Habets [87] (see also I.T. Kiguradze [186]).

Proposition A-4.1 can be found in N. Rouche and J. Mawhin [263, vol. 2]. Proposition A-4.3 is used in P. Habets and P. Torres [152]. Proposition A-4.5 comes from J. Mawhin [211]. Its extension to the singular Dirichlet case as in Proposition A-4.4 appears in C. De Coster, M.R. Grossinho, and P. Habets [87]. Proposition A-4.6 is the parallel for the Dirichlet problem of a result that can be found in J. Mawhin [210].

The maximum principle is by now classical. A good account of it can be found in M.H. Protter and H.F. Weinberger [257]. The results presented here rely also on the Krein-Rutman theorem.

The anti-maximum principle was first clarified with the works of P. Clément and L.A. Peletier [70] and P. Hess [162]. The results presented here are due to A. Cabada, P. Habets, and S. Lois [51] for the Neumann case and the extension for the periodic problem can be found in M. Cherpion [63].

Appendix

1 Degree Theory

Definition 1.1. *Let X and Y be real normed vector spaces. The operator $T : E \subset Y \to X$ is said to be* completely continuous *if*
(a) T is continuous;
(b) for all bounded set $B \subset E$, $T(B)$ is relatively compact, i.e. $\overline{T(B)}$ is compact.

In case $X = \mathcal{C}([a,b],\mathbb{R})$, the Arzelá-Ascoli Theorem gives a relative compactness criterion.

Theorem 1.1. (Arzelá-Ascoli Theorem) *Let $I \subset \mathbb{R}$ be a closed bounded interval. A set $A \subset \mathcal{C}(I,\mathbb{R})$ is relatively compact if*
(a) A is equibounded, i.e. $\forall t \in I$, $\exists K > 0$, $\forall u \in A$, $|u(t)| < K$;
(b) A is equicontinuous, i.e. $\forall t_0 \in I$, $\forall \epsilon > 0$, $\exists \delta > 0$, $\forall u \in A$, $\forall t \in I$, $|t - t_0| < \delta$ implies $|u(t) - u(t_0)| < \epsilon$.

Proof. See [255, p.136].

Definition 1.2. (Leray-Schauder degree) *Let X be a real normed vector space and I be the identity on X. A* degree *is an application, which associates to any open bounded set $\Omega \subset X$ and any completely continuous operator $T : \Omega \to X$ such that*

$$\forall u \in \partial\Omega, \quad u \neq Tu,$$

an integer $\deg(I - T, \Omega)$ *with the following properties :*
(i) Normalization – *If $0 \in \Omega$ then* $\deg(I, \Omega) = 1$*, while if $0 \notin \bar{\Omega}$ then* $\deg(I, \Omega) = 0$;

(ii) Additivity – *If* Ω_1 *and* Ω_2 *are disjoint open bounded subsets of* X *such that* $0 \notin (I - T)(\partial\Omega_1 \cup \partial\Omega_2)$ *then*

$$\deg(I - T, \Omega_1 \cup \Omega_2) = \deg(I - T, \Omega_1) + \deg(I - T, \Omega_2);$$

(iii) Invariance with respect to an homotopy – *If* $H : [0,1] \times \overline{\Omega} \to X$ *is completely continuous and such that* $0 \notin (I - H)([0,1] \times \partial\Omega)$, *then*

$$\deg(I - H(0,\cdot), \Omega) = \deg(I - H(1,\cdot), \Omega).$$

The integer $\deg(I - T, \Omega)$ *is called the* topological degree *of* $I - T$ *with respect to* Ω.

Such a degree is unique and coincide with Leray-Schauder degree whose construction can be found in [205]. This degree also satisfies the following fundamental properties.

Proposition 1.2. *Let* X *be a real normed vector space,* Ω *an open bounded subset of* X *and* $T : \Omega \to X$ *a completely continuous operator such that* $0 \notin (I - T)(\partial\Omega)$.

Then the following properties hold :

(iv) *If* $0 \notin (I - T)(\overline{\Omega})$, *then* $\deg(I - T, \Omega) = 0$;

(v) *If* $\deg(I - T, \Omega) \neq 0$, *then there exists* $u \in \Omega$ *such that* $u = Tu$;

(vi) Excision – *If* $A \subset \Omega$ *is such that* $0 \notin (I - T)(\overline{A})$ *then*

$$\deg(I - T, \Omega) = \deg(I - T, \Omega \setminus \overline{A}).$$

Proof. See [205, p. 66].

Theorem 1.3. *Let* X *be a real normed vector space and* $T : B[0,R] \subset X \to B[0,R]$ *be a completely continuous operator such that* $0 \notin (I - T)(\partial B(0,R))$.

Then $\deg(I - T, B(0,R)) = 1$ *and* T *has at least one fixed point in* $B(0,R)$.

Proof. Using the properties of the degree, we see right away that for $\lambda \in [0,1]$

$$\deg(I - T, B(0,R)) = \deg(I - \lambda T, B(0,R)) = \deg(I, B(0,R)) = 1. \qquad \square$$

As a corollary, we can prove Schauder's Theorem.

Theorem 1.4. (Schauder's Theorem) *Let* X *be a real normed space and* $T : B[0,R] \subset X \to B[0,R]$ *be a completely continuous operator.*

Then there exists at least one fixed point $u^* \in B[0,R]$ *of* T.

Proof. Notice that there exists either a fixed point of T on $\partial B(0,R)$ or Theorem 1.3 applies. $\square$

A second application is the odd mapping theorem.

Theorem 1.5. (Odd Mapping Theorem) *Let X be a Banach space, $\Omega \subset X$ be a bounded, open set, symmetric with respect to $0 \in \Omega$, and $T : \overline{\Omega} \to X$ be a completely continuous operator such that $0 \notin (I - T)(\partial\Omega)$ and*

$$\forall u \in \partial\Omega, \quad T(u) = -T(-u).$$

Then $\deg(I - T, \Omega)$ *is odd.*

Proof. [205, p. 66].

Theorem 1.6. (Linear Mapping Theorem) *Let X be a Banach space, $\Omega \subset X$ be a bounded open set such that $0 \in \Omega$ and $T : \overline{\Omega} \to X$ be a completely continuous linear operator such that $I - T$ is an homeomorphism. Then*

$$|deg(I - T, \Omega)| = 1.$$

Proof. See [205, p. 66] or [263, vol. 2 p. 185].

Given two spaces $X \subset E$, it is known that if the application $T : E \to X$ is completely continuous and X is closed in E, the degree in E can be computed from the degree in X, i.e. $\deg_E(I - T, \Omega) = \deg_X(I - T, \Omega \cap X)$ (see [103, p.59]). The following theorem gives such a result without assuming X to be closed.

Theorem 1.7. (Reduction Theorem) *Let $X \subset E$ be two continuously included Banach spaces, $T : E \to X$ be a completely continuous operator and let $\Omega_X \subset \Omega_E$ be open, bounded subsets respectively of X and E. Assume that for any $u \in \mathrm{adh}_E\Omega_E \setminus \Omega_X$, $u \neq Tu$.*

Then

$$\deg_E(I - T, \Omega_E) = \deg_X(I - T, \Omega_X).$$

Proof. Claim 1 : There exists $\delta > 0$ such that

$$\inf\{\|u - Tu\|_X \mid u \in \mathrm{adh}_E\Omega_E \setminus \Omega_X\} > \delta. \tag{1.1}$$

Otherwise there exists $(u_n)_n \subset \mathrm{adh}_E\Omega_E \setminus \Omega_X$ such that $u_n - Tu_n \overset{X}{\to} 0$. As $(u_n)_n$ is bounded in E and $T : E \to X$ is completely continuous we have, going to a subsequence if necessary, $Tu_n \overset{X}{\to} v$. Hence $u_n \overset{X}{\to} v$ and by the continuous injection of X into E, $u_n \overset{E}{\to} v$. It follows that $v \in \mathrm{adh}_E\Omega_E \setminus \Omega_X$ and $v = Tv$ which contradicts the assumptions.

Claim 2 : There exists $\delta' > 0$ *such that*

$$\inf\{\|u - Tu\|_E \mid u \in \partial_E\Omega_E\} > \delta'.$$

The proof is similar to the proof of Claim 1.

Claim 3 : There exists a finite dimensional subspace $X_1 \subset X$ *and a completely continuous application* $F : E \to X_1$ *such that*

$$\sup_{\mathrm{adh}_E\Omega_E} \|Fu - Tu\|_X \leq \delta \quad \text{and} \quad \sup_{\partial_E\Omega_E} \|Fu - Tu\|_E \leq \delta'.$$

Let $c > 0$ be such that for any $u \in X$, $\|u\|_E \leq c\|u\|_X$ and $\eta = \min\{\delta, \frac{1}{c}\delta'\}$. We deduce then from [103, Proposition 8.1, p. 55] that there exists a finite dimensional subspace $X_1 \subset X$ and a completely continuous application $F : \mathrm{adh}_E\Omega_E \to X_1$ such that

$$\sup_{\mathrm{adh}_E\Omega_E} \|Fu - Tu\|_X \leq \eta \leq \delta. \tag{1.2}$$

We also have

$$\sup_{\partial_E\Omega_E} \|Fu - Tu\|_E \leq c \sup_{\partial_E\Omega_E} \|Fu - Tu\|_X \leq c\eta \leq \delta'.$$

Conclusion – As $\|u - Fu\|_E \geq \|u - Tu\|_E - \|Tu - Fu\|_E > 0$ for $u \in \partial_E\Omega_E$ and $\|u - Fu\|_X \geq \|u - Tu\|_X - \|Tu - Fu\|_X > 0$ for $u \in \partial_X\Omega_X$, it is easy to deduce now using [103, Theorems 8.1, p.57, and 8.7, p.59] that

$$\deg_E(I - T, \Omega_E) = \deg_E(I - F, \Omega_E) = \deg_{X_1}(I - F, X_1 \cap \Omega_E)$$

and

$$\deg_X(I - T, \Omega_X) = \deg_X(I - F, \Omega_X) = \deg_{X_1}(I - F, X_1 \cap \Omega_X).$$

We deduce from (1.1) and (1.2), that $\|u - Fu\|_X > 0$ for all $u \in \mathrm{adh}_E\Omega_E \setminus \Omega_X$ so that using the excision property we have

$$\begin{aligned} \deg_X(I - T, \Omega_X) &= \deg_{X_1}(I - F, X_1 \cap \Omega_X) \\ &= \deg_{X_1}(I - F, X_1 \cap \Omega_E) = \deg_E(I - T, \Omega_E). \end{aligned}$$ □

2 Variational Methods

Definition 2.1. *We say that the functional* $\phi : X \to \mathbb{R}$ satisfies the Palais-Smale Condition *if for every sequence* $(u_n)_n \subset X$ *such that* $\phi(u_n)$ *is bounded and* $\nabla\phi(u_n) \to 0$ *there exists a subsequence that converges to some function* $u \in X$.

Theorem 2.1. *Let $f : [a,b] \times \mathbb{R} \to \mathbb{R}$ be an L^1-Carathéodory function and*

$$\phi : H_0^1(a,b) \to \mathbb{R}, u \to \int_a^b [\frac{u'^2(t)}{2} - F(t,u(t))]\, dt,$$

with $F(t,s) = \int_0^s f(t,x)\, dx$. Then, for every sequence $(u_n)_n \subset H_0^1(a,b)$ with $\phi(u_n)$ bounded, $\nabla\phi(u_n) \to 0$ and $\|u_n\|_{H_0^1} \leq C$, there exists a subsequence $(u_{n_k})_k$ and $u \in H_0^1(a,b)$ such that $u_{n_k} \to u$ in $H_0^1(a,b)$.

Proof. See [259, Prop B.35]. □

Theorem 2.2. *Let X be a reflexive Banach space and let $\phi : X \to \mathbb{R}$ be weakly lower semi-continuous and coercive. Then ϕ has a minimum.*

Proof. See [219, Thm 1.1].

Definition 2.2. *Let X be a Banach space, $\phi \in C^1(X,\mathbb{R})$ and $c \in \mathbb{R}$. A critical point $u_0 \in X$ of ϕ with $\phi(u) = c$ is of* mountain-pass type *if for every neighbourhood W of u_0 the topological space $W \cap \phi^c$, where $\phi^c = \{u \in X \mid \phi(u) < c\}$, is nonempty and not path-connected.*

Theorem 2.3. (Mountain Pass Theorem) *Let X be a Banach space and let $\phi \in C^1(X,\mathbb{R})$ satisfy the Palais-Smale condition. Suppose*
(a) there exist $e_1 \in X$ and constants ρ, γ such that for all $u \in \partial B(e_1,\rho)$, $\phi(u) \geq \gamma > \phi(e_1)$;
(b) there exist $e_2 \in X \setminus B(e_1,\rho)$ such that $\phi(e_2) < \gamma$.
Then ϕ has a critical value $c \geq \gamma$ characterized by

$$c = \inf_{\varphi \in \Gamma} \max_{t \in [0,1]} \phi(\varphi(t)),$$

where

$$\Gamma = \{\varphi \in C([0,1],X) \mid \varphi(0) = e_1, \quad \varphi(1) = e_2\}.$$

Moreover, if $K_c = \{u \in X \mid \phi(u) = c,\ \nabla\phi(u) = 0\}$ consists of isolated points then K_c contains a critical point of mountain pass type.

Proof. See [219, Thm 4.10] and [165]. □

Theorem 2.4. (Characterization of minimizer) *Let X be a Banach space and let $\phi \in C^1(X,\mathbb{R})$ satisfy the Palais-Smale condition. Suppose that $u_0 \in X$ is a local minimum, i.e. there exists $\epsilon > 0$ such that*

$$\phi(u_0) \leq \phi(u) \quad \textit{for } \|u - u_0\| \leq \epsilon.$$

Then given $0 < \epsilon_0 \leq \epsilon$, the following alternative holds:

(a) either there exists $0 < \gamma < \epsilon_0$ *such that*

$$\inf\{\phi(u) \mid \|u - u_0\| = \gamma\} > \phi(u_0),$$

(b) or, for each $\gamma \in]0, \epsilon_0[$, ϕ *has a local minimum at a point* u_γ *with* $\|u_\gamma - u_0\| = \gamma$ *and* $\phi(u_\gamma) = \phi(u_0)$.

Proof. See [99, Thm 5.10]. □

Corollary 2.5. *Let* X *be a Banach space and* $\phi \in C^1(X, \mathbb{R})$ *satisfy the Palais-Smale condition. If* ϕ *has two local minima, then* ϕ *has a third critical point.*

Proof. This result is an application of Theorem 2.3 and 2.4. □

Theorem 2.6. *Let* H *be a Hilbert space,* $\phi \in C^2(H, \mathbb{R})$ *and assume that* $\nabla\phi(u) = u - Tu$ *with* T *a compact operator. Suppose further that for all critical point* u_0 *of* ϕ*, the first eigenvalue* λ_1 *of* $d^2\phi(u_0) \in L(H)$ *is either positive or simple. Let* $u_0 \in H$ *and* $\epsilon > 0$ *be such that* u_0 *is the only critical point of* ϕ *in* $B(u_0, \epsilon)$.
(a) If u_0 *is of mountain-pass type,*

$$\deg(\nabla\phi, B(u_0, \epsilon)) = -1.$$

(b) If u_0 *is a local minimum,*

$$\deg(\nabla\phi, B(u_0, \epsilon)) = 1.$$

Proof. See [165] and [18]. See also [60, corollary 2-3.1]. □

Corollary 2.7. *Let* $f : [a, b] \times \mathbb{R} \to \mathbb{R}$ *be an* L^1*-Carathéodory function which is* C^1 *in the second variable and*

$$\phi : H_0^1(a, b) \to \mathbb{R}, u \to \int_a^b [\frac{u'^2(t)}{2} - F(t, u(t))]\, dt,$$

with $F(t, s) = \int_0^s f(t, x)\, dx$.

Let $u_0 \in H_0^1(a, b)$ *and* $\epsilon > 0$ *be such that* u_0 *is the only critical point of* ϕ *in* $B(u_0, \epsilon)$.
(a) If u_0 *is of mountain-pass type,*

$$\deg(\nabla\phi, B(u_0, \epsilon)) = -1.$$

(b) If u_0 *is a local minimum,*

$$\deg(\nabla\phi, B(u_0, \epsilon)) = 1.$$

Proof. First observe that $\nabla\phi(u) = u - KN(u)$, where

$$N : H_0^1(a,b) \to L^1(a,b), u \mapsto f(\cdot,u),$$
$$K : L^1(a,b) \to H_0^1(a,b), h \mapsto Kh$$

and Kh is defined to be the unique solution of

$$-u'' = h(t), \quad u(a) = 0,\ u(b) = 0.$$

Moreover $d^2\phi(u_0)(v) = v - K\frac{\partial f}{\partial u}(\cdot, u_0(\cdot))v$. If λ_1 and φ_1 are the first eigenvalue and the corresponding eigenfunction of $d^2\phi(u_0)$, then

$$\begin{aligned} -\varphi_1'' &= \mu_1 \tfrac{\partial f}{\partial u}(t, u_0(t))\varphi_1, \\ \varphi_1(a) &= 0,\ \varphi_1(b) = 0, \end{aligned}$$

with $\mu_1 = (1-\lambda_1)^{-1}$. Observe that, if $\lambda_1 \le 0$ we have

$$\int_a^b \tfrac{\partial f}{\partial u}(t, u_0(t))\varphi_1^2(t)\, dt > 0$$

and we can apply [207] to conclude that λ_1 is simple. The result follows then from Theorem 2.6. □

3 Spectral Results

3.1 Abstract Spectral Result

Theorem 3.1. (Abstract Spectral Theorem) *Let H be an infinite dimensional Hilbert space with scalar product $\tilde{a}(\cdot,\cdot)$ and let $\tilde{b} : H \times H \to \mathbb{R}$ be a bilinear, symmetric, positive definite form (i.e. for all $u \in H \setminus \{0\}$, $\tilde{b}(u,u) > 0$) such that $\tilde{b}(u_n,u_n)$ converges to $\tilde{b}(u,u)$, for any sequence $(u_n)_n$ that converges weakly to u in H.*

Then the problem

find $\lambda \in \mathbb{R}$ and $\varphi \in H \setminus \{0\}$ such that for every $v \in H$, $\tilde{a}(\varphi, v) = \lambda\tilde{b}(\varphi, v)$,

has a sequence of solutions $(\lambda_n, \varphi_n)_{n\ge1}$. The λ_n are called the eigenvalues *and φ_n the corresponding* eigenfunctions, *and are such that*

(i) $0 < \lambda_1 \le \lambda_2 \le \dots$ *and* $\lim_{n\to\infty} \lambda_n = \infty$;

(ii) the sequence $(\varphi_n)_{n\ge1}$ is an orthonormal Hilbertian basis of H and is orthogonal with respect to $\tilde{b}(\cdot,\cdot)$; moreover, for every $u \in H$,

$$\tilde{a}(u,u) = \sum_{i=1}^{\infty} \tilde{a}(u,\varphi_i)^2 \quad \textit{and} \quad \tilde{b}(u,u) = \sum_{i=1}^{\infty} \frac{1}{\lambda_i}\tilde{a}(u,\varphi_i)^2;$$

(iii)
$$\lambda_n = \min_{X_n \in \mathcal{E}_n} \Big[\max_{u \in X_n \setminus \{0\}} \frac{\tilde{a}(u,u)}{\tilde{b}(u,u)} \Big],$$
where $\mathcal{E}_n$ is the set of n-dimensional subspaces of H.

Proof. See [307, section 4.5], see also [83, vol 5].

The next subsections apply this result to the separated and the periodic boundary conditions.

3.2 Separated Boundary Conditions

In this section, we consider the eigenvalue problem

$$\begin{aligned} u'' + p(t)u' + q(t)u + \lambda r(t)u &= 0, \\ a_1 u(a) - a_2 u'(a) &= 0, \\ b_1 u(b) + b_2 u'(b) &= 0. \end{aligned} \tag{3.1}$$

Theorem 3.2. *Let p, q, $r \in L^1(a,b)$ be such that $r(t) > 0$ a.e. on $[a,b]$ and let a_1, $b_1 \in \mathbb{R}$, a_2, $b_2 \in \mathbb{R}^+$ satisfy $a_1^2 + a_2^2 > 0$ and $b_1^2 + b_2^2 > 0$. Write*

$$P(t) = \exp(\int_a^t p(s)\,ds), \quad Q(t) = q(t)P(t) \quad \text{and} \quad R(t) = r(t)P(t)$$

and define H and $a(\cdot,\cdot)$ as follows :
(a) if $a_2 = 0$, $b_2 = 0$,
$$H = H_0^1(a,b) \text{ and } a(u,v) = \int_a^b (Pu'v' - Quv)\,dt;$$
(b) if $a_2 = 0$, $b_2 \neq 0$,
$$H = \{u \in H^1(a,b) \mid u(a) = 0\}$$
and
$$a(u,v) = \int_a^b (Pu'v' - Quv)\,dt + P(b)\tfrac{b_1}{b_2}u(b)v(b);$$
(c) if $a_2 \neq 0$, $b_2 = 0$,
$$H = \{u \in H^1(a,b) \mid u(b) = 0\}$$
and
$$a(u,v) = \int_a^b (Pu'v' - Quv)\,dt + P(a)\tfrac{a_1}{a_2}u(a)v(a);$$
(d) if $a_2 \neq 0$, $b_2 \neq 0$,
$$H = H^1(a,b)$$
and
$$a(u,v) = \int_a^b (Pu'v' - Quv)\,dt + P(b)\tfrac{b_1}{b_2}u(b)v(b) + P(a)\tfrac{a_1}{a_2}u(a)v(a).$$
At last, define $b(u,v) = \int_a^b Ruv\,dt$.

Then the problem (3.1) *has a sequence of simple eigenvalues* $(\lambda_n)_{n\geq 1}$ *and a sequence of corresponding eigenfunctions* $(\varphi_n)_{n\geq 1}$ *such that*

(i) for some $M \geq 0$, $-M < \lambda_1 < \lambda_2 < \dots$ *and* $\lim_{n\to\infty} \lambda_n = \infty$;

(ii) the sequence $(\varphi_n)_{n\geq 1}$ *is orthogonal for* $a(\cdot,\cdot)$ *and* $b(\cdot,\cdot)$, *and total for* H *with the scalar product* $(\cdot,\cdot)_{H^1}$;

(iii)
$$\lambda_n = \min_{X_n \in \mathcal{E}_n} \Big[\max_{u\in X_n\setminus\{0\}} \frac{a(u,u)}{b(u,u)} \Big],$$
where $\mathcal{E}_n$ *is the set of* n*-dimensional subspaces of* H.

Proof. Problem (3.1) is equivalent to

$$\begin{aligned} (P(t)u')' + Q(t)u + \lambda R(t)u &= 0,\\ a_1 u(a) - a_2 u'(a) &= 0,\\ b_1 u(b) + b_2 u'(b) &= 0, \end{aligned} \tag{3.2}$$

which has a variational structure.

Let us concentrate on the fourth case, $a_2 \neq 0$, $b_2 \neq 0$, the other ones being similar.

Claim 1 : there exist an $M > 0$ *and* $\epsilon > 0$ *such that for all* $u \in H$,

$$a(u,u) + Mb(u,u) \geq \epsilon \int_a^b (u'^2 + u^2)\,dt. \tag{3.3}$$

Assume on the contrary that, for every n, there exists a $u_n \in H^1(a,b)$ such that

$$\int_a^b (Pu_n'^2 - Qu_n^2)\,dt + P(b)\tfrac{b_1}{b_2}u_n^2(b) + P(a)\tfrac{a_1}{a_2}u_n^2(a) + n\int_a^b Ru_n^2\,dt < \tfrac{1}{n}\|u_n\|_{H^1}^2. \tag{3.4}$$

Without loss of generality, we can assume that $\|u_n\|_{H^1} = 1$ and hence, going to subsequences, $(u_n)_n$ converges weakly in $H^1(a,b)$ and strongly in $C([a,b])$ to some $u \in H^1(a,b)$. Dividing (3.4) by n and going to the limit, we obtain

$$\int_a^b Ru^2\,dt \leq 0,$$

which implies using the positivity of R that $u = 0$. Hence, we deduce from (3.4) that

$$\limsup_{n\to\infty} \Big[\int_a^b Pu_n'^2\,dt + n\int_a^b Ru_n^2\,dt\Big] \leq 0$$

and from the positivity of P and R

$$0 \leq \liminf_{n\to\infty}[\int_a^b Pu_n'^2\,dt + n\int_a^b Ru_n^2\,dt].$$

It follows that

$$\lim_{n\to\infty}\int_a^b Pu_n'^2\,dt = 0 \quad \text{and} \quad \lim_{n\to\infty} n\int_a^b Ru_n^2\,dt = 0.$$

As $P \in C([a,b])$ satisfies $P(t) \geq K$ on $[a,b]$ for some $K > 0$, we obtain the contradiction $\|u_n\|_{H^1} \to 0$.

The weak problem – Problem (3.2) is equivalent to the weak problem

$$\begin{aligned}&\textit{find } \tilde{\lambda} \in \mathbb{R} \textit{ and } \tilde{\varphi} \in H \setminus \{0\} \textit{ such that,}\\ &\textit{for every } v \in H,\ a(\tilde{\varphi}, v) + Mb(\tilde{\varphi}, v) = \tilde{\lambda}b(\tilde{\varphi}, v),\end{aligned} \tag{P}$$

i.e. $(\tilde{\lambda}, \tilde{\varphi})$ solves (P) if and only if $(\tilde{\lambda} - M, \tilde{\varphi})$ solves (3.2). It is clear now that Theorem 3.1 applies to problem (P).

Claim (i) – Existence of the eigenvalues follows from Theorem 3.1. Further these are simple. Otherwise, let φ and ψ be two eigenfunctions corresponding to λ_n. Then every solution u of

$$u'' + p(t)u' + q(t)u + \lambda_n r(t)u = 0,$$

is of the form $u = A\varphi + B\psi$ and satisfies $a_1u(a) - a_2u'(a) = 0$. This fact contradicts the existence of a solution for Cauchy problem with initial conditions such that $a_1u(a) - a_2u'(a) \neq 0$.

Claim (ii) – The sequence $(\tilde{\varphi}_n)_n$, given by Theorem 3.1, is orthogonal with respect to $a(\cdot,\cdot) + Mb(\cdot,\cdot)$ and $b(\cdot,\cdot)$. Hence it is also orthogonal with respect to $a(\cdot,\cdot)$. From Theorem 3.1 and Claim 1, such a sequence is total for H with $(\cdot,\cdot)_{H^1}$.

Claim (iii) – Notice that

$$\begin{aligned}\lambda_n = \tilde{\lambda}_n - M &= \min_{X_n\in\mathcal{E}_n}\Big[\max_{u\in X_n\setminus\{0\}} \frac{a(u,u) + Mb(u,u)}{b(u,u)}\Big] - M\\ &= \min_{X_n\in\mathcal{E}_n}\Big[\max_{u\in X_n\setminus\{0\}} \frac{a(u,u)}{b(u,u)}\Big].\end{aligned}$$

□

Our next result concerns the monotonicity of the eigenvalues.

Theorem 3.3. (Monotonicity of the eigenvalues) *Let a_1, $b_1 \in \mathbb{R}$, a_2, $b_2 \in \mathbb{R}^+$ satisfy $a_1^2 + a_2^2 > 0$ and $b_1^2 + b_2^2 > 0$. Let p, q, $\tilde{q}$, r, $\tilde{r} \in L^1(a,b)$, with $r(t) > 0$ and $\tilde{r}(t) > 0$ a.e. on $[a,b]$. Denote $(\lambda_n)_{n\geq 1}$ (resp. $(\tilde{\lambda}_n)_{n\geq 1}$) the* *sequence of eigenvalues of* (3.1) *corresponding to* p, q *and* r *(resp.* p, $\tilde{q}$ *and* $\tilde{r}$*).*

(i) If, for a.e. $t \in [a,b]$, $\tilde{q}(t) \leq q(t)$, $\tilde{r}(t) \leq r(t)$, *then, for every* $n \geq 1$ *such that* $\lambda_n \geq 0$, *we have* $\tilde{\lambda}_n \geq \lambda_n$.

(ii) If, for a.e. $t \in [a,b]$, $\tilde{q}(t) \leq q(t)$, $\tilde{r}(t) \leq r(t)$, *and either* $\tilde{q} \lneqq q$ *or* $\tilde{r} \lneqq r$, *then, for every* $n \geq 1$ *such that* $\lambda_n \geq 0$, *we have* $\tilde{\lambda}_n > \lambda_n$.

(iii) If, for a.e. $t \in [a,b]$, $\tilde{q}(t) \leq q(t)$, $\tilde{r}(t) = r(t)$, *then, for every* $n \geq 1$, $\tilde{\lambda}_n \geq \lambda_n$ *and if moreover* $\tilde{q} \lneqq q$ *then, for every* $n \geq 1$, $\tilde{\lambda}_n > \lambda_n$.

Proof. Let $a(\cdot,\cdot)$ and $b(\cdot,\cdot)$ (resp. $\tilde{a}(\cdot,\cdot)$ and $\tilde{b}(\cdot,\cdot)$) be defined from p, q and r (resp. p, $\tilde{q}$ and $\tilde{r}$) as in Theorem 3.2. If $\lambda_n \geq 0$, we deduce from the variational characterization of eigenvalues (Theorem 3.2 (iii)) that

$$0 \leq \lambda_n = \min_{X_n \in \mathcal{E}_n} \Big[\max_{u \in X_n \setminus \{0\}} \frac{a(u,u)}{b(u,u)} \Big].$$

Therefore, for any n-dimensional subspace X_n,

$$\max_{u \in X_n \setminus \{0\}} \frac{a(u,u)}{b(u,u)} \geq 0$$

and there exists a $v \in X_n$ such that $a(v,v) \geq 0$.

Let $X_n \in \mathcal{E}_n$ be fixed. For any $u \in X_n$ such that $a(u,u) \geq 0$, we compute

$$\frac{a(u,u)}{b(u,u)} \leq \frac{\tilde{a}(u,u)}{\tilde{b}(u,u)} \leq \max_{u \in X_n \setminus \{0\}} \frac{\tilde{a}(u,u)}{\tilde{b}(u,u)}.$$

In particular, $\max_{u \in X_n \setminus \{0\}} \frac{\tilde{a}(u,u)}{\tilde{b}(u,u)} \geq \frac{a(v,v)}{b(v,v)} \geq 0$ and

$$\frac{a(u,u)}{b(u,u)} \leq \max_{u \in X_n \setminus \{0\}} \frac{\tilde{a}(u,u)}{\tilde{b}(u,u)}$$

holds for all $u \in X_n \setminus \{0\}$.

It follows then that

$$\lambda_n = \min_{X_n \in \mathcal{E}_n} \Big[\max_{u \in X_n \setminus \{0\}} \frac{a(u,u)}{b(u,u)} \Big] \leq \min_{X_n \in \mathcal{E}_n} \Big[\max_{u \in X_n \setminus \{0\}} \frac{\tilde{a}(u,u)}{\tilde{b}(u,u)} \Big] = \tilde{\lambda}_n.$$

Proofs of Claims (ii) and (iii) follow now easily. □

The next propositions concern the positivity of the first eigenfunction.

Proposition 3.4. *Under the assumptions of Theorem 3.2, the eigenfunction* φ_1 *corresponding to the first eigenvalue* λ_1 *of* (3.1) *does not change sign. More precisely, we can choose* φ_1 *in such a way that*

(i) $\varphi_1 > 0$ *on* $]a, b[$;
(ii) $\varphi_1(a) > 0$ *if* $a_2 > 0$ *and* $\varphi_1'(a) > 0$ *if* $a_2 = 0$;
(iii) $\varphi_1(b) > 0$ *if* $b_2 > 0$ *and* $\varphi_1'(b) < 0$ *if* $b_2 = 0$.

Proof. Assume $\varphi_1^+ = \max\{\varphi_1, 0\}$ and $\varphi_1^- = \max\{-\varphi_1, 0\}$ are nonzero functions. Recall that for positive numbers A, B, C and D, we have either $\frac{A+B}{C+D} \geq \frac{A}{C}$ or $\frac{A+B}{C+D} \geq \frac{B}{D}$. Hence, we can write

$$\lambda_1 = \min_{H\setminus\{0\}} \frac{a(u,u)}{b(u,u)} = \frac{a(\varphi_1,\varphi_1)}{b(\varphi_1,\varphi_1)} \geq \min\Big(\frac{a(\varphi_1^+,\varphi_1^+)}{b(\varphi_1^+,\varphi_1^+)}, \frac{a(\varphi_1^-,\varphi_1^-)}{b(\varphi_1^-,\varphi_1^-)}\Big) \geq \lambda_1.$$

It follows that φ_1^+ and φ_1^- are also eigenfunctions corresponding to λ_1. As eigenvalues are simple, $\varphi_1 = \varphi_1^+$ or $\varphi_1 = \varphi_1^-$.

If $\varphi_1(t_0) = 0$ for some $t_0 \in]a, b[$ then $\varphi_1'(t_0) = 0$ and φ_1 is the solution of the Cauchy problem

$$u'' + p(t)u' + q(t)u + \lambda_1 r(t)u = 0,$$
$$u(t_0) = 0, \ u'(t_0) = 0$$

which implies $\varphi_1 = 0$, a contradiction. It is now easy to conclude. □

Proposition 3.5. *Under the assumptions of Theorem 3.2, if φ is an eigenfunction of* (3.1) *that does not change sign, then φ is an eigenfunction corresponding to the first eigenvalue λ_1.*

Proof. Let φ_n, $n \neq 1$, be an eigenfunction of (3.1) that does not change sign. We compute

$$0 = b(\varphi_1, \varphi_n) = \int_a^b R\varphi_1\varphi_n \, dt,$$

which is impossible since $R\varphi_1 > 0$ a.e. on $[a, b]$. □

Proposition 3.6. *Under the assumptions and with the notations of Theorem 3.2, if φ is such that $\frac{a(\varphi,\varphi)}{b(\varphi,\varphi)} = \lambda_1$ then φ is an eigenfunction of* (3.1) *corresponding to the first eigenvalue λ_1.*

Proof. Observe that, by the characterization of the first eigenvalue λ_1, φ is a minimum of $a(u, u) - \lambda_1 b(u, u)$ and hence satisfies for all $v \in H$, $a(\varphi, v) = \lambda_1 b(\varphi, v)$, i.e. φ is an eigenfunction of (3.1) corresponding to λ_1. □

3.3 The Periodic Problem

Consider the periodic eigenvalue problem

$$\begin{aligned} u'' + p(t)u' + q(t)u + \lambda r(t)u = 0, \\ u(a) = u(b), \ u'(a) = u'(b). \end{aligned} \tag{3.5}$$

Theorem 3.7. *Let p, q, $r \in L^1(a,b)$ be such that $r(t) > 0$ a.e. on $[a,b]$. Write*

$$P(t) = \exp\left(\int_a^t p(s)\,ds\right), \quad Q(t) = q(t)P(t) \quad \text{and} \quad R(t) = r(t)P(t).$$

Define

$$H = \{u \in H^1(a,b) \mid u(a) = u(b)\},$$

$$a(u,v) = \int_a^b (Pu'v' - Quv)\,dt \qquad \text{and} \qquad b(u,v) = \int_a^b Ruv\,dt.$$

Then the problem (3.5) has a sequence of eigenvalues $(\lambda_n)_{n\geq 1}$ and a sequence of corresponding eigenfunctions $(\varphi_n)_{n\geq 1}$ such that

(i) for some $M \geq 0$, $-M < \lambda_1 \leq \lambda_2 \leq \ldots$ and $\lim_{n\to\infty} \lambda_n = \infty$;

(ii) the sequence $(\varphi_n)_{n\geq 1}$ is orthogonal for $a(\cdot,\cdot)$ and $b(\cdot,\cdot)$, and total for H with the scalar product $(\cdot,\cdot)_{H^1}$;

(iii)
$$\lambda_n = \min_{X_n \in \mathcal{E}_n} \left[\max_{u \in X_n \setminus \{0\}} \frac{a(u,u)}{b(u,u)} \right],$$
where $\mathcal{E}_n$ is the set of n-dimensional subspaces of H.

Proof. As in the proof of Theorem 3.2, we consider the weak problem

find $\lambda \in \mathbb{R}$ and $u \in H \setminus \{0\}$ such that for every $v \in H$, $a(u,v) = \lambda b(u,v)$

and apply Theorem 3.1 to the shifted equation

$$a(u,v) + Mb(u,v) = \tilde{\lambda} b(u,v),$$

where $M \geq 0$ is such that the scalar product $a(u,v) + Mb(u,v)$ is equivalent to $(\cdot,\cdot)_{H^1}$. □

As for the separated boundary conditions, the eigenvalues satisfy monotonicity properties the proof of which paraphrases the argument used in Theorem 3.3.

Theorem 3.8. (Monotonicity of the eigenvalues) *Let p, q, $\tilde{q}$, r, $\tilde{r} \in L^1(a,b)$, with $r(t) > 0$ and $\tilde{r}(t) > 0$ a.e. on $[a,b]$. Denote $(\lambda_n)_{n\geq 1}$ (resp. $(\tilde{\lambda}_n)_{n\geq 1}$) the sequence of eigenvalues of (3.5) corresponding to p, q and r (resp. p, $\tilde{q}$ and $\tilde{r}$).*

(i) If, for a.e. $t \in [a,b]$, $\tilde{q}(t) \leq q(t)$, $\tilde{r}(t) \leq r(t)$, then for every $n \geq 1$ such that $\lambda_n \geq 0$, we have $\tilde{\lambda}_n \geq \lambda_n$.

(ii) *If, for a.e.* $t \in [a,b]$, $\tilde{q}(t) \le q(t)$, $\tilde{r}(t) \le r(t)$, *and either* $\tilde{q} \lneqq q$ *or* $\tilde{r} \lneqq r$, *then, for every* $n \ge 1$ *such that* $\lambda_n \ge 0$, *we have* $\tilde{\lambda}_n > \lambda_n$.

(iii) *If, for a.e.* $t \in [a,b]$, $\tilde{q}(t) \le q(t)$, $\tilde{r}(t) = r(t)$, *then, for every* $n \ge 1$, $\tilde{\lambda}_n \ge \lambda_n$ *and if moreover* $\tilde{q} \lneqq q$ *then, for every* $n \ge 1$, $\tilde{\lambda}_n > \lambda_n$.

The following propositions deal with the properties of the first eigenfunction. These are proved as the corresponding propositions for boundary value problems with separated boundary conditions.

Proposition 3.9. *Under the assumptions of Theorem 3.7, the eigenfunction* φ_1 *corresponding to the first eigenvalue* λ_1 *of* (3.5) *does not change sign. More precisely, we can choose* φ_1 *in such a way that* $\varphi_1 > 0$ *on* $[a,b]$.

Proposition 3.10. *Under the assumptions of Theorem 3.7, if* φ *is an eigenfunction of* (3.5) *that does not change sign, then* φ *is an eigenfunction corresponding to the first eigenvalue* λ_1.

Proposition 3.11. *Under the assumptions of Theorem 3.7, if* φ *is such that* $\frac{a(\varphi,\varphi)}{b(\varphi,\varphi)} = \lambda_1$, *then* φ *is an eigenfunction of* (3.5) *corresponding to the first eigenvalue* λ_1.

3.4 The Dirichlet Problem

In this section, we collect several results concerning the eigenvalue problem

$$\begin{aligned} u'' + \mu\, r(t)u = 0, \\ u(a) = 0,\ u(b) = 0, \end{aligned} \tag{3.6}$$

where $r \in \mathcal{A}$.

Lemma 3.12. *Let* $r \in \mathcal{A}$, $r(t) > 0$ *a.e. on* $[a,b]$ *and* $\mu \in \mathbb{R}$.
Then the equation

$$u'' + \mu\, r(t)u = 0, \tag{3.7}$$

has a solution $u_0 \in C^1([a,b[,\mathbb{R})$ *such that*

$$u_0(a) = 0, \qquad u_0'(a) = 1 \tag{3.8}$$

and a solution $u_1 \in C^1(]a,b],\mathbb{R})$ *such that*

$$u_1(b) = 0, \qquad u_1'(b) = 1.$$

Proof. Let $T : C([a,a+\epsilon]) \to C([a,a+\epsilon])$ be defined by

$$Tv(a) = 1$$

and

$$Tv(t) = 1 - \frac{\mu}{t-a}\int_a^t (t-s)(s-a)r(s)v(s)\,ds \text{ for } t \in]a, a+\epsilon].$$

For ϵ small enough, T is a contraction since

$$\begin{aligned}\|Tv - Tw\|_\infty &\le |\mu|\, \| \int_a^t \tfrac{(t-s)(s-a)}{t-a} r(s)(v(s) - w(s))\,ds\|_\infty \\ &\le |\mu| \int_a^{a+\epsilon} (s-a)r(s)\,ds\, \|v - w\|_\infty.\end{aligned}$$

Hence, by Banach's fixed point Theorem, there exists a v_0 defined on $[a, a+\epsilon]$ such that $v_0 = Tv_0$. It follows that $u_0(t) = (t-a)v_0(t)$ satisfies the Cauchy problem (3.7)-(3.8) on $[a, a+\epsilon]$. We compute also

$$u_0'(t) = 1 - \mu \int_a^t (s-a)r(s)v_0(s)\,ds \in \mathcal{C}([a, a+\epsilon]).$$

Notice at last that since $r \in L^1(T_1, T_2)$ for every $a < T_1 < T_2 < b$, we can extend u_0 as a $\mathcal{C}^1$-solution on $[a, b[$.

The existence of u_1 follows from the same argument. □

Notice that for all $u \in H_0^1(a, b)$,

$$\begin{aligned}|u(t)| &\le \min(\int_a^t |u'(s)|\,ds, \int_t^b |u'(s)|\,ds) \\ &\le \min(\sqrt{t-a}(\int_a^t u'^2(s)\,ds)^{1/2}, \sqrt{b-t}(\int_t^b u'^2(s)\,ds)^{1/2}) \\ &\le C\sqrt{t-a}\sqrt{b-t}\,\|u\|_{H_0^1}.\end{aligned} \tag{3.9}$$

This formula implies that for $r \in \mathcal{A}$, $r(t) > 0$ a.e. on $[a, b]$,

$$\int_a^b |ruv|\,dt \le C^2 \int_a^b r(t)(b-t)(t-a)\,dt \|u\|_{H_0^1}\|v\|_{H_0^1} < \infty.$$

Hence, we can consider the following eigenvalue problem:

$$\text{find } \mu \in \mathbb{R} \text{ and } \psi \in H_0^1(a, b) \setminus \{0\} \text{ such that, for any } v \in H_0^1(a, b), \quad \int_a^b \psi'(t)v'(t)\,dt = \mu \int_a^b r(t)\psi(t)v(t)\,dt. \tag{3.10}$$

Proposition 3.13. *Let $r \in \mathcal{A}$, $r(t) > 0$ a.e. on $[a, b]$. Then the problem (3.6) has sequences of simple eigenvalues $(\mu_n)_{n\ge1}$ and eigenfunctions $(\psi_n)_{n\ge1}$ such that*
(i) $0 < \mu_1 < \mu_2 < \ldots < \mu_n \le \ldots$ and $\lim_{n\to\infty} \mu_n = \infty$;
(ii) the sequence $(\psi_n)_{n\ge1}$ is total in $H_0^1(a, b)$ for the scalar product $(\cdot,\cdot)_{H^1}$.

Proof. The proof follows from Theorem 3.1 with

$$\tilde{a}(u,v) = \int_a^b u'(t)v'(t)\,dt \quad \text{and} \quad \tilde{b}(u,v) = \int_a^b r(t)u(t)v(t)\,dt,$$

provided we prove that

$$u_n \overset{H_0^1}{\rightharpoonup} u \text{ implies } \int_a^b ru_n^2\,dt \to \int_a^b ru^2\,dt.$$

Observe that, $u_n \overset{H_0^1}{\rightharpoonup} u$ implies $u_n(t) \to u(t)$ for all $t \in]a,b[$. Hence, for a.e. $t \in]a,b[$,

$$r(t)u_n^2(t) \to r(t)u^2(t).$$

On the other hand, if $u_n \overset{H_0^1}{\rightharpoonup} u$, there exists a $K > 0$ such that for all n, $\|u_n\|_{H_0^1} \leq K$. We deduce from (3.9) that

$$r(t)u_n^2(t) \leq C^2K^2\,r(t)\,(t-a)(b-t) \in L^1(a,b).$$

Hence, the result follows from Lebesgue's Dominated Convergence Theorem.

Notice at last that eigenvalues are simple, i.e. $\mu_i \neq \mu_j$ if $i \neq j$ since the corresponding eigenfunctions satisfy a Cauchy problem with initial conditions $\psi(a) = 0$, $\psi'(a) = A$. □

We state now some bounds and sign properties of the eigenfunctions.

Proposition 3.14. *Assume $r \in \mathcal{A}$, $r(t) > 0$ a.e. on $[a,b]$. Let μ_n and ψ_n be the eigenvalues and eigenfunctions of* (3.6). *Then we have*
(i) $\psi_n \in W^{2,1}(a,b)$ and, for some $K_n > 0$,

$$|\psi_n(t)| \leq K_n(t-a)(b-t);$$

(ii) ψ_1 has no interior zero.

Proof. Let u_0 and u_1 be the solutions of (3.7) with $\mu = \mu_n$ given by Lemma 3.12. Let $\varphi \in C^\infty([a,b])$ be such that $\varphi(t) = 1$ if $t \in [a,(a+b)/3]$ and $\varphi(t) = 0$ if $t \in [2(a+b)/3, b]$. Notice that as $r \in \mathcal{A}$ and ψ_n satisfies (3.9), we have

$$\psi_n'' = -\mu_n r\psi_n \in \{u \in L^1_{loc}(a,b) \mid \int_a^b u(t)\,\sqrt{(t-a)(b-t)}\,dt < \infty\}.$$

Hence, we can write

$$-\int_{a+\delta}^{b-\delta} \mu_n r\psi_n\varphi u_0\,ds = \int_{a+\delta}^{b-\delta} \psi_n''\varphi u_0\,ds = \psi_n'\varphi u_0|_{a+\delta}^{b-\delta} - \int_{a+\delta}^{b-\delta} \psi_n'(\varphi u_0)'\,ds.$$

Going to the limit as δ goes to zero and using (3.10), we obtain

$$\lim_{\delta\to 0}\psi_n'(a+\delta)u_0(a+\delta)=0.$$

As

$$\frac{d}{dt}(\psi_n' u_0-\psi_n u_0')=0,$$

we have for any $t\in[a,b]$,

$$\begin{aligned}\psi_n'(t)u_0(t)-\psi_n(t)u_0'(t)&=\lim_{\delta\to 0}(\psi_n'(a+\delta)u_0(a+\delta)-\psi_n(a+\delta)u_0'(a+\delta))\\&=0\end{aligned}$$

which proves that ψ_n is a multiple of u_0. Similarly, ψ_n is a multiple of u_1. This result proves $\psi_n\in W^{2,1}(a,b)$ and for some $K_n>0$

$$|\psi_n(t)|\le K_n(t-a)(b-t).$$

The proof that ψ_1 has no interior zero follows the proof of Proposition 3.4. □

Lemma 3.15. *Let $r\in\mathcal{A}$, $r(t)>0$ a.e. on $[a,b]$ and ψ_1 be the first eigenfunction of* (3.6)*. Then, for all $u\in W^{2,\mathcal{A}}(a,b)$ with $u(a)=0$, $u(b)=0$, we have*

$$\int_a^b u''\psi_1\,dt=\int_a^b \psi_1''u\,dt.$$

Proof. Observe that by integration by parts, we have

$$\int_{a+\delta}^{b-\delta} u''\psi_1\,dt=u'\psi_1|_{a+\delta}^{b-\delta}-\psi_1'u|_{a+\delta}^{b-\delta}+\int_{a+\delta}^{b-\delta}\psi_1''u\,dt.$$

The thesis holds if we prove that

$$\lim_{\delta\to 0}\psi_1(a+\delta)u'(a+\delta)=0\ \text{ and }\ \lim_{\delta\to 0}\psi_1(b-\delta)u'(b-\delta)=0.$$

Let us consider the first equality; the second one is similar. Compute

$$\begin{aligned}|\psi_1(a+\delta)u'(a+\delta)|&\le[\int_{a+\delta}^{t_0}|u''(t)|\,dt+|u'(t_0)|]\,K_1\delta(b-a-\delta)\\&\le\int_{a+\delta}^{t_0}(t-a)|u''(t)|\,dt\;K_1(b-a-\delta)\\&\qquad+|u'(t_0)|\,K_1\delta(b-a-\delta).\end{aligned}$$

Hence, for any $\epsilon>0$, we can take t_0 and δ small enough so that the second member is bounded by ϵ, which ends the proof. □

Proposition 3.16. *Let $r \in \mathcal{A}$, $r(t) > 0$ a.e. on $[a, b]$. If ψ is an eigenfunction of (3.6) that does not change sign, then ψ is an eigenfunction corresponding to the first eigenvalue μ_1.*

Proof. The proof is similar to the one of Proposition 3.5. □

4 Inequalities

This section concerns inequalities which are used to obtain *a priori* bounds on solutions.

Proposition 4.1. *Let the function $u \in H^1(a, b)$ be such that $u(a) = u(b)$ and $\int_a^b u(t)dt = 0$. Then*

$$\|u\|_\infty \leq \sqrt{\frac{b-a}{12}}\|u'\|_{L^2}.$$

Proof. Let $u \in H^1(a, b)$ satisfy the assumptions of the proposition. Such a function has a Fourier serie which reads

$$u(t) = \sum_{n=1}^{\infty}(a_n \cos \tfrac{2\pi nt}{b-a} + b_n \sin \tfrac{2\pi nt}{b-a}),$$

where a_n and $b_n \in \mathbb{R}$. We also have

$$u'(t) = \frac{2\pi}{b-a}\sum_{n=1}^{\infty} n(-a_n \sin \tfrac{2\pi nt}{b-a} + b_n \cos \tfrac{2\pi nt}{b-a}),$$

and

$$\|u'\|_{L^2}^2 = \frac{2\pi^2}{b-a}\sum_{n=1}^{\infty} n^2(a_n^2 + b_n^2).$$

We compute then

$$|u(t)| \leq \sum_{n=1}^{\infty} \tfrac{1}{n}\sqrt{n^2(a_n^2 + b_n^2)} \leq \sqrt{\sum_{n=1}^{\infty} \tfrac{1}{n^2}}\ \sqrt{\sum_{n=1}^{\infty} n^2(a_n^2 + b_n^2)}$$
$$\leq \frac{\pi}{\sqrt{6}}\ \frac{\sqrt{b-a}}{\sqrt{2}\ \pi}\ \|u'\|_{L^2}.$$ □

Let us notice that Proposition 4.1 is optimal as equality is obtained for

$$u(t) = (t - \frac{a+b}{2})^2 - \frac{(b-a)^2}{12}.$$

Proposition 4.2. *Let $u \in W^{2,1}(a,b)$ be such that $u(a) = u(b)$, $u'(a) = u'(b)$ and $\int_a^b u(t)\,dt = 0$. Then*

$$\|u\|_\infty \leq \frac{b-a}{12}\|u''\|_{L^1}.$$

Proof. Using Proposition 4.1 and an integration by part, we compute

$$\|u\|_\infty^2 \leq \frac{b-a}{12}\|u'\|_{L^2}^2 = \frac{b-a}{12}\left|\int_a^b u''(t)u(t)\,dt\right| \leq \frac{b-a}{12}\|u''\|_{L^1}\|u\|_\infty$$

and the result follows. □

Proposition 4.3. *Let $u \in H^2(a,b)$ be such that $u(a) = u(b)$, $u'(a) = u'(b)$ and $\int_a^b u(t)\,dt = 0$. Then*

$$\|u\|_\infty \leq \frac{(b-a)\sqrt{b-a}}{12\sqrt{5}}\|u''\|_{L^2}.$$

Proof. The proof is similar to the one of Proposition 4.1. Let $u \in H^2(a,b)$ satisfy the assumptions of the proposition. Its Fourier serie reads

$$u(t) = \sum_{n=1}^{\infty}(a_n \cos \tfrac{2\pi n t}{b-a} + b_n \sin \tfrac{2\pi n t}{b-a}),$$

where a_n and $b_n \in \mathbb{R}$. We also have

$$u''(t) = -\left(\frac{2\pi}{b-a}\right)^2 \sum_{n=1}^{\infty} n^2(a_n \cos \tfrac{2\pi n t}{b-a} + b_n \sin \tfrac{2\pi n t}{b-a}),$$

and

$$\|u''\|_{L^2}^2 = \left(\frac{2}{b-a}\right)^3 \pi^4 \sum_{n=1}^{\infty} n^4(a_n^2 + b_n^2).$$

We compute then

$$\begin{aligned}|u(t)| &\leq \sum_{n=1}^{\infty} \tfrac{1}{n^2}\sqrt{n^4(a_n^2+b_n^2)} \leq \sqrt{\sum_{n=1}^{\infty}\tfrac{1}{n^4}}\sqrt{\sum_{n=1}^{\infty} n^4(a_n^2+b_n^2)}\\ &\leq \frac{\pi^2}{3\sqrt{10}}\,\frac{(b-a)\sqrt{b-a}}{2\sqrt{2}\,\pi^2}\|u''\|_{L^2} = \frac{(b-a)\sqrt{b-a}}{12\sqrt{5}}\|u''\|_{L^2}.\end{aligned}$$ □

For the next proposition, recall that if $d \in \mathcal{A}$ satisfies $d(t) \gneq 0$ a.e. on $[a,b]$, the problem

$$\begin{aligned} u'' + \mu d(t)u &= 0, \\ u(a) = 0, \ u(b) &= 0, \end{aligned} \tag{4.1}$$

has a first eigenvalue $\mu_1(d)$. Denote by $\psi_1(t;d)$ the corresponding eigenfunction. Recall that $\psi_1(.;d)$ is one-signed and we can assume $\psi_1(.;d) > 0$ on $]a,b[$.

Proposition 4.4. *Let $d \in \mathcal{A}$ be such that $d \gneq 0$, μ_1 and ψ_1 be the corresponding first eigenvalue and eigenfunction of* (4.1).

Then there exists a constant $K > 0$ such that for all $u \in W^{2,\mathcal{A}}(a,b)$ with $u(a) = 0$, $u(b) = 0$ and $\int_a^b u'\psi_1' dt = 0$, we have

$$\|u\|_\infty \le K \int_a^b |u'' + \mu_1 du| \psi_1 \, dt.$$

Proof. Let $u \in W^{2,\mathcal{A}}(a,b)$ be such that $u(a) = 0$, $u(b) = 0$, $\int_a^b u'\psi_1' \, dt = 0$. This function solves the problem

$$x'' + \mu_1 d(t)x = F(t), \quad x(a) = 0, \ x(b) = 0, \tag{4.2}$$

where

$$F(t) := u''(t) + \mu_1 d(t)u(t).$$

Part 1 – Fundamental solutions of the homogeneous equation. The function ψ_1 solves

$$x'' + \mu_1 d(t)x = 0$$

on $[a,b]$. To obtain a second linearly independent solution v, consider the Cauchy problem

$$\begin{aligned} v'' + \mu_1 d(t)v &= 0, \\ v(\tfrac{a+b}{2}) = 0, \ v'(\tfrac{a+b}{2}) &= -k, \end{aligned} \tag{4.3}$$

where $k = 1/\psi_1(\frac{a+b}{2}) > 0$. Let us prove that its solution exists and is bounded on $[a,b]$. Assume there exists $T \in](a+b)/2, b[$ such that $v < 0$ on $](a+b)/2, T[$, $v(T) = 0$ and $v'(T) \ge 0$. This assumption leads to the contradiction

$$0 = \int_{(a+b)/2}^T \frac{d}{dt}(\psi_1' v - v'\psi_1) \, dt = -v'(T)\psi_1(T) - k\,\psi_1((a+b)/2) < 0.$$

Hence, v is negative on $](a+b)/2, b[$. Since v is a solution of (4.3), v is convex on this interval thus bounded. In a similar way, we prove that v is bounded on $[a, (a+b)/2]$.

Part 2 – The solutions of (4.2). As v is bounded on $[a,b]$, we can define, using Proposition 3.14,

$$w(t) = \Big(\int_a^t \frac{s-a}{b-a} v(s)F(s)\,ds - \int_t^b \frac{b-s}{b-a} v(s)F(s)\,ds\Big)\,\psi_1(t) \\ +\Big(\int_t^b \psi_1(s)F(s)\,ds\Big)\,v(t).$$

This function solves (4.2) and

$$u(t) = w(t) - \frac{\int_a^b w'(s)\psi_1'(s)\,ds}{\int_a^b \psi_1'^2(s)\,ds}\,\psi_1(t). \tag{4.4}$$

Part 3 – Bounds on $u(t)$. Notice that as ψ_1 is positive and concave, we have, for all $t \in [a,b]$,

$$\begin{aligned} &\frac{s-a}{b-a}\psi_1(t) \le \frac{s-a}{t-a}\psi_1(t) \le \psi_1(s), \quad \text{if } s < t, \\ &\frac{b-s}{b-a}\psi_1(t) \le \frac{b-s}{b-t}\psi_1(t) \le \psi_1(s), \quad \text{if } s > t. \end{aligned}$$

It follows that

$$\begin{aligned} |w(t)| &\le \int_a^b \psi_1(s)|v(s)||F(s)|\,ds + \int_t^b \psi_1(s)|F(s)|\,ds\,|v(t)| \\ &\le 2\,\|v\|_\infty \int_a^b \psi_1(s)|F(s)|\,ds \le C_1 \int_a^b \psi_1(s)|F(s)|\,ds. \end{aligned}$$

Hence, by (4.4) and since ψ_1 is an eigenfunction corresponding to μ_1,

$$\begin{aligned} |u(t)| &\le |w(t)| + C_2|\psi_1(t)|\,\mu_1 \int_a^b d(s)\psi_1(s)w(s)\,ds \\ &\le (1 + C_2\|\psi_1\|_\infty\mu_1 \int_a^b d(s)\psi_1(s)\,ds)\,\|w\|_\infty \\ &\le K \int_a^b \psi_1(s)|F(s)|\,ds, \end{aligned}$$

which proves the result. □

In a similar way, we can extend Proposition 4.4 to the separated boundary conditions.

Proposition 4.5. *Let a_1, $b_1 \in \mathbb{R}$, a_2, $b_2 \in \mathbb{R}^+$ with $a_1^2 + a_2^2 > 0$, $b_1^2 + b_2^2 > 0$. Let $q \in L^1(a,b)$ and define λ_1 and φ_1 to be the first eigenvalue and eigenfunction of*

$$\begin{aligned} u'' + q(t)u + \lambda u &= 0, \\ a_1u(a) - a_2u'(a) &= 0, \\ b_1u(b) + b_2u'(b) &= 0. \end{aligned}$$

Then there exists a $K > 0$ such that for all $u \in W^{2,1}(a,b)$ with $a_1 u(a) - a_2 u'(a) = 0$, $b_1 u(b) + b_2 u'(b) = 0$ and $\int_a^b (u'\varphi_1' - qu\varphi_1)\,dt = 0$, we have

$$\|u\|_\infty \le K \int_a^b |u'' + q(t)u + \lambda_1 u|\varphi_1\,dt.$$

Proposition 4.6. *Let $\gamma \in L^1(a,b)$ be such that $\gamma(t) \lneqq (\frac{\pi}{b-a})^2$.*
Then there exists $\varepsilon > 0$ such that for all $u \in H_0^1(a,b)$, we have

$$\int_a^b (u'^2 - \gamma u^2)\,dt \ge \varepsilon \|u\|_{H^1}^2.$$

Proof. Assume by contradiction there exists a sequence $(u_n)_n \subset H_0^1(a,b)$ so that

$$\|u_n\|_{H^1} = 1 \quad \text{and} \quad \lim_{n\to\infty} \int_a^b ({u_n'}^2 - \gamma(t)u_n^2)\,dt = 0.$$

Recall that $\int_a^b ({u_n'}^2 - \gamma(t)u_n^2)\,dt \ge 0$. Let $u_n(t) = \bar{u}_n \sin \pi\frac{t-a}{b-a} + \tilde{u}_n(t)$, where

$$\int_a^b \tilde{u}_n(t) \sin \pi\tfrac{t-a}{b-a}\,dt = 0.$$

We compute then

$$\int_a^b ({u_n'}^2 - \gamma(t)u_n^2)\,dt \ge \int_a^b ({u_n'}^2 - (\tfrac{\pi}{b-a})^2 u_n^2)\,dt \ge \tfrac{3\pi^2}{4\pi^2+(b-a)^2} \|\tilde{u}_n\|_{H^1}^2.$$

It follows that $\tilde{u}_n \overset{H^1}{\to} 0$. Going to a subsequence, we can assume $\bar{u}_n \to \bar{u}$ so that $u_n \overset{H^1}{\to} \bar{u} \sin \pi\frac{t-a}{b-a}$, with $\bar{u} \ne 0$. Hence, we obtain the contradiction

$$0 = \lim_{n\to\infty} \int_a^b ({u_n'}^2 - \gamma(t)u_n^2)\,dt = \bar{u}^2 \int_a^b (\tfrac{\pi^2}{(b-a)^2} - \gamma(t)) \sin^2 \pi\tfrac{t-a}{b-a}\,dt \ne 0. \qquad \square$$

5 Maximum Principle

Our first result is a maximum principle which is valid for $W^{2,1}$-solutions.

Theorem 5.1. (Maximum principle) *Let p, $q \in L^1(a,b)$, $q \le 0$ and assume $u \in W^{2,1}(a,b)$, $u(t) \not\equiv 0$, is such that*

$$Lu = -(u'' + pu' + qu) \ge 0, \ \text{a.e. on } [a,b], \ u(a) \ge 0, \ u(b) \ge 0.$$

Then $u(t) > 0$ if $t \in]a,b[$, $u(a) = 0$ implies $u'(a) > 0$ and $u(b) = 0$ implies $u'(b) < 0$.

Proof. Claim 1 – *$u \geq 0$ on $[a,b]$.* If u takes negative values on $]a,b[$, we can find $t_0 \in [a,b[$ and $t_1 \in]a,b]$ such that

$$u(t) < 0 \text{ on }]t_0,t_1[,\ u(t_0) = u(t_1) = 0. \tag{5.1}$$

Let

$$P(t) = \int_{t_0}^{t} p(s)\,ds \tag{5.2}$$

and notice that a.e. in $[t_0,t_1]$,

$$\frac{d}{dt}[u'(t)e^{P(t)}] = -[(Lu)(t) + q(t)u(t)]e^{P(t)} \leq 0.$$

Hence, for any $t \in]t_0,t_1[$,

$$u'(t)e^{P(t)} \leq u'(t_0) \leq 0$$

and as $u(t_0) = u(t_1) = 0$, we have $u(t) = 0$ on $[t_0,t_1]$ which contradicts (5.1).

Claim 2 – *$u > 0$ on $]a,b[$.* Suppose now that $u(t) \not\equiv 0$ is nonnegative. If u is not positive on $]a,b[$, there exist t_0 and t_1 in $]a,b[$ such that $u(t_0) = 0$ and $u(t_1) > 0$. Assume $t_0 < t_1$, define $z \in W^{2,1}(a,b)$ to be the solution of the Cauchy problem

$$\begin{aligned} z'' + p(t)z' + q(t)z &= -1, \\ z(t_0) = 0,\ z'(t_0) &= -1, \end{aligned}$$

and consider the function

$$w = u + \epsilon z,$$

where $\epsilon > 0$ is small enough so that $w(t_1) > 0$. Notice that

$$Lw > 0 \text{ on } [t_0,t_1],\ w(t_0) = 0,\ w(t_1) > 0.$$

From Claim 1, $w(t) \geq 0$ on $[t_0,t_1]$, but this inequality contradicts the fact that $w(t_0) = 0$ and $w'(t_0) = -\epsilon < 0$. The proof is similar if $t_1 < t_0$.

Claim 3 – *$u(a) = 0$ implies $u'(a) > 0$; $u(b) = 0$ implies $u'(b) < 0$.* Suppose $u(a) = 0$. Let $t_0 = a$ and define z and $w = u + \epsilon z$ as in Claim 2. We know that $w(t) \geq 0$ on $[a,t_1]$. Hence

$$w'(a) = u'(a) - \epsilon \geq 0, \text{ i.e. } u'(a) > 0.$$

Similarly, we prove that $u'(b) < 0$ if $u(b) = 0$. □

The above maximum principle can be adapted to deal with the separated boundary value problem.

Theorem 5.2. *Let a_1, $b_1 \in \mathbb{R}$, a_2, $b_2 \in \mathbb{R}^+$, $a_1^2 + a_2^2 > 0$ and $b_1^2 + b_2^2 > 0$, p, $q \in L^1(a,b)$ such that the first eigenvalue λ_1 of (3.1) with $r \equiv 1$ satisfies $\lambda_1 > 0$. Assume $u \in W^{2,1}(a,b)$ is a nontrivial function such that*

$$\begin{aligned} Lu = -(u'' + pu' + qu) \geq 0,\ \text{a.e. on } [a,b],\\ a_1u(a) - a_2u'(a) \geq 0,\ b_1u(b) + b_2u'(b) \geq 0. \end{aligned}$$

Then $u(t) > 0$ if $t \in]a,b[$, $u(a) = 0$ implies $u'(a) > 0$ and $u(b) = 0$ implies $u'(b) < 0$.

Proof. Claim 1 – *$u \geq 0$ on $[a,b]$.* Assume u takes negative values. Let $[t_0, t_1]$ be a maximum interval on which u is nonpositive and define

$$\begin{aligned} v(t) &= u(t), && \text{if } t \in [t_0, t_1],\\ &= 0, && \text{otherwise.} \end{aligned}$$

Observe that $v \in H$ with H defined as in Theorem A-3.2 and that

$$-\int_a^b [(P(t)u')' + Q(t)u]v\, dt \leq 0,$$

where $P(t) = \exp(\int_a^t p(s)\, ds)$ and $Q(t) = q(t)P(t)$. Hence, we have the contradiction

$$0 \geq a(v,v) \geq \lambda_1 \int_a^b v^2(t)\, dt > 0,$$

where $a(\cdot,\cdot)$ is defined in Theorem A-3.2.

Claim 2 – *$u(t) > 0$ if $t \in]a,b[$, $u(a) = 0$ implies $u'(a) > 0$ and $u(b) = 0$ implies $u'(b) < 0$.* Observe that, using Claim 1,

$$-(u'' + pu' + \min\{q,0\}u) \geq 0,$$

and we conclude by Theorem 5.1. □

A similar result holds for the periodic boundary value problem.

Theorem 5.3. *Let p, $q \in L^1(a,b)$ be such that the first eigenvalue λ_1 of (3.5) with $r \equiv 1$ satisfies $\lambda_1 > 0$. Assume $u \in W^{2,1}(a,b)$ is a nontrivial function such that*

$$Lu = -(u'' + pu' + qu) \geq 0,\ \text{a.e. on } [a,b],\ u(a) = u(b),\ u'(a) \leq u'(b).$$

Then $u > 0$ on $[a,b]$.

Proof. The proof of this theorem repeats the argument used to prove Theorem 5.2. □

6 Anti-maximum Principle

Our first result concerns the periodic problem

$$\begin{aligned} &u'' + 2p|u'| + qu = \sigma(t), \\ &u(a) = u(b), \quad u'(a) = u'(b) + A, \end{aligned} \tag{6.1}$$

where $p \in \mathbb{R}$, $q > 0$, $\sigma \in L^1(a,b)$ and $A \in \mathbb{R}$.

To investigate this problem, we consider first the Cauchy problem

$$\begin{aligned} &v'' + 2p|v'| + qv = 0, \\ &v(0) = 1, \quad v'(0) = 0, \end{aligned} \tag{6.2}$$

whose solution can be described as follows:

(i) if $q < p^2$, $v(t)$ is an even function defined by

$$v(t) = e^{pt}\left(\cosh(\sqrt{p^2-q}\,t) - \frac{p}{\sqrt{p^2-q}}\sinh(\sqrt{p^2-q}\,t)\right), \quad t \geq 0;$$

(ii) if $q = p^2$, $v(t)$ is even and

$$v(t) = e^{pt}(1-pt), \quad t \geq 0;$$

(iii) if $q > p^2$, $v(t)$ is a periodic function of period $\frac{2\pi}{\sqrt{q-p^2}}$ defined by

$$v(t) = e^{p|t|}\left(\cos(\sqrt{q-p^2}\,|t|) - \frac{p}{\sqrt{q-p^2}}\sin(\sqrt{q-p^2}\,|t|)\right),$$
$$t \in [-\tfrac{\pi}{\sqrt{q-p^2}}, \tfrac{\pi}{\sqrt{q-p^2}}].$$

Note that if $q \leq p^2$, we have $v'(t) < 0$ for all $t > 0$ and $v'(t) > 0$ for all $t < 0$. Also, if $q > p^2$, we obtain $v'(t) > 0$ for $t \in]-\frac{\pi}{\sqrt{q-p^2}}, 0[$ and $v'(t) < 0$ for $t \in]0, \frac{\pi}{\sqrt{q-p^2}}[$.

Proposition 6.1. *Let $p \in \mathbb{R}$, $q > 0$, $\sigma \in L^1(a,b)$ and $A \in \mathbb{R}$. Suppose that problem* (6.2) *has a solution v such that $v \geq 0$ in $[-(b-a)/2, (b-a)/2]$.*

Then for all $A \geq 0$ and $\sigma \geq 0$, every solution u of problem (6.1) *is such that $u \geq 0$ on $[a,b]$.*

Proof. Claim 1 – A solution $u \neq 0$ of problem (6.1) *must take positive values.* Let $u \leq 0$ be a nontrivial solution of (6.1). As $u(a) = u(b)$, there exists $t_0 \in]a,b[$ such that $u'(t_0) = 0$.

Let us prove that $u'(t) \le 0$ on $[a, t_0[$. If not, there exists $[t_1, t_2] \subset [a, t_0]$ such that $u'(t) > 0$ on $[t_1, t_2[$ and $u'(t_2) = 0$. Hence,

$$\frac{d}{dt}(u'e^{2pt}) = (u'' + 2pu')e^{2pt} = (\sigma - qu)e^{2pt} \ge 0, \quad \forall t \in [t_1, t_2],$$

and we deduce $0 = u'(t_2)e^{2pt_2} \ge u'(t_1)e^{2pt_1} > 0$, which is a contradiction. Similarly, we prove $u'(t) \ge 0$ on $]t_0, b]$. We deduce now from the boundary conditions that $u'(a) = u'(b) = 0$.

As $u'(b) = 0$, the above arguments imply $u'(t) \le 0$ on $[a, b]$ and $u'(a) = 0$ implies $u'(t) \ge 0$. Hence $u'(t) \equiv 0$ and we deduce from the equation the contradiction $0 > qu = \sigma \ge 0$.

Claim 2 – Problem (6.1) *has no solution which changes sign.* Let u be a solution of (6.1) that changes sign. Extending u by periodicity if necessary, we can find some $t_1 \in [a, b]$ and $t_2 \in]t_1, t_1 + (b-a)/2]$ so that $u(t_1)u(t_2) < 0$, $u'(t_1) = 0$ and $u'(t_2) = 0$. Assume $u(t_1) < 0 < u(t_2)$ and choose $s_1 \in [t_1, t_2]$ such that $u(s_1) < 0$ and $u'(s_1) > 0$. Define then $s_2 := \sup\{t \ge s_1 \mid u' > 0 \text{ on } [s_1, t]\}$. Clearly $s_2 \le t_2$ and $u'(s_2) = 0$. Define now $\bar{u}(t) = u(t)e^{pt}$, $\bar{v}(t) = v(t - s_2)e^{pt}$, where v is the solution of (6.2), and compute

$$\frac{d}{dt}(\bar{u}\bar{v}' - \bar{u}'\bar{v}) = -\sigma e^{pt}\bar{v} \le 0 \quad \text{on } [s_1, s_2].$$

This inequality gives the contradiction

$$0 \ge (\bar{u}\bar{v}' - \bar{u}'\bar{v})\Big|_{s_1}^{s_2} = [-u(s_1)v'(s_1 - s_2) + v(s_1 - s_2)u'(s_1)]e^{2ps_1} > 0.$$

The proof is similar if $u(t_1) > 0 > u(t_2)$. □

Next, for all $p \ge 0$, we introduce the first positive zero of the solution v of (6.2), which reads

$$\theta(q,p) = \begin{cases} \frac{1}{\sqrt{p^2-q}} \operatorname{arctanh} \frac{\sqrt{p^2-q}}{p}, & \text{if } 0 < q < p^2, \\ \frac{1}{p}, & \text{if } 0 < q = p^2, \\ \frac{1}{\sqrt{q-p^2}} \left(\frac{\pi}{2} - \arctan \frac{p}{\sqrt{q-p^2}} \right), & \text{if } 0 \le p^2 < q. \end{cases} \tag{6.3}$$

Notice that comparing solutions of (6.2) for different values of p, it is easy to see that $\theta(q, p)$ is a decreasing function of p.

If $p < 0$ and $q \le p^2$, the function v is positive in $\mathbb{R}$ and if $q > p^2$ the first positive zero is given by

$$\chi(q,p) = \frac{1}{\sqrt{q-p^2}} \left(\frac{\pi}{2} + \arctan \frac{|p|}{\sqrt{q-p^2}} \right). \tag{6.4}$$

Proposition 6.1 and the definitions of $\theta(q,p)$ and $\chi(q,p)$ imply the following theorem.

Theorem 6.2. (Anti-maximum principle) *Let $p \in \mathbb{R}$, $q > 0$, $\sigma \in L^1(a,b)$ and $A \in \mathbb{R}$. Suppose u is a solution of problem* (6.1) *for some $A \geq 0$ and $\sigma \geq 0$.*

Then, any of the conditions
(i) $p \geq 0$ and $\theta(q,p) \geq \frac{b-a}{2}$,
(ii) $p < 0$ and $q \leq p^2$ or
(iii) $p < 0$, $q > p^2$ and $\chi(q,p) \geq \frac{b-a}{2}$
imply $u \geq 0$ on $[a,b]$.

As a corollary, we obtain the following result in case $p = 0$.

Corollary 6.3. *The problem*

$$
\begin{gathered}
u'' + qu = \sigma(t), \\
u(a) - u(b) = 0, \ u'(a) - u'(b) = A,
\end{gathered}
$$

has a positive solution for any $A \geq 0$ and $\sigma \in L^1(a,b)$, $\sigma(t) \geq 0$, $\sigma \not\equiv 0$, if $0 < q \leq (\frac{\pi}{b-a})^2$.

We can extend the above results to the Neumann problem. The proofs are essentially the same as in the periodic case. Consider the problem

$$
\begin{aligned}
&u'' + 2p|u'| + qu = \sigma(t), \\
&u'(a) = A, \quad u'(b) = B,
\end{aligned}
\tag{6.5}
$$

where $p \in \mathbb{R}$, $q > 0$, $\sigma \in L^1(a,b)$, $A \in \mathbb{R}$ and $B \in \mathbb{R}$.

Proposition 6.4. *Let $p \in \mathbb{R}$, $q > 0$, $\sigma \in L^1(a,b)$ and A, $B \in \mathbb{R}$. Suppose that problem* (6.2) *has a solution v such that $v \geq 0$ in $[0, b-a]$ and $v' < 0$ in $]0, b-a]$.*

Then for all $A \geq 0$, $B \leq 0$ and $\sigma \geq 0$, every solution u of problem (6.5) *is such that $u \geq 0$ on $[a,b]$.*

Proof. Claim 1 – A solution $u \neq 0$ of problem (6.5) *must take positive values.* See Claim 1 in the proof of Proposition 6.1.

Claim 2 – Problem (6.5) *has no solution which changes sign.* The argument is similar to the one used in the proof of Proposition 6.1. In case there exist $t_1 < t_2$ such that $u(t_1) < 0 < u(t_2)$ we choose $s_1 \in [t_1, t_2]$ such that $u(s_1) < 0$ and $u'(s_1) > 0$ and take $s_2 := \sup\{t \geq s_1 \mid u' > 0 \text{ on } [s_1, t]\}$. We define then $\bar{u}(t) = u(t)e^{pt}$, $\bar{v}(t) = v(t - s_2)e^{pt}$, where v is the solution of (6.2), and come to a contradiction using a similar argument. The same reasoning applies if $u(t_1) > 0 > u(t_2)$. □

We obtain the following result which follows from Proposition 6.4 and the definitions of $\theta(q,p)$ and $\chi(q,p)$.

Theorem 6.5. *Let* $p \in \mathbb{R}$, $q > 0$, $\sigma \in L^1(a,b)$ *and* A, $B \in \mathbb{R}$. *Suppose* u *is a solution of problem* (6.5) *for some* $A \geq 0$, $B \leq 0$ *and* $\sigma \geq 0$.
Then, any of the conditions
(i) $p \geq 0$ *and* $\theta(q,p) \geq b-a$,
(ii) $p < 0$ *and* $q \leq p^2$ *or*
(iii) $p < 0$, $q > p^2$ *and* $\chi(q,p) \geq b-a$
imply $u \geq 0$ *on* $[a,b]$.

Bibliography

[1] **A. Adje**, *Sur et sous solutions dans les équations différentielles discontinues avec conditions aux limites non linéaires*, Dissertation Doctorale, Université Catholique de Louvain, Louvain-la-Neuve, 1987.

[2] **A. Adje**, *Sur et sous-solutions généralisées et problèmes aux limites du second ordre*, Bull. Soc. Math. Belg. 42 ser.B (1990), 347-368.

[3] **L. Aguinaldo and K. Schmitt**, *On the boundary value problem* $u'' + u = \alpha u^-$, $u(0) = 0 = u(\pi)$, Proc. A.M.S. 68 (1978), 64-68.

[4] **S. Ahmad, A.C. Lazer, and J.L. Paul**, *Elementary critical point theory and perturbations of elliptic boundary value problems at resonance*, Indiana Univ. Math. J. 25 (1976), 933-944.

[5] **K. Akô**, *On the Dirichlet problem for quasi-linear elliptic differential equations of the second order*, J. Math. Soc. Japan 13 (1961), 45-62.

[6] **K. Akô**, *Subfunctions for ordinary differential equations*, J. Fac. Sci. Univ. Tokyo 12 (1965), 17-43.

[7] **K. Akô**, *Subfunctions for ordinary differential equations II*, Funkcialaj Ekvacioj 10 (1967), 145-162.

[8] **K. Akô**, *Subfunctions for ordinary differential equations III*, Funkcialaj Ekvacioj 11 (1968), 111-129.

[9] **K. Akô**, *Subfunctions for ordinary differential equations IV*, Funkcialaj Ekvacioj 11 (1968), 185-195.

[10] **K. Akô**, *Subfunctions for ordinary differential equations V*, Funkcialaj Ekvacioj 12 (1969), 239-249.

[11] **K. Akô**, *Subfunctions for ordinary differential equations VI*, J. Fac. Sci. Univ. Tokyo 16 (1969), 149-156.

[12] **K. Akô and T. Kusano**, *On bounded solutions of second order elliptic differential equations*, J. Fac. Sci. Univ. Tokyo 9 (1964), 29-37.

[13] **H. Amann**, *On the existence of positive solutions of nonlinear elliptic boundary value problems*, Indiana Univ. Math. J. 21 (1971), 125-146.

[14] **H. Amann**, *Existence of multiple solutions for nonlinear elliptic boundary value problems*, Indiana Univ. Math. J. 21 (1972), 925-935.

[15] **H. Amann**, *On the number of solutions of nonlinear equations in ordered Banach spaces*, J. Funct. Anal. 11 (1972), 346-384.

[16] **H. Amann**, *Existence and multiplicity theorems for semi-linear elliptic boundary value problems*, Math. Z. 150 (1976), 281-295.

[17] **H. Amann**, *Fixed point equations and nonlinear eigenvalue problems in ordered Banach spaces*, SIAM Review 18 (1976), 620-709.

[18] **H. Amann**, *A note on degree theory for gradient mappings*, Proc. A.M.S. 85 (1982), 591-595.

[19] **H. Amann, A. Ambrosetti, and G. Mancini**, *Elliptic equations with noninvertible Fredholm linear part and bounded nonlinearities*, Math. Z. 158 (1978), 179-194.

[20] **H. Amann and M. Crandall**, *On some existence theorems for semi-linear elliptic equations*, Indiana Univ. Math. J. 27 (1978), 779-790.

[21] **H. Amann and P. Hess**, *A multiplicity result for a class of elliptic boundary value problems*, Proc. Royal Soc. Edinburgh 84A (1979), 145-151.

[22] **A. Ambrosetti and P. Hess**, *Pairs of solutions for some nonlinear elliptic equations*, Boll. U.M.I. 16A (1979), 588-592.

[23] **A. Ambrosetti and G. Mancini**. *Theorems of existence and multiplicity for nonlinear elliptic problems with noninvertible linear part*, Ann. Scuola Norm. Sup. Pisa Serie IV,5 (1978), 15-28.

[24] **A. Ambrosetti and G. Prodi**, *On the inversion of some differentiable mappings with singularities between Banach spaces*, Ann. Mat. Pure Appl. 93 (1972) 231-247.

[25] **B.N. Babkin**, *Solution of a boundary value problem for an ordinary differential equation of second order by Caplygin's method*, Prikl. Math. Meh. Akad. Nauk. SSSR 18 (1954), 239-242.

[26] **P.B. Bailey, L.F. Shampine, and P.E. Waltman**, *Nonlinear two point boundary value problems*, Academic Press, New York, 1968.

[27] **I. Barbalat**, *Applications du principe topologique de T. Wazewski aux équations différentielles du second ordre*, Ann. Polon. Math 5 (1958), 303-317.

[28] **T. Bartsch and Z.Q. Wang**, *On the existence of sign-changing solutions for semilinear Dirichlet problems*, Top. Meth. Nonlin. Anal. 7 (1996), 115-131.

[29] **J.V. Baxley**, *A singular nonlinear boundary value problem: membrane response of a spherical cap*, SIAM J. Appl. Math. 48 (1988), 497-505.

[30] **A. Bellen**, *Monotone methods for periodic solutions of second order scalar functional differential equations*, Numer. Math. 42 (1983), 15-30.

[31] **K. Ben-Naoum and C. De Coster**, *On the existence and multiplicity of positive solutions of the p-Laplacian separated boundary value problem*, Diff. Int. Equ. 10 (1997), 1093-1112.

[32] **H. Berestycki**, *Le nombre de solutions de certains problèmes semilinéaires elliptiques*, J. Funct. Anal. 40 (1981), 1-29.

[33] **M.S. Berger and E. Podolak**, *On the solutions of a nonlinear Dirichlet problem*, Indiana Univ. Math. J. 24 (1975), 837-846.

[34] **S.R. Bernfeld and J. Chandra**, *Minimal and maximal solutions of nonlinear boundary value problems*, Pacific J. Math. 71 (1977), 13-20.

[35] **S. Bernstein**, *Sur certaines équations différentielles ordinaires du second ordre*, C.R. Acad. Sci. Paris 138 (1904), 950-951.

[36] **S. Bernstein**, *Sur les équations du calcul des variations*, Ann. Sci. Ecole Norm. Sup. 29 (1912), 431-485.

[37] **L.E. Bobisud, D. O'Regan, and W.D. Royalty**, *Solvability of some nonlinear boundary value problems*, Nonlinear Anal. T.M.A. 12 (1988), 855-869.

[38] **L. Boccardo, F. Murat, and J.P. Puel**, *Existence de solutions faibles pour des équations elliptiques quasi-linéaires à croissance quadratique*, Nonlinear partial differential equations and their applications, Collège de France Seminar, vol. IV (Ed: H. Brezis and J.L. Lions), Research Notes in Math. 84, Pitman, London (1983), 19-73.

[39] **L. Boccardo, F. Murat, and J.P. Puel**, *Résultats d'existence pour certains problèmes elliptiques quasilinéaires*, Ann. Scuola Norm. Sup. Pisa 11 (1984), 213-225.

[40] **Y.P. Boglaev**, *The two-point problem for a class of ordinary differential equations with a small parameter coefficient of the derivative*, U.S.S.R. Comput. Math. Math. Phys. 10 (1970), 191-204.

[41] **D. Bonheure and C. De Coster**, *Forced singular oscillators and the method of lower and upper solutions*, Topological Methods in Nonlinear Analysis 22 (2003), 297-317.

[42] **D. Bonheure, C. Fabry, and D. Smets**, *Periodic solutions of forced isochronous oscillators at resonance*, Discrete Cont. Dyn. Syst. 8 (2002), 907-930.

[43] **H. Brezis**, *Analyse fonctionnelle: Théorie et applications*, Masson, Paris, 1983.

[44] **H. Brezis and L. Nirenberg**, *H^1 versus C^1 local minimizers*, C. R. Acad. Sci. Paris 317 (1993), 465-472.

[45] **H. Brezis and R.E.L. Turner**, *On a class of superlinear elliptic problems*, Comm. Partial Diff. Equ. 2 (1977), 601-614.

[46] **N.I. Briš**, *On boundary value problems for the equation $\epsilon y'' = f(x, y, y')$ for small ϵ* (in russian), Dokl. Akad. Nauk. SSR 95 (1954), 429-432.

[47] **K.J. Brown and H. Budin**, *Multiple positive solutions for a class of nonlinear boundary value problems*, J. Math. Anal. and Appl. 60 (1977), 329-338.

[48] **K.J. Brown and H. Budin**, *On the existence of positive solutions for a class of semilinear elliptic boundary value problems*, SIAM J. Math. Anal. 10 (1979), 875-883.

[49] **V.F. Butuzov, N.N. Nefedov, and K.R. Schneider**, *Singularly perturbed problems in case of exchange of stabilities*, J. Math. Sciences 121 (2004), 1973-2079.

[50] **A. Cabada**, *El metodo monotono para problemas de frontera*, Ph. D. Thesis, Univ. Santiago de Compostela, 1992.

[51] **A. Cabada, P. Habets, and S. Lois**, *Monotone method for the Neumann problem with lower and upper solutions in the reverse order*, Applied Mathematics and Computation 117 (2001), 1-14.

[52] **A. Cabada, P. Habets, and R. Pouso**, *Optimal existence conditions for Φ-Laplacian equations with upper and lower solutions in the reversed order*, J. Differential Equations 166 (2000), 385-401.

[53] **A. Cabada and L. Sanchez**, *A positive operator approach to the Neumann problem for a second order ordinary differential equation*, J. Math. Anal. Appl. 204 (1996), 774-785.

[54] **A. Cañada and P. Drábek**, *On semilinear problems with nonlinearities depending only on derivatives*, SIAM J. Math. Anal. 27 (1996), 543-557.

[55] **S. Carl**, *A combined variational-monotone iterative method for elliptic boundary value problems with discontinuous nonlinearity*, Appl. Anal. 43 (1992), 21-45.

[56] **A. Castro**, *Periodic solutions of the forced pendulum equation*, Differential Equations (Ed: S. Ahmad, M. Keener, A. Lazer), Academic Press, New York 1980, 149-160.

[57] **A. Castro, J.B. Garner, and R. Shivaji**, *Existence results for classes of sublinear semipositone problems*, Results in Math. 23 (1993), 214-220.

[58] **K.C. Chang**, *A variant mountain pass lemma*, Scientia Sinica (Series A) 26 (1983), 1241-1255.

[59] **K.C. Chang**, *Variational methods and sub- and super-solutions*, Scientia Sinica (Series A) 26 (1983), 1256-1265.

[60] **K.C. Chang**, *Infinite dimensional Morse theory and multiple solution problems*, Birkhäuser, Boston, 1993.

[61] **K.W. Chang and F.A. Howes**, *Nonlinear singular perturbation phenomena theory and application*, Applied Mathematical Sciences 56, Springer-Verlag, Berlin, 1984.

[62] **S.A. Chaplygin**, *Complete works III*, Akademia Nauk SSSR, Leningrad, 1935.

[63] **M. Cherpion**, *La méthode des sous et sur-solutions: itérations monotones et problèmes singuliers*, Thèse de doctorat, Louvain-la-Neuve (Belgique), 2002.

[64] **M. Cherpion, C. De Coster, and P. Habets**, *Monotone iterative methods for boundary value problems*, Differential and Integral Equations 12 (1999), 309-338.

[65] **M. Cherpion, C. De Coster, and P. Habets**, *A constructive monotone iterative method for second order BVP in presence of lower and upper solutions*, Applied Math. and Comp. 123 (2001), 75-91.

[66] **R. Chiappinelli, J. Mawhin, and R. Nugari**, *Generalized Ambrosetti-Prodi conditions for nonlinear two-point boundary value problems*, J. Diff. Equ. 69 (1987), 422-434.

[67] **Y. Choquet-Bruhat and J. Leray**, *Sur le problème de Dirichlet, quasilinéaire, d'ordre 2*, C.R. Acad. Sci. Paris 274 (1972), 81-85.

[68] **J.A. Cid**, *On extremal solutions in Schauder's theorem with applications to differential equations*, Bull. Belgian Math. Soc. 11 (2004), 15-20.

[69] **F.H. Clarke**, *Optimization and nonsmooth analysis*, Université de Montréal, Centre de Recherches Mathématiques, Montréal, 1989.

[70] **P. Clément and L.A. Peletier**, *An anti-maximum principle for second order elliptic operators*, J. Differential Equations 34 (1979), 218-229.

[71] **E.A. Coddington and N. Levinson**, *A boundary value problem for a nonlinear differential equation with a small parameter*, Proc. A.M.S. 3 (1952), 73-81.

[72] **D.S. Cohen**, *Positive solutions of a class of nonlinear eigenvalue problems*, J. Math. Mech. 17 (1967), 209-215.

[73] **L. Collatz and J. Schröder**, *Einschliessen der Lösungen von Randwertaufgaben*, Num. Math. 1 (1959), 61-72.

[74] **M. Conti, L. Merizzi, and S. Terracini**, *Remarks on variational methods and lower-upper solutions*, NoDEA Nonlinear Differential Equations Appl. 6 (1999), 371-393.

[75] **F.J.S.A. Correa**, *On pairs of positive solutions for a class of sub-superlinear elliptic problems*, Diff. and Int. Equations 5 (1992), 387-392.

[76] **D.G. Costa and J.V.A. Goncalves**, *On the existence of positive solutions for a class of non-selfadjoint elliptic boundary value problems*, Applicable Analysis 31 (1989), 309-320.

[77] **R. Courant and D. Hilbert**, *Methods of Mathematical Physics, vol. II: Partial Differential Equations*, Interscience Publishers, New York - London, 1962.

[78] **M. Cuesta and J.P. Gossez**, *A variational approach to nonresonance with respect to the Fučík spectrum*, Nonlinear Anal. T.M.A. 19 (1992), 487-500.

[79] **M. Cuesta, J.P. Gossez, and P. Omari** *Nonresonance to the right of the first eigenvalue for the one-dimensional p-laplacian*, Nonlinear Anal. 38 (1999), 481-496.

[80] **B. Dacorogna**, *Direct methods in the calculus of variations*, Springer-Verlag, Berlin, 1989.

[81] **E.N. Dancer**, *On the ranges of certain weakly nonlinear elliptic partial differential equations*, J. Math. Pures Appl. 57 (1978), 351-366.

[82] **E.N. Dancer and Y. Du**, *A note on multiple solutions of some semilinear elliptic problems*, J. Math. Anal. Appl. 211 (1997), 626-640.

[83] **R. Dautray and J.-L.Lions**, *Analyse mathématique et calcul numérique pour les sciences et les techniques*, Masson, Paris, 1988.

[84] **C. De Coster**, *Pairs of positive solutions for the one-dimensional p-laplacian*, Nonlinear Anal. T.M.A. 23 (1994), 669-681.

[85] **C. De Coster**, *La méthode des sur et sous solutions dans l'étude de problèmes aux limites*, Thèse de doctorat, Louvain-la-Neuve (Belgique), 1994.

[86] **C. De Coster**, *Lower and upper solutions for singular derivative dependent Dirichlet problem*, Math. Inequal. Appl. 4 (2001), 377-396.

[87] **C. De Coster, M.R. Grossinho, and P. Habets**, *On pairs of positive solutions for a singular boundary value problem*, Applicable Analysis 59 (1995), 241-256.

[88] **C. De Coster and P. Habets**, *A two parameters Ambrosetti-Prodi problem*, Portugaliae Math. 53 (1996), 279-303.

[89] **C. De Coster and P. Habets**, *Upper and lower solutions in the theory of ODE boundary value problems : classical and recent results*, in "Nonlinear analysis and boundary value problems for ordinary differential equations" (Ed: F. Zanolin), C.I.S.M. Courses and Lectures 371, Springer, New York (1993) 1-79.

[90] **C. De Coster and P. Habets**, *Existence and multiplicity of positive solutions of the Ginzburg-Landau boundary value problem*, J. Comp. Appl. Math. 113 (2000), 317-327.

[91] **C. De Coster and P. Habets**, *An overview of the method of lower and upper solutions for ODEs*, in "Nonlinear analysis and its Applications to Differential Equations" (M.R. Grossinho, M. Ramos, C. Rebelo and L. Sanchez eds), Progress in Nonlinear Differential Equations and their Applications vol. 43, Birkhauser, Boston (2001) 3-22.

[92] **C. De Coster and P. Habets**, *The lower and upper solution method for boundary value problems*, in "Handbook of Differential Equations, Ordinary Differential Equations" Vol. 1 (Eds: A. Cañada, P. Drábek, and A. Fonda), Elsevier Science B.V., North Holland, Amsterdam et al. (2004), 69-160.

[93] **C. De Coster and M. Henrard**, *Existence and localization of solution for elliptic problem in presence of lower and upper solutions without any order*, J. Differential Equations 145 (1998), 420-452.

[94] **C. De Coster and P. Omari**, *Unstable periodic solutions for a parabolic problem in presence of non-well-ordered lower and upper solutions*, J. Funct. Anal. 175 (2000), 52-88.

[95] **C. De Coster and M. Tarallo**, *Foliations, associated reductions and lower-upper solutions*, Calculus of Variations and P.D.E. 15 (2002), 25-44.

[96] **D.G. de Figueiredo**, *Positive solutions of semilinear elliptic problems*, in "Differential Equations", Proc. Sao Paolo, 1981, L.N.M. 957, Springer-Verlag, Berlin (1982), 34-87.

[97] **D.G. de Figueiredo**, *Positive solutions for some classes of semilinear elliptic problems*, Proceedings of Symposia in Pure Mathematics 45 (1986), 371-379.

[98] **D.G. de Figueiredo**, *On the existence of multiple ordered solutions of nonlinear eigenvalue problems*, Nonlinear Anal. T.M.A. 11 (1987), 481-492.

[99] **D.G. de Figueiredo**, *Lectures on the Ekeland variational principle with applications and detours*, Tata Institute of Fundamental Research, Lectures on Math. and Phys. 81, Springer Verlag, Berlin-New York, 1989.

[100] **D.G. de Figueiredo and P.L. Lions**, *On pairs of positive solutions for a class of semilinear elliptic problems*, Indiana Univ. Math. J. 34 (1985), 591-606.

[101] **D.G. de Figueiredo and W.N. Ni**, *Perturbations of second order linear elliptic problems by nonlinearities without Landesman-Lazer condition*, Nonlinear Anal. T.M.A. 3 (1979), 629-634.

[102] **D.G. de Figueiredo and S. Solimini**, *A variational approach to superlinear elliptic problems*, Comm. Partial Diff. Equations 9 (1984), 699-717.

[103] **K. Deimling**, *Nonlinear functional analysis*, Springer-Verlag, Berlin, 1985.

[104] **M.A. del Pino, R. Manasevich, and A. Montero**, *T-periodic solutions for some second order differential equations with singularities*, Proc. Roy. Soc. Edimburgh Sect. A 120 (1992), 231-243.

[105] **J. Deuel and P. Hess**, *A criterion for the existence of solutions of non-linear elliptic boundary value problems*, Proc. Royal Society of Edinburgh 74A (1974/75), 49-54.

[106] **C.L. Dolph**, *Nonlinear integral equations of the Hammerstein type*, Trans. A.M.S. 66 (1949), 289-386.

[107] **D. Dunninger**, *Existence of positive solutions for fourth order nonlinear problems*, Boll. U.M.I. (7) 1-B (1987), 1129-1138.

[108] **H. Epheser**, *Über die Existenz der Lösungen von Randwertaufgaben mit gewöhnlichen, nichtlinearen Differentialgleichungen zweiter Ordnung*, Math. Zeitschr. 61 (1955), 435-454.

[109] **L.H. Erbe**, *Nonlinear boundary value problems for second order differential equations*, J. Diff. Equ. 7 (1970), 459-472.

[110] **L.H. Erbe, S. Hu, and H. Wang**, *Multiple positive solutions of some boundary value problems*, J. Math. Anal. Appl. 184 (1994), 640-648.

[111] **L.H. Erbe and H. Wang**, *On the existence of positive solutions of ordinary differential equations*, Proc. A.M.S. 120 (1994), 743-748.

[112] **A. Erdélyi**, *On a nonlinear boundary value problem involving a small parameter*, J. Austral. Math. Soc. 2 (1962), 425-439.

[113] **C. Fabry and P. Habets**, *The Picard boundary value problem for nonlinear second order vector differential equations*, J. Diff. Equ. 42 (1981), 186-198.

[114] **C. Fabry and P. Habets**, *Upper and lower solutions for second-order boundary value problems with nonlinear boundary conditions*, Nonlinear Anal. T.M.A. 10 (1986), 985-1007.

[115] **C. Fabry, J. Mawhin, and M.N. Nkashama**, *A multiplicity result for periodic solutions of forced nonlinear second order ordinary differential equations*, Bull. London Math. Soc. 18 (1986), 173-180.

[116] **M.L. Fernandes, P. Omari, and F. Zanolin**, *On the solvability of a semilinear two-point BVP around the first eigenvalue*, Diff. and Int. Equations 2 (1989), 63-79.

[117] **A. Fonda**, *Periodic solutions of scalar second order differential equations with a singularity*, Academie Royale de Belgique, Classe des Sciences, Mémoires, 1993.

[118] **A. Fonda**, *On the existence of periodic solutions for scalar second order differential equations when only the asymptotic behaviour of the potential is known*, Proc. A.M.S. 119 (1993), 439-445.

[119] **A. Fonda, J.P. Gossez, and F. Zanolin**, *On a nonresonance condition for a semilinear elliptic problem*, Diff. Int. Equ. 4 (1991), 945-951.

[120] **A. Fonda and M. Ramos**, *Large-amplitude subharmonic oscillations for scalar second-order differential equations with asymmetric nonlinearities*, J. Diff. Equ. 109 (1994), 354-372.

[121] **S. Fučík**, *Solvability of nonlinear equations and boundary value problems*, Reidel, Dordrecht, 1980.

[122] **R.E. Gaines**, *A Priori bounds and upper and lower solutions for nonlinear second-order boundary-value problems*, J. Diff. Equ. 12 (1972), 291-312.

[123] **R.E. Gaines and J. Mawhin**, *Coincidence degree and nonlinear differential equations*, Lectures Notes in Math. 568, Springer-Verlag, Berlin, 1977.

[124] **J.A. Gatica, V. Oliker, and P. Waltman**, *Singular nonlinear boundary value problems for second order ordinary differential equations*, J. Diff. Equ. 79 (1989), 62-78.

[125] **M. Gaudenzi and P. Habets**, *Existence and multiplicity of positive solutions for boundary value problems of 2d order ODE*, Topological Methods in Nonlinear Analysis 14 (1999), 131-150.

[126] **M. Gaudenzi, P. Habets, and F. Zanolin**, *Positive solutions of superlinear boundary value problems with singular indefinite weight*, Bull. Belgian Math. Soc. 9 (2002), 607-619.

[127] **G.V. Gendzhoyan**, *On two-sided Chaplygin approximations to the solution of the two point boundary value problem*, Izv. SSR Jiz Mate Nauk 17 (1964), 21-27.

[128] **H. Gingold and S. Rosenblat**, *Differential equations with moving singularities*, SIAM J. Math. Anal. 7 (1976), 942-957.

[129] **J.P. Gossez**, *Some nonlinear differential equations with resonance at the first eigenvalue*, Confer. Sem. Mat. Univ. Bari 167 (1979), 355-389.

[130] **J.P. Gossez and P. Omari**, *Periodic solutions of a second order ordinary differential equation: a necessary and sufficient condition for nonresonance*, J. Diff. Equ. 94 (1991), 67-82.

[131] **J.P. Gossez and P. Omari**, *A necessary and sufficient condition of nonresonance for a semilinear Neumann problem*, Proc. A.M.S. 114 (1992), 433-442.

[132] **J.P. Gossez and P. Omari**, *Non-ordered lower and upper solutions in semilinear elliptic problems*, Comm. P.D.E. 19 (1994), 1163-1184.

[133] **J.P. Gossez and P. Omari**, *On a semilinear elliptic Neumann problem with asymmetric nonlinearities*, Trans. A.M.S. 347 (1995), 2553-2562.

[134] **J.L. Gouze and K.P. Hadeler**, *Monotone flows and order intervals*, Nonlinear World 1 (1994), 23-34.

[135] **A. Granas, R.B. Guenther, and J.W. Lee**, *Topological transversality II. Applications to the Neumann problem for $y'' = f(t, y, y')$*, Pacific J. Math. 104 (1983), 95-109.

[136] **L.J. Grimm and K. Schmitt**, *Boundary value problems for delay-differential equations*, Bull. A.M.S. 74 (1968), 997-1000.

[137] **L.J. Grimm and K. Schmitt**, *Boundary value problem for differential equations with deviating arguments*, Aequationes Mathematicae 4 (1970), 176-190.

[138] **V.V. Gudkov, A. Klokov, and A.J. Lepin**, *Two points boundary value problems for ordinary differential equations*, (in russian) Zinatne Riga, 1973.

[139] **V.V. Gudkov and A.J. Lepin**, *On necessary and sufficient conditions for the solvability of certain boundary-value problems for a second-order ordinary differential equation*, Dokl. Akad. Nauk S.S.S.R. 210 (1973), 800-803.

[140] **D. Guo**, *A fixed point theorem of decreasing operators and its applications*, (in chinese), Kexue Tongbao 29 (1984), 189.

[141] **D. Guo and V. Lakshmikantham**, *Nonlinear problems in abstract cones*, Academic Press, Inc., 1988.

[142] **Z. Guo**, *Solvability of some singular nonlinear boundary value problems and existence of positive radial solutions of some nonlinear elliptic problems*, Nonlinear Anal. T.M.A. 16 (1991), 781-790.

[143] **S. Haber and N. Levinson**, *A boundary value problem for a singularly perturbed differential equation*, Proc. A.M.S. 6 (1955), 866-872.

[144] **P. Habets and M. Laloy**, *Perturbations singulières de problèmes aux limites: Les sur- et sous-solutions*, Séminaire de Mathématique Appliquée et Mécanique 76, U.C.L., 1974.

[145] **P. Habets and P. Omari**, *Existence and localization of solutions of second order elliptic problems using lower and upper solutions in the reversed order*, Top. Meth. Nonlinear Anal. 8 (1996), 25-56.

[146] **P. Habets and P. Omari**, *Positive solutions of an indefinite prescribed mean curvature problem on a general domain*, Advanced Nonlinear Studies 4 (2004), 1-13.

[147] **P. Habets, P. Omari, and F. Zanolin**, *Nonresonance conditions on the potential with respect to the Fučík spectrum for the periodic boundary value problem*, Rocky Mountain J. Math. 25 (1995), 1305-1340.

[148] **P. Habets and R. Pouso**, *Examples of non existence of solution in presence of upper and lower solutions*, ANZIAM J. 44 (2003), 591-594.

[149] **P. Habets and L. Sanchez**, *Periodic solutions of some Liénard equations with singularities*, Proc. A.M.S. 109 (1990), 1035-1044.

[150] **P. Habets and L. Sanchez**, *A two-point problem with nonlinearity depending only on the derivative*, SIAM J. Math. Anal. 28 (1997), 1205-1211.

[151] **P. Habets and K. Schmitt**, *Nonlinear boundary value problems for systems of differential equations*, Archiv der Math. 40 (1983), 441-446.

[152] **P. Habets and P. Torres**, *Some multiplicity for periodic solutions of a Rayleigh differential equation*, Dyn. of Contin. Discrete Impuls. Syst. Ser. A Math. Anal. 8 (2001), 335-351.

[153] **P. Habets and F. Zanolin**, *Upper and lower solutions for a generalized Emden-Fowler equation*, J. Math. Anal. Appl. 181 (1994), 684-700.

[154] **P. Habets and F. Zanolin**, *Positive solutions for a class of singular boundary value problem*, Boll. U.M.I. (7) 9-A (1995), 273-286.

[155] **G. Harris**, *The influence of boundary data on the number of solutions of boundary value problems with jumping nonlinearities*, Trans. A.M.S. 321 (1990), 417-464.

[156] **W.A. Harris**, *Application of the method of differential inequalities in singular perturbation problems*, in "New developments in differential equations", W. Eckhaus (Ed.), North-Holland, Amsterdam (1976), 111-116.

[157] **J.W. Heidel**, *A second-order nonlinear boundary value problem*, J. Math. Anal. Appl. 48 (1974), 493-503.

[158] **S. Heikkilä and V. Lakshmikantham**, *Monotone iterative techniques for discontinuous nonlinear differential equations*, Monographs and textbooks in pure and applied mathematics 181, Marcel Dekker, Inc., New York, 1994.

[159] **P. Hess**, *On the solvability of nonlinear elliptic boundary value problems*, Indiana Univ. Math. J. 25 (1976), 461-466.

[160] **P. Hess**, *On a second-order nonlinear elliptic boundary value problem*, Nonlinear Anal, A collection of papers in honor of E. Rothe, Ed: L. Cesari, R. Kannan and H. Weinberger, Academic Press, New York (1978), 99-107.

[161] **P. Hess**, *On multiple positive solutions of nonlinear elliptic eigenvalue problems*, Comm. Partial Diff. Equ. 6 (1981) 951-961.

[162] **P. Hess**, *An antimaximum principle for linear elliptic equations with an indefinite weight function*, J. Diff. Equ. 41 (1981), 369-374.

[163] **I. Hirai and K. Akô**, *On generalized Peano's theorem concerning the Dirichlet problem for semi-linear elliptic differential equations*, Proc. Japan Acad. 36 (1960), 480-485.

[164] **H. Höfer**, *Variational and topological methods in partially ordered Hilbert spaces*, Math. Ann. 261 (1982), 493-514.

[165] **H. Höfer**, *A note on the topological degree at a critical point of the Mountain-Pass type*, Proc. A.M.S. 90 (1984), 309-315.

[166] **F.A. Howes**, *Singular perturbations and differential inequalities*, PhD. Thesis, University of Southern California, 1974.

[167] **F.A. Howes**, *Singular perturbations and differential inequalities*, Memoirs A.M.S. 168 (1976).

[168] **F.A. Howes**, *Some classical and nonclassical singular perturbation problems* , Funkcialaj Ekvacioj 19 (1976), 113-132.

[169] **F.A. Howes**, *Singularly perturbed boundary value problems with angular limiting solutions*, Trans. A.M.S. 241 (1978), 155-182.

[170] **F.A. Howes**, *Some old and new results on singularly perturbed boundary value problems*, in "Singular Perturbations and Asymptotics", R.E. Meyer - S.V. Parter (Ed.), Academic Press, New York (1980), 41-85.

[171] **V.C. Hutson**, *Boundary value problems for differential difference equations*, J. Diff. Equ. 36 (1980), 363-373.

[172] **R. Iannacci and M.N. Nkashama**, *Nonlinear two point boundary value problems at resonance without Landesman-Lazer condition*, Proc. A.M.S. 106 (1989), 943-952.

[173] **L.K. Jackson**, *Subfunctions and second-order ordinary differential inequalities*, Advances in Math. 2 (1967), 307-363.

[174] **J. Janus and J. Myjak**, *A generalized Emden-Fowler equation with a negative exponent*, Nonlinear Anal. T.M.A. 23 (1994), 953-970.

[175] **L. Kantorovich**, *The method of successive approximations for functional equations*, Acta Math. 71 (1939), 63-97.

[176] **J.K. Kazdan and R.J. Kramer**, *Invariant criteria for existence of solutions to second-order quasilinear elliptic equations*, Comm. Pure Appl. Math. 31 (1978), 619-645.

[177] **J.L. Kazdan and F.W. Warner**, *Remarks on some quasilinear elliptic equations*, Comm. Pure Applied Math. 28 (1975), 567-597.

[178] **H.B. Keller and D.S. Cohen**, *Some positone problems suggested by non-linear heat generation*, J. Math. Mech. 16 (1967), 1361-1376.

[179] **H.B. Keller and E. Reiss**, *Iterative solutions for the nonlinear bending of circular plates*, Comm. Pure Appl. Math. 9 (1958), 273-292.

[180] **J.L. Kelley**, *General Topology*, Graduate Texts in Mathematics 27, Springer Verlag, New York, 1955.

[181] **M. Khavanin and V. Lakshmikantham**, *The method of mixed monotony and second order boundary value problems*, J. Math. Anal. Appl. 120 (1986), 737-744.

[182] **A.Ya. Khokhryakov and B.M. Arkhipov**, *The differential inequality theorem for a periodic boundary value problem with a second order ODE*, Diff. Urav. 1 (1965), 335-345.

[183] **I.T. Kiguradze**, *A priori estimates for derivatives of bounded functions satisfying second-order differential inequalities*, Differentsial'nye Uravneniya 3 (1967), 1043-1052.

[184] **I.T. Kiguradze**, *On periodic solutions of nonlinear second order differential equations*, in "Proc. IV Int. Conf. Nonlinear Oscill." Prague (1967), 175-180.

[185] **I.T. Kiguradze**, *Some singular boundary value problems for ordinary nonlinear second order differential equations*, Differentsial'nye Uravneniya 4 (1968), 1753-1773.

[186] **I.T. Kiguradze**, *On a singular boundary value problem*, J. Math. Anal. and Appl. 30 (1970), 475-489.

[187] **I.T. Kiguradze and B.L. Shekhter**, *Singular boundary-value problems for ordinary second-order differential equations*, Itogi Nauki i Tekhniki, Seriya Sovremennye Problemy Matematiki, Noveishie Dostizheniya 30 (1987), 105-201, translated in J. Soviet Math. 43 (1988), 2340-2417.

[188] **D. Kinderlehrer and G. Stampacchia**, *An introduction to variational inequalities and their applications*, Academic Press, New York, 1980.

[189] **K. Klingelhöfer**, *Nonlinear boundary value problems with simple eigenvalue of the linear part*, Arch. Rat. Mech. Anal. 37 (1970), 381-398.

[190] **H.W. Knobloch**, *Zwei Kriterien für die Existenz periodischer Lösungen von Differentialgleichungen zweiter Ordnung*, Archiv. der Math. 14 (1963), 182-185.

[191] **H.W. Knobloch**, *Eine neue Methode zur Approximation periodischer Lösungen nicht-linearer Differentialgleichungen zweiter Ordnung*, Math. Zeitschr. 82 (1963), 177-197.

[192] **Bongsoo Ko**, *The third solution of semilinear elliptic boundary value problems and applications to singular perturbation problems*, J. Diff. Equ. 101 (1993), 1-14.

[193] **Y.S. Kolesov**, *Schauder's principle and the stability of periodic solutions*, Soviet Math. Dokl. 10 (1969), 1290-1293.

[194] **Y.S. Kolesov**, *Periodic solutions of quasilinear parabolic equations of second order*, Trans. Moscow Math. Soc. 21 (1970), 114-146.

[195] **M.A. Krasnosel'skii**, *Positive solutions of operator equations*, Noordhoff, Groeningen, 1964.

[196] **M.A. Krasnosel'skii, G.M. Vainniko, P.P. Zabreiko, Ya.B. Rutitskii, and V.Ya. Stetsenko**, *Approximate solution of operator equations*, Wolters-Noordhoff, 1972.

[197] **T. Kura**, *The weak supersolution-subsolution method for second order quasilinear elliptic equations*, Hiroshima Math. J. 19 (1989), 1-36.

[198] **E.M. Landesman and A.C. Lazer**, *Nonlinear perturbations of linear elliptic boundary value problems at resonance*, J. Math. Mech. 19 (1970), 609-623.

[199] **A.C. Lazer and S. Solimini**, *On periodic solutions of nonlinear differential equations with singularities*, Proc. A.M.S. 99 (1987), 109-114.

[200] **S.J. Li and T. Wang**, *Mountain pass theorem in order intervals and multiple solutions for semilinear elliptic Dirichlet problems*, J. Anal. Math. 81 (2000), 373-396.

[201] **P.L. Lions**, *On the existence of positive solutions of semilinear elliptic equations*, SIAM Review 24 (1982), 441-467.

[202] **Z. Liu**, *Positive solutions of a class of nonlinear elliptic eigenvalue problems*, Math. Z. 242 (2002), 663-686.

[203] **Z. Liu and J. Sun**, *Calculus of variations and super and sub-solutions in reversed order*, Acta Math. Sinica (in chinese) 37 (1994), 512-514.

[204] **Z. Liu and J. Sun**, *Invariant sets of descending flow in critical point theory with applications to nonlinear differential equations*, J. Diff. Equ. 172 (2001), 257-299.

[205] **N.G. Lloyd**, *Degree theory*, Cambridge Univ. Press, Cambridge, 1978.

[206] **A.G. Lomtatidze**, *Positive solutions of boundary value problems for second order ordinary differential equations with singular points*, Differentsial'nye Uravneniya 23 (1987), 1685-1692.

[207] **A. Manes and A.M. Micheletti**, *Un'estensione della teoria variazionale classica degli autovalori per operatori ellittici del secondo ordine*, Boll. U.M.I. 7 (1973), 285-301.

[208] **R. Martins**, *Existence of periodic solutions for second-order differential equations and the strong force condition*, preprint (2000).

[209] **J. Mawhin**, *Nonlinear functional analysis and periodic solutions of ordinary differential equations*, Summer School "Difford 74", Stara Lesna, Czechoslovaquia (1974), 37-60.

[210] **J. Mawhin**, *Compacité, monotonie et convexité dans l'étude de problèmes aux limites semi-linéaires*, Séminaire d'analyse moderne 19, Université de Sherbrooke, 1981.

[211] **J. Mawhin**, *Boundary value problems with nonlinearities having infinite jumps*, Comment. Math. Univ. Carolin. 25 (1984) 401-414.

[212] **J. Mawhin**, *Remarks on the preceding paper of Ahmad and Lazer on periodic solutions*, Boll. U.M.I. 3A (1984), 229-238.

[213] **J. Mawhin**, *Points fixes, points critiques et problèmes aux limites*, Sém. de Math. Supérieures, Univ. Montréal, 1985.

[214] **J. Mawhin**, *Problèmes de Dirichlet variationnels non linéaires*, Sém. de Math. Supérieures, Univ. Montréal, 1987.

[215] **J. Mawhin**, *On a differential equation for the periodic motions of a satellite around its center of mass*, in "Asymptotic Methods of Mathematical Physics", Kiev Naukova Dumka (1988),150-157.

[216] **J. Mawhin**, *Some remarks on semilinear problems at resonance where the nonlinearity depends only on the derivatives*, Acta Math. Inf. Univ. Ostraviensis 2 (1994), 61-69.

[217] **J. Mawhin**, *Boundary value problems for nonlinear ordinary differential equations: from successive approximations to topology*, in "Development of Mathematics, 1900-1950" (ed: J.P. Pier), Birkhauser, Basel (1994), 445-478.

[218] **J. Mawhin and J.R. Ward**, *Nonuniform nonresonance conditions at the two first eigenvalues for periodic solutions of forced Liénard and Duffing equations*, Rocky Mountain J. Math. 12 (1982), 643-654.

[219] **J. Mawhin and M. Willem**, *Critical point theory and Hamiltonian systems*, Springer-Verlag, Berlin, 1989.

[220] **E.J. McShane**, *Integration*, Princeton University Press, Princeton, N.J., 1944.

[221] **W. Mlak**, *Parabolic differential inequalities and Chaplighin's method*, Ann. Polon. Math. 8 (1960), 139-152.

[222] **W. Mlak**, *An example of the equation* $u_t = u_{xx} + f(x,t,u)$ *with distinct maximum and minimum solutions of a mixed problem*, Ann. Polon. Math. 13 (1963), 101-103.

[223] **M. Müller**, *Über das Fundamentaltheorem in der Theorie der gewöhnlichen Differentialgleichungen*, Math. Z. 26 (1926), 619-649.

[224] **M. Nagumo**, *Über die Differentialgleichung* $y'' = f(t,y,y')$, Proc. Phys-Math. Soc. Japan 19 (1937), 861-866.

[225] **M. Nagumo**, *Über das Verhalten des Integrale von* $\lambda y'' + f(x,y,y',\lambda) = 0$ *für* $\lambda \to 0$, Proc. Phys-Math. Soc. Japan 21 (1939), 529-534.

[226] **M. Nagumo**, $y'' = f(t,y,y')$ *no Kyôkaichi Mondai ni tsuite I, II*, Kansû Hôteishiki 30 (1941), 36-46; 17 (1942), 50-52.

[227] **M. Nagumo**, *On the periodic solution of an ordinary differential equation of second order*, Zenkoku Shijou Suugaku Danwakai (1944), 54-61.

[228] **M. Nagumo**, *On principally linear elliptic differential equations of the second order*, Osaka Math. J. 6 (1954), 207-229.

[229] **J.J. Nieto**, *Nonlinear second order boundary value problems with Carathéodory function*, Applicable Analysis 34 (1989), 111-128.

[230] **F.I. Njoku, P. Omari, and F. Zanolin**, *Multiplicity of positive radial solutions of a quasilinear elliptic problem in a ball*, Adv. Differential Equations 5 (2000), 1545-1570.

[231] **F.I. Njoku and F. Zanolin**, *Positive solutions for two points BVP's : existence and multiplicity results*, Nonlinear Analysis, T.M.A., 13 (1989), 1329-1338.

[232] **F. Obersnel and P. Omari**, *Old and new results for first order periodic ODEs without uniqueness: a comprehensive analysis via lower and upper solutions*, preprint.

[233] **H. Okamura**, $y'' = f(t,y,y')$ *ni tsuite I, II, III*, Kansu Huteishiki 27 (1941), 27-35; 30 (1941), 14-19; 31 (1942), 32-40.

[234] **P. Omari**, *A monotone method for constructing extremal solutions of 2nd order scalar BVPs*, Appl. Math. Comput. 18 (1986), 257-275.

[235] **P. Omari**, *Non-ordered lower and upper solutions and solvability of the periodic problem for the Liénard and the Rayleigh equations*, Rend. Ist. Mat. Univ. Trieste 20 (1988), 54-64.

[236] **P. Omari and M. Trombetta**, *Remarks on the lower and upper solutions method for second and third-order periodic boundary value problems*, Appl. Math. Comp. 50 (1992), 1-21.

[237] **P. Omari and W. Ye**, *Necessary and sufficient conditions for the existence of periodic solutions of second order ordinary differential equations with singular nonlinearities*, Diff. Int. Equ. 8 (1995), 1843-1858.

[238] **P. Omari and W. Ye**, *Periodic solutions of a second order ordinary differential equation with an effective damping*, Funkcialaj Ekvacioj 38 (1995), 71-80.

[239] **P. Omari and F. Zanolin**, *Infinitely many solutions of a quasilinear elliptic problem with an oscillatory potential*, Comm. P.D.E. 21 (1996), 721-733.

[240] **P. Omari and F. Zanolin**, *An elliptic problem with arbitrary small positive solutions*, in "Proc. Conf. Nonlinear Differential Equations (Coral Gables, FL 1999)", Electron. J. Diff. Equ. Conf. 5, Southwest Texas States Univ., San Marcos, TX (2000), 301-308.

[241] **R. Ortega**, *Some applications of the topological degree to stability theory*, in "Topological methods in differential equations and inclusions (Montréal, PQ, 1994)", NATO Adv. Sci. Inst. Ser. C Math. Phys. Sci. 472, Kluwer Acad. Publ, Dordrecht (1995), 377-409.

[242] **R. Ortega and M. Tarallo**, *Almost periodic upper and lower solutions*, J. Diff. Equ. 193 (2003), 343-358.

[243] **S.V. Parter**, *Midly nonlinear elliptic partial differential equations and their numerical solution. I*, Numerische Mathematik 7 (1965), 113-128.

[244] **G. Peano**, *Sull'integrabilità delle equazioni differenziali di primo ordine*, Atti Acad. Torino 21 (1885), 677-685.

[245] **A. Pelczar**, *On invariant points of monotone transformations in partially ordered spaces*, Ann. Polon. Math. 17 (1965), 49-53.

[246] **O. Perron**, *Ein neuer Existenzbeweis für die Integrale der Differentialgleichung $y' = f(x, y)$*, Math. Ann. 76 (1915), 471-484.

[247] **O. Perron**, *Eine neue Behandlung des ersten Randwerproblems für* $\Delta u = 0$, Math. Z. 18 (1923), 42-54.

[248] **R.R. Phelps**, *Convex functions, monotone operators and differentiability*, L.N. Math. 1364, Springer-Verlag, Berlin Heidelberg, 1989.

[249] **E. Picard**, *Mémoire sur la théorie des équations aux dérivées partielles et la méthode des approximations successives*, J. de Math. 6 (1890), 145-210.

[250] **E. Picard**, *Sur l'application des méthodes d'approximations successives à l'étude de certaines équations différentielles ordinaires*, J. de Math. 9 (1893), 217-271.

[251] **E. Picard**, *Sur un exemple d'approximations successives divergentes*, Comptes Rendus 118 (1894), 899-902.

[252] **E. Picard**, *Traité d'analyse, III*, Gauthier-Villars, Paris, 1896.

[253] **E. Picard**, *Sur certains exemples singuliers d'approximations successives*, Comptes Rendus 126 (1898), 497-500.

[254] **E. Picard**, *Sur un exemple d'approximations successives divergentes*, Bull. Soc. Math. 28 (1900), 137-143.

[255] **L.C. Piccinini, G. Stampacchia, and G. Vidossich**, *Ordinary differential equations in* $\mathbb{R}^n$*: Problems and methods*, Applied Math. Sciences 39, Springer-Verlag, Berlin, 1984.

[256] **G. Prodi**, *Teoremi di esistenza per equazioni alle derivate parziali non lineari di tipo parabolico, Nota I e II*, Rend. Ist. Lombardo 86 (1953), 1-47.

[257] **M.H. Protter and H.F. Weinberger**, *Maximum principles in differential equations*, Prentice Hall, Englewood Cliffs, 1967.

[258] **P. Rabinowitz**, *Pairs of positive solutions of nonlinear elliptic partial differential equations*, Indiana Univ. Math. J. 23 (1973/1974), 173-186.

[259] **P. Rabinowitz**, *Minimax methods in critical point theory with applications to differential equations*, C.B.M.S. - A.M.S. 65, 1986.

[260] **I. Rachůnková and M. Tvrdý**, *Nonlinear systems of differential inequalities and solvability of certain boundary value problems*, J. of Inequal. and Appl. 6 (2001), 199-226.

[261] **I. Rachůnková, M. Tvrdý, and I. Vrkoč**, *Existence of nonnegative and nonpositive solutions for second order periodic boundary value problems*, J. Diff. Equ. 176 (2001), 445-469.

[262] **A. Rosenblatt**, *Sur les théorèmes de M. Picard dans la théorie des problèmes aux limites des équations différentielles ordinaires non linéaires*, Bull. Sc. Math. 57 (1933), 100-106.

[263] **N. Rouche and J. Mawhin**, *Equations différentielles ordinaires*, Masson, Paris, 1973.

[264] **F.Z. Sadyrbaev**, *Lyapunov functions and the solvability of the first boundary-value problem for ordinary second-order differential equations*, Diff. Equ. 16 (1980), 387-391.

[265] **L. Sanchez**, *Positive solutions for a class of semilinear two-point BVP's*, Bull. Austr. Math. Soc. 45 (1992), 439-451.

[266] **T. Satō**, *Sur l'équation aux dérivées partielles $\Delta z = f(x, y, z, p, q)$*, Compositio Math. 12 (1954), 157-177.

[267] **T. Satō**, *Sur l'équation aux dérivées partielles $\Delta z = f(x, y, z, p, q)$ II*, Compositio Math. 14 (1959), 152-171.

[268] **D.H. Sattinger**, *Monotone methods in nonlinear elliptic and parabolic boundary value problems*, Indiana Univ. Math. J. 21 (1972), 979-1000.

[269] **K. Schmitt**, *Periodic solutions of nonlinear second order differential equations*, Math. Zeitschr. 98 (1967), 200-207.

[270] **K. Schmitt**, *Bounded solutions of nonlinear second order differential equations*, Duke Math. J. 36 (1969), 237-244.

[271] **K. Schmitt**, *A nonlinear boundary value problem*, J. Diff. Equ. 7 (1970), 527-537.

[272] **K. Schmitt**, *A note on periodic solutions of second order ordinary differential equations*, SIAM J. Appl. Math. 21 (1971), 491-494.

[273] **K. Schmitt**, *Intermediate value theorems for periodic functional differential equations*, in "Equations différentielles et fonctionnelles non linéaires", P. Janssens, J. Mawhin, N. Rouche (Ed.), Hermann, Paris, (1973), 65-78.

[274] **K. Schmitt**, *Applications of variational equations to ordinary differential equations - Multiple solutions of boundary value problems*, J. Diff. Equ. 17 (1975), 154-186.

[275] **K. Schmitt**, *Boundary value problems for quasilinear second order elliptic equations*, Nonlinear Anal. T.M.A. 2 (1978), 263-309.

[276] **K. Schrader**, *Solutions of second order ordinary differential equations*, J. Diff. Equ. 4 (1968), 510-518.

[277] **K. Schrader**, *Existence theorems for second order boundary value problems*, J. Diff. Equ. 5 (1969), 572-584.

[278] **J. Schröder**, *Anwendung von Fixpunktsätzen bei der numerischen Behandlung nichtlinearer Gleichungen in halbgeordneten Räumen*, Arch. Rat. Mech. Anal. 4 (1959/60), 177-192.

[279] **G. Scorza Dragoni**, *Sugli integrali dei sistemi di equazioni differenziali*, Rend. Istituto Lombardo di Scienze e Lettere 64 (1931), 659-682.

[280] **G. Scorza Dragoni**, *Il problema dei valori ai limiti studiato in grande per gli integrali di una equazione differenziale del secondo ordine*, Giornale di Mat (Battaglini) 69 (1931), 77-112.

[281] **G. Scorza Dragoni**, *Il problema dei valori ai limiti studiato in grande per le equazioni differenziali del secondo ordine*, Math. Ann. 105 (1931), 133-143.

[282] **G. Scorza Dragoni**, *Su un problema di valori ai limite per le equazioni differenziali ordinarie del secondo ordine*, Rend. Semin. Mat. R. Univ. Roma 2 (1938), 177-215. Aggiunta, ibid, 253-254.

[283] **G. Scorza Dragoni**, *Elementi uniti di transformazioni funzionali e problemi di valori ai limiti*, Rend. Semin. Mat. R. Univ. Roma 2 (1938), 255-275.

[284] **G. Scorza Dragoni**, *Intorno a un criterio di esistenza per un problema di valori ai limiti*, Rend. Semin. R. Accad. Naz. Lincei 28 (1938), 317-325.

[285] **V. Šeda**, *Antitone operators and ordinary differential equations*, Czech. Math. J. 31 (1981), 531-553.

[286] **V. Šeda**, *On some non-linear boundary value problems for ordinary differential equations*, Archivum Mathematicum (Brno) 25 (1989), 207-222.

[287] **E. Serra and M. Tarallo**, *A reduction method for periodic solutions of second order subquadratic equations*, Adv. Diff. Equ. 3 (1998), 199-226.

[288] **E. Serra and M. Tarallo**, *A unified approach to boundary value problems at resonance with the first eigenvalue*, in "Dynamical systems and differential equations, Vol II", Discrete Contin. Dynam. Systems Added Volume II (1998), 182-195.

[289] **L.F. Shampine**, *Some nonlinear eigenvalue problems*, J. Math. Mech. 17 (1968), 1065-1072.

[290] **L.F. Shampine and G.M. Wing**, *Existence and uniqueness of solutions of a class of nonlinear elliptic boundary value problems*, J. Math. and Mech. 19 (1970), 971-979.

[291] **S.N. Slugin**, *A modification of the abstract analogue to Chaplygin's method*, Dokl. Akad. Nauk SSSR 120 (1958), 256-258.

[292] **G. Stampacchia**, *Le problème de Dirichlet pour les équations elliptiques du second ordre à coefficients discontinus*, Ann. Inst. Fourier 15 (1965), 189-258.

[293] **C.A. Stuart**, *Maximal and minimal solutions of elliptic differential equations with discontinuous non-linearities*, Math. Z. 163 (1978), 239-249.

[294] **M. Struwe**, *Variational Methods: Applications to nonlinear partial differential equations and hamiltonian systems*, Springer-Verlag, Berlin, 1990.

[295] **J. Szarski**, *Differential Inequalities*, Monografie Matematyczne 43, PWN-Polish Scientific Publishers, Warszawa, 1965.

[296] **S.D. Taliaferro**, *A nonlinear singular boundary value problem*, Nonlinear Anal. T.M.A. 3 (1979), 897-904.

[297] **A. Tarski**, *A lattice-theoretical fixpoint theorem and its applications*, Pacific J. Math. 5 (1955), 285-309.

[298] **A. Tineo**, *Existence theorems for a singular two-point Dirichlet problem*, Nonlinear Anal. T.M.A. 19 (1992), 323-333.

[299] **F. Tomi**, *Über semilineare elliptische Differentialgleichungen zweiter Ordnung*, Math. Z. 111 (1969), 350-366.

[300] **L. Tonelli**, *Sull'equazione differenziale* $y'' = f(x, y, y')$, Ann. Scuola Norm. Sup. Pisa, Sc. Fis. Mat. 8 (1939), 75-88.

[301] **G.M. Troianiello**, *On solutions to quasilinear parabolic unilateral problems*, Boll. U.M.I. 1-B (1982), 535-552.

[302] **G.M. Troianiello**. *Elliptic differential equations and obstacle problems.* Plenum Press, New York, 1987.

[303] **V.V. Vasin**, *Iterative regularization of the monotonic operator equations of the first kind in partially ordered spaces*, Dokl. Math. 51 (1995), 180-183.

[304] **W. Walter**, *Differential and integral inequalities*, Springer Verlag, New York, 1970.

[305] **M.X. Wang, J.J. Nieto, and A. Cabada**, *Monotone method for nonlinear second order periodic boundary value problem with Carathéodory functions*, Ann. Pol. Math. 58 (1993), 221-235.

[306] **H. Weyl**, *On the differential equations of the simplest boundary layer problem*, Annals of Math. 43 (1942), 381-407.

[307] **M. Willem**, *Analyse harmonique réelle*, Hermann, Paris, 1995.

[308] **J.B.M. Xavier**, *A priori estimates and multiplicity results for equations of the form* $-\Delta u = f(x, u, Du)$, Dyn. Syst. Appl. 3 (1994), 489-500.

[309] **E. Zeidler**, *Nonlinear functional analysis and its applications I: Fixed point theorems*, Springer-Verlag, New York, 1986.

[310] **M. Zhang**, *A relationship between the periodic and the Dirichlet BVPs of singular differential equations*, Proc. Royal Soc. Edinburgh 128A (1998), 1099-1114.

Index

Mathematics in Science and Engineering
Edited by C.K. Chui, Stanford University

Recent titles:
I. Podlubny, *Fractional Differential Equations*
E. Castillo, A. Iglesias, R. Ruíz-Cobo, *Functional Equations in Applied Sciences*
V. Hutson, J.S. Pym, M.J. Cloud, *Applications of Functional Analysis and Operator Theory (Second Edition)*
V. Lakshmikantham and S.K. Sen, *Computational Error and Complexity in Science and Engineering*
T.A. Burton, V*olterra Integral and Differential Equations (Second Edition)*
E.N. Chukwu, *A Mathematical Treatment of Economic Cooperation and Competition Among Nations: with Nigeria, USA, UK, China and Middle East Examples*
Wei-Bin Zhang, *Discrete Dynamical Systems, Bifurcations and Chaos in Economics*